Encyclopaedia

of

PARTIAL DIFFERENTIAL EQUATION

Encyclopaedia of PARTIAL DIFFERENTIAL EQUATION

Volume 1

Dr. Rakesh Kumar Pandey

ANMOL PUBLICATIONS PVT. LTD.
NEW DELHI - 110 002 (INDIA)

ANMOL PUBLICATIONS PVT. LTD.
Regd. Office: 4360/4, Ansari Road, Daryaganj,
New Delhi-110 002 (India)
Ph.: 23278000, 23261597
Branch Office: No. 1015, Ist Main Road, BSK IIIrd Stage
IIIrd Phase, IIIrd Block,
Bangalore-560 085 (India)
Tel.: 080-41723429
Visit us at: www.anmolpublications.com

Encyclopaedia of Partial Differential Equation

First Edition, 2009
ISBN 978-81-261-4107-4 (Set)

PRINTED IN INDIA

Printed at Mehra Offset Press, Delhi.

Contents

Preface

Encyclopaedia of Partial Differential Equation is a unique Encyclopaedia among itself, aiming to provide a fairly complete account of the basic concepts required to build a strong foundation for a student endeavoring to study this subject. The Analytical Approach to the major concepts makes the book highly self-contained and comprehensive guide that succeeds in making the concepts easily understandable. All the Elementary Principles and Fundamental Concepts have been explained rigorously, leaving no scope for Confusion. The focus throughout the text has been on presenting the subject matter in a well-knit manner and lucid style, so that even a student with average Mathematical Skill would find it accessible to himself. In addition, the book provides numerous well-graded solved examples, generally set in various university and competitive examinations, which will facilitate easy understanding besides acquainting the students with a variety of questions.

The Principal objective of the book is solving Boundary Value Problems involving partial digfferential equations. Seperation of variable receives the greatest attention because it is widely used in applications and because it provides a uniform method for solving importanvt cases of Heat, Wave and Potential Equatioins. one technique is not enough of course. The wave equation is developed in parallel with the series solution, and the distrubuted-source solution is constructed for the Heat Equation. In addition, there are chapters on Periodic and Orthogonal Functions, Laplace Transforms, Infinite Series, Series Solutions to Differential Equations, Exact Differential Equations, Bernoulli Differential Equations, Nonhomogeneous Differential Equations, Boundary Value, Intervals of Validity, Solutions to Systems.

Author

List of Symbols

$\pm \quad \not\subset \quad \subset \quad \supset \quad \subseteq \quad \in \quad \notin \quad \partial$

$\neq \quad \equiv \quad \approx \quad \angle \quad \nabla \quad \Sigma \quad \Pi \quad \Omega$

$\phi \quad \alpha \quad \beta \quad \chi \quad \varepsilon \quad \gamma \quad \eta \quad \lambda$

$\mu \quad \nu \quad \perp \quad \pi \quad \theta \quad \rho \quad \sigma \quad \tau$

$\Re \quad \omega \quad \xi \quad \psi \quad \zeta \quad \Upsilon \quad \exists \quad \int$

$\iint \quad \iiint \quad \oint \quad \times \quad \sqrt{\ } \quad \therefore \quad \because \quad \Delta$

$\geq \quad \leq \quad \Rightarrow \quad \Downarrow \quad \Leftrightarrow \quad \infty \quad \propto \quad \mathbb{R}$

$\mathbb{Z} \quad \mathbb{N} \quad \mathbb{C} \quad \mathbb{Q}$

Chapter 1

Introduction to Differential Equation

In this chapter we can get some basic definitions and concepts out of the way. Most of the definitions and concepts introduced here can be introduced without any real knowledge of how to solve differential equations. Most of them are terms that we'll use throughout a class so getting them out of the way right at the beginning is a good idea.During an actual class I tend to hold off on a couple of the definitions and introduce them at a later point when we actually start solving differential equations. The reason for this is mostly a time issue. In this class time is usually at a premium and some of the definitions/concepts require a differential equation and/or its solution so I use the first couple differential equations that we will solve to introduce the definition or concept.

DIFFERENTIAL EQUATION

The first definition that we should cover should be that of differential equation. A differential equation is any equation which contains derivatives, either ordinary derivatives or partial derivatives. There is one differential equation that everybody probably knows, that is Newton's Second Law of Motion. If an object of mass m is moving with acceleration a and being acted on with force F then Newton's Second Law tells us.

$$F = m.$$

To see that this is in fact a differential equation we need to rewrite it a little. First, remember that we can rewrite the acceleration, a, in one of two ways.

$$a - \frac{dv}{dt} \quad \text{OR} \quad a = \frac{d^2u}{dt^2}.$$

Where v is the velocity of the object and u is the position function of the object at any time t. We should also remember at this point that the force, F may also be a function of time, velocity, and/or position.

So, with all these things in mind Newton's Second Law can now be written as a differential equation in terms of either the velocity, v, or the position, u, of the object as follows.

$$m\frac{dv}{dt} = F(t,v)$$

$$m\frac{d^2u}{dt^2} = F\left(t,u,\frac{du}{dt}\right).$$

So, here is our first differential equation. We will see both forms of this in later chapters.

Here are a few more examples of differential equations.

$$ay'' + by' + cy = g(t)$$

$$\sin(y)\frac{d^2y}{dx^2} = (1-y)\frac{dy}{dx} + y^2e^{-5y}$$

$$y^{(4)} + 10y'' - 4y' + 2y = \cos(t)$$

$$\alpha^2\frac{\partial^2 u}{\partial x^2} = \frac{\partial u}{\partial t}$$

$$a^2u_{xx} = u_u$$

$$\frac{\partial^3 u}{\partial^2 x \partial t} = 1 + \frac{\partial u}{\partial y}.$$

ORDINARY AND PARTIAL DIFFERENTIAL EQUATIONS

A differential equation is called an ordinary differential equation, abbreviated by ode, if it has ordinary derivatives in it. Likewise, a differential equation is called a partial differential equation, abbreviated by pde, if it has differential derivatives in it. The vast majority of these notes will deal with ode's. The only exception to this will be the last chapter in which we'll take a brief look at a common and basic solution technique for solving pde's.

LINEAR DIFFERENTIAL EQUATIONS

A linear differential equation is any differential equation that can be written in the following form.

$$a_n(t)y^{(n)}(t) + a_{n-1}(t)y^{(n-1)}(t) + \cdots + a_1(t)y'(t) + a_0(t)y(t) = g(t).$$

The important thing to note about linear differential equations is that there are no products of the function, $y(t)$, and its derivatives and neither the function or its derivatives occur to any power other than the first power.

The coefficients $a_0(t),\ldots, a_n(t)$ and $g(t)$ can be zero or non-zero functions, constant or non-constant functions, linear or non-linear functions. Only the function, $y(t)$, and its derivatives are used in determining if a differential equation is linear. If a differential equation cannot be written in the form, then it is called a non-linear differential equation. All the other are linear differential equations. We can't classify since we do not know what form the function F has. These could be either linear or non-linear depending on F.

SOLUTION

A solution to a differential equation on an interval $\alpha < 1 < \beta$ is any function $y(t)$ which satisfies the differential equation in question on the interval $\alpha < 1 < \beta$. It is important to note that solutions are often accompanied by intervals and these intervals can impart some important information about the solution. Consider the following example.

Example: Show that $y(x) = x^{-\frac{3}{2}}$ is a solution to $4x^2y'' + 12xy' + 3y = 0$ for $x > 0$.

Solution: We'll need the first and second derivative to do this.

$$y'(x) = -\frac{3}{2}x^{-\frac{5}{2}} \qquad y''(x) = -\frac{15}{4}x^{-\frac{7}{2}}.$$

Plug these as well as the function into the differential equation.

$$4x^2\left(\frac{15}{4}x^{-\frac{3}{2}}\right) + 12x\left(-\frac{3}{2}x^{-\frac{5}{2}}\right) + 3\left(x^{\frac{3}{2}}\right) = 0$$

$$15x^{-\frac{3}{2}} - 18x^{-\frac{3}{2}} + 3x^{-\frac{3}{2}} = 0$$

$$0 = 0$$

So, $y(x) = x^{-\frac{3}{2}}$ does satisfy the differential equation and hence is a solution. Why then did I include the condition that ? I did not use this condition anywhere in the work showing that the function would satisfy the differential equation.

To see why recall that

$$y(x) = x^{-\frac{3}{2}} = \frac{1}{\sqrt{x^3}}.$$

In this form it is clear that we'll need to avoid at the least as this would give division by zero.

Also, there is a general rule of thumb that we're going to run with in this class. This rule of thumb is: Start with real numbers, end with real numbers. In other words, if our differential equation only contains real numbers then we don't want solutions that give complex numbers. So, in order to avoid complex numbers we will also need to avoid negative values of x.

So, we saw in the last example that even though a function may symbolically satisfy a differential equation, because of certain restrictions brought about by the solution we cannot use all values of the independent variable and hence, must make a restriction on the independent variable. This will be the case with many solutions to differential equations.

In the last example. note that there are in fact many more possible solutions to the differential equation given. For instance all of the following are also solutions

$$y(x) = x^{-\frac{1}{2}}$$

$$y(x) = -9x^{-\frac{3}{2}}$$

$$y(x) = 7x^{-\frac{1}{2}}$$

$$y(x) = -9x^{-\frac{3}{2}} + 7x^{-\frac{1}{2}}.$$

I'll leave the details to you to check that these are in fact solutions. Given these examples can you come up with any other solutions to the differential equation? There are in fact an infinite number of solutions to this differential equation.

So, given that there are an infinite number of solutions to the differential equation in the last example (provided you believe me when I say that anyway....) we can ask a natural question. Which is the solution that we want or does it matter which solution we use? This question leads us to the next definition in this section.

INITIAL CONDITION(S)

Initial Condition(s) are a condition, or set of conditions, on the solution that will allow us to determine which solution that we are after. Initial conditions (often abbreviated i.c.'s when I'm feeling lazy...) are of the form,

$$y(t_0) = y_0 \quad \text{and/or} \quad y^{(k)}(t_0) = y_k.$$

So, in other words, initial conditions are values of the solution and/or its derivative(s) at specific points. As we will see eventually, solutions to "nice enough" differential equations are unique and hence only one solution will meet the given conditions.

The number of initial conditions that are required for a given differential equation will depend upon the order of the differential equation as we will see.

Example: $y(x) = x^{-\frac{3}{2}}$ is a solution to $4x^2y'' + 12xy' + 3y = 0$, $y(4) = \frac{1}{8}$, and $y'(4) = \frac{3}{64}$.

Solution: As we saw in previous example the function is a solution and we can then note that

$$y(4) = 4^{-\frac{3}{2}} = \frac{1}{\left(\sqrt{4}\right)^3} = \frac{1}{8}$$

$$y'(4) = 4^{-\frac{5}{2}} = -\frac{3}{2}\frac{1}{\left(\sqrt{4}\right)^5} = -\frac{3}{64}$$

and so this solution also meets the initial conditions of $y(4) = \frac{1}{8}$ and $y'(4) = -\frac{3}{64}$. In fact, $y(x) = x^{-\frac{3}{2}}$ is the only solution to this differential equation that satisfies these two initial conditions.

INITIAL VALUE PROBLEM

An Initial Value Problem (or IVP) is a differential equation along with an appropriate number of initial conditions.

Example: The following is an IVP.

$$4x^2y'' + 12xy' + 3y = 0 \qquad y(4) = \frac{1}{8},\ y'(4) = -\frac{3}{64}.$$

Example: Here's another IVP.

$$2ty' + 4y = 3 \qquad y(1) = -\ 4.$$

As I noted earlier the number of initial condition required will depend on the order of the differential equation.

Interval of Validity

The interval of validity for an IVP with initial condition(s)

$$y(t_0) = y_0 \qquad \text{and/or} \qquad y^{(s)}(t_0) = y_k$$

is the largest possible interval on which the solution is valid and contains t_0 $t_0 t_0$. These are easy to define, but can be difficult to find, so I'm going to put off saying anything more about these until we get into actually solving differential equations and need the interval of validity.

GENERAL SOLUTION

The general solution to a differential equation is the most general form that the solution can take and doesn't take any initial conditions into account.

Example: $y(t) = \frac{3}{4} + \frac{c}{t^2}$ is the general solution to

$$2ty' + 4y = 3.$$

I'll leave it to you to check that this function is in fact a solution to the given differential equation. In fact, all solutions to this differential equation will be in this form. This is one of the first differential equations that you will learn how to solve and you will be able to verify this shortly for yourself.

ACTUAL SOLUTION

The actual solution to a differential equation is the specific solution that not only satisfies the differential equation, but also satisfies the given initial condition(s).

Example: What is the actual solution to the following IVP?

$$2ty' + 4y = 3 \quad y(1) = -4.$$

Solution: This is actually easier to do that it might at first appear. From the previous example we already know (well that is provided you believe my solution to this example...) that all solutions to the differential equation are of the form.

$$y(t) = \frac{3}{4} + \frac{c}{t^2}.$$

All that we need to do is determine the value of c that will give us the solution that we're after. To find this all we need do is use our initial condition as follows.

$$-4 = y(1) = \frac{3}{4} + \frac{c}{1^2} \quad \Rightarrow \quad c = -4 - \frac{3}{4} = -\frac{19}{4}.$$

So, the actual solution to the IVP is.

$$y(t) = \frac{3}{4} - \frac{19}{4t^2}.$$

From this last example we can see that once we have the general solution to a differential equation finding the actual solution is nothing more than applying the initial condition(s) and solving for the constant(s) that are in the general solution.

IMPLICIT/EXPLICIT SOLUTION

In this case it's easier to define an explicit solution, then tell you what an implicit solution isn't, and then give you an example to show you the difference. So, that's what I'll do.

An explicit solution is any solution that is given in the form $y = y(t)$. In other words, the only place that y actually shows up is once on the left side and only raised to the first power. An implicit solution is any solution that isn't in explicit form. Note that it is possible to have either general implicit/explicit solutions and actual implicit/explicit solutions.

Example: $y^2 = t^2 - 3$ is the actual implicit solution to

$$y' = \frac{t}{y}, \qquad y(2) = -1.$$

At this point I will ask that you trust me that this is in fact a solution to the differential equation. You will learn how to get this solution in a later section. The point of this example is that since there is a y^2 on the left side instead of a single $y(t)$ this is not an explicit solution!

Example: Find an actual explicit solution to $y' = \frac{t}{y}, \quad y(2) = -1$.

Solution: We already know from the previous example that an implicit solution to this IVP is $y^2 = t^3 - 3$. To find the explicit solution all we need to do is solve for $y(t)$.

$$y(t) = \pm\sqrt{t^2 - 3}.$$

Now, we've got a problem here. There are two functions here and we only want one and in fact only one will be correct! We can determine the correct function by reapplying the initial condition. Only one of them will satisfy the initial condition.

In this case we can see that the "-" solution will be the correct one. The actual explicit solution is then

$$y(t) = -\sqrt{t^2 - 3}.$$

In this case we were able to find an explicit solution to the differential equation. It should be noted however that it will not always be possible do find an explicit solution. Also, note that in this case we were only able to get the explicit actual solution because we had the initial condition to help us determine which of the two functions would be the correct solution. We've now gotten most of the basic definitions out of the way and so we can move onto other topics.

DIRECTION FIELDS

This topic is given its own section for a couple of reasons. First, understanding direction fields and what they tell us about a differential equation and its solution is important and can be introduced without any knowledge of how to solve a differential equation and so can be done here before we get into solving them. So, having some information about the solution to a differential equation without actually having the solution is a nice idea that needs some investigation.

Next, since we need a differential equation to work with this is a good section to show you that differential equations occur naturally in many cases and how we get them. Almost every physical situation that occurs in nature can be described with an appropriate differential equation. The differential equation may be easy or difficult to arrive at depending on the situation and the assumptions that are made about the situation and we may not every able to solve it, however it will exist.

The process of describing a physical situation with a differential equation is called modeling. We will be looking at modeling several times throughout this class.

One of the simplest physical situations to think of is a falling object. So let's consider a falling object with mass *m* and derive a differential equation that, when solved, will give us the velocity of the object at any time, *t*. We will assume that only gravity and air resistance will act upon the object as it falls. Below is a figure showing the forces that will act upon the object.

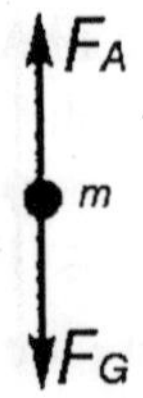

Before defining all the terms in this problem we need to set some conventions. We will assume that forces acting in the downward direction are positive forces while forces that act in the upward direction are negative. Likewise, we will assume that an object moving downward (*i.e.* a falling object) will have a positive velocity.

Now, let's take a look at the forces shown in the diagram above. F_G is the force due to gravity and is given by F_G where *g* is the acceleration due to gravity. In this class I use $g = 9.8$ m/s^2 or $g = 32$ ft/s^2 depending on whether we will use the metric or British system. F_A is the force due to air resistance and for this example we will assume that it is proportional to the velocity, *v*, of the mass. Therefore the force due to air resistance is then given by $F_A = -\gamma v$, where $\gamma > 0$. Note that the "—" is required to get the correct sign on the force. Both γ and *v* are positive and the force is acting upward and hence must be negative. The "—" will give us the correct sign and hence direction for this force.

Recall from the previous section that Newton's Second Law of motion can be written as

$$m\frac{dv}{dt} = F(t,v)$$

where *F(t,v)* is the sum of forces that act on the object and may be a function of the time *t* and the velocity of the object, *v*. For our situation we will have two forces acting on the object gravity, $F_G = mg$. acting in the downward direction and hence will be positive, and air resistance, $F_A = -\gamma v$, acting in the upward direction and hence will be negative. Putting all of this together into Newton's Second Law gives the following.

$$m\frac{dv}{dt} = mg - \gamma v.$$

To simplify the differential equation let's divide out the mass, *m*.

$$\frac{dv}{dt} = g - \frac{\gamma v}{m}.$$

This then is a first order linear differential equation that, when solved, will give the velocity, v (in m/s), of a falling object of mass m that has both gravity and air resistance acting upon it.

In order to look at direction fields (that is after all the topic of this section....) it would be helpful to have some numbers for the various quantities in the differential equation. So, let's assume that we have a mass of 2 kg and that $\gamma = 0.392$. Plugging this gives the following differential equation.

$$\frac{dv}{dt} = 9.8 - 0.196v.$$

Let's take a geometric view of this differential equation. Let's suppose that for some time, t, the velocity just happens to be v = 30 m/s. Note that we're not saying that the velocity ever will be 30 m/s. All that we're saying is that let's suppose that by some chance the velocity does happen to be 30 m/s at some time t. So, if the velocity does happen to be 30 m/s at some time t we can plug $v = 30$ to get.

$$\frac{dv}{dt} - 3.92.$$

Recall from your Calculus I course that a positive derivative means that the function in question, the velocity in this case, is increasing, so if the velocity of this object is ever 30m/s for any time t the velocity must be increasing at that time.

Also, recall that the value of the derivative at a particular value of t gives the slope of the tangent line to the graph of the function at that time, t. So, if for some time t the velocity happens to be 30 m/s the slope of the tangent line to the graph of the velocity is 3.92.

We could continue in this fashion and pick different values of v and compute the slope of the tangent line for those values of the velocity. However, let's take a slightly more organized approach to this. Let's first identify the values of the velocity that will have zero slope or horizontal tangent lines. These are easy enough to find. All we need to do is set the derivative equal to zero and solve for v.

In the case of our example we will have only one value of the velocity which will have horizontal tangent lines, v = 50 m/s. What this means is that IF (again, there's that word if), for some time t, the velocity happens to be 50 m/s then the tangent line at that point will be horizontal. What the slope of the tangent line is at times before and after this point is not known yet and has no bearing on the slope at this particular time, t.

Now, let's get some tangent lines and hence arrows for our graph for some other values of v. At this point the only exact slope that is useful to us is where the slope horizontal. So instead of going after exact slopes for the rest of the graph we are only going to go after general trends in the slope. Is the slope increasing or decreasing? How fast is the slope increasing or decreasing? For this example those types of trends are very easy to get.

So, if we have $v = 50$, we know that the tangent lines will be horizontal. We denote this on an axis system with horizontal arrows pointing in the direction of increasing t at the level of $v = 50$ as shown in the following figure.

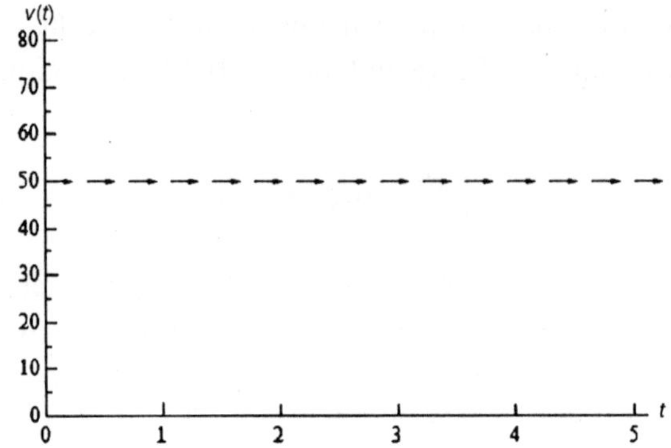

This means that it can only change sign if it first goes through zero. So, if the derivative will change signs (no guarantees that it will) it will do so at $v = 50$ and the only place that it may change sign is $v = 50$. This means that for $v > 50$ the slope of the tangent lines to the velocity will have the same sign. Likewise, for $v < 50$ the slopes will also have the same sign. The slopes in these ranges may have (and probably will) have different values, but we do know what their signs must be.

Let's start by looking at $v < 50$. We saw earlier that if $v = 30$ the slope of the tangent line will be 3.92, or positive. Therefore, for all values of $v < 50$ we will have positive slopes for the tangent lines. We can see that as v approaches 50, always staying less than 50, the slopes of the tangent lines will approach zero and hence flatten out. If we move v away from 50, staying less than 50, the slopes of the tangent lines will become steeper. If you want to get an idea of just how steep the tangent lines become you can always pick specific values of v and compute values of the derivative. For instance, we know that at $v = 30$ the derivative is 3.92 and so arrows at this point should have a slope of around 4. Using this information we can now add in some arrows for the region below $v = 50$ as shown in the graph below.

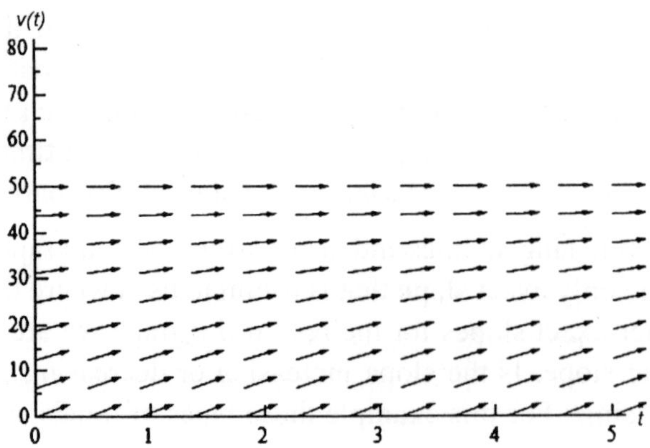

Now, let's look at $v > 50$. The first thing to do is to find out if the slopes are positive or negative. We will do this the same way that we did in the last bit, *i.e.* pick a value of v, the derivative is positive or negative. Note, that you should NEVER assume that the derivative will change signs where the derivative is zero. It is easy enough to check so you should always do so.

We need to check the derivative so let's use $v = 60$. The slope of the tangent line as -1.96, or negative. Therefore, for all values of $v > 50$ we will have negative slopes for the tangent lines. As with $v < 50$, we can see that as v approaches 50, always staying greater than 50, the slopes of the tangent lines will approach zero and flatten out. While moving v away from 50 again, staying greater than 50, the slopes of the tangent lines will become steeper. We can now add in some arrows for the region above $v = 50$ as shown in the graph below.

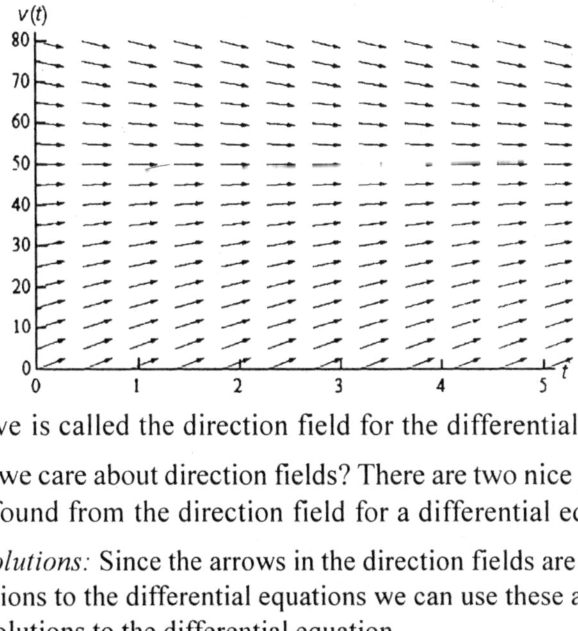

This graph above is called the direction field for the differential equation.

So, just why do we care about direction fields? There are two nice pieces of information that can be readily found from the direction field for a differential equation.

1. *Sketch of solutions:* Since the arrows in the direction fields are in fact tangents to the actual solutions to the differential equations we can use these as guides to sketch the graphs of solutions to the differential equation.
2. *Long Term Behavior:* In many cases we are less interested in the actual solutions to the differential equations as we are in how the solutions behave as t increases. Direction fields, if we can get our hands on them, can be used to find information about this long term behavior of the solution.

So, back to the direction field for our differential equation. Suppose that we want to know what the solution that has the value $v(0) = 30$ looks like. We can go to our direction field and start at 30 on the vertical axis. At this point we know that the solution is increasing and that as it increases the solution should flatten out because the velocity will be approaching the value of $v = 50$. So we start drawing an increasing solution and when we hit an arrow we just make sure that we stay parallel to that arrow. This gives us the figure below.

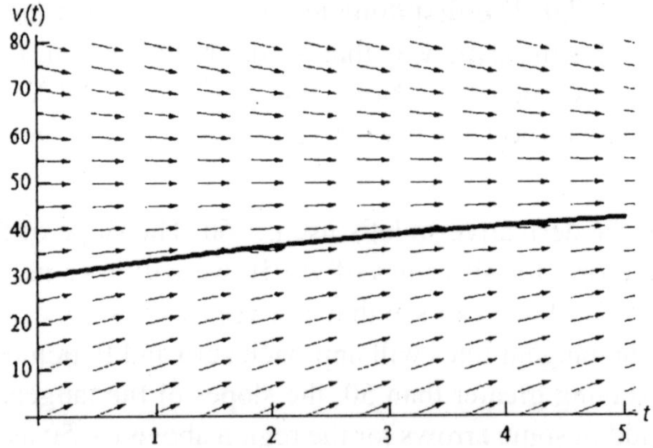

To get a better idea of how all the solutions are behaving, let's put a few more solutions in. Adding some more solutions gives the figure below. The set of solutions that we've graphed below is often called the family of solution curves or the set of integral curves. The number of solutions that is plotted when plotting the integral curves varies. You should graph enough solution curves to illustrate how solutions in all portions of the direction field are behaving.

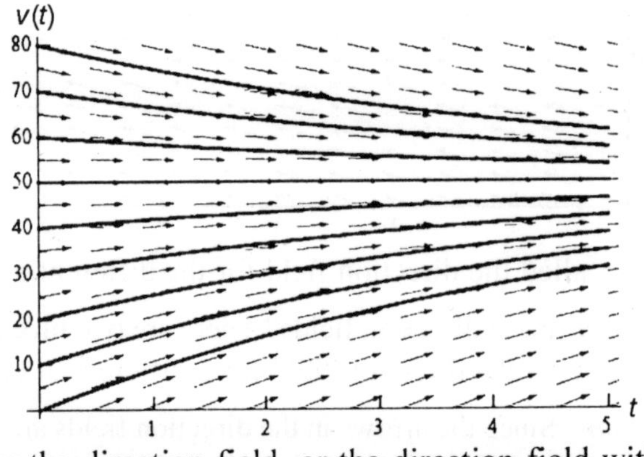

Now, from either the direction field, or the direction field with the solution curves sketched in we can see the behavior of the solution as t increases. For our falling object, it looks like all of the solutions will approach $v = 50$ as t increases.

We will often want to know if the behavior of the solution will depend on the value of $v(0)$. In this case the behavior of the solution will not depend on the value of $v(0)$, but that is probably more of the exception than the rule so don't expect that.

Let's take a look at a more complicated example.

Example: Sketch the direction field for the following differential equation. Sketch the set of integral curves for this differential equation. Determine how the solutions behave as and if this behavior depends on the value of $y(0)$ describe this dependency.

$$y' = (y^2 - y - 2)(1 - y)^2.$$

Solution: First, do not worry about where this differential equation came from. To be honest, I just make it up. It may, or may not describe an actual physical situation.

This differential equation looks somewhat more complicated than the falling object example from above. However, with the exception of a little more work, it is not much more complicated. The first step is to determine where the derivative is zero.

$$0 = (y^2 - y - 2)(1 - y)^2$$

$$0 = (y - 2)(y + 1)(1 - y)^2.$$

We can now see that we have three values of y in which the derivative, and hence the slope of tangent lines, will be zero. The derivative will be zero at y = -1, 1, and 2. So, let's start our direction field with drawing horizontal tangents for these values. This is shown in the figure below.

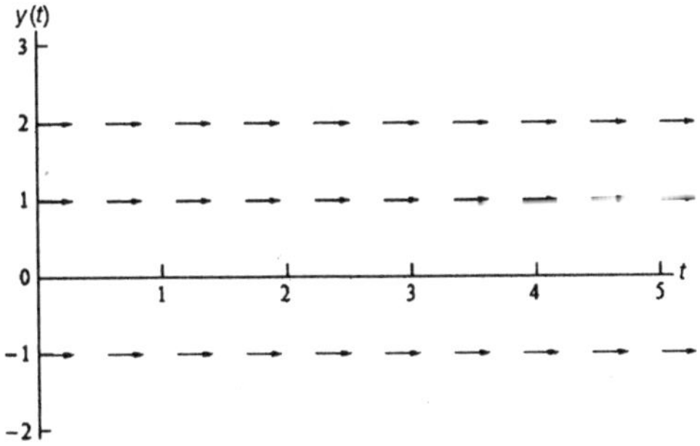

Now, we need to add arrows to the four regions that the graph is now divided into. For each of these regions I will pick a value of y in that region and plug it into the right hand side of the differential equation to see if the derivative is positive or negative in that region. Again, to get an accurate direction fields you should pick a few more over values over the whole range to see how the arrows are behaving over the whole range.

$y < -1$

In this region we can use $y = -2$ as the test point. At this point we have $y' = 36$. So, tangent lines in this region will have very steep and positive slopes. Also as $y \to -1$ the slopes will flatten out while staying positive. The figure below shows the direction fields with arrows in this region.

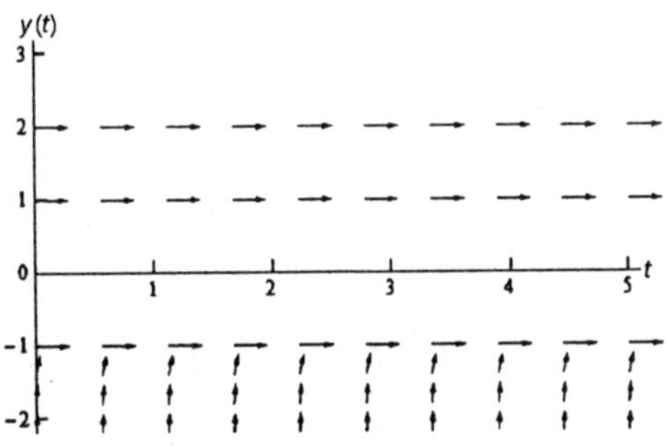

$-1 < y < 1$

In this region we can use $y = 0$ as the test point. At this point we have $y' = -2$. Therefore, tangent lines in this region will have negative slopes and apparently not be very steep. So what do the arrows look like in this region?

As $y \to 1$ staying less that 1 of course, the slopes should be negative and approach zero. As we move away from 1 and towards -1 the slopes will start to get steeper (and stay negative), but eventually flatten back out, again staying negative, as $y \to -1$ since the derivative must approach zero at that point. The figure below shows the direction fields with arrows added to this region.

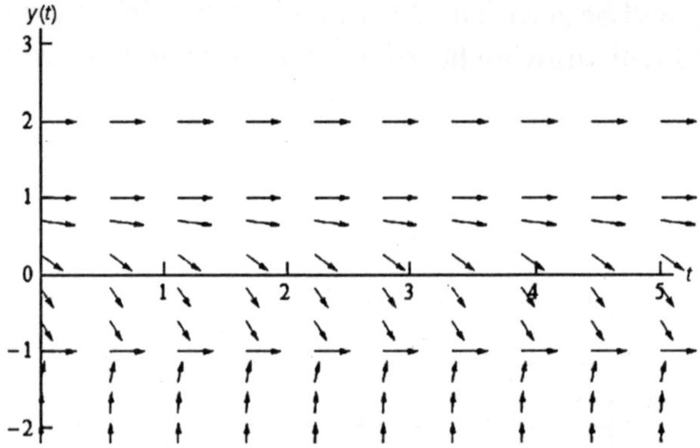

$1 < y < 2$

In this region we will use $y = 1.5$ as the test point. At this point we have $y' = -0.3125$. Tangent lines in this region will also have negative slopes and apparently not be as steep as the previous region.

Arrows in this region will behave essentially the same as those in the previous region. Near $y = 1$ and $y = 2$ the slopes will flatten out and as we move from one to the other the slopes will get somewhat steeper before flattening back out. The figure below shows the direction fields with arrows added to this region.

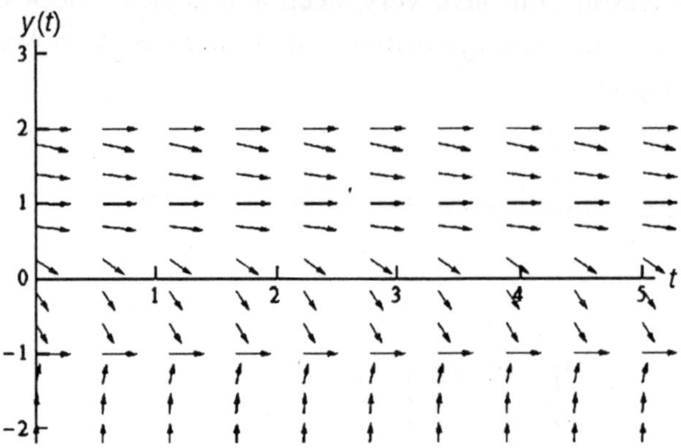

$y > 2$

In this last region we will use $y = 3$ as the test point. At this point we have $y' = 16$. So, as we saw in the first region tangent lines will start out fairly flat near $y = 2$ and then as we move way from $y = 2$ they will get fairly steep. The complete direction field for this differential equation is shown below.

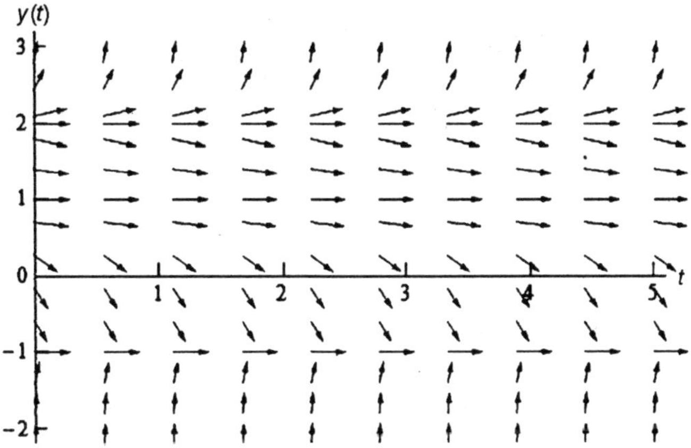

Here is the set of integral curves for this differential equation. Note that due to the steepness of the solutions in the lowest region and the software used to generate these images I was unable to include more than one solution curve in this region.

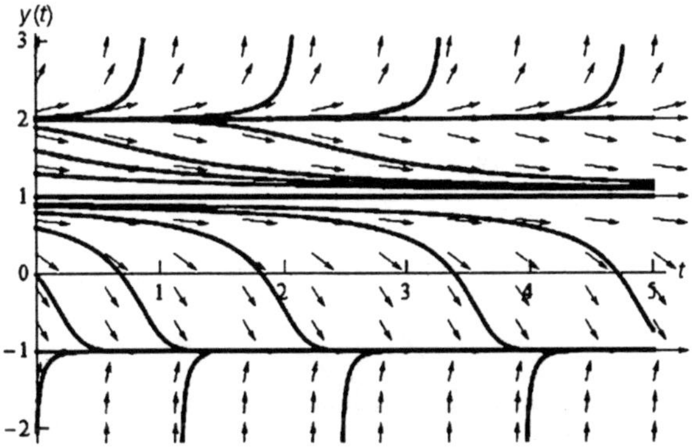

Finally, let's take a look at long term behavior of all solutions. Unlike the first example, the long term behavior in this case will depend on the value of y at $t = 0$. By examining either of the previous two figures we can arrive at the following behavior of solutions as $t \to \infty$.

Value of $y(0)$ Behavior as $t \to \infty$

$$y(0) < 1 \quad y \to -1$$

$$1 \le y(0) < 2 \quad y \to 1$$

$$y(0) = 2 \quad y \to 2$$

$$y(0) > 2 \quad y \to \infty$$

Do, not forget to acknowledge what the horizontal solutions are doing. This is often the most missed portion of this kind of problem.

In both of the examples that we've worked to this point the right hand side of the derivative has only contained the function and NOT the independent variable. When the right hand side of the differential equation contains both the function and the independent variable the behavior can be much more complicated and sketching the direction fields by hand can be very difficult. Computer software is very handy in these cases. In some cases they aren't too difficult to do by hand however. Let's take a look at the following example.

Example: Sketch the direction field for the following differential equation. Sketch the set of integral curves for this differential equation.

$$y' = y - x$$

Solution: To sketch direction fields for this kind of differential equations we first identify places where the derivative will be constant. To do this we set the derivative in the differential equation equal to a constant, say c. This gives us a family of equations, called isoclines, that we can plot and on each of these curves the derivative will be a constant value of c.

Notice that in the previous examples we looked at the isocline for $c = 0$ to get the direction field started. For our case the family of isoclines is.

$$c = y - x.$$

The graph of these curves for several values of c is shown below.

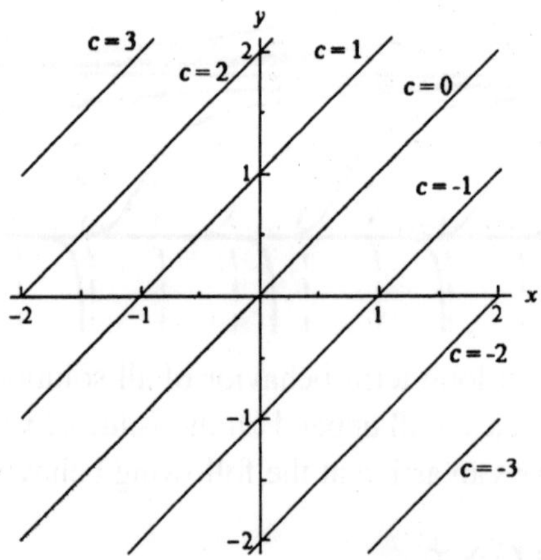

Now, on each of these lines, or isoclines, the derivative will be constant and will have a value of c. On the $c = 0$ isocline the derivative will always have a value of zero and hence the tangents will all be horizontal. On the $c = 1$ isocline the tangents will always have a slope of 1, on the $c = -2$ isocline the tangents will always have a slope of -2, *etc.* Below is a few tangents put in for each of these isoclines.

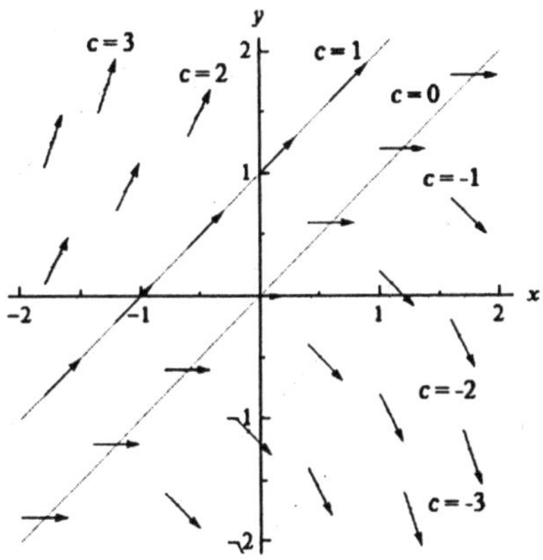

To add more arrows for those areas between the isoclines start at say, $c = 0$ and move up to $c = 1$ and as we do that we increase the slope of the arrows (tangents) from 0 to 1. This is shown in the figure below.

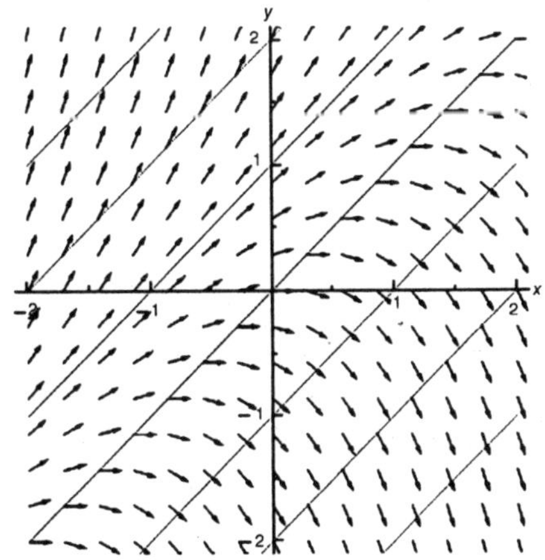

We can then add in integral curves as we did in the previous examples. This is shown in the figure below.

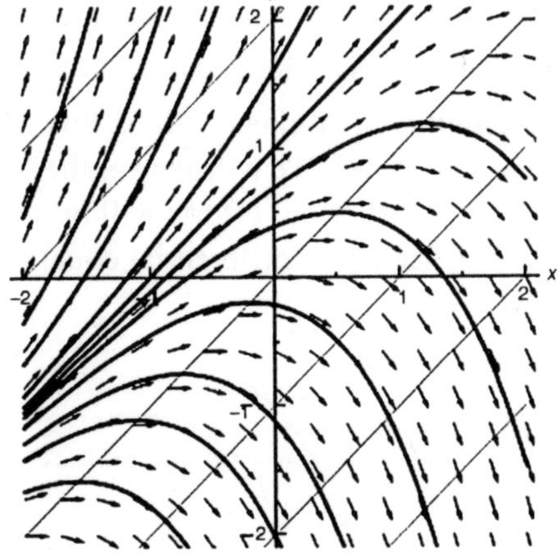

FINAL THOUGHTS

Before moving on to learning how to solve differential equations I want give a few final thoughts. Any differential equations course will concern itself with answering one or more of the following questions.

1. *Given a differential equation will a solution exist*? Not all differential equations will have solutions so it's useful to know ahead of time if there is a solution or not. If there isn't a solution why waste our time trying to find something that doesn't exist?

   This question is usually called the existence question in a differential equations course.

2. *If a differential equation does have a solution how many solutions are there*? As we will see eventually, it is possible for a differential equation to have more than one solution. We would like to know how many solutions there will be for a given differential equation.

   There is a sub question here as well. What condition(s) on a differential equation are required to obtain a single unique solution to the differential equation?

   Both this question and the sub question are more important than you might realize. Suppose that we derive a differential equation that will give the temperature distribution in a bar of iron at any time t. If we solve the differential equation and end up with two (or more) completely separate solutions we will have problems. Consider the following situation to see this.

   If we subject 10 identical iron bars to identical conditions they should all exhibit the same temperature distribution. So only one of our solutions will be accurate, but we will have no way of knowing which one is the correct solution.

   It would be nice if, during the derivation of our differential equation, we could make sure that our assumptions would give us a differential equation that upon solving will yield a single unique solution.

   This question is usually called the uniqueness question in a differential equations course.

3. *If a differential equation does have a solution can we find it*? This may seem like an odd question to ask and yet the answer is not always yes. Just because we know that a solution to a differential equations exists does not mean that we will be able to find it.

   In a first course in differential equations (such as this one) the third question is the question that we will concentrate on. We will answer the first two equations for special, and fairly simple, cases, but most of our efforts will be concentrated on answering the third question for as wide a variety of differential equations as possible.

Chapter 2

First Order Differential Equations

In this chapter we will look at solving first order differential equations. The most general first order differential equation can be written as,

$$\frac{dy}{dt} = f(y,t)$$

What we will do instead is look at several special cases and see how to solve those. We will also look at some of the theory behind first order differential equations as well as some applications of first order differential equations. Below is a list of the topics discussed in this chapter.

LINEAR DIFFERENTIAL EQUATIONS

The first special case of first order differential equations that we will look is the linear first order differential equation. In this case, unlike most of the first order cases that we will look at, we can actually derive a formula for the general solution. The general solution is derived below. However, I would suggest that you do not memorize the formula itself. Instead of memorizing the formula you should memorize and understand the process that I'm going to use to derive the formula. Most problems are actually easier to work by using the process instead of using the formula.

So, let's see how to solve a linear first order differential equation. Remember as we go through this process that the goal is to arrive at a solution that is in the form $y = y(t)$. It's sometimes easy to lose sight of the goal as we go through this process for the first time. In order to solve a linear first order differential equation we MUST start with the differential equation in the form shown below. If the differential equation is not in this form then the process we're going to use will not work.

$$\frac{dy}{dt} + P(t)y = g(t)$$

Where both *p(t)* and *g(t)* are continuous functions.μ (t) Recall that a quick and dirty definition of a continuous function is that a function will be continuous provided you can draw the graph from left to right without ever picking up your pencil/pen. In other words, a function is continuous if there are no holes or breaks in it.

Now, we are going to assume that there is some magical function somewhere out there in the world,, called an integrating factor. Do not, at this point, worry about what this function is or where it came from. We will figure out what $\mu(t)$is once we have the formula for the general solution in hand.

So, now that we have assumed the existence of $\mu(t)$multiply everything by $\mu(t)$. This will give.

$$\mu(t)\frac{dy}{dt}+\mu(t)p(t)y=\mu(t)g(t)$$

Now, this is where the magic of $\mu(t)$ comes into play. We are going to assume that whatever $\mu(t)$is, it will satisfy the following.

$$\mu(t)p(t)=\mu'(t)$$

Again do not worry about how we can find a $\mu(t)$that will satisfy (3). As we will see, provided $p(t)$ is continuous we can find it. So substituting (3) into (2) we now arrive at.

$$\mu(t)\frac{dy}{dt}+\mu'(t)y=\mu(t)g(t)$$

At this point we need to recognize that the left side of (4) is nothing more than the following product rule.

$$\mu(t)\frac{dy}{dt}+\mu'(t)y=\left(\mu(t)y(t)\right)'$$

So we can replace the left side of Equation with this product rule. Upon doing this (4) becomes

$$\left(\mu(t)y(t)\right)'=\mu(t)g(t)$$

Now, recall that we are after $y(t)$. We can now do something about that. All we need to do is integrate both sides then use a little algebra and we'll have the solution. So, integrate both sides to get.

$$\int\left(\mu(t)y(t)\right)'dt=\int\mu(t)g(t)$$

$$\mu(t)y(t)+c=\int\mu(t)g(t)dt$$

Note the constant of integration, c, from the left side integration is included here. It is vitally important that this be included. If it is left out you will get the wrong answer every time.

The final step is then some algebra to solve for the solution, $y(t)$.

$$\mu(t)y(t) = \int \mu(t)g(t)dt - c$$

$$y(t) = \frac{\int \mu(t)g(t)dt - c}{\mu(t)}$$

Now, from a notational standpoint we know that the constant of integration, c, is an unknown constant and so to make our life easier we will absorb the minus sign in front of it into the constant and use a plus instead. This will NOT affect the final answer for the solution. So with this change we have.

$$y(t) = \frac{\int \mu(t)g(t)dt = c}{\mu(t)}$$

Again, changing the sign on the constant will not affect our answer. If you choose to keep the minus sign you will get the same value of c as I do except it will have the opposite sign. Upon plugging in c we will get exactly the same answer.

There is a lot of playing fast and loose with constants of integration in this section, so you will need to get used to it. When we do this we will always to try to make it very clear what is going on and try to justify why we did what we did. So, now that we've got a general solution we need to go back and determine just what this magical function $\mu(t)$ $\mu(t)$ $\mu(t)$ is. This is actually an easier process that you might think. We'll start with (3).

$$\mu(t)p(t) = \mu'(t)$$

Divide both sides by,

$$\frac{\mu'(t)}{\mu(t)} = (t)$$

Now, hopefully you will recognize the left side of this from your Calculus I class as nothing more than the following derivative.

$$(\text{In } \mu(t))' = p(t)$$

As with the process above all we need to do is integrate both sides to get.

$$In\mu(t) + k = \int p(t)dt$$

$$In\mu(t) = \int p(t)dt + k$$

You will notice that the constant of integration from the left side, k, had been moved to the right side and had the minus sign absorbed into it again as we did earlier. Also note that we're using k here because we've already used c and in a little bit we'll have both of them in the same equation. So, to avoid confusion we used different letters to represent the fact that they will, in all probability, have different values.

Exponentiate both sides to get $\mu(t)$ out of the natural logarithm.

$$\mu(t) = e^{\int p(t)dt + k}$$

Now, it's time to play fast and loose with constants again. It is inconvenient to have the k in the exponent so we're going to get it out of the exponent in the following way.

$$\begin{aligned}\mu(t) &= e^{\int p(t)dt + k} \\ &= e^k e^{\int p(t)dt} \qquad \text{Recall } x^{a+b} = x^a x^b!\end{aligned}$$

Now, let's make use of the fact that k is an unknown constant. If k is an unknown constant then so is $e^k e^k$ e^kso we might as well just rename it k and make our life easier. This will give us the following.

$$\mu(t) = ke^{\int P(t)dt}$$

So, we now have a formula for the general solution, and a formula for the integrating factor. We do have a problem however. We've got two unknown constants and the more unknown constants we have the more trouble we'll have later on. Therefore, it would be nice if we could find a way to eliminate one of them (we'll not be able to eliminate both....).

This is actually quite easy to do. First, substitute and rearrange the constants.

$$\begin{aligned}y(t) &= \frac{\int ke^{\int p(t)dt} g(t)dt + c}{ke^{\int p(t)dt}} \\ &= \frac{k\int e^{\int p(t)dt} g(t)dt + c}{ke^{\int p(t)dt}} \\ &= \frac{\int e^{\int p(t)dt} g(t)dt + \frac{c}{k}}{e^{\int P(t)dt}}\end{aligned}$$

So, Equation can be written in such a way that the only place the two unknown constants show up is a ratio of the two. Then since both c and k are unknown constants so is the ratio of the two constants.

Therefore we'll just call the ratio c and then drop k out of since it will just get absorbed into c eventually.

The solution to a linear first order differential equation is then

$$y(t) = \frac{\int (t)g(t)dt + c}{\mu(t)}$$

where,

$$\mu(t) = e^{\int p(t)dt}$$

Now, the reality is that is not as useful as it may seem. It is often easier to just run through the process that got rather than using the formula. We will not use this formula in any of my examples. We will need to use regularly, as that formula is easier to use than the process to derive it.

Solution Process: The solution process for a first order linear differential equation is as follows.

Put the differential equation in the correct initial form.

1. Find the integrating factor, μ (*t*).
2. Multiply everything in the differential equation by μ (*t*)and verify that the left side becomes the product rule (μ (*t*) y (*t*))and write it as such.
3. Integrate both sides, make sure you properly deal with the constant of integration.
4. Solve for the solution *y(t)*.

Let's work a couple of examples. Let's start by solving the differential equation that we derived back in the Direction Field section.

Example: Find the solution to the following differential equation.

$$\frac{dv}{dt} = 9.8 - 0.196v$$

Solution: First we need to get the differential equation in the correct form.

$$\frac{dv}{dt} + 0.196v = 9.8$$

From this we can see that *p(t)* = 0.196 and μ (*t*) so is then.

$$\mu(t) = e^{\int 0.196dt} = e^{0.196t}$$

Note that officially there should be a constant of integration in the exponent from the integration.

However, we can drop that for exactly the same reason that we dropped the *k*. Now multiply all the terms in the differential equation by the integrating factor and do some simplification.

$$e^{0.196t}\frac{dv}{dt}+0.196e^{0.196t}v=9.8e^{0.196t}$$

$$\left(e^{0.196t}v\right)'=9.8e^{0.196t}$$

Integrate both sides and don't forget the constants of integration that will arise from both integrals.

$$\int\left(e^{0.196}v\right)'dt=\int 9.8e^{0.196t}dt$$

$$e^{0.196t}v+k=50e^{0.196t}+c$$

Okay. It's time to play with constants again. We can subtract k from both sides to get.

$$e^{0.196t}v=50e^{0.196t}+c-k$$

Both c and k are unknown constants and so the difference is also an unknown constant. We will therefore write the difference as c. So, we now have

$$e^{0.196t}v=50e^{0.196t}+c$$

From this point on we will only put one constant of integration down when we integrate both sides knowing that if we had written down one for each integral, as we should, the two would just end up getting absorbed into each other.

The final step in the solution process is then to divide both sides by $e^{0.196t}$ or to multiply both sides by $e^{-0.196t}$. Either will work, but I usually prefer the multiplication route. Doing this gives the general solution to the differential equation.

$$v(t)=50+ce^{-0.196t}$$

From the solution to this example we can now see why the constant of integration is so important in this process. Without it, in this case, we would get a single, constant solution, $v(t) = 50$. With the constant of integration we get infinitely many solutions, one for each value of c.

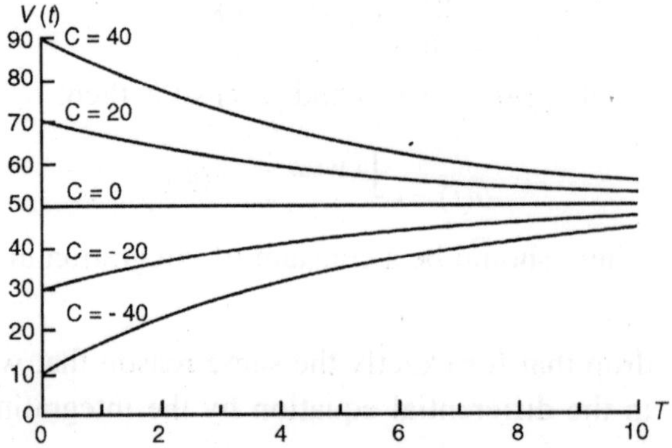

Back in the direction field section where we first derived the differential equation used in the last example we used the direction field to help us sketched some solutions. Let's see if we got them correct. To sketch some solutions all we need to do is to pick different values of c to get a solution. Several of these are shown in the graph below.

So, it looks like we did pretty good sketching the graphs back in the direction field section.

Now, recall from the Definitions section that the Initial Condition(s) will allow us to zero in on a particular solution. Solutions to first order differential equations (not just linear as we will see) will have a single unknown constant in them and so we will need exactly one initial condition to find the value of that constant and hence find the solution that we were after. The initial condition for first order differential equations will be of the form

$$y(t_0) = y_0$$

Recall as well that a differential equation along with a sufficient number of initial conditions is called an Initial Value Problem (IVP).

Example: Solve the following IVP.

$$\frac{dv}{dt} = 9.8 - 0.196v \qquad v(0) = 48$$

Solution: To find the solution to an IVP we must first find the general solution to the differential equation and then use the initial condition to identify the exact solution that we are after. So, since this is the same differential equation as we looked at in Example 1, we already have its general solution.

$$v = 50 + ce^{-0.196t}$$

Now, to find the solution we are after we need to identify the value of c that will give us the solution we are after. To do this we simply plug in the initial condition which will give us an equation we can solve for c. So let's do this

$$48 = v(0) = 50 + c \Rightarrow \quad c = -2$$

So, the actual solution to the IVP is.

$$v = 50 - 2e^{-0.196t}$$

A graph of this solution can be seen in the figure above.

Let's do a couple of examples that are a little more involved.

Example: Solve the following IVP.

$$\cos(x)y' + \sin(x)y = 2\cos^3(x)\sin(x) - 1 \quad y\left(\frac{\pi}{4}\right) = 3\sqrt{2}, \; 0 \le x < \frac{\pi}{2}$$

Solution: Rewrite the differential equation to get the coefficient of the derivative a one.

$$y'+\frac{\sin(x)}{\cos(x)}y = 2\cos^2(x)\sin(x) - \frac{01}{\cos(x)}$$

$$y'+\tan(x)y = 2\cos^2(x)\sin(x) - \sec(x)$$

Now find the integrating factor.

$$\mu(t) = e^{\int \tan x dx} = e^{\text{In}\,|\sec x|} = e^{\text{In}\,\sec x} = \sec x$$

Can you do the integral? If not rewrite tangent back into sines and cosines and then use a simple substitution. Note that we could drop the absolute value bars on the secant because of the limits on x. In fact, this is the reason for the limits on x.

Also note that we made use of the following fact.

$$\mathrm{e}^{\text{In}\,f(x)} = f(x)$$

This is an important fact that you should always remember for these problems. We will want to simplify the integrating factor as much as possible in all cases and this fact will help with that simplification.

Now back to the example. Multiply the integrating factor through the differential equation and verify the left side is a product rule. Note as well that we multiply the integrating factor through the rewritten differential equation and NOT the original differential equation. Make sure that you do this. If you multiply the integrating factor through the original differential equation you will get the wrong solution!

$$\sec(x)y'+(x)\tan(x)y = 2\sec(x)\cos^2(x)\sin(x) - \sec^2(x)$$

$$(\sec(x)y)' = 2\cos(x)\sin(x) - \sec^2(x)$$

Integrate both sides.

$$\int(\sec(x)y(x))'dx = \int 2\cos(x)\sin(x) - \sec^2(x)dx$$

$$\sec(x)y(x) = 1\int \sin(2x) - \sec^2(x)dx$$

$$\sec(x)y(x) = -\frac{1}{2}\cos(2x) - \tan(x) + c$$

Note the use of the trig formula $\sin(2\theta) = 2\sin\theta\cos\theta$ that made the integral easier. Next, solve for the solution.

$$y(x) = -\frac{1}{2}\cos(x)\cos(2x) - \cos(x)\tan(x) + c\cos(x)$$

$$= -\frac{1}{2}\cos(x)\cos(2x) - \sin(x) + c\cos(x)$$

Finally, apply the initial condition to find the value of c.

$$3\sqrt{2} = y\left(\frac{\pi}{4}\right) = -\frac{1}{2}\cos\left(\frac{\pi}{4}\right)\cos\left(\frac{\pi}{2}\right) - \sin\left(\frac{\pi}{4}\right) + c\cos\left(\frac{\pi}{4}\right)$$

$$3\sqrt{2} = -\frac{\sqrt{2}}{2} + c\frac{\sqrt{2}}{2}$$

$$c = 7$$

The solution is then.

$$y(x) = -\frac{1}{2}\cos(x)\cos(2x) - \sin(x) + 7\cos(x)$$

Below is a plot of the solution.

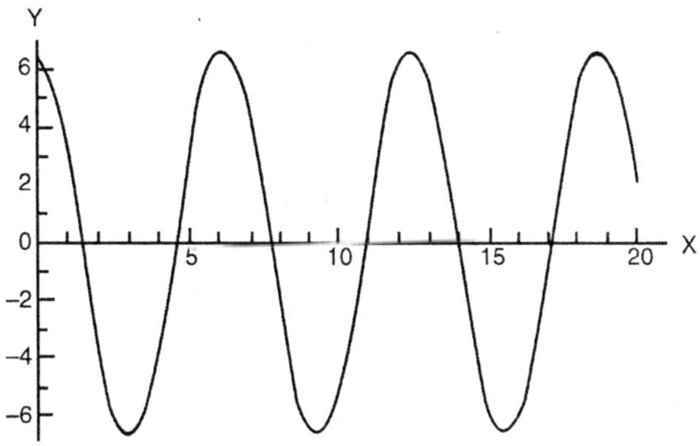

Example: Find the solution to the following IVP.

Solution: First, divide through by the t to get the differential equation into the correct form.

$$y' + \frac{2}{t}y = t - 1 + \frac{1}{t}$$

Now let's get the integrating factor, $\mu(t)$.

$$\mu(t) = e^{\int \frac{2}{t}dt} = e^{2\ln|t|}$$

Now, we need to simplify $\mu(t)$. However, we can't use (11) yet as that requires a coefficient of one in front of the logarithm. So, recall that

$$\ln x^r = r \ln x$$

and rewrite the integrating factor in a form that will allow us to simplify it.

$$\mu(t) = e^{2\ln|t|} = e^{\ln|t|^2} = |t|^2 = t^2$$

We were able to drop the absolute value bars here because we were squaring the t, but often they can't be dropped so be careful with them and don't drop them unless you know that you can. Often the absolute value bars must remain.

Now, multiply the rewritten differential equation (remember we can't use the original differential equation here...) by the integrating factor.

$$\left(t^2 y\right)' = t^3 - t^2 + t$$

Integrate both sides and solve for the solution.

$$t^2 y = \int t^3 - t^2 + t dt$$
$$= \frac{1}{4}t^4 - \frac{1}{3}t^3 + \frac{1}{2}t^2 + c$$
$$y(t) = \frac{1}{4}t^2 - \frac{1}{3}t + \frac{1}{2} + \frac{c}{t^2}$$

Finally, apply the initial condition to get the value of c.

$$\frac{1}{2} = y(1) = \frac{1}{4} - \frac{1}{3} + \frac{1}{2} + c \quad \Rightarrow \quad c = \frac{1}{12}$$

The solution is then,

$$y(t) = \frac{1}{4}t^2 - \frac{1}{3}t + \frac{1}{2} + \frac{1`}{12t^2}$$

Here is a plot of the solution.

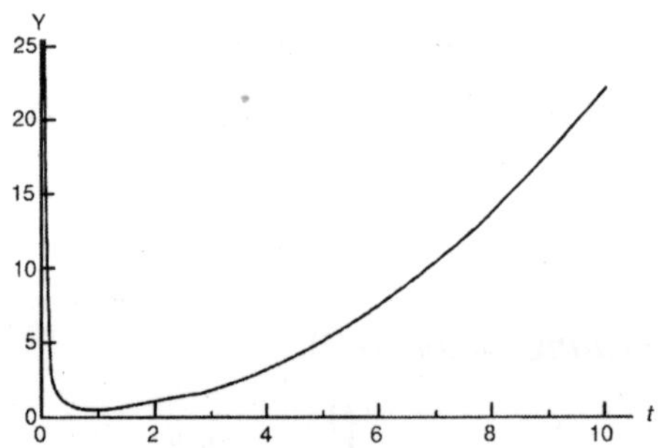

Example: Find the solution to the following IVP.

$$ty' - 2y = t^5 \sin(2t) - t^3 + 4r^4 \qquad y(\pi) = \frac{3}{2}\pi^4$$

Solution: First, divide through by t to get the differential equation in the correct form.

$$y' - \frac{2}{t}y = t^4 \sin(2t) - t^2 + 4t^3$$

Now that we have done this we can find the integrating factor, $\mu(t)$.

$$\mu(t) = e^{\int \frac{2}{t} dt} = e^{-2\ln|t|}$$

Do not forget that the "-" is part of $p(t)$. Forgetting this minus sign can take a problem that is very easy to do and turn it into a very difficult, if not impossible problem so be careful! Now, we just need to simplify this as we did in the previous example.

$$\mu(t) = e^{-2\text{In}|t|} = e^{\text{In}|t|^{-2}} = |t|^{-2} = t^{-2}$$

Again, we can drop the absolute value bars since we are squaring the term.

Now multiply the differential equation by the integrating factor (again, make sure it's the rewritten one and not the original differential equation).

$$(t^2 y)' = t^2 \sin(2t) - 1 + 4t$$

Integrate both sides and solve for the solution.

$$t^{-2} y(t) = \int t^2 \sin(2t)dt + \int -1 + 4t dt$$

$$t^{-2} y(t) = -\frac{1}{2}t^2 \cos(2t) + \frac{1}{2}t\sin(2t) + \frac{1}{4}\cos(2t) - t + 2t^2 + c$$

$$y(t) = -\frac{1}{2}t^4 \cos(2t) + \frac{1}{2}t^3 \sin(2t) + \frac{1}{4}t^2 \cos(2t) - t^3 + 2t^4 + ct^2$$

Apply the initial condition to find the value of c.

$$\frac{3}{2}\pi^4 = y(\pi) = -\frac{1}{2}\pi^4 + \frac{1}{4}\pi^2 - \pi^3 + 2\pi^4 + c\pi^2 = \frac{3}{2}\pi^4 - \pi^3 + \frac{1}{4}\pi^2 + c\pi^2$$

$$\pi^3 - \frac{1}{4}\pi^2 = c\pi^2$$

$$c = \pi - \frac{1}{4}$$

The solution is then

$$y(t) = -\frac{1}{2}t^4 \cos(2t) + \frac{1}{2}t^3 \sin(2t) + \frac{1}{4}t^2 \cos()2t - t^3 + 2t^4 + \left(\pi - \frac{1}{4}\right)t^2$$

Below is a plot of the solution.

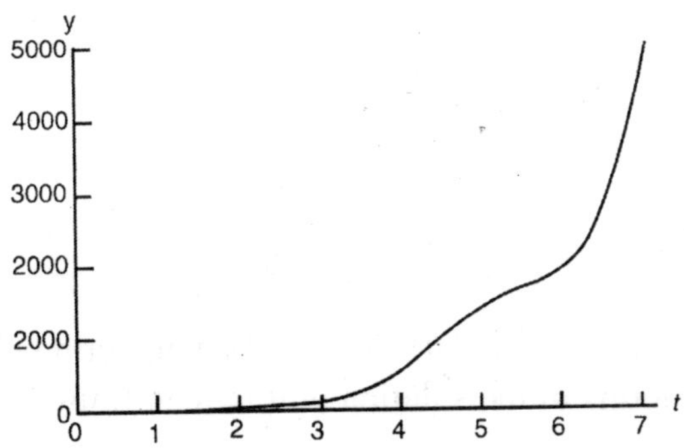

Let's work one final example that looks more at interpreting a solution rather than finding a solution.

Example: Find the solution to the following IVP and determine all possible behaviors of the solution as $t \to \infty$. If this behavior depends on the value of y_0 give this dependence.

$$2y' - y = 4\sin(3t) \quad y(0) = y_0$$

Solution: First, divide through by a 2 to get the differential equation in the correct form.

$$y' - \frac{1}{2}y = 2\sin(3t)$$

Now find $\mu(t)$.

$$\mu(t) = e^{\int \frac{1}{2}dt} = e^{\frac{1}{2}}$$

Multiply $\mu(t)$ through the differential equation and rewrite the left side as a product rule.

$$\left(e^{\frac{t}{2}}y\right)' = 2e^{\frac{t}{2}}\sin(3t)$$

Integrate both sides and solve for the solution.

$$e^{\frac{t}{2}}y = \int 2e^{\frac{t}{2}}\sin(3t)dt + c$$

$$e^{\frac{t}{2}}y = -\frac{24}{37}e^{\frac{t}{2}}\cos(3t) - \frac{4}{37}e^{\frac{t}{2}}\sin(3t) + c$$

$$y(t) = -\frac{24}{37}\cos(3t) - \frac{4}{37}\sin(3t) + ce^{\frac{t}{2}}$$

Apply the initial condition to find the value of c and note that it will contain y_0 as we don't have a value for that.

$$y_0 = y(0) = -\frac{24}{37} + c \quad \Rightarrow \quad c = y_0 + \frac{24}{37}$$

So the solution is

$$y(t) = -\frac{24}{37}\cos(3t) - \frac{4}{37}\sin(3t) + \left(y_0 + \frac{24}{37}\right)e^{\frac{t}{2}}$$

Now that we have the solution, let's look at the long term behavior (*i.e* $t \to \infty$) of the solution. The first two terms of the solution will remain finite for all values of t. It is the last term that will determine the behavior of the solution. The exponential will always go to infinity as $t \to \infty$, however depending on the sign of the coefficient c (yes we've already found it, but for ease of this discussion we'll continue to call it c). The following table gives the long term behavior of the solution for all values of c.

Range of c	Behavior of solution as $t \to \infty$
$c < 0$	$y(t) \to \infty$
$c = 0$	$y(t)$ remains finite
$c > 0$	$y(t) \to \infty$

This behavior can also be seen in the following graph of several of the solutions.

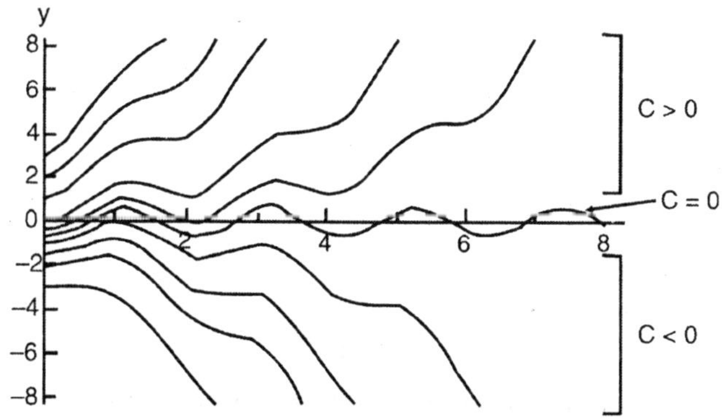

Now, because we know how c relates to y_0 we can relate the behavior of the solution to y_0. The following table give the behavior of the solution in terms of y_0 instead of c.

Range of y_0	Behavior of solution as $t \to \infty$
$y_0 < -\frac{24}{37}$	$y(t) \to -\infty$
$y_0 = -\frac{24}{37}$	$y(t)$ remains finite
$y_0 > -\frac{24}{37}$	$y(t) \to -\infty$

Note that for $y_0 = -\frac{24}{37}$ the solution will remain finite. That will not always happen.

Investigating the long term behavior of solutions is sometimes more important than the solution itself. Suppose that the solution above gave the temperature in a bar of metal. In this case we would want the solution(s) that remains finite in the long term. With this investigation we would now have the value of the initial condition that will give us that solution and more importantly values of the initial condition that we would need to avoid so that we didn't melt the bar.

SEPARABLE DIFFERENTIAL EQUATIONS

We are now going to start looking at nonlinear first order differential equations. The first type of nonlinear first order differential equations that we will look at is separable differential equations.

A separable differential equation is any differential equation that we can write in the following form.

$$N(y)\frac{dy}{dx} = M(x)$$

Note that in order for a differential equation to be separable all the y's in the differential equation must be multiplied by the derivative and all the x's in the differential equation must be on the other side of the equal sign.

Solving separable differential equation is fairly easy. We first rewrite the differential equation as the following

$$N(y)dy = M(x)dx$$

Then you integrate both sides.

$$\int N(y)dy = \int M(x)dx$$

So, after doing the integrations you will have an implicit solution that you can hopefully solve for the explicit solution, *y(x)*. Note that it won't always be possible to solve for an explicit solution.

Recall from the Definitions section that an implicit solution is a solution that is not in the form y = y (x)while an explicit solution has been written in that form.

We will also have to worry about the interval of validity for many of these solutions. Recall that the interval of validity was the range of the independent variable, x in this case, on which the solution is valid. In other words, we need to avoid division by zero, complex numbers, logarithms of negative numbers or zero, *etc.* Most of the solutions that we will get from separable differential equations will not be valid for all values of x.

Let's start things off with a fairly simple example so we can see the process without getting lost in details of the other issues that often arise with these problems.

Example: Solve the following differential equation and determine the interval of validity for the solution.

$$\frac{dy}{dx} = 6y^2x \qquad y(1) = \frac{1}{25}$$

Solution: It is clear, hopefully, that this differential equation is separable. So, let's separate the differential equation and integrate both sides. As with the linear first order officially we will pick up a constant of integration on both sides from the integrals on each side of the equal sign. The two can be moved to the same side an absorbed into each other. We will use the convention that puts the single constant on the side with the x's.

$$y^{-2}dy = 6x\,dx$$
$$\int y^{-2}dy = \int 6x\,dx$$
$$-\frac{1}{y} = 3x^2 + c$$

So, we now have an implicit solution. This solution is easy enough to get an explicit solution, however before getting that it is usually easier to find the value of the constant at this point. So apply the initial condition and find the value of c.

$$-\frac{1}{1/25} = 3(1)^2 + c \qquad c = -28$$

Plug this into the general solution and then solve to get an explicit solution.

$$-\frac{1}{y} = 3x^2 - 28$$
$$y(x) = \frac{1}{28 - 3x^2}$$

Now, as far as solutions go we've got the solution. We do need to start worrying about intervals of validity however.

Recall that there are two conditions that define an interval of validity. First, it must be a continuous interval with no breaks or holes in it.

Second it must contain the value of the independent variable in the initial condition, $x = 1$ in this case.

So, for our case we've got to avoid two values of x. Namely, $x \neq \pm\sqrt{\frac{28}{3}} \approx \pm 3.05505$ since these will give us division by zero. This gives us three possible intervals of validity.

$$-\infty < x < -\sqrt{\frac{28}{3}} \qquad -\sqrt{\frac{28}{3}} < x < \sqrt{\frac{28}{3}} \qquad \sqrt{\frac{28}{3}} < x < \infty$$

However, only one of these will contain the value of x from the initial condition and so we can see that

$$-\sqrt{\frac{28}{3}} < x < \sqrt{\frac{28}{3}}$$

must be the interval of validity for this solution.

Here is a graph of the solution.

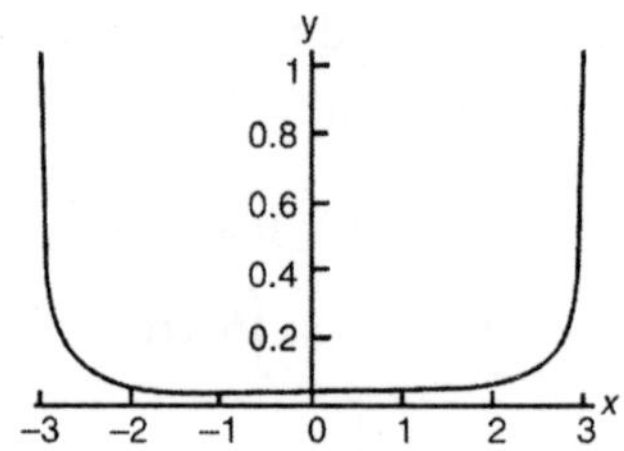

Note that this does not say that either of the other two intervals listed above can't be the interval of validity for any solution. With the proper initial condition either of these could have been the interval of validity. We'll leave it to you to verify the details of the following claims. If we use an initial condition of

$$y(-4) = -\frac{1}{20}$$

we will get exactly the same solution however in this case the interval of validity would be the first one.

$$-\infty < x < -\sqrt{\frac{28}{3}}$$

Likewise, if we use

$$y(6) = -\frac{1}{80}$$

as the initial condition we again get exactly the same solution and in this case the third interval becomes the interval of validity.

$$\sqrt{\frac{28}{3}} < x < \infty$$

So, simply changing the initial condition a little can give any of the possible intervals.

Example: Solve the following IVP and find the interval of validity for the solution.

$$y' = \frac{3x^2 + 4x - 4}{2y - 4} \qquad y(1) = 3$$

Solution: This differential equation is clearly separable, so let's put it in the proper form and then integrate both sides.

$$(2y - 4)dy = (3x^2 + 4x - 4)dx$$

$$\int (2y - 4)dy = \int (3x^2 + 4x - 4)dx$$

$$y^2 - 4y = x^3 + 2x^2 - 4x + c$$

We now have our implicit solution, so as with the first example let's apply the initial condition at this point to determine the value of c.

$$(3)^2 - 4(3) = (1)^3 + 2(1)^2 - 4(1) + c \qquad c = -2$$

The implicit solution is then

$$y^2 - 4y = x^3 + 2x^2 - 4x - 2$$

We now need to find the explicit solution. This is actually easier than it might look and you already know how to do it. First we need to rewrite the solution a little

$$y^2 - 4y - (x^3 + 2x^2 - 4x - 2) = 0$$

To solve this all we need to recognize is that this is quadratic in y and so we can use the quadratic formula to solve it. However, unlike quadratics you are used to, at least some of the "constants" will not actually be constant, but will in fact involve x's.

So, upon using the quadratic formula on this we get.

$$y(x) = \frac{4 \pm \sqrt{16 - 4(1)\left(-(x^3 + 2x^2 - 4x - 2)\right)}}{2}$$
$$= \frac{4 \pm \sqrt{16 + 4(x^3 + 2x^2 - 4x - 2)}}{2}$$

Next, notice that we can factor a 4 out from under the square root (it will come out as a 2…) and then simplify a little.

$$y(x) = \frac{4 \pm 2\sqrt{4 + \left(x^3 + 2x^2 - 4x - 2\right)}}{2}$$
$$= 2 \pm \sqrt{x^3 + 2x^2 - 4x + 2)}$$

We are almost there. Notice that we've actually got two solutions here (the "$\pm$ $\pm\pm$") and we only want a single solution. In fact, only one of the signs can be correct. So, to figure out which one is correct we can reapply the initial condition to this. Only one of the signs will give the correct value so we can use this to figure out which one of the signs is correct. Plugging $x = 1$ into the solution gives.

$$3 = y(1) = 2 \pm \sqrt{1 + 2 - 4 + 2} = 2 \pm 1 = 3,1$$

In this case it looks like the "+" is the correct sign for our solution. Note that it is completely possible that the "" could be the solution so don't always expect it to be one or the other.

The explicit solution for our differential equation is.

$$y(x) = 2 + \sqrt{x^3 + 2x^2 - 4x + 2}$$

To finish the example out we need to determine the interval of validity for the solution. If we were to put a large negative value of x in the solution we would end up with complex values in our solution and we want to avoid complex numbers in our solutions here. So, we will need to determine which values of x will give real solutions. To do this we will need to solve the following inequality.

$$x^3 + 2x^2 - 4x + 2 \geq 0$$

In other words, we need to make sure that the quantity under the radical stays positive.

Using a computer algebra system like Maple or Mathematica we see that the left side is zero at $x = 3.36523$ as well as two complex values, but we can ignore complex values for interval of validity computations. Finally a graph of the quantity under the radical is shown below.

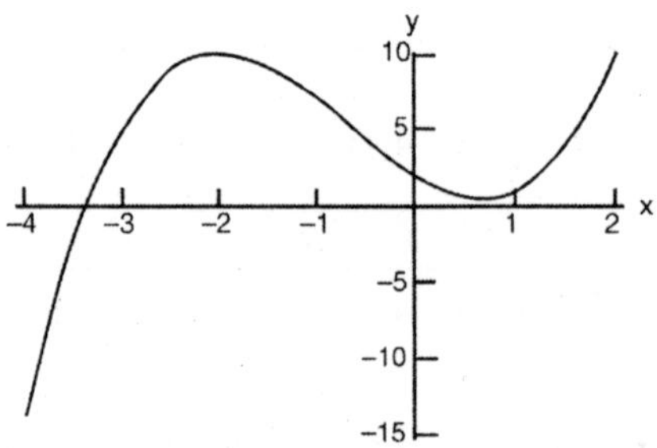

So, in order to get real solutions we will need to require $x \geq -3.36523$ because this is the range of x's for which the quantity is positive. Notice as well that this interval also contains the value of x that is in the initial condition as it should.

Therefore, the interval of validity of the solution is $x \geq -3.36523$.

Here is graph of the solution.

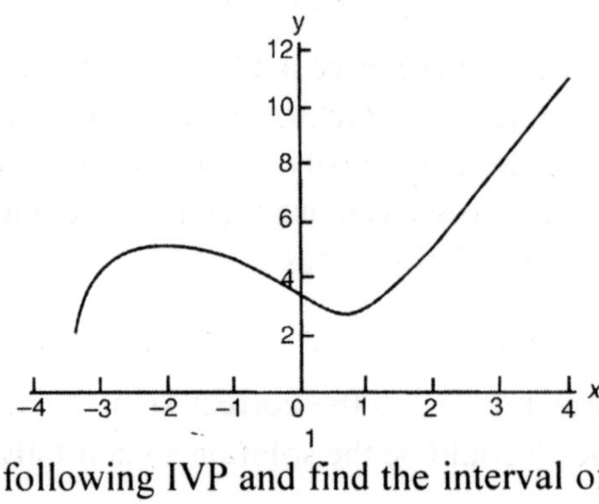

Example: Solve the following IVP and find the interval of validity of the solution.

$$y' = \frac{xy^3}{\sqrt{1+x^2}} \qquad y(0) = -1$$

Solution: First separate and then integrate both sides.

$$y^{-3}dy = x(1+x^2)^{-\frac{1}{2}}dx$$

$$\int y^{-3}dy = \int x(1+x^2)^{\frac{1}{2}}dx$$

$$-\frac{1}{2y^2} = \sqrt{1=x^2}+c$$

Apply the initial condition to get the value of c.

$$-\frac{1}{2} = \sqrt{1+c} \qquad c = -\frac{3}{2}$$

The implicit solution is then,

$$\frac{1}{2y^2} = \sqrt{1+x^2} - \frac{3}{2}$$

Now let's solve for $y(x)$.

$$\frac{1}{y^2} = 3 - 2\sqrt{1+x^2}$$

$$y^2 = \frac{1}{3-2\sqrt{1+x^2}}$$

$$y(x) = \pm\frac{1}{\sqrt{3-2\sqrt{1+x^2}}}$$

Reapplying the initial condition shows us that the "—" is the correct sign. The explicit solution is then,

$$y(x) = -\frac{1}{\sqrt{3-2\sqrt{1+x^2}}}$$

Let's get the interval of validity. That's easier than it might look for this problem. First, since $1+x^2 \geq 0$ the "inner" root will not be a problem. Therefore all we need to worry about is division by zero and negatives under the "outer" root. We can take care of both be requiring

$$3 - 2\sqrt{1+x^2} > 0$$

$$3 > 2\sqrt{1+x^2}$$

$$9 > 4(1+x^2)$$

$$\frac{9}{4} > 1 + x^2$$

$$\frac{5}{4} > x^2$$

Note that we were able to square both sides of the inequality because both sides of the inequality are guaranteed to be positive in this case. Finally solving for x we see that the only possible range of x's that will not give division by zero or square roots of negative numbers will be,

$$-\frac{\sqrt{5}}{2} < x < \frac{\sqrt{5}}{2}$$

and nicely enough this also contains the initial condition $x=0$. This interval is therefore our interval of validity.

Here is a graph of the solution.

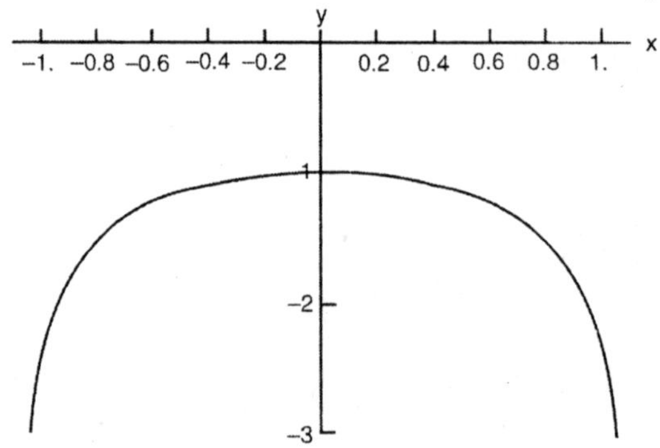

Example: Solve the following IVP and find the interval of validity of the solution.

$$y' = e^{-y}(2x-4) \qquad y(5)=0$$

Solution:

This differential equation is easy enough to separate, so let's do that and then integrate both sides.

$$e^y dy = (2x-4)dx$$
$$\int e^y dy = \int (2x-4)dx$$
$$e^y = x^2 - 4x + c$$

Applying the initial condition gives

$$1 = 25 - 20 + c \qquad c = -4$$

This then gives an implicit solution of.

$$e^y = x^2 - 4x - 4$$

We can easily find the explicit solution to this differential equation by simply taking the natural log of both sides.

$$x^2 - 4x - 4 > 0$$

Finding the interval of validity is the last step that we need to take. Recall that we can't plug negative values or zero into a logarithm, so we need to solve the following inequality

The quadratic will be zero at the two points $x = 2 \pm 2\sqrt{2}$. A graph of the quadratic (shown below) shows that there are in fact two intervals in which we will get positive values of the polynomial and hence can be possible intervals of validity.

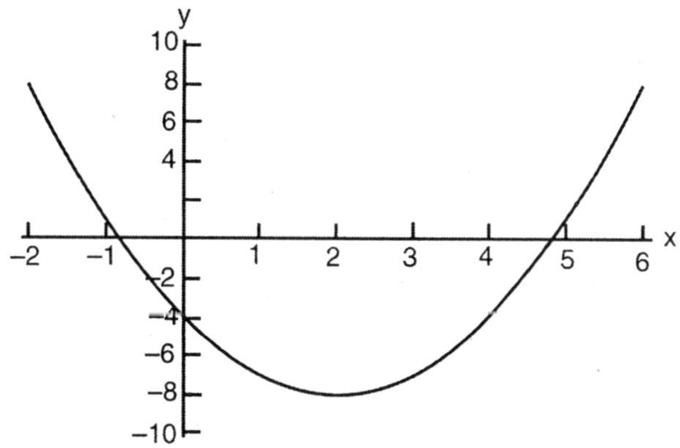

So, possible intervals of validity are

$$-\infty < x < 2 - 2\sqrt{2}$$

$$2 + 2\sqrt{2} < x < \infty$$

From the graph of the quadratic we can see that the second one contains $x = 5$, the value of the independent variable from the initial condition. Therefore the interval of validity for this solution is.

$$2 + 2\sqrt{2} < x < \infty$$

Here is a graph of the solution.

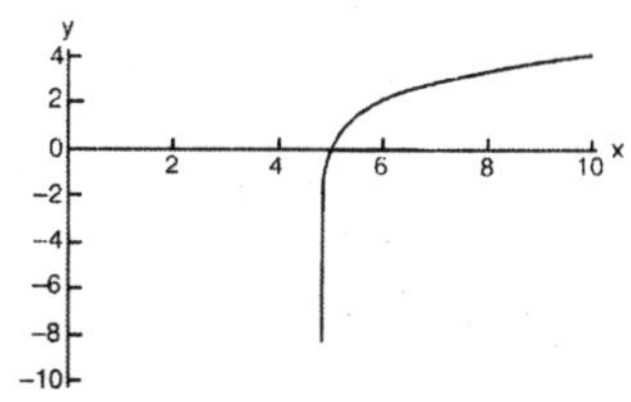

Example: Solve the following IVP and find the interval of validity for the solution.

$$\frac{dr}{d\theta} = \frac{r^2}{\theta} \qquad r(1) = 2$$

Solution: This is actually a fairly simple differential equation to solve. I'm doing this one mostly because of the interval of validity.

So, get things separated out and then integrate.

$$\frac{1}{r^2}dr = \frac{1}{\theta}d\theta$$

$$\int\frac{1}{r^2}dr = \int\frac{1}{\theta}d\theta$$

$$-\frac{1}{r} = \text{In}|\theta| + c$$

Now, apply the initial condition to find *c*.

$$-\frac{1}{2} = \text{In}(1) + c \qquad c = -\frac{1}{2}$$

So, the implicit solution is then,

$$-\frac{1}{r} = \text{In}|\theta| - \frac{1}{2}$$

Solving for *r* gets us our explicit solution.

$$r = \frac{1}{\frac{1}{2} - \text{In}|\theta|}$$

Now, there are two problems for our solution here. First we need to avoid è= 0 because of the natural log. Notice that because of the absolute value on the èwe don't need to worry about èbeing negative. We will also need to avoid division by zero. In other words, we need to avoid the following points.

$$\frac{1}{2} - \text{In}|\theta| = 0$$

$$\text{In}|\theta| = \frac{1}{2} \qquad \text{exponentiate both sides}$$

$$|\theta| = e^{\frac{1}{2}}$$

$$\theta = \pm\sqrt{e}$$

So, these three points break the number line up into four portions, each of which could be an interval of validity.

$$-\infty < \theta < -\sqrt{e}$$

$$\_\sqrt{e} < \theta < 0$$

$$0 < \theta < \sqrt{e}$$

$$\sqrt{e} < \theta < \infty$$

The interval that will be the actual interval of validity is the one that contains è= 1. Therefore, the interval of validity is $0 < \theta < \sqrt{e}$.

Here is a graph of the solution.

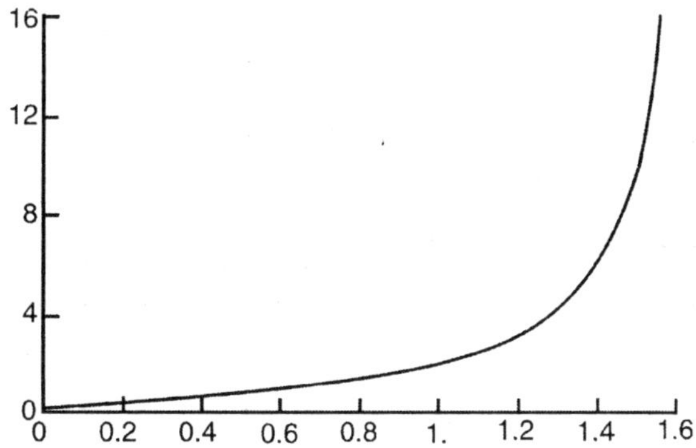

Example: Solve the following IVP.

$$\frac{dy}{dt} = e^{y-t}\sec(y)(1+t^2) \qquad y(0)=0$$

Solution: This problem will require a little work to get it separated and in a form that we can integrate, so let's do that first.

$$\frac{dy}{dt} = \frac{e^{y}e^{-t}}{\cos(y)}(1+t^2)$$

$$e^{-y}\cos(y)dy = e^{-t}(1+t^2)dt$$

Now, with a little integration by parts on both sides we can get an implicit solution.

$$\int e^{-y}\cos(y)dy = \int e^{-t}(1+t^2)dt$$

$$\frac{e^{-y}}{2}(\sin(y)-\cos(y)) = -e^{-t}(t^2+2t+3)+c$$

Applying the initial condition gives.

$$\frac{1}{2}(-1) = -(3)+c \qquad c = \frac{5}{2}$$

Therefore, the implicit solution is.

$$\frac{e^{y}}{2}(\sin(y)-\cos(y)) = -e^{-t}\left(t^2+2t+3\right)+\frac{5}{2}$$

It is not possible to find an explicit solution for this problem and so we will have to leave the solution in its implicit form. Finding intervals of validity from implicit solutions can often be very difficult so we will also not bother with that for this problem.

Chapter 3

Second Order Differential Equations

In the previous chapter we looked at first order differential equations. In this chapter we will move on to second order differential equations. Just as we did in the last chapter we will look at some special cases of second order differential equations that we can solve. Unlike the previous chapter however, we are going to have to be even more restrictive as to the kinds of differential equations that we'll look at. This will be required in order for us to actually be able to solve them.

BASIC CONCEPTS

In this chapter we will be looking exclusively at linear second order differential equations. The most general linear second order differential equation is in the form.

$$p(t)y^{n} + q(t)y' + r(t)y = g(t).$$

In fact, we will rarely look at non-constant coefficient linear second order differential equations. In the case where we assume constant coefficients we will use the following differential equation.

$$ay^{n} + by' + cy = g(t).$$

Where possible we will use the equation just to make the point that certain facts, theorems, properties, and/or techniques can be used with the non-constant form. It can be fairly difficult to solve second order non-constant coefficient differential equations.

Initially we will make our life easier by looking at differential equations with $g(t) = 0$. When $g(t) = 0$ we call the differential equation homogeneous and when $g(t) \neq 0$we call the differential equation nonhomogeneous.

So, let's start thinking about how to go about solving a constant coefficient, homogeneous, linear, second order differential equation.Here is the general constant coefficient, homogeneous, linear, second order differential equation.

$$ay^{n} + by' + cy = 0.$$

It's probably best to start off with an example. This example will lead us to a very important fact that we will use in every problem from this point on. The example will also give us clues into how to go about solving these in general.

Example: Determine some solutions to

$$y''-9y=0.$$

Solution: We can get some solutions here simply by inspection. We need functions whose second derivative is 9 times the original function. One of the first functions that I can think of that comes back to itself after two derivatives is an exponential function and with proper exponents the 9 will get taken care of as well.

So, it looks like the following two functions are solutions.

$$y(t)=e^{3t} \quad \text{and} \quad y(t)=e^{-3t}.$$

We'll leave it to you to verify that these are in fact solutions.

These two functions are not the only solutions to the differential equation however.Any of the following are also solutions to the differential equation.

$$y(t)=9e^{3t} \qquad y(t)=123e^{3t}$$

$$y(t)=56e^{-3t} \qquad y(t)=\frac{14}{9}e^{-3t}$$

$$y(t)=7e^{3t}-6e^{-3t} \qquad y(t)=-92e^{3t}016e^{-3t}$$

In fact if you think about it any function that is in the form

$$y(t)=c_1e^{3t}+c_2e^{-3t}$$

will be a solution to the differential equation.

This example leads us to a very important fact that we will use in practically every problem in this chapter.

PRINCIPLE OF SUPERPOSITION

If $y_1(t)$ and $y_2(t)$are two solutions to a linear, homogeneous differential equation then so is

$$y(t)=c_1y_1(t)+c_2y_2(t).$$

Note that we didn't include the restriction of constant coefficient or second order in this. This will work for any linear homogeneous differential equation.

If we further assume second order and one other condition (which we'll give in a second) we can go a step further.

If $y_1(t)$and $y_2(t)$are two solutions to a linear, second order homogeneous differential equation and they are "nice enough" then the general solution to the linear, second order differential equation.

So, just what do we mean by "nice enough"? We'll hold off on that until a later section. At this point you'll hopefully believe it when we say that specific functions are "nice enough".

So, if we now make the assumption that we are dealing with a linear, second order differential equation. The next question that we can ask is how to find the constants c_1 and c_2. Since we have two constants it makes sense, hopefully, that we will need two equations, or conditions, to find them.

One way to do this is to specify the value of the solution at two distinct points, or,

$$y(t) = y_0 \qquad y(t_1) = y_1.$$

These are typically called boundary values and are not really the focus of this course so we won't be working with them.

Another way to find the constants would be to specify the value of the solution and its derivative at a particular point. Or,

$$y(t_0) = y_0 \qquad y(t_1) = y_1.$$

These are the two conditions that we'll be using here. As with the first order differential equations these will be called initial conditions.

$$y(t_0) = y_0 \qquad y'(t_1) = y'_1.$$

Example: Solve the following IVP.

$$y'' - 9y = 0 \qquad y(0) = 2 \quad y'(0) = -1.$$

Solution: First, the two functions

$$y(t) = e^{3t} \quad \text{and} \quad y(t) = e^{-3t}$$

are "nice enough" for us to form the general solution to the differential equation. At this point, please just believe this. You will be able to verify this for yourself in a couple of sections. The general solution to our differential equation is then

$$y(t) = c_1 e^{-3t} + c_2 e^{3t}.$$

Now all we need to do is apply the initial conditions. This means that we need the derivative of the solution.

$$y'(t) = -3c_1 e^{-3t} + 3c_2 e^{3t}.$$

Plug in the initial conditions

$$2 = y(0) = c_1 + c_2$$

$$-1 = y'(0) - 3c_1 + 3c_2$$

This gives us a system of two equations and two unknowns that can be solved. Doing this yields

$$c_1 = \frac{7}{6} \qquad c_2 = \frac{5}{6}.$$

The solution to the IVP is then,

$$y(t) = \frac{7}{6}\,\mathrm{e}^{-3t} + = \frac{5}{6}e^{3t}.$$

Up to this point we've only looked at a single differential equation and we got its solution by inspection. For a rare few differential equations we can do this. However, for the vast majority of the second order differential equations out there we will be unable to do this. So, we would like a method for arriving at the two solutions we will need in order to form a general solution that will work for any linear, constant coefficient, second order differential equation. This is easier than it might initially look.

We will use the solutions we found in the first example as a guide. All of the solutions in this example were in the form

$$y(t) = \mathrm{e}^{rt}.$$

Note, that we didn't include a constant in front of it since we can literally include any constant that we want and still get a solution. The important idea here is to get the exponential function. Once we have that we can add on constants to our hearts content.

So, let's assume that all solutions to

$$ay'' + by' + cy = 0$$

will be of the form

$$y(t) = e^{rt}.$$

To see if we are correct all we need to do is plug this into the differential equation and see what happens. So, let's get some derivatives and then plug in.

$$y'(t) = re^{rt} \qquad\qquad y''(t) = r^2 e^{rt}$$

$$a(r^2 e^{rt}) + b(re^{rt}) + c(e^{rt}) = 0$$

$$e^{rt}(ar^{rt} + br + c) = 0.$$

So, the following must be true

$$e^{rt}(ar^2 + br + c) = 0.$$

This can be reduced further by noting that exponentials are never zero. Therefore, it will be a solution to the equation provided r is a solution to

$$ar^2 + br + c = 0.$$

This equation is typically called the characteristic equation.

Okay, so how do we use this to find solutions to a linear, constant coefficient, second order differential equation? First write down the characteristic equation, for the differential equation. This will be a quadratic equation and so we should expect two roots, r_1 and r_2. Once we have these two roots we have two solutions to the differential equation.

$$y_1(t) = e^{r_1 t} \quad \text{and} \quad y_2(t) = e^{r_2 t}.$$

Let's take a look at a quick example.

Example: Find two solutions to

$$y'' - 9y = 0.$$

Solution: This is the same differential equation that we looked at in the first example. This time however, let's not just guess. Let's go through the process as out lined above to see the functions that we guess above are the same as the functions the process gives us. First write down the characteristic equation for this differential equation and solve it.

$$r^2 - 9 = 0 \quad \Rightarrow \quad r = \pm 3.$$

The two roots are 3 and -3. Therefore, two solutions are

$$y_1(t) = e^{3t} \quad \text{and} \quad y_2(t) = e^{-3t}.$$

These match up with the first guesses that we made in the first example.

You'll notice that we neglected to mention whether or not the two solutions listed are in fact "nice enough" to form the general solution. This was intentional. We have three cases that we need to look at and this will be addressed differently in each of these cases. So, what are the cases? As we previously noted the characteristic equation is quadratic and so will have two roots, r_1 and r_2. The roots will have three possible forms. These are

Real, distinct roots, $r_1 \neq r_2$.

1. Complex root, $r_{1,2} = \lambda \pm \mu i$.
2. Double roots, $r_1 = r_2 = r$.

The next three sections will look at each of these in some more depth, including giving forms for the solution that will be "nice enough" to get a general solution.

REAL, DISTINCT ROOTS

It's time to start solving constant coefficient, homogeneous, linear, second order differential equations. So, let's recap how we do this from the last section. We start with the differential equation.

$$ay''+by'+cy=0.$$

Write down the characteristic equation.

$$ar^2+br+c=0.$$

Solve the characteristic equation for the two roots, r_1 and r_2. This gives the two solutions

$$y_1(t)=e^{r_1 t} \quad \text{and} \quad y_2(t)=e^{r_2 t}.$$

Now, if the two roots are real and distinct (*i.e.* $r_1 \neq r_2$) it will turn out that these two solutions are "nice enough" to form the general solution

$$y(t)=c_1 e^{r_1 t} + c_2 e^{r_2 t}.$$

As with the last section, we'll ask that you believe us when we say that these are "nice enough". You will be able to prove this easily enough once we reach a later section. With real, distinct roots there really isn't a whole lot to do other than work a couple of examples so let's do that.

Example: Solve the following IVP.

$$y''+11y'+24y=0 \qquad y(0)=0 \qquad y'(0)=-7.$$

Solution: The characteristic equation is

$$r^2+11r+24=0$$
$$(r+8)(r+3)=0$$

Its roots are $r_1=-8$ and $r_2=-3$ and so the general solution and its derivative is.

$$y(t)=c_1 e^{-8t}+c_2 e^{-3t}$$
$$y'(t)=-8c_1 e^{-8t}-3c_2 e^{-3t}.$$

Now, plug in the initial conditions to get the following system of equations.

$$0=y(0)=c_1+c_2$$
$$-7=y'(0)=-8c_1-3c_2.$$

Solving this system gives $c_1=\frac{7}{5}$ and $c_2=\frac{7}{5}$. The actual solution to the differential equation is then

$$y(t)=\frac{7}{5}e^{-8t}-\frac{7}{5}e^{-3t}.$$

Example: Solve the following IVP

$$y''+3y'-10y=0 \qquad y(0)=4 \quad y'(0)=-2.$$

Solution: The characteristic equation is

$$r^2+3r-10=0$$

$$(r+5)(r-2)=0.$$

Its roots are $r_1=-5$ and $r_2=2$ and so the general solution and its derivative is.-

$$y(t)=c_1e^{-5t}+c_1e^{2t}$$

$$y'(t)=-5c_1e^{-5t}+2c_2e^{2t}.$$

Now, plug in the initial conditions to get the following system of equations.

$$4=y(0)=c_1+c_2$$

$$-2=y'(0)=-5c_1+2c_2.$$

Solving this system gives $c_1=\frac{10}{7}$ and $c_2=\frac{18}{7}$. The actual solution to the differential equation is then

$$y(t)=\frac{10}{7}e^{-5t}+\frac{18}{7}e^{2t}.$$

Example: Solve the following IVP.

$$3y''+2y'-8y=0 \qquad y(0)=-6 \qquad y'(0)=-18.$$

Solution: The characteristic equation is

$$3r^2+2r-8=0$$

$$(3r-4)(r+2)=0.$$

Its roots are $r_1=$ and $r_2=-2$ and so the general solution and its derivative is.-

$$y(t)=c_1e^{\frac{4t}{3}}+c_2e^{-2t}$$

$$y'(t)=\frac{4}{3}c_1e^{\frac{4t}{3}}-2c_2e^{-2t}.$$

Now, plug in the initial conditions to get the following system of equations.

$$-6=y(0)=c_1+c_2$$

$$-18=y'(0)=\frac{4}{3}c_1-2c_2.$$

Solving this system gives $c_1 = -9$ and $c_2 = 3$. The actual solution to the differential equation is then.

$$y(t) = -9e^{\frac{4t}{3}} + 3e^{-2t}.$$

Example: Solve the following IVP

$$4y'' - 5y' = 0 \qquad y(-2) = 0 \qquad y'(-2) = 7.$$

Solution: The characteristic equation is

$$4r^2 - 5r = 0$$
$$r(4r - 5) = 0.$$

The roots of this equation are $r_1 = 0$ and $r_2 = \frac{5}{4}$. Here is the general solution as well as its derivative.

$$y(t) = c_1 e^0 + c_2 e^{\frac{5t}{4}} = c_1 + c_2 e^{\frac{5t}{4}}$$
$$y'(t) = \frac{5}{4} c_2 e^{\frac{5t}{4}}.$$

Up to this point all of the initial conditions have been at and this one isn't. Don't get to locked into initial conditions always being at and you just automatically use that instead of the actual value for a given problem. So, plugging in the initial conditions gives the following system of equations to solve.

$$0 = y(-2) = c_1 + c_2 e^{-\frac{5}{2}}$$
$$7 = y'(-2) = \frac{5}{2} c_2 e^{-\frac{5}{2}}.$$

Solving this gives.

$$c_1 = -\frac{28}{5} \qquad c_2 = -\frac{28}{5} e^{\frac{5}{2}}.$$

The solution to the differential equation is then.

$$y(t) = -\frac{28}{5} + \frac{28}{5} e^{\frac{5t}{4}} e^{\frac{5t}{4}} = -\frac{28}{5} + \frac{28}{5} e^{\frac{5t}{4} + \frac{5}{2}}.$$

In a differential equations class most instructors (including me....) tend to use initial conditions at $t = 0$ because it makes the work a little easier for the students as they are trying to learn the subject. However, there is no reason to always expect that this will be the case, so do not start to always expect initial conditions at $t = 0$!

Let's do one final example to make another point that you need to be made aware of.

Example: Find the general solution to the following differential equation.

$$y''-6y'-2y=0.$$

Solution: The characteristic equation is.

$$r^2-6r-2=0.$$

The roots of this equation are.

$$r_{1,2}=3\pm\sqrt{11}.$$

Now, do NOT get excited about these roots they are just two real numbers.

$$r_1=3+\sqrt{11} \quad \text{and} \quad r_1=3-\sqrt{11}.$$

Admittedly they are not as nice looking as we may be used to, but they are just real numbers. Therefore, the general solution is

$$y(t)=c_1e^{(3+\sqrt{11})}+c_2e^{(3-\sqrt{11})t}.$$

If we had initial conditions we could proceed as we did in the previous two examples although the work would be somewhat messy and so we aren't going to do that for this example.

The point of the last example is make sure that you don't get to used to "nice", simple roots. In practice roots of the characteristic equation will generally not be nice, simple integers or fractions so don't get too used to them!

COMPLEX ROOTS

In this section we will be looking at solutions to the differential equation

$$ay''+by'+cy=0$$

in which roots of the characteristic equation,

$$ar^2+br+c=0$$

are complex roots in the form $r_{1,2}=\lambda\pm\mu i$.

Now, recall that we arrived at the characteristic equation by assuming that all solutions to the differential equation will be of the form

$$y(t)=e^{rt}.$$

Plugging our two roots into the general form of the solution gives the following solutions to the differential equation.

$$y_1(t)=e^{(\lambda+\mu i)t} \quad \text{and} \quad y_2(t)=e^{(\lambda+\mu i)t}.$$

Now, these two functions are "nice enough" (there's those words again... we'll get around to defining them eventually) to form the general solution. We do have a problem however. Since we started with only real numbers in our differential equation we would like our solution to only involve real numbers. The two solutions above are complex and so we would like to get our hands on a couple of solutions ("nice enough" of course...) that are real.

To do this we'll need Euler's Formula.

$$e^{i\theta} = \cos\theta + i\sin\theta.$$

A nice variant of Euler's Formula that we'll need is.

$$e^{-i\theta} = \cos(-\theta) + i\sin(-\theta) = \cos\theta - i\sin\theta.$$

Now, split up our two solutions into exponentials that only have real exponents and exponentials that only have imaginary exponents. Then use Euler's formula, or its variant, to rewrite the second exponential.

$$y_1(t) = e^{\lambda t}e^{i\mu t} = e^{\lambda t}(\cos(\mu t) + i\sin(\mu t))$$

$$y_2(t) = e^{\lambda t}e^{-i\mu t} = e^{\lambda t}(\cos(\mu t) - i\sin(\mu t)).$$

This doesn't eliminate the complex nature of the solutions, but it does put the two solutions into a form that we can eliminate the complex parts.

Recall from the basics section that if two solutions are "nice enough" then any solution can be written as a combination of the two solutions. In other words,

$$y(t) = c_1y_1(t) + c_2y_2(t)$$

will also be a solution.

Using this let's notice that if we add the two solutions together we will arrive at.

$$y_1(t) + y_2(t) = 2e^{\lambda t}\cos(\mu t).$$

This is a real solution and just to eliminate the extraneous 2 let's divide everything by a 2. This gives the first real solution that we're after.

$$u(t) = \frac{1}{2}y_1 + \frac{1}{2}y_2(t) = e^{\lambda t}\cos(\mu t).$$

Note that this is just equivalent to taking

$$c_1 = c_2 = \frac{1}{2}.$$

Now, we can arrive at a second solution in a similar manner. This time let's subtract the two original solutions to arrive at.

$$y_1(t) - y_2(t) = 2ie^{\lambda t}\sin(\mu t).$$

On the surface this doesn't appear to fix the problem as the solution is still complex. However, upon learning that the two constants, c_1 and c_2 can be complex numbers we can arrive at a real solution by dividing this by $2i$. This is equivalent to taking

$$c_1 = \frac{1}{2i} \quad \text{and} \quad c_2 = \frac{1}{2i}.$$

Our second solution will then be

$$v(t) = \frac{1}{2i} y_1(t) - \frac{1}{2i} y_2(t) = e^{\lambda t} \sin(\mu t).$$

We now have two solutions (we'll leave it to you to check that they are in fact solutions) to the differential equation.

$$u(t) = e^{\lambda t} \cos(\mu t) \quad \text{and} \quad v(t) = e^{\lambda t} \sin(\mu t).$$

It also turns out that these two solutions are "nice enough" to form a general solution. So, if the roots of the characteristic equation happen to be $r_{1,2} = \lambda \pm \mu i$the general solution to the differential equation is.

$$y(t) = c_1 e^{\lambda t} \cos(\mu t) + c_2 e^{\lambda t} \sin(\mu t).$$

Let's take a look at a couple of examples now.

Example: Solve the following IVP.

$$y'' - 4y' + 9y = 0 \quad y(0) = 0 \quad y'(0) = -8.$$

Solution: The characteristic equation for this differential equation is.

$$r^2 - 4r + 9 = 0.$$

The roots of this equation are $r_{1,2} = 2 \pm \sqrt{5}i$. The general solution to the differential equation is then.

$$y(t) = c_1 e^{2t} \cos\left(\sqrt{5}t\right) + c_2 e^{2t} \sin\left(\sqrt{5}t\right).$$

Now, you'll note that we didn't differentiate this right away as we did in the last section. The reason for this is simple. While the differentiation is not terribly difficult, it can get a little messy. So, first looking at the initial conditions we can see from the first one that if we just applied it we would get the following.

$$0 = y(0) = c_1.$$

In other words, the first term will drop out in order to meet the first condition. The makes the solution, along with its derivative

$$y(t) = c_2 e^{2t} \sin\left(\sqrt{5}t\right)$$
$$y'(t) = 2c_2 e^{2t} \sin\left(\sqrt{5}t\right) + \sqrt{5} c_2 e^{2t} \cos\left(\sqrt{5}t\right).$$

A much nicer derivative than if we'd done the original solution. Now, apply the second initial condition to the derivative to get.

$$-8 = y'(0) = \sqrt{5}c_2 \quad \Rightarrow \quad c_2 = -\frac{8}{\sqrt{5}}.$$

The actual solution is then.

$$y(t) = -\frac{8}{\sqrt{5}}e^{2t}\sin\left(\sqrt{5}t\right).$$

Example: Solve the following IVP.

$$y''-8y'+17y=0 \qquad y(0)=-4 \quad y'(0)=-1.$$

Solution: The characteristic equation this time is.

$$r^2-8r+17=0.$$

The roots of this are $r_{1,2} = 4 \pm i$. The general solution as well as its derivative is

$$y(t) = c_1e^{4t}\cos(t) + c_2e^{4t}\sin(t)$$

$$y'(t) = 4c_1e^{4t}\cos(t) - c_2e^{4t}\sin(t) + 4c_2e^{4t}\sin(t) + c_2e^{4t}\cos(t).$$

Notice that this time we will need the derivative from the start as we won't be having one of the terms drop out. Applying the initial conditions gives the following system.

$$-4 = y(0) = c_1$$

$$-1 = y'(0) = 4c_1 + c_2.$$

Solving this system gives $c_1 = -4$and $c_2 = 15$. The actual solution to the IVP is then.

$$y(t) = -4e^{4t}\cos(t) + 15e^{4t}\sin(t).$$

Example: Solve the following IVP.

$$4y''+24y'+37y=0 \qquad y(\pi)=1 \quad y'(\pi)=0.$$

Solution: The characteristic equation this time is.

$$4r^2+24r+37=0.$$

The roots of this are $r_{1,2} = -3 \pm \frac{1}{2}i$. The general solution as well as its derivative is

$$y(t) = c_1e^{-3t}\cos\left(\frac{t}{2}\right) + c_2e^{-3t}\sin\left(\frac{t}{2}\right)$$

$$y'(t) = -3c_1e^{-3t}\cos\left(\frac{t}{2}\right) - \frac{c_1}{2}e^{-3t}\sin\left(\frac{t}{2}\right) - 3c_2e^{-3t}\sin\left(\frac{t}{2}\right) + \frac{c_2}{2}e^{-3t}\cos\left(\frac{t}{2}\right).$$

Applying the initial conditions gives the following system.

$$1 = y(\pi) = c_1 e^{-3\pi} \cos\left(\frac{\pi}{2}\right) + c_2 e^{-3\pi} \sin\left(\frac{\pi}{2}\right) = c_2 e^{-3\pi}$$

$$0 = y'(\pi) = -\frac{c_1}{2} e^{-3\pi} - 3c_2 e^{-3\pi}.$$

Do not forget to plug the $t =$ ðinto the exponential! This is one of the more common mistakes that students make on these problems. Also, make sure that you evaluate the trig functions as much as possible in these cases. It will only make your life simpler. Solving this system gives

$$c_1 = -6e^{3\pi} \qquad c_2 = e^{3\pi}.$$

The actual solution to the IVP is then.

$$y(t) = -6e^{3\pi} e^{-3t} \cos\left(\frac{t}{2}\right) + e^{3\pi} e^{-3t} \sin\left(\frac{t}{2}\right)$$

$$y(t) = -6e^{-3(t-\pi)} \cos\left(\frac{t}{2}\right) + e^{-3(t-\pi)} \sin\left(\frac{t}{2}\right).$$

Let's do one final example before moving on to the next topic.

Example: Solve the following IVP.

$$y'' + 16y = 0 \qquad y\left(\frac{\pi}{2}\right) = -10 \qquad y'\left(\frac{\pi}{2}\right) = 3.$$

Solution: The characteristic equation for this differential equation and its roots are.

$$r^2 + 16 = 0 \Rightarrow r = \pm 4i.$$

The general solution to this differential equation and its derivative is.

$$y(t) = c_1 \cos(4t) + c_2 \sin(4t)$$

$$y'(t) = -4c_1 \sin(4t) + 4c_2 \cos(4t).$$

Plugging in the initial conditions gives the following system.

$$-10 = y\left(\frac{\pi}{2}\right) = c_1 \qquad c_1 = -10$$

$$3 = y'\left(\frac{\pi}{2}\right) = 4c \qquad c_2 = \frac{3}{4}.$$

So, the constants drop right out with this system and the actual solution is.

$$y(t) = -10\cos(4t) + \frac{3}{4}\sin(4).$$

REPEATED ROOTS

In this section we will be looking at the last case for the constant coefficient, linear, homogeneous second order differential equations. In this case we want solutions to

$$ay^{n} + by' + cy = 0$$

where solutions to the characteristic equation

$$ar^2 + br + c = 0$$

are double roots $r_1 = r_2 = r$.

This leads to a problem however. Recall that the solutions are

$$y_1(t) = e^{r_1 t} = e^{n} \qquad y_2(t) = e^{r_2 t} = e^{rt}.$$

These are the same solution and will NOT be "nice enough" to form a general solution. I do promise that I'll define "nice enough" eventually! So, we can use the first solution, but we're going to need a second solution.

Before finding this second solution let's take a little side trip. The reason for the side trip will be clear eventually. From the quadratic formula we know that the roots to the characteristic equation are,

$$r_{1,2} = \frac{-b \pm \sqrt{b^2 - 4ac}}{2a}.$$

In this case, since we have double roots we must have

$$b^2 - 4ac = 0.$$

This is the only way that we can get double roots and in this case the roots will be

$$r_{1,2} = \frac{-b}{2a}.$$

So, the one solution that we've got is

$$y_1(t) = e^{\frac{bt}{2a}}.$$

To find a second solution we will use the fact that a constant times a solution to a linear homogeneous differential equation is also a solution. If this is true then *maybe* we'll get lucky and the following will also be a solution

$$y_2(t) = v(t)y_1(t) = v(t)e^{\frac{bt}{2a}}$$

with a proper choice of *v(t)*. To determine if this in fact can be done, let's plug this back into the differential equation and see what we get. We'll first need a couple of derivatives.

$$y_2'(t) = v'e^{\frac{bt}{2a}} - \frac{b}{2a}ve^{-\frac{bt}{2a}}$$

$$y_2''(t) = v''e^{\frac{bt}{2a}} - \frac{b}{2a}v'e^{-\frac{bt}{2a}} - \frac{b}{2a}v'e^{-\frac{bt}{2a}} + \frac{b^2}{4a}ve^{-\frac{bt}{2a}}$$

$$= v''e^{\frac{bt}{2a}} - \frac{b}{2a}v'e^{-\frac{bt}{2a}} + \frac{b^2}{4a^2}ve^{-\frac{bt}{2a}}.$$

We dropped the *(t)* part on the *v* to simplify things a little for the writing out of the derivatives. Now, plug these into the differential equation.

$$a\left(v''e^{\frac{bt}{2a}} - \frac{b}{a}v'e^{-\frac{bt}{2a}} + \frac{b^2}{4a^2}ve^{-\frac{bt}{2a}}\right) + b\left(v'e^{\frac{bt}{2a}} \frac{b}{2a}ve^{\frac{bt}{2a}}\right) + c\left(ve^{-\frac{bt}{2a}}\right) = 0.$$

We can factor an exponential out of all the terms so let's do that. We'll also collect all the coefficients of *v* and its derivatives.

$$e^{-\frac{bt}{2a}}\left(av'' + (-b+b)v' + \left(\frac{b^2}{4a} - \frac{b^2}{2a} + c\right)v\right) = 0$$

$$e^{-\frac{bt}{2a}}\left(av'' + \left(-\frac{b^2}{4a} + c\right)v\right) = 0$$

$$e^{-\frac{bt}{2a}}\left(av'' - \frac{1}{4a}\left(b^2 - 4ac\right)v\right) = 0.$$

Now, because we are working with a double root we know that that the second term will be zero. Also exponentials are never zero. Therefore, equation will be a solution to the differential equation provided *v(t)* is a function that satisfies the following differential equation.

$$av'' = 0 \quad \text{OR} \quad v'' = 0.$$

We can drop the *a* because we know that it can't be zero. If it were we wouldn't have a second order differential equation! So, we can now determine the most general possible form that is allowable for *v(t)*.

$$v' = \int v''dt = c \qquad v = \int v'dt = ct + k.$$

This is actually more complicated than we need and in fact we can drop both of the constants from this. To see why this is let's go ahead and use this to get the second solution. The two solutions are then

$$y_1(t) = e^{-\frac{bt}{2a}} \qquad y_2(t) = (ct+k)e^{-\frac{bt}{2a}}.$$

Eventually you will be able to show that these two solutions are "nice enough" to form a general solution. The general solution would then be the following.

$$\begin{aligned} y(t) &= c_1 e^{-\frac{bt}{2a}} + c_2(ct+k)\, e^{\frac{bt}{2a}} \\ &= c_1 e^{-\frac{bt}{2a}} + (c_2 ct + c_2 k)\, e^{-\frac{bt}{2a}} \\ &= (c_1 + c_2 k)\, e^{-\frac{bt}{2a}} + c_2 ct e^{-\frac{bt}{2a}}. \end{aligned}$$

Notice that we rearranged things a little. Now, c, k, c_1, and c_2 are all unknown constants so any combination of them will also be unknown constants. In particular, $c_1 + c_2 k$ and $c_2 c$ are unknown constants so we'll just rewrite them as follows.

$$y(t) = c_1 e^{-\frac{bt}{2a}} + c_2 t e^{-\frac{bt}{2a}}.$$

So, if we go back to the most general form for $v(t)$ we can take $c=1$ and $k=0$ and we will arrive at the same general solution. Let's recap. If the roots of the characteristic equation are $r_1 = r_2 = r$, then the general solution is then

$$y(t) = c_1 e^{rt} + c_2 t e^{rt}.$$

Now, let's work a couple of examples.

Example: Solve the following IVP.

$$y'' - 4y' + 4y = 0 \qquad y(0) = 12 \qquad y'(0) = -3.$$

Solution: The characteristic equation and its roots are.

$$r^2 - 4r + 4 = (r-2)^2 = 0 \qquad r_{1,2} = 2.$$

The general solution and its derivative are

$$\begin{aligned} y(t) &= c_1 e^{2t} + c_2 t e^{2t} \\ y'(t) &= 2c_1 e^{2t} + c_2 e^{2t} + 2c_2 t e^{2t}. \end{aligned}$$

Don't forget to product rule the second term! Plugging in the initial conditions gives the following system.

$$\begin{aligned} 12 &= y(0) = c_1 \\ -3 &= y'(0) = 2c_1 + c_2. \end{aligned}$$

This system is easily solve to get $c_1 = 12$ and $c_2 = -27$. The actual solution to the IVP is then.

$$y(t) = 12e^{2t} - 27te^{2t}.$$

Example: Solve the following IVP.

$$16y'' - 40y' + 25y = 0 \quad y(0) = 3 \quad y'(0) = -\frac{9}{4}.$$

Solution: The characteristic equation and its roots are.

$$16r^2 - 40r + 25 = (4r - 5) = 0 \qquad r_{1,2} = \frac{5}{4}.$$

The general solution and its derivative are

$$y(t) = c_1 e^{\frac{5t}{4}} + c_2 t e^{\frac{5t}{4}}$$

$$y'(t) = \frac{5}{4} c_1 e^{\frac{5t}{4}} + c_2 e^{\frac{5t}{4}} + \frac{5}{4} c_2 t e^{\frac{5t}{4}}.$$

Don't forget to product rule the second term! Plugging in the initial conditions gives the following system.

$$3 = y(0) = c_1$$

$$-\frac{9}{4} = y'(0) = \frac{5}{4} c_1 + c_2.$$

This system is easily solve to get $c_1 = 3$ and $c_2 = -6$. The actual solution to the IVP is then.

$$y(t) = 3e^{\frac{5t}{4}} - 6te^{\frac{5t}{4}}.$$

Example: Solve the following IVP

$$y'' + 14y' + 49y = 0 \qquad y(-4) = -1 \qquad y'(-4) = 5.$$

Solution: The characteristic equation and its roots are.

$$r^2 + 14r + 49 = (r + 7)^2 = 0 \qquad r_{1,2} = -7.$$

The general solution and its derivative are

$$y(t) = c_1 e^{-7t} + c_2 t e^{-7t}$$

$$y'(t) = -7c_1 e^{-7t} + c_2 t e^{-7t} - 7c_2 t e^{-7t}.$$

Plugging in the initial conditions gives the following system of equations.

$$-1 = y(-4) = c_1 e^{28} - 4c_2 e^{28}$$

$$5 = y'(-4) = -7c_1 e^{28} + c_2 e + 28c_2 e^{28} = -7c_1 e^{28} + 29c_2 e^{28}.$$

Solving this system gives the following constants.

$$c_1 = -9e^{-28} \qquad c_2 = -2e^{-28}.$$

The actual solution to the IVP is then.

$$y(t) = -9e^{-28} e^{-7t} - 2te^{-28} e^{-7t}$$

$$y(t) = -9e^{-7(t+4)} - 2te^{-7(t+4)}.$$

REDUCTION OF ORDER

We're now going to take a brief detour and look at solutions to non-constant coefficient, second order differential equations of the form.

$$p(t)y''+q(t)y'+r(t)y=0.$$

In general, finding solutions to these kinds of differential equations can be much more difficult than finding solutions to constant coefficient differential equations. However, if we already know one solution to the differential equation we can use the method that we used in the last section to find a second solution. This method is called reduction of order.

Let's take a quick look at an example to see how this is done.

Example: Find the general solution to

$$2t^2y''+ty'-3y=0.$$

given that $y_1(t)=t^{-1}$ is a solution.

Solution: Reduction of order requires that a solution already be known. Without this known solution we won't be able to do reduction order.

Once we have this first solution we will then assume that a second solution will have the form

$$y_2(t)=v(t)y_1(t)$$

for a proper choice of *v(t)*. To determine the proper choice, we plug the guess into the differential equation and get a new differential equation that can be solved for *v(t)*.

So, let's do that for this problem. Here is the form of the second solution as well as the derivatives that we'll need.

$$y_2(t)=t^{-1}v \qquad y_2'(t)=-t^{-2}v+r^{-1}v' \qquad y_2''(t)=2t^{-3}v-2t^{-2}v'+t^{-1}v''.$$

Plugging these into the differential equation gives

$$2t^2(2t^{-3}v-2t^{-2}v'+t^{-1}v'')+t(-t^{-2}v+t^{-1}v')-3(t^{-1}v)=0.$$

Rearranging and simplifying gives

$$2tv''+(-4+1)v'+(4t^{-1}-t^{-1}-3t^{-1})v=0$$

$$2tv''-3v'=0.$$

Note that upon simplifying the only terms remaining are those involving the derivatives of v. The term involving v drops out. If you've done all of your work correctly this should always happen. Sometimes, as in the repeated roots case, the first derivative term will also drop out. So, in order to be a solution then v must satisfy

$$2tv''-3v'=0.$$

This appears to be a problem. In order to find a solution to a second order non-constant coefficient differential equation we need to solve a different second order non-constant coefficient differential equation.

However, this isn't the problem that it appears to be. Because the term involving the v drops out we can actually solve and we can do it with the knowledge that we already have at this point. We will solve this by making the following change of variable.

$$w = v' \quad \Rightarrow \quad w' = v''$$

With this change of variable becomes

$$2tw' - 3w = 0$$

and this is a linear, first order differential equation that we can solve. This also explains the name of this method. We've managed to reduce a second order differential equation down to a first order differential equation.

This is a fairly simple first order differential equation so I'll leave the details of the solving to you. If you need a refresher on solving linear, first order differential equations go back to the second chapter and check out that section. The solution to this differential equation is

$$w(t) = ct^{\frac{3}{2}}.$$

Now, this is not quite what we were after. However, we can now find this. Recall our change of variable.

$$v' = w.$$

With this we can easily solve for *v(t)*.

$$v(t) = \int w\,dt = \int ct^{\frac{3}{2}}dt = \frac{2}{5}ct^{\frac{5}{2}} + k.$$

This is the most general possible *v(t)* that we can use to get a second solution. So, just as we did in the repeated roots section, we can choose the constants to be anything we want so choose them to clear out all the extraneous constants. In this case we can use

$$c = \frac{5}{2} \qquad k = 0.$$

Using these gives the following for *v(t)* and for the second solution.

$$v(t) = t^{\frac{5}{2}} \quad \Rightarrow \quad y_2(t) = t^{-1}\left(t^{\frac{5}{2}}\right) = t^{\frac{3}{2}}.$$

Then general solution will then be,

$$y(t) = c_1 t^{-1} + c_2 t^{\frac{3}{2}}.$$

If we had been given initial conditions we could then differentiate, apply the initial conditions and solve for the constants.

Reduction of order, the method used in the previous example can be used to find second solutions to differential equations. However, this does require that we already have a solution and often finding that first solution is a very difficult task and often in the process of finding the first solution you will also get the second solution without needing to resort to reduction of order. So, for those cases when we do have a first solution this is a nice method for getting a second solution.

Let's do one more example.

Example: Find the general solution to

$$t^2 y''+2ty'-2y=0$$

given that $y_1(t) = t$ is a solution.

Solution: The form for the second solution as well as its derivatives are,

$$y_2(t)=tv \quad y_2'(t)=v+tv' \quad y_2''(t)=2v'+tv'.$$

Plugging these into the differential equation gives,

$$t^2(2v'+tv'')+2t(v+tv')-2(tv)=0=0.$$

Rearranging and simplifying gives the differential equation that we'll need to solve in order to determine the correct v that we'll need for the second solution.

$$t^3v''+4t^2v'=0.$$

Next use the variable transformation as we did in the previous example.

$$w=v' \quad \Rightarrow \quad w'=v''.$$

With this change of variable the differential equation becomes

$$t^3w'+4t^2w=0$$

and this is a linear, first order differential equation that we can solve. We'll leave the details of the solution process to you.

$$w(t)=ct^{-4}.$$

Now solve for $v(t)$.

$$v(t)=\int w\,dt=\int ct^{-4}dt=-\frac{1}{3}ct^{-3}+k.$$

As with the first example we'll drop the constants and use the following $v(t)$

$$v(t)=t^{-3} \quad \Rightarrow \quad y_2(t)=t(t^{-3})=t^{-2}.$$

Then general solution will then be,

$$y(t) = c_1 t + \frac{c_2}{t^2}.$$

On a side note, both of the differential equations in this section were of the form,

$$t^2 y'' + \alpha t\, y' + \beta y = 0.$$

These are called Euler differential equations and are fairly simple to solve directly for both solutions. To see how to solve these directly take a look at the Euler Differential Equation section.

FUNDAMENTAL SETS OF SOLUTIONS

The time has finally come to define "nice enough". We've been using this term throughout the last few sections to describe those solutions that could be used to form a general solution and it is now time to officially define it.

First, because everything that we're going to be doing here only requires linear and homogeneous we won't require constant coefficients in our differential equation. So, let's start with the following IVP.

$$p(t)y'' + q(t)y' + r(t)y = 0$$

$$y(t_0) = y_0 \qquad y'(t_0) = y_0'.$$

Let's also suppose that we have already found two solutions to this differential equation, $y_1(t)$ and $y_2(t)$. We know from the Principle of Superposition that

$$y(t) = c_1 y_1(t) + c_2 y_2(t).$$

Will also be a solution to the differential equation. What we want to know is whether or not it will be a general solution.

$$y(t_0) = y_0 \qquad y'(t_0) = y_0'.$$

This will also imply that any solution to the differential equation can be written in this form. So, let's see if we can find constants that will satisfy these conditions. First differentiate and plug in the initial conditions.

$$y_0 = y(t_0) = c_1 y_1(t_0) + c_2 t_2(t_0)$$

$$y_0' = y'(t_0) = c_1 y_1'(t_0) + c_2 t_2'(t_0).$$

Since we are assuming that we've already got the two solutions everything in this system is technically known and so this is a system that can be solved for c_1 and c_2. This can be done in general using Cramer's Rule. Using Cramer's Rule gives the following solution.

$$c_1 = \frac{\begin{vmatrix} y_0 & y_0(t_0) \\ y_0' & y_2'(t_0) \end{vmatrix}}{\begin{vmatrix} y_1(t_0) & y_0(t_0) \\ y_1'(t_0) & y_2'(t_0) \end{vmatrix}} \qquad c_2 = \frac{\begin{vmatrix} y_1(t_0) & y_0 \\ y_1'(t_0) & y_0' \end{vmatrix}}{\begin{vmatrix} y_1(t_0) & y_2(t_0) \\ y_1'(t_0) & y_2'(t_0) \end{vmatrix}}$$

where,

$$\begin{vmatrix} a & b \\ c & d \end{vmatrix} = ad - bc$$

is the determinant of a 2x2 matrix. If you don't know about determinants that is okay, just use the formula that we've provided above.

Now, equation will give the solution to the system. Note that in practice we generally don't use Cramer's Rule to solve systems, we just proceed in a straightforward manner and solve the system using basic algebra techniques. So, why did we use Cramer's Rule here then?

We used Cramer's Rule because we can use to develop a condition that will allow us to determine when we can solve for the constants. All three (yes three, the denominators are the same!) of the quantities are just numbers and the only thing that will prevent us from actually getting a solution will be when the denominator is zero.

The quantity in the denominator is called the Wronskian and is denoted as

$$W(f,g)(t) = \begin{vmatrix} f(t) & g(t) \\ f'(t) & g'(t) \end{vmatrix} = f(t)g'(t) - g(t)f'(t).$$

When it is clear what the functions and/or t are we often just denote the Wronskian by W. Let's recall what we were after for here. We wanted to determine when two solutions would be nice enough to form a general solution. The two solutions will form a general solution if they satisfy the general initial conditions given and we can see from Cramer's Rule we can see that they will satisfy the initial conditions provided the Wronskian isn't zero. Or,

$$W(y_1, y_2)(t_0) = \begin{vmatrix} y_1(t_0) & y_2(t_0) \\ y_1'(t_0) & y_2'(t_0) \end{vmatrix} = y_1(t_0)y_2'(t_0) - y_2(t_0)y_1'(t_0) \neq 0.$$

So, suppose that $y_1(t)$ and $y_2(t)$ are two solutions and that $W(y_1, y_2)(t) \neq 0$. Then the two solutions are called a fundamental set of solutions and the general solution to equation is

$$t(y) = c_1 y_1(t) + c_2 y_2(t).$$

We know now what "nice enough" means. Two solutions are "nice enough" if they are a fundamental set of solutions.

So, let's check one of the claims that we made in a previous section. We'll leave the other two to you to check if you'd like to.

Example: Back in the complex root section we made the claim that

$$y_1(t) = e^{\lambda t}\cos(\mu t) \qquad \text{and} \qquad y_2(t) = e^{\lambda t}\sin(\mu t)$$

where a fundamental set of solutions. Prove that they in fact are.

Solution: So, to prove this we will need to take find the Wronskian for these two solutions and show that it isn't zero.

$$W = \begin{vmatrix} e^{\lambda t}\cos(\mu t) & e^{\lambda t}\sin(\mu t) \\ \lambda e^{\lambda t}\cos(\mu t) - \mu e^{\lambda t}\sin(\mu t) & \lambda e^{\lambda t}\sin(\mu t) + \mu e^{\lambda t}\cos(\mu t) \end{vmatrix}$$

$$= e^{\lambda t}\cos(\mu t)(\lambda e^{\lambda t}\sin(\mu t) + \mu e^{\lambda t}\cos(\mu t)) - e^{\lambda t}\sin(\mu t)(\lambda e^{\lambda t}\cos(\mu t) - \mu e^{\lambda t}\sin(\mu t))$$

$$= \mu e^{2\lambda t}\cos^2(\mu t) + \mu e^{2\lambda t}\sin^2(\mu t)$$

$$= \mu e^{2\lambda t}(\cos^2(\mu t) + \sin^2(\mu t))$$

$$= \mu e^{2\lambda t}.$$

Now, the exponential will never be zero and $\mu \neq 0$(if it were we wouldn't have complex roots!) and so $W \neq 0$. Therefore, these two solutions are in fact a fundamental set of solutions and so the general solution in this case is.

$$y(t) = c_1 e^{\lambda t}\cos(\mu t) + c_2 e^{\lambda t}\sin(\mu t).$$

Example: In the first example that we worked in the Reduction of Order section we found a second solution to

$$2t^2 y'' + ty' - 3y = 0.$$

Show that this second solution, along with the given solution, form a fundamental set of solutions for the differential equation.

Solution: The two solutions from that example are

$$y_1(t) = t^{-1} \qquad y_2(t) = t^{\frac{3}{2}}.$$

Let's compute the Wronskian of these two solutions.

$$W = \begin{vmatrix} t^{-1} & t^{\frac{3}{2}} \\ -t^{-2} & \frac{3}{2}t^{\frac{1}{2}} \end{vmatrix} = \frac{3}{2}t^{-\frac{3}{2}} - \left(-t^{-\frac{3}{2}}\right) = \frac{5}{2}t^{-\frac{1}{2}} = \frac{5}{2\sqrt{t}}.$$

So, the Wronskian will never be zero. Note that we can't plug $t = 0$ into the Wronskian. This would be a problem in finding the constants in the general solution, except that we also can't plug $t = 0$ into the solution either and so this isn't the problem that it might appear to be. So, since the Wronskian isn't zero for any t the two solutions form a fundamental set of solutions and the general solution is

$$y(t) = c_1 t^{-1} + c_2 t^{\frac{3}{2}}$$

as we claimed in that example.

To this point we've found a set of solutions then we've claimed that they are in fact a fundamental set of solutions. Of course, you can now verify all those claims that we've made, however this does bring up a question. How do we know that for a given differential equation a set of fundamental solutions will exist? The following theorem answers this question.

Theorem: Consider the differential equation

$$y'' + p(t)y' + q(t)y = 0.$$

where $p(t)$ and $q(t)$ are continuous functions on some interval I. Choose t_0 to be any point in the interval I. Let $y_1(t)$ be a solution to the differential equation that satisfies the initial conditions.

$$y(t_0) = 1 \qquad y'(t_0) = 0.$$

Let $y_2(t)$ be a solution to the differential equation that satisfies the initial conditions.

$$y(t_0) = 0 \qquad y'(t_0) = 1.$$

Then $y_1(t)$ and $y_2(t)$ form a fundamental set of solutions for the differential equation. It is easy enough to show that these two solutions form a fundamental set of solutions. Just compute the Wronskian.

$$W(y_1, y_2)(t_0) = \begin{vmatrix} y_1(t_0) & y_2(t_0) \\ y_1'(t_0) & y_2'(t_0) \end{vmatrix} = \begin{vmatrix} 1 & 0 \\ 0 & 1 \end{vmatrix} = 1 - 0 = 1 \neq 0.$$

So, fundamental sets of solutions will exist provided we can solve the two IVP's given in the theorem.

Example: Use the theorem to find a fundamental set of solutions for

$$y'' + 4y' + 3y = 0$$

using $t_0 = 0$.

Solution: Using the techniques from the first part of this chapter we can find the two solutions that we've been using to this point.

$$y(t) = e^{-3t} \qquad y(t) = e^{-t}.$$

These do form a fundamental set of solutions as we can easily verify. However, they are NOT the set that will be given by the theorem. Neither of these solutions will satisfy either of the two sets of initial conditions given in the theorem. We will have to use these to find the fundamental set of solutions that is given by the theorem.

We know that the following is also solution to the differential equation.

$$y(t) = c_1 e^{-3t} + c_2 e^{-t}.$$

So, let's apply the first set of initial conditions and see if we can find constants that will work.

$$y(0) = 1 \qquad y'(0) = 0.$$

We'll leave it to you to verify that we get the following solution upon doing this.

$$y_1(t) = -\frac{1}{2}e^{-3t} + \frac{3}{2}e^{-t}.$$

Likewise, if we apply the second set of initial conditions,

$$y(0) = 0 \qquad y'(0) = 1$$

we will get

$$y_2(t) = -\frac{1}{2}e^{-3t} + \frac{1}{2}e^{-t}.$$

According to the theorem these should form a fundament set of solutions. This is easy enough to check.

$$W = \begin{vmatrix} -\frac{1}{2}e^{-3t} + \frac{1}{2}e^{-t} & -\frac{1}{2}e^{-3t} + \frac{1}{2}e^{-t} \\ \frac{1}{2}e^{-3t} - \frac{3}{2}e^{-t} & -\frac{1}{2}e^{-3t} - \frac{1}{2}e^{-t} \end{vmatrix}$$

$$= \left(-\frac{1}{2}e^{-3t} + \frac{3}{2}e^{-t}\right)\left(\frac{3}{2}e^{-3t} - \frac{1}{2}e^{-t}\right) - \left(-\frac{1}{2}e^{-3t} + \frac{3}{2}e^{-t}\right)\left(\frac{3}{2}e^{-3t} - \frac{3}{2}e^{-t}\right)$$

$$= e^{-4t} \neq 0.$$

So, we got a completely different set of fundamental solutions from the theorem that what we've been using up to this point. This is not a problem. There are an infinite number of pairs of functions that we could use as a fundamental set of solutions for this problem.

So, which set of fundamental solutions should we use? Well, if we use the ones that we originally found, the general solution would be,

$$y(t) = c_1 e^{-3t} + c_2 e^{-t}.$$

Whereas, if we used the set from the theorem the general solution would be,

$$y(t) = c_1\left(-\frac{1}{2}e^{-3t} + \frac{3}{2}e^{-t}\right) + c_2\left(-\frac{1}{2}e^{-3t} + \frac{3}{2}e^{-t}\right).$$

This would not be very fun to work with when it came to determining the coefficients to satisfy a general set of initial conditions. So, which set of fundamental solutions should we use? We should always try to use the set that is the most convenient to use for a given problem.

MORE ON THE WRONSKIAN

In the previous section we introduced the Wronskian to help us determine whether two solutions were a fundamental set of solutions. In this section we will look at another application of the Wronskian as well as an alternate method of computing the Wronskian.

Let's start with the application. We need to introduce a couple of new concepts first.

Given two non-zero functions *f(x)* and *g(x)* write down the following equation.-

$$c\,f(x) + kg(x) = 0.$$

Notice that $c = 0$ and $k = 0$ will make (1) true for all x regardless of the functions that we use.

Now, if we can find non-zero constants c and k for which will also be true for all x then we call the two functions linearly dependent. On the other hand if the only two constants for which is true are $c = 0$ and $k = 0$ then we call the functions linearly independent.

Example: Determine if the following sets of functions are linearly dependent or linearly independent.

1. $f(x) = 9\cos(2x) \qquad g(x) = 2\cos^2(x) - 2\sin^2(x)$
2. $f(t) = 2t^2 \qquad g(t) = t^4.$

Solution:

1. $f(x) = 9\cos(2x) \qquad g(x) = 2\cos^2(x) - 2\sin^2(x).$

We'll start by writing down (1) for these two functions.

$$c(9\cos(2x) + k(2\cos^2(x) - 2\sin^2(x)) = 0.$$

We need to determine if we can find non-zero constants c and k that will make this true for all x or if $c = 0$ and $k = 0$ are the only constants that will make this true for all x. This is often a fairly difficult process. The process can be simplified with a good intuition for this kind of thing, but that's hard to come by, especially if you haven't done many of these kinds of problems.

In this case the problem can be simplified by recalling

$$\cos^2(x) - \sin^2(x) = \cos(2x).$$

Using this fact our equation becomes.

$$9c\cos(2x) + 2k\cos(2x) = 0$$

$$(9c + 2k)\cos(2x) = 0.$$

With this simplification we can see that this will be zero for any pair of constants c and k that satisfy

$$9c + 2k = 0.$$

Among the possible pairs on constants that we could use are the following pairs.

$$c = 1, \qquad k = -\frac{9}{2}$$

$$c = \frac{2}{9}, \qquad k = -1$$

$$c = -2 \qquad k = 9$$

$$c = -\frac{7}{6} \qquad k = -\frac{21}{4} \text{ etc.}$$

As I'm sure you can see there are literally thousands of possible pairs and they can be made as "simple" or as "complicated" as you want them to be. So, we've managed to find a pair of non-zero constants that will make the equation true for all x and so the two functions are linearly dependent.

2. $f(t) = 2t^2 \qquad g(t) = t^4.$

As with the last part, we'll start by writing for these functions.

$$2ct^2 + kt^4 = 0.$$

In this case there isn't any quick and simple formula to write one of the functions in terms of the other as we did in the first part. So, we're just going to have to see if we can find constants. We'll start by noticing that if the original equation is true, then if we differentiate everything we get a new equation that must also be true. In other words, we've got the following system of two equations in two unknowns.

$$2ct^2 + kt^4 = 0$$

$$4ct + 4kt^3 = 0.$$

We can solve this system for c and k and see what we get. We'll start by solving the second equation for c.

$$c = -kt^2.$$

Now, plug this into the first equation.

$$2(-kt^2)t^2 + kt^4 = 0$$

$$-kt^4 = 0.$$

Recall that we are after constants that will make this true for all t. The only way that this will ever be zero for all t is if $k = 0$! So, if $k = 0$ we must also have $c = 0$.

Therefore, we've shown that the only way that

$$2ct^2 + kt^4 = 0$$

will be true for all t is to require that $c = 0$ and $k = 0$. The two functions therefore, are linearly independent.

As we saw in the previous examples determining whether two functions are linearly independent or linearly dependent can be a fairly involved process. This is where the Wronskian can help.

FACT

Given two functions $f(x)$ and $g(x)$ that are differentiable on some interval I.

1. If $W(f, g)(x_0) \neq 0$ for some x_0 in I, then $f(x)$ and $g(x)$ are linearly independent on the interval I.
2. If $f(x)$ and $g(x)$ are linearly dependent on I then $W(f,g)(x) = 0$ for all x in the interval I.

Be very careful with this fact. It DOES NOT say that if $W(f,g)(x) = 0$ then $f(x)$ and $g(x)$ are linearly dependent! In fact it is possible for two linearly independent functions to have a zero Wronskian!

This fact is used to quickly identify linearly independent functions and functions that are liable to be linearly dependent.

Example: Verify the fact using the functions from the previous example.

Solution: (a) $f(x) = 9\cos(2x) \quad g(x) = 2\cos^2(x) - 2\sin^2(x)$.

In this case if we compute the Wronskian of the two functions we should get zero since we have already determined that these functions are linearly dependent.

$$W = \begin{vmatrix} 9\cos(2x) & 2\cos^2(x) - 2\sin^2(x) \\ -18\sin(2x) & -4\cos(x)\sin(x) - 4\sin(x)\cos(x) \end{vmatrix}$$

$$= \begin{vmatrix} 9\cos(2x) & 2\cos(2x) \\ -18\sin(2x) & -2\sin(2x) - 2\sin(2x) \end{vmatrix} = \begin{vmatrix} 9\cos(2x) & 2\cos(2x) \\ -18\sin(2x) & -4\sin(2x) \end{vmatrix}$$

$$= -36\cos(2x)\sin(2x) - (-36\cos(2x)\sin(2x)) = 0.$$

So, we get zero as we should have. Notice the heavy use of trig formulas to simplify the work!

2. $f(t) = 2t^2 \quad g(t) = t^4$.

Here we know that the two functions are linearly independent and so we should get a non-zero Wronskian.

$$W = \begin{vmatrix} 2t^2 & t^4 \\ 4t & 4t^3 \end{vmatrix} = 8t^5 - 4t^5 = 4t^5.$$

The Wronskian is non-zero as we expected provided ≠ 0. This is not a problem. As long as the Wronskian is not identically zero for all t we are okay.

Example: Determine if the following functions are linearly dependent or linearly independent.

1. $f(t) = \cos t \quad g(t) = \sin t$
2. $f(x) = 6^x \quad g(x) = 6^{x+2}$.

Solution: 1. Now that we have the Wronskian to use here let's first check that. If its non-zero then we will know that the two functions are linearly independent and if its zero then we can be pretty sure that they are linearly dependent.-

$$W = \begin{vmatrix} \cos t & \sin t \\ -\sin t & \cos t \end{vmatrix} = \cos^2 t + \sin^2 t = 1 \neq 0.$$

So, by the fact these two functions are linearly independent. Much easier this time around!

2. We'll do the same thing here as we did in the first part. Recall that

$$(a^x)' = a^x \ln a.$$

Now compute the Wronskian.

$$W = \begin{vmatrix} 6^x & 6x+2 \\ 6^x \ln 6 & 6^{x+2} \ln 6 \end{vmatrix} = 6^x 6^{x+2} \ln 6 - 6^{x+2} 6^x \ln 6 = 0.$$

Now, this does not say that the two functions are linearly dependent! However, we can guess that they probably are linearly dependent. To prove that they are in fact linearly dependent we'll need to write and see if we can find non-zero c and k that will make it true for all x.-

$$c6^x + k6^{x+2} = 0$$

$$c6^x + k6^x 6^2 = 0$$

$$c6^x + 36k6^x = 0$$

$$(c + 36k)6^x = 0.$$

So, it looks like we could use any constants that satisfy

$$c + 36k = 0$$

to make this zero for all x. In particular we could use

$$c = 36 \qquad k = -1$$

$$c = -36 \qquad k = 1$$

$$c = 9 \qquad k = -\frac{1}{4} \quad etc.$$

We have non-zero constants that will make the equation true for all x. Therefore, the functions are linearly dependent.

Before proceeding to the next topic in this section let's talk a little more about linearly independent and linearly dependent functions. Let's start off by assuming that $f(x)$ and $g(x)$ are linearly dependent. So, that means there are non-zero constants c and k so that -

$$cf(x) + kg(x) = 0$$

is true for all x.

Now, we can solve this in either of the following two ways.

$$f(x) = -\frac{k}{c}g(x) \quad \text{OR} \quad g(x) = -\frac{c}{k}f(x).$$

Note that this can be done because we know that c and k are non-zero and hence the divisions can be done without worrying about division by zero.

So, this means that two linearly dependent functions can be written in such a way that one is nothing more than a constant time the other. Go back and look at both of the sets of linearly dependent functions that we wrote down and you will see that this is true for both of them. Two functions that are linearly independent can't be written in this manner and so we can't get from one to the other simply by multiplying by a constant.

Next, we don't want to leave you with the impression that linear independence and linear dependence is only for two functions. We can easily extend the idea to as many functions as we'd like.

Let's suppose that we have n non-zero functions, $f_1(x), f_2(x), \ldots, f_n(x)$. Write down the following equation.

$$c_1 f_1(x) + c_2 f_2(x) + \cdots + c_n f_n(x) = 0.$$

If we can find constants c_1, c_2, ..., c_n with at least two non-zero so that the equation is true for all x then we call the functions linearly dependent. If, on the other hand, the only constants that make it true for x are $c_1 = 0, c_2 = 0, \ldots, c_n = 0$ then we call the functions linearly independent.-

Note that unlike the two function case we can have some of the constants be zero and still have the functions be linearly dependent.

In this case just what does it mean for the functions to be linearly dependent? Well, let's suppose that they are. So, this means that we can find constants, with at least two non-zero so that it is true for all x. For the sake of argument let's suppose that c_1 is one of the non-zero constants. This means that we can do the following.

$$c_1 f_1(x) + c_2 f_2(x) + \cdots + c_n f_n(x) = 0$$

$$c_1 f_1(x) = (c_2 f_2(x) + \cdots + c_n f_n(x))$$

$$f_1(x) = -\frac{1}{c_1}(c_2, f_2(x) + \cdots + c_n f_n(x)).$$

In other words, if the functions are linearly dependent then we can write at least one of them in terms of the other functions.

Okay, let's move on the other topic of this section. There is an alternate method of computing the Wronskian. The following theorem gives this alternate method.

ABEL'S THEOREM

If $y_1(t)$ and $y_2(t)$ are two solutions to

$$y'' + p(t)y' + q(t)y = 0$$

then the Wronskian of the two solutions is

$$W(y_1, y_2)(t) = W(y_1, y_2)(t_0)e^{-\int_{t_0}^{t} p(x)dx}$$

for some t_0.

Because we don't know the Wronskian and we don't know t_0 this won't do us a lot of good apparently. However, we can rewrite this as

$$W(y_1, y_2)(t) = ce^{-\int p(t)dt}$$

where the original Wronskian sitting in front of the exponential is absorbed into the c and the evaluation of the integral at t_0 will put a constant in the exponential that can also be brought out and absorbed into the constant c. If you don't recall how to do this go back and take a look at the linear, first order differential equation section as we did something similar there.

With this rewrite we can compute the Wronskian up to a multiplicative constant, which isn't too bad. Notice as well that we don't actually need the two solutions to do this. All we need is the coefficient of the first derivative from the differential equation (provided the coefficient of the second derivative is one of course...).

Let's take a look at a quick example of this.

Example: Without solving, determine the Wronskian of two solutions to the following differential equation.

$$t^4 y'' - 2t^3 y' - t^8 y = 0.$$

Solution: The first thing that we need to do is divide the differential equation by the coefficient of the second derivative as that needs to be a one. This gives us

$$y'' - \frac{2}{t} y' - t^4 y = 0.$$

Now, using (3) the Wronskian is

$$W = ce^{\int -\frac{2}{t} dt} = ce^{2\ln t} = ce^{\ln t^2} = ct^2.$$

Chapter 4

Higher Order Differential Equations

In this chapter we're going to take a look at higher order differential equations. This chapter will actually contain more than most text books tend to have when they discuss higher order differential equations.

We will definitely cover the same material that most text books do here. However, in all the previous chapters all of our examples were 2nd order differential equations or 2 x 2 systems of differential equations.

So, in this chapter we're also going to do a couple of examples here dealing with 3rd order or higher differential equation with Laplace transforms and series as well as a discussion of some larger systems of differential equations.

Basic Concepts for nth Order Linear Equations

We'll start this chapter off with the material that most text books will cover in this chapter. We will take the material from the Second Order chapter and expand it out to n^{th} order linear differential equations. As we'll see almost all of the 2nd order material will very naturally extend out to n^{th} order with only a little bit of new material.

So, let's start things off here with some basic concepts for n^{th} order linear differential equations. The most general n^{th} order linear differential equation is,

$$P_n(t)\,y^{(n)} + P_{n-1}(t)\,y^{(n-1)} + \cdots + P_1(t)\,y' + P_0(t)\,y = G(t)$$

where you'll hopefully recall that,

$$y^{(m)} = \frac{d^m y}{dx^m}$$

Many of the theorems and ideas for this material require $y^{(m)}$ that has a coefficient of 1 and so if we divide out by $P_n(t)$ we get,

$$y^{(n)} + P_{n-1}(t)\,y^{(n-1)} + \cdots + P_1(t)\,y' + P_0(t)\,y = g(t)$$

As we might suspect an IVP for an nth order differential equation will require the following n initial conditions.

$$y(y_0) = \overline{y}(t_0) = \overline{y}, \cdots, y^{(n-1)}(t_0) = \overline{y}_{n-1}$$

Theorem: Suppose the functions $p_0, p_1, \cdots, p_{n-1}$ and $g(t)$ are all continuous in some open interval I containing $t_0$then there is a unique solution to the IVP given and the solution will exist for all t in I.

This theorem is a very natural extension of a similar theorem we saw in the 1st order material.

Next we need to move into a discussion of the nth order linear homogeneous differential equation,

$$y^{(n)} + p_{n-1}(t)\, y^{(n-1)} + \cdots + p_1(t)\, y' + p_0(t)\, y = 0$$

Let's suppose that we know $y_1(t), y_2(t), \cdots, y_n(t)$ are all solutions then by the an extension of the Principle of Superposition we know that

$$y(t) = c_1 y_1(t) + c_2 y_2(t) + \cdots + c_n y_n(t)$$

will also be a solution. The real question here is whether or not this will form a general solution.

In order for this to be a general solution then we will have to be able to find constants $c_1, c_2, \cdots, c_n$ for any choice of (in the interval I from Theorem 1) and any choice of $\overline{y}_1, \overline{y}_2, \cdots \overline{y}_n$. Or, in other words we need to be able to find $c_1, c_2, \cdots, c_n$ that will solve,

$$\begin{aligned} c_1 y_1(t_0) + c_2 y_2(t_0) + \cdots + c_n y_n(t_0) &= \overline{y}_0 \\ c_1 y_1'(t_0) + c_2 y_2'(t_0) + \cdots + c_n y_n'(t_0) &= \overline{y}_0 \\ &\vdots \\ c_1 y_1^{(n-1)}(t_0) + c_2 y_2^{(n-1)}(t_0) + \cdots + c_n y_n^{(n-1)}(y_0) &= \overline{y}_{n-1} \end{aligned}$$

Just as we did for 2nd order differential equations, we can use Cramer's Rule to solve this and the denominator of each the answers will be the following determinant of an $n \times n$ matrix.

$$\begin{vmatrix} y_1 & y_2 & \cdots & y_n \\ y_1' & y_2' & \cdots & y_n' \\ \vdots & \vdots & \ddots & \vdots \\ y_1^{(n-1)} & y_2^{(n-1)} & \cdots & y_n^{(n-1)} \end{vmatrix}$$

As we did back with the 2nd order material we'll define this to be the Wronskian and denote it by,

$$W\left(y_1, y_2, \cdots y_n\right)(t) = \begin{vmatrix} y_1 & y_2 & \cdots & y_n \\ y_1' & y_2' & \cdots & y_n' \\ \vdots & \vdots & \ddots & \vdots \\ y_1^{(n-1)} & y_2^{(n-1)} & \cdots & y_n^{(n-1)} \end{vmatrix}$$

Now that we have the definition of the Wronskian out of the way we need to get back to the question at hand. Because the Wronskian is the denominator in the solution to each of the c_1 we can see that we'll have a solution provided it is not zero for any value of $t = t_0$ that we chose to evaluate the Wronskian at. The following theorem summarizes all this up.

Theorem: Suppose the functions $p_0, p_1, \cdots, p_{n-1}$ are all continuous on the open interval I and further suppose that $y_1(t), y_2(t), \cdots y_n(t)$ are all solutions. If $W\left(y_1, y_2 \cdots y_n\right)(t) \neq 0$ for every t in I then $y_1(t), y_2(t), \cdots y_n(t)$ form a Fundamental Set of Solutions and the general solution is,

$$y(t) = c_1 y_1(t) + c_2 y_2(t) + \cdots + c_n y_n(t)$$

Recall as well that if a set of solutions form a fundamental set of solutions then they will also be a set of linearly independent functions. We'll close this section off with a quick reminder of how we find solutions to the nonhomogeneous differential equation. We first need the nth order version of a theorem we saw back in the 2nd order material.

Theorem: Suppose that $Y_1(t)$ and $Y_2(t)$ are two solutions and that $y_1(t), y_2(t), \ldots y_n(t)$ are a fundamental set of solutions to the homogeneous differential equation then,

$$Y_1(t) - Y_2(t)$$

is a solution and it can be written as

$$Y_1(t) - Y_2(t) = c_1 y_1 = c_1 y_1(t) + c_2 y_2(t) + \cdots + c_n y_n(t)$$

Now, just as we did with the 2nd order material if we let $Y(t)$ be the general solution and if we let $Y_P(t)$ be any solution then using the result of this theorem we see that we must have,

$$Y(t) = c_1 y_1(t) + c_2 y_2(t) + \cdots + c_n y_n(t) + Y_P(t) = y_c(t) + Y_P(t)$$

where, $y_c(t) = c_1 y_1(t) + c_2 y_2(t) + \cdots + c_n y_n(t)$ is called the complimentary solution and $Y_P(t)$ is called a particular solution.

Over the course of the next couple of sections we'll discuss the differences in finding the complimentary and particular solutions for n^{th} order differential equations in relation to what we know about 2nd order differential equations. We'll see that, for the most part, the methods are the same. The amount of work involved however will often be significantly more.

HOMOGENEOUS LINEAR HIGHER ORDER DIFFERENTIAL EQUATIONS

As with 2nd order differential equations we can't solve a nonhomogeneous differential equation unless we can first solve the homogeneous differential equation. We'll also need to restrict ourselves down to constant coefficient differential equations as solving non-constant coefficient differential equations is quite difficult and so we won't be discussing them here. Likewise, we'll only be looking at linear differential equations.

So, let's start off with the following differential equation,

$$a_n y^{(n)} + a_{n-1} y^{(n-1)} + \cdots + a_1 y' + a_0 y = 0$$

Now, assume that solutions to this differential equation will be in the form $y(t) = e^{rt}$ and plug this into the differential equation and with a little simplification we get,

$$e^{rt}\left(a_n r^n + a_{n-1} r^{n-1} + \cdots + a_1 r + a_0\right) = 0$$

and so in order for this to be zero we'll need to require that

$$a_n r^n + a_{n-1} r^{n-1} + \cdots + a_1 r + a_0 = 0$$

This is called the characteristic polynomial/equation and its roots/solutions will give us the solutions to the differential equation. We know that, including repeated roots, an nth degree polynomial (which we have here) will have n roots. So, we need to go through all the possibilities that we've got for roots here.

This is where we start to see differences in how we deal with n^{th} order differential equations versus 2nd order differential equations. There are still the three main cases: real distinct roots, repeated roots and complex roots (although these can now also be repeated as well see). In 2nd order differential equations each differential equation could only involve one of these cases. Now, however, that will not necessarily be the case. We could very easily have differential equations that contains each of these cases.

For instance suppose that we have an 9th order differential equation. The complete list of roots could have 3 roots which only occur once in the list (*i.e.* real distinct roots), a root with multiplicity 4 (*i.e.* occurs 4 times in the list) and a set of complex conjugate roots (recall that because the coefficients are all real complex roots will always occur in conjugate pairs).

So, for each nth order differential equation we'll need to form a set of n linearly independent functions (*i.e.* a fundamental set of solutions) in order to get a general solution. In the work that follows we'll discuss the solutions that we get from each case but we will leave it to you to verify that when we put everything together to form a general solution that we do indeed get a fundamental set of solutions. Recall that in order to this we need to verify that the Wronskian is not zero.

So, let's get started with the work here. Let's start off by assuming that in the list of roots of the characteristic equation we have $r_1 r_2, \ldots, r_k$ and they only occur once in the list. The solution from each of these will then be,

$$e^{r_1 t}, e^{r_2 t}, \cdots, e^{r_k t}$$

There's nothing really new here for real distinct roots.

Now let's take a look at repeated roots. The result here is a natural extension of the work we saw in the 2nd order case. Let's suppose that r is a root of multiplicity k (*i.e.* r occurs k times in the list of roots). We will then get the following k solutions to the differential equation,

$$e^{rt}, te^{rt}, \cdots, t^{k-1}e^{rt}$$

So, for repeated roots we just add in a t for each of the solutions past the first one until we have a total of k solutions. Again, we will leave it to you to compute the Wronskian to verify that these are in fact a set of linearly independent solutions.

Finally we need to deal with complex roots. The biggest issue here is that we can now have repeated complex roots for 4th order or higher differential equations. We'll start off by assuming that occurs only once in the list of roots. In this case we'll get the standard two solutions,

$$e^{\lambda t}\cos(\mu t) \qquad e^{\lambda t}\sin(\mu t)$$

Now let's suppose that has a multiplicity of k (*i.e.* they occur k times in the list of roots). In this case we can use the work from the repeated roots above to get the following set of $2k$ complex-valued solutions,

$$e^{(\lambda+\mu i)t}, te^{(\lambda+\mu i)t}, \cdots, t^{k-1}e^{(\lambda+\mu i)t}$$

$$e^{(\lambda-\mu i)t}, te^{(\lambda-\mu i)t}, \cdots, t^{k-1}e^{(\lambda-\mu i)t}$$

The problem here of course is that we really want real-valued solutions. So, recall that in the case where they occurred once all we had to do was use Euler's formula on the first one and then take the real and imaginary part to get two real valued solutions.

We'll do the same thing here and use Euler's formula on the first set of complex-valued solutions above, split each one into its real and imaginary parts to arrive at the following set of $2k$ real-valued solutions.

$$e^{\lambda t}\cos(\mu t), e^{\lambda t}\sin(\mu t), te^{\lambda t}\sin(\mu t), \cdots, t^{k-1}e^{\lambda t}\cos(\mu t), t^{k-1}e^{\lambda t}\sin(\mu t)$$

Once again we'll leave it to you to verify that these do in fact form a fundamental set of solutions.

Before we work a couple of quick examples here we should point out that the characteristic polynomial is now going to be at least a 3rd degree polynomial and finding the roots of these by hand is often a very difficult and time consuming process and in many cases if the roots are not rational (*i.e.* in the form $\frac{p}{g}$) it can be almost impossible to find them all by hand. To see a process for determining all the rational roots of a polynomial check out the Finding Zeroes of Polynomials page in my Algebra notes. In practice however, we usually use some form of computation aid such as Maple or Mathematica to find all the roots. So, let's work a couple of example here to illustrate at least some of the ideas discussed here.

Example: Solve the following IVP.

$$y^{(3)} - 5y'' - 22y'' + 56y = 0 \qquad y(0) = 1 \quad y'(0) = -2 \quad y''(0) = -4$$

Solution: The characteristic equation is,

$$r^3 - 5r^3 - 22r + 56 = (r+4)(r-2)(r-7) = 0 \quad \Rightarrow \quad r_1 = -4, r_2 = 2, r_3 = 7$$

So we have three real distinct roots here and so the general solution is,

$$y(t) = c_1 e^{-4t} + c_2 e^{2t} + c_3 e^{7t}$$

Differentiating a couple of times and applying the initial conditions gives the following system of equations that we'll need to solve in order to find the coefficients.

$$\begin{aligned} 1 &= y(0) = c_1 + c_2 + c_3 & & c_1 = \frac{14}{33} \\ -2 &= y'(0) = -4c_1 + 2c_2 + 7c_3 & \Rightarrow & c_2 = \frac{13}{15} \\ -4 &= y''(0) = 16c_1 + 4c_2 + 49c_3 & & c_3 = -\frac{16}{55} \end{aligned}$$

The actual solution is then,

$$y(t) = \frac{14}{33}e^{-4t} + \frac{13}{15}e^{2t} - \frac{16}{55}e^{7t}$$

So, outside of needing to solve a cubic polynomial (which we left the details to you to verify) and needing to solve a system of 3 equations to find the coefficients (which we also left to you to fill in the details) the work here is pretty much identical to the work we did in solving a 2nd order IVP.

Because the initial condition work is identical to work that we should be very familiar with to this point with the exception that it involved solving larger systems we're going to not bother with solving IVP's for the rest of the examples. The main point of this section is the new ideas involved in finding the general solution to the differential equation anyway and so we'll concentrate on that for the remaining examples.

Also note that we'll not be showing very much work in solving the characteristic polynomial. We are using computational aids here and would encourage you to do the same here. Solving these higher degree polynomials is just too much work and would obscure the point of these examples.

So, let's move into a couple of examples where we have more than one case involved in the solution.

Example: Solve the following differential equation.

$$2y^{(4)} + 11y^{(3)} + 18y^{n} + 4y' - 8y = 0$$

Solution:

The characteristic equation is,

$$2r^4 + 11r^3 + 18r^2 + 4r - 8 = (2r-1)(r+2)^3 = 0$$

So, we have two roots here, $r_1 = \frac{1}{2}$ and $r_2 = -2$ which is multiplicity of 3.Remember that we'll get three solutions for the second root and after the first we add t's only the solution until we reach three solutions.

The general solution is then,

$$y(t) = c_1 e^{\frac{1}{2}t} + c_2 e^{2t} + c_3 t e^{2t} + c_4 t^2 e^{2t}$$

Example: Solve the following differential equation.

$$y^{(5)} + 12y^{(4)} + 104y^{(3)} + 408y' + 1156y' = 0$$

Solution: The characteristic equation is,

$$y^5 + 12r^4 + 104r^3 + 408r^2 + 1156r = r\left(r^2 + 6r + 34\right)^2 = 0$$

So, we have one real root $r = 0$and a pair of complex roots $r = -3 \pm 5i$each with multiplicity 2. So, the solution for the real root is easy and for the complex roots we'll get a total of 4 solutions, 2 will be the *normal* solutions and two will be the normal solution each multiplied by t.

The general solution is,

$$y(t)=c_1+c_2e^{-3t}\cos(5t)+c_3e^{-3t}\sin(5t)+c_4te^{-3t}\cos(5t)+c_5te^{-3t}\sin(5t)$$

Let's now work an example that contains all three of the basic cases just to say that we that we've got one work here.

Example: Solve the following differential equation.

$$y^{(5)}-15y^{(4)}+84y^{(3)}-220y''+275y'-125y=0$$

Solution: The characteristic equation is

$$r^5-15r^4+84r^3-220r^2+275r-125=(r-1)(r-5)^2\left(r^2-4r+5\right)=0$$

In this case we've got one real distinct root, $r = 1$, and double root,$r = 5$, and a pair of complex roots, $r = 2 \pm$ i that only occur once.

The general solution is then,

$$y(t)=c_1e^t+c_2e^{5t}+c_3te^{5t}+c_4e^{2t}\cos(t)+c_5e^{2t}\sin(t)$$

We've got one final example to work here that on the surface at least seems almost too easy. The problem here will be finding the roots as well see.

Example: Solve the following differential equation.

$$y^{(4)}+16y=0$$

Solution: The characteristic equation is

$$r^4+16=0$$

So, a really simple characteristic equation. However, in order to find the roots we need to compute the fourth root of -16 and that is something that most people haven't done at this point in their mathematical career. We'll just give the formula here for finding them, but if you're interested in seeing a little more about this you might want to check out the Powers and Roots section of my Complex Numbers Primer.

The 4 (and yes there are 4!) 4th roots of -16 can be found by evaluating the following,

$$\sqrt[4]{-16}=(-16)^{\frac{1}{4}}=\sqrt[4]{16}e^{\left(\frac{\pi}{4}+\frac{\pi k}{2}\right)}=2\left(\cos\left(\frac{\pi}{4}+\frac{\pi k}{2}\right)+i\cos\left(\frac{\pi}{4}+\frac{\pi k}{2}\right)\right)\qquad k=0,1,2,3$$

Note that each value of k will give a distinct 4th root of -16. Also, note that for the 4th root (and ONLY the 4th root) of any negative number all we need to do is replace the

16 in the above formula with the absolute value of the number in question and this formula will work for those as well.

Here are the 4th roots of -16.

$$k=0:2\left(\cos\left(\frac{\pi}{4}\right)+i\sin\left(\frac{\pi}{4}\right)\right)=2\left(\frac{1}{\sqrt{2}}+\frac{1}{\sqrt{2}}i\right)=\sqrt{2}+i\sqrt{2}$$

$$k=1:2\left(\cos\left(\frac{3\pi}{4}\right)+i\sin\left(\frac{3\pi}{4}+\frac{\pi k}{2}\right)\right)=2\left(-\frac{1}{\sqrt{2}}+\frac{1}{\sqrt{2}}i\right)=-\sqrt{2}+i\sqrt{2}$$

$$k=2:2\left(\cos\left(\frac{5\pi}{4}\right)+i\sin\left(\frac{5\pi}{4}\right)\right)=2\left(-\frac{1}{\sqrt{2}}-\frac{1}{\sqrt{2}}i\right)=-\sqrt{2}-i\sqrt{2}$$

$$k=3:2\left(\cos\left(\frac{7\pi}{4}\right)+i\sin\left(\frac{7\pi}{4}\right)\right)=2\left(\frac{1}{\sqrt{2}}-\frac{1}{\sqrt{2}}i\right)=\sqrt{2}-i\sqrt{2}$$

So, we have two sets of complex roots: $r\sqrt{2}\pm i\sqrt{2}$ and $r=-\sqrt{2}\pm i\sqrt{2}$. The general solution is,

$$Y(t)=c_1e^{\sqrt{2}t}\cos\left(\sqrt{2}t\right)+c_2e^{\sqrt{2}t}\sin\left(\sqrt{2}t\right)+c_3e^{-\sqrt{2}t}\cos\left(\sqrt{2}t\right)+c_4e^{-\sqrt{2}t}\sin\left(\sqrt{2}t\right)$$

So, we've worked a handful of examples here of higher order differential equations that should give you a feel for how these work in most cases.

There are of course a great many different kinds of combinations of the basic cases than what we did here and of course we didn't work any case involving 6th order or higher, but once you've got an idea on how these work it's pretty easy to see that they all work pretty in pretty much the same manner. The biggest problem with the higher order differential equations is that the work in solving the characteristic polynomial and the system for the coefficients on the solution can be quite involved.

UNDETERMINED COEFFICIENTS

We now need to start looking into determining a particular solution for n^{th} order differential equations. The two methods that we'll be looking at are the same as those that we looked at in the 2nd order chapter. In this section we'll look at the method of Undetermined Coefficients and this will be a fairly short section. With one small extension, which we'll see in the lone example in this section, the method is identical to what we saw back when we were looking at undetermined coefficients in the 2nd order differential equations chapter. Given the differential equation,

$$y^{(n)}+p_{n-1}(t)y^{(n-1)}+\cdots+p_1(t)y'+p_0(t)y=g(t)$$

if $g(t)$ is an exponential function, polynomial, sine, cosine, sum/difference of one of these and/or a product of one of these then we guess the form of a particular solution using the same guidelines that we used in the 2nd order material. We then plug the guess into the differential equation, simplify and set the coefficients equal to solve for the constants.

The one thing that we need to recall is that we first need the complimentary solution prior to making our guess for a particular solution. If any term in our guess is in the complimentary solution then we need to multiply the portion of our guess that contains that term by a t. This is where the one extension to the method comes into play. With a 2nd order differential equation the most we'd ever need to multiply by is. With higher order differential equations this may need to be more than t^2.

The work involved here is almost identical to the work we've already done and in fact it isn't even that much more difficult unless the guess is particularly messy and that makes for more mess when we take the derivatives and solve for the coefficients. Because there isn't much difference in the work here we're only going to do a single example in this section illustrating the extension. So, let's take a look at the lone example we're going to do here.

Example: Solve the following differential equation.

$$y^{(3)} - 12y'' + 48y' - 64y = 12 - 32e^{-8t} + 2e^{4t}$$

Solution: We first need the complimentary solution so the characteristic equation is,

$$r^3 - 12r^2 + 48r - 64 = (r-4)^3 = \tilde{0} \qquad \Rightarrow \qquad r = 4\left(\text{Multiplicity } 3\right)$$

We've got a single root of multiplicity 3 so the complimentary solution is,

$$y_c(t) = c_1 e^{4t} + c_2 t e^{4t} + c_3 t^2 e^{4t}$$

Now, our first guess for a particular solution is,

$$Y_p = A + Be^{-8t} + Ce^{4t}$$

Notice that the last term in our guess is in the complimentary solution so we'll need to add one at least one t to the third term in our guess. Also notice that multiplying the third term by either t or t^2will result in a new term that is still in the complimentary solution and so we'll need to multiply the third term by t^3in order to get a term that is not contained in the complimentary solution.

Our final guess is then,

$$Y_p = A + Be^{-8t} + Ct^3 e^{4t}$$

Now all we need to do is take three derivatives of this, plug this into the differential equation and simplify to get (we'll leave it to you to verify the work here),

$$-64A - 1728Be^{-8t} + 6Ce^{4t} = 12 - 32e^{-8t} + 2e^{4t}$$

Setting coefficients equal and solving gives,

$$t^0 : -64A = 12 \qquad A = -\frac{3}{16}$$

$$e^{-8t} : -1728B = -32 \quad \Rightarrow \quad B = \frac{1}{54}$$

$$e^{4t} : 6C = 2 \qquad C\frac{1}{3}$$

A particular solution is then,

$$Y_p = -\frac{3}{16} + \frac{1}{54}e^{-8t} + \frac{1}{3}t^3 e^{4t}$$

The general solution to this differential equation is then,

$$y(t) = c_1 e^{4t} + c_2 t e^{4t} + c_3 t^2 e^{4t} - \frac{3}{16} + \frac{1}{54}e^{-8t} + \frac{1}{3}t^3 e^{4t}$$

Okay, we've only worked one example here, but remember that we mentioned earlier that with the exception of the extension to the method that we used in this example the work here is identical to work we did the 2nd order material.

VARIATION OF PARAMETERS

We now need to take a look at the second method of determining a particular solution to a differential equation. As we did when we first saw Variation of Parameters we'll go through the whole process and derive up a set of formulas that can be used to generate a particular solution. However, as we saw previously when looking at 2nd order differential equations this method can lead to integrals that are not easy to evaluate. So, while this method can always be used, unlike Undetermined Coefficients, to at least write down a formula for a particular solution it is not always going to be possible to actually get a solution.

So let's get started on the process. We'll start with the differential equation,

$$y^{(n)} + p_{n-1}(t)y^{(n-1)} + \cdots + p_1(t)y' + p_0(t)y = g(t)$$

and assume that we've found a fundamental set of solutions, $y_1(t), y_2(t), \ldots, y_n(t)$, for the associated homogeneous differential equation.

Because we have a fundamental set of solutions to the homogeneous differential equation we now know that the complimentary solution is,

$$y_1(t), c_1 y_2(t) + c_2 y_2(t) + \cdots + c_n y_n(t)$$

The method of variation of parameters involves trying to find a set of new functions, $u_1(t), u_2(t), \ldots, u_n(t)$ so that,

$$Y(t) = u_1(t)y_1(t) + u_2(t)y_2(t) + \cdots + u_n(t)y_n(t)$$

will be a solution to the nonhomogeneous differential equation. In order to determine if this is possible, and to find the $u_i(t)$ if it is possible, we'll need a total of n equations involving the unknown functions that we can (hopefully) solve.

One of the equations is easy. The guess, will need to satisfy the original differential equation. So, let's start taking some derivatives and as we did when we first looked at variation of parameters we'll make some assumptions along the way that will simplify our work and in the process generate the remaining equations we'll need.

The first derivative is,

$$Y'(t) = u_1y_1' + u_2y_2' + \cdots + u_ny_n' + u_1'y_1 + u_2'y_2 + \cdots + u_n'y_n$$

Note that we rearranged the results of the differentiation process a little here and we dropped the (t)part on the u and y to make this a little easier to read. Now, if we keep differentiating this it will quickly become unwieldy and so $u_1'y_1 + u_2'y_2 + \cdots + u_n'y_n = 0$ let's make as assumption to simplify things here. Because we are after the $u_i(t)$ we should probably try to avoid letting the derivatives on these become too large. So, let's make the assumption that,

$$u_1'y_1 + u_2'y_2 + \cdots + u_n'y_n = 0$$

The natural question at this point is does this even make sense to do? The answer is, if we end up with a system of n equation that we can solve for the $u_i(t)$ then yes it does make sense to do. Of course, the other answer is, we wouldn't be making this assumption if we didn't know that it was going to work. However to accept this answer requires that you trust us to make the correct assumptions so maybe the first answer is the best at this point.

At this point the first derivative is,

$$Y'(t) = u_1y_1' + u_2y_2' + \cdots + u_ny_n'$$

and so we can now take the second derivative to get,

$$Y''(t) = u_1y_1'' + u_2y_2'' + \cdots + u_ny_n'' + u_1'y_1'' + u_2'y_2'' + \cdots + u_n'y_n''$$

This looks an awful lot alike the original first derivative prior to us simplifying it so let's again make a simplification. We'll again want to keep the derivatives on the $u_i(t)$ to a minimum so this time let's assume that,

$$u_1'y_1'' + u_2'y_2'' + \cdots + u_n'y_n'' = 0$$

and with this assumption the second derivative becomes,

$$Y''(t) = u_1y_1'' + u_2y_2'' + \cdots + u_ny_n''$$

Hopefully you're starting to see a pattern develop here. If we continue this process for the first $n = 1$derivatives we will arrive that the following formula for these derivatives.

$$Y^{(k)}(t) = u_1 y_1{}^{(k)} + u_2 y_2{}^{(k)} + \cdots + u_n y_n{}^{(k)} \qquad k = 1, 2, \ldots, n-1$$

To get to each of these formulas we also had to assume that,

$$u_1' y_1{}^{(k)} + u_2' y_2{}^{(k)} + \cdots + u_n' y_n{}^{(k)} = 0 \qquad k = 0, 1, \ldots n-2$$

and recall that the 0th derivative of a function is just the function itself. So, for example, $y_2^{(0)}(t) = y_2(t)$.

Notice as well that the set of assumptions actually give us n–1 equations in terms of the derivatives of the unknown functions: $u_1(t), u_2(t), \ldots, u_n(t)$.

All we need to do then is finish generating the first equation we started this process to find. To do this we'll need one more derivative of the guess.

Differentiating the $(n-1)^{\text{st}}$ derivative, to get the nth derivative gives,

$$Y^{(n)}(t) = u_1 y_1{}^{(n)} + u_2 y_2{}^{(n)} + \ldots + u_n y_n^{(n)} + u_1' y_2^{(n-1)} + u_2' y_2^{(n-1)} + \cdots + u_n' y_n^{(n-1)}$$

This time we'll also not be making any assumptions to simplify this but instead just plug this along with the derivatives given

$$\begin{aligned} u_1 y_1^{(n)} + u_2 y_2^{(n)} + \cdots + u_n y_n^{(n)} + u_1' y_1^{(n-1)} + \cdots + u_n' y_n^{(n-1)} &+ \\ p_{n-1}(t)\left[u_1 y_1^{(n-1)} + u_2 y_2^{(n-1)} + \cdots + u_n y_n^{(n-1)}\right] &+ \\ &\vdots \\ p_1(t)\left[u_1 y_1' + u_2 y_2' + \cdots + u_n y_n'\right] &+ \\ p_0(t)\left[u_1 y_1 + u_2 y_2 + \cdots + u_n y_n\right] &= g(t) \end{aligned}$$

Next, rearrange this a little to get,

$$\begin{aligned} &u_1\left[y_1^{(n)} + p_{n-1}(t) y_1^{(n-1)} + \cdots + p_1(t) y_1' + p_0(t) y_1\right] + \\ &u_2\left[y_2^{(n)} + p_{n-1}(t) y_2^{(n-1)} + \cdots + p_1(t) y_2' + p_0(t) y_2\right] + \\ &\vdots \\ &u_n\left[y_n^{(n)} + p_{n-1}(t) y_1^{(n-1)} + \cdots + p_1(t) y_n' + p_0(t) y_n\right] + \\ &u_1' y_1^{(n-1)} + u_2' y_2^{(n-1)} + \cdots + u_n' y_n^{(n-1)} = g(t) \end{aligned}$$

Recall that $y_1(t), y_2(t), \dots, y_n(t)$ are all solutions to the homogeneous differential equation and so all the quantities in the [] are zero and this reduces down to,

$$u_1' y_1^{(n-1)} + u_2' y_2^{(n-1)} + \cdots + u_n' y_n^{(n-1)} = g(t)$$

So this equation, along with those given, give us the *n* equations that we needed. Let's list them all out here for the sake of completeness.

$$\begin{aligned} u_1' y_1 + u_2' y_2 + \cdots + u_n' y_n &= 0 \\ u_1' y_1' + u_2' y_2' + \cdots + u_n' y_n' &= 0 \\ u_1' y_1'' + u_2' y_2'' + \cdots + u_n' y_n'' &= 0 \\ &\vdots \\ u_1' y_1^{(n-2)} + u_2' y_2^{(n-2)} + \cdots + u_n' y_n^{(n-2)} &= 0 \\ u_1' y_1^{(n-1)} + u_2' y_2^{(n-1)} + \cdots + u_n' y_n^{(n-1)} &= g(t) \end{aligned}$$

So, we've got *n* equations, but notice that just like we got when we did this for 2nd order differential equations the unknowns in the system are not $u_1(t), u_2(t), \dots, u_n(t)$ but instead they are the derivatives, $u_1'(t), u_2'(t), \dots, u_n'(t)$. This isn't a major problem however. Provided we can solve this system we can then just integrate the solutions to get the functions that we're after. Also, recall that the $y_1(t), y_2(t), \dots, y_n(t)$ are assumed to be known functions and so they along with their derivatives (which appear in the system) are all known quantities in the system.

Now, we need to think about how to solve this system. If there aren't too many equations we can just solve it directly if we want to. However, for large *n* (and it won't take much to get large here) that could be quite tedious and prone to error and it won't work at all for general *n* as we have here.

The best solution method to use at this point is then Cramer's Rule. We've used Cramer's Rule several times in this course, but the best reference for our purposes here is when we used it when we first defined Fundamental Sets of Solutions back in the 2nd order material.

Upon using Cramer's Rule to solve the system the resulting solution for each will be a quotient of two determinants of *n* x *n* matrices. The denominator of each solution will be the determinant of the matrix of the known coefficients,

$$\begin{vmatrix} y_1 & y_2 & \cdots & y_n \\ y_1' & y_2' & \cdots & y_n' \\ \vdots & \vdots & \ddots & \vdots \\ y_1^{(n-1)} & y_2^{(n-1)} & \cdots & y_n^{(n-1)} \end{vmatrix} = W(y_1, y_2, \cdots y_n)(t)$$

This however, is just the Wronskian of $y_1(t), y_2(t), \ldots, y_n(t)$ as noted above and because we have assumed that these form a fundamental set of solutions we also know that the Wronskian will not be zero. This in turn tells us that the system above is in fact solvable and all of the assumptions we apparently made out of the blue above did in fact work. The numerators of the solution for u_i' will be the determinant of the matrix of coefficients with the ith column replaced with the column $(0,0,0,\ldots,0,g(t))$. For example, the numerator for the first one, u_i' is,

$$\begin{vmatrix} 0 & y_2 & \cdots & y_n \\ 0 & y_2' & \cdots & y_n' \\ \vdots & \vdots & \ddots & \vdots \\ g(t) & y_2^{(n-1)} & \cdots & y_n^{(n-1)} \end{vmatrix}$$

Now, by a nice property of determinants if we factor something out of one of the columns of a matrix then the determinant of the resulting matrix will be the factor times the determinant of new matrix. In other words, if we factor out of this matrix we arrive at,

$$\begin{vmatrix} 0 & y_2 & \cdots & y_n \\ 0 & y_2' & \cdots & y_n' \\ \vdots & \vdots & \ddots & \vdots \\ g(t) & y_2^{(n-1)} & \cdots & y_n^{(n-1)} \end{vmatrix} = g(t) \begin{vmatrix} 0 & y_2 & \cdots & y_n \\ 0 & y_2' & \cdots & y_n' \\ \vdots & \vdots & \ddots & \vdots \\ 1 & y_2^{(n-1)} & \cdots & y_n^{(n-1)} \end{vmatrix}$$

We did this only for the first one, but we could just as easily done this with any of the n solutions. So, let W_i represent the determinant we get by replacing the ith column of the Wronskian with the column (0,0,0,…,0,1) and the solution to the system can then be written as,

$$u_1' = \frac{g(t)W_1(t)}{W(t)}, \quad u_2' = \frac{g(t)W_2(t)}{W(t)}, \ldots, u_n' = \frac{g(t)W_2(t)}{W(t)}$$

Wow! That was a lot of effort to generate and solve the system but we're almost there. With the solution to the system in hand we can now integrate each of these terms to determine just what the unknown functions, $u_1(t), u_2(t), \cdots, u_n(t)$ we've after all along are.

$$u_1 = \int \frac{g(t)W_1(t)}{W(t)}dt, \qquad u_2 = \int \frac{g(t)W_2(t)}{W(t)}dt, \ldots, u_n = \int \frac{g(t)W_n(t)}{W(t)}dt$$

Finally, a particular solution is then given by,

$$Y(t) = y_1(t)\int \frac{g(t)W_1(t)}{W(t)}dt + y_2(t)\int \frac{g(t)W_2(t)}{W(t)}dt + \cdots + y_n(t)\int \frac{g(t)W_n(t)}{W(t)}dt$$

We should also note that in the derivation process here we assumed that the coefficient of

We should also note that in the derivation process here we assumed that the coefficient of the $y(n)$ term was a one and that has been factored into the formula above. If the coefficient of this term is not one then we'll need to make sure and divide it out before trying to use this formula.

Before we work an example here we really should note that while we can write this formula down actually computing these integrals may be all but impossible to do.

Okay let's take a look at a quick example.

Example: Solve the following differential equation.

$$y(3)-2y''-21y'-18y=3+4e^{-t}$$

Solution: The characteristic equation is,

$$r^3-2r^2-21r-18=(r+3)(r+1)(r-6)=0 \qquad \Rightarrow \qquad r_1=-3, r_2=-1, r_3=6$$

So we have three real distinct roots here and so the general solution is,

$$y_0(t)=c_1e^{-3t}+c_2e^{-t}+c_3e^{6t}$$

Okay, we've now got several determinants to compute. We'll leave it to you to verify the following determinant computations.

$$W=\begin{vmatrix} e^{-3t} & e^{-t} & e^{6t} \\ -3e^{-3t} & -e^{-t} & 6e^{6t} \\ 9e^{-3t} & e^{-t} & 36e^{6t} \end{vmatrix}=126e^{2t} \qquad W_1=\begin{vmatrix} 0 & e^{-t} & e^{6t} \\ 0 & -e^{-t} & 6e^{6t} \\ 1 & e^{-t} & 36e^{6t} \end{vmatrix}=7e^{5t}$$

$$W_2=\begin{vmatrix} e^{-3t} & 0 & e^{6t} \\ -3e^{-3t} & 0 & 6e^{6t} \\ 9e^{-3t} & 1 & 36e^{6t} \end{vmatrix}=-9e^{3t} \qquad W_3=\begin{vmatrix} e^{-3t} & e^{-t} & 0 \\ -3e^{-3t} & -e^{-t} & 0 \\ 9e^{-3t} & e^{-t} & 1 \end{vmatrix}=2e^{4t}$$

Now, given that $g(t)=3+4e^{-t}$ we can compute each of the u_i. Here are those integrals.

$$u_1=\int\frac{(3+4e^{-t})(7e^{5t})}{126e^{2t}}dt=\frac{1}{18}\int 3e^{3t}+4e^{2t}dt=\frac{1}{18}\left(e^{3t}+2e^{2t}\right)$$

$$u_2=\int\frac{(3+4e^{-t})(9e^{3t})}{126e^{2t}}dt=-\frac{1}{14}\int 3e^{t}+4dt=-\frac{1}{14}\left(3e^{t}+4t\right)$$

$$u_3=\int\frac{(3+4e^{-t})(2e^{-4t})}{126e^{2t}}dt=\frac{1}{63}\int 3e^{-6t}+4e^{-7t}dt=\frac{1}{63}\left(-\frac{1}{2}e^{-6t}-\frac{4}{7}e^{-7t}\right)$$

Note that we didn't include the constants of integration in each of these because including them would just have introduced a term that would get absorbed into the complimentary solution just as we saw when we were dealing with 2nd order differential equations.

Finally, a particular solution for this differential equation is then,

$$\begin{aligned} Y_P &= u_1 y_1 + u_2 y_2 + u_3 y_3 \\ &= \frac{1}{18}\left(e^{3t} + 2e^{2t}\right)e^{-3t} - \frac{1}{14}\left(3e^{t} + 4t\right)e^{-t} + \frac{1}{63}\left(-\frac{1}{2}e^{-6t} - \frac{4}{7}e^{-7t}\right)e^{6t} \\ &= -\frac{1}{6} + \frac{5}{49}e^{-t} - \frac{2}{7}te^{-t} \end{aligned}$$

The general solution is then,

$$y(t) = c_1 e^{-3t} + c_2 e^{-t} + c_3 e^{6t} - \frac{1}{6} + \frac{5}{49}e^{-t} - \frac{2}{7}te^{-t}$$

We're only going to do a single example in this section to illustrate the process more than anything so with that we'll close out this section

LAPLACE TRANSFORMS

There really isn't all that much to this section. All we're going to do here is work a quick example using Laplace transforms for a 3rd order differential equation so we can say that we worked at least one problem for a differential equation whose order was larger than 2. Everything that we know from the Laplace Transforms chapter is still valid.

The only new bit that we'll need here is the Laplace transform of the third derivative. We can get this from the general formula that we gave when we first started looking at solving IVP's with Laplace transforms. Here is that formula,

$$\mathcal{L}\{y'''\} = s^3 Y(s) - s^2 y(0) - sy'(0) - y''(0)$$

Example: Solve the following IVP.

$$y''' - 4y'' = 4t + 3u_6(t)e^{30-5t}, \qquad y(0) = -3\ y'(0) = 1\ y''(0) = 4$$

Solution: As always we first need to make sure the function multiplied by the Heaviside function has been properly shifted.

$$y''' - 4y'' = 4t + 3u_6(t)e^{-5(t-6)}$$

It has been properly shifted and we can see that we're shifting e^{-5t}.

All we need to do now is take the Laplace transform of everything, plug in the initial conditions and solve for. Doing all of this gives,

$$s^3Y(s)-s^2y(0)-sy'(0)-y''(0)-4\left(s^2Y(s)-sy(0)-y'(0)\right)=\frac{4}{s^2}+\frac{3e^{-6s}}{s+5}$$

$$\left(s^3-4s^2\right)Y(s)+3s^2-13s=\frac{4}{s^2}+\frac{3e^{-6s}}{s+5}$$

$$\left(s^3-4s^2\right)Y(s)=\frac{4}{s^2}-3s^2+13s+\frac{3e^{-6s}}{s+5}$$

$$\left(s^3-4s^2\right)Y(s)=\frac{4-3s^4+13s^3}{s^2}+\frac{3e^{-6s}}{s+5}$$

$$Y(s)=\frac{4-3s^4+13s^3}{s^4(s-4)}+\frac{3e^{-6s}}{s^2(s-4)(s+5)}$$

$$Y(s)=F(s)+3e^{-6s}G(s)$$

Now we need to partial fraction and inverse transform *F(s)* and *G(s)*. We'll leave it to you to verify the details.

$$F(s)=\frac{4-3s^4+13s^3}{s^4(s-4)}=\frac{\frac{17}{64}}{s-4}-\frac{\frac{209}{64}}{s}-\frac{\frac{1}{16}}{s^2}-\frac{\frac{1}{4}\left(\frac{2!}{2!}\right)}{s^3}-\frac{1\left(\frac{3!}{3!}\right)}{s^4}$$

$$f(t)=\frac{17}{64}e^{4t}-\frac{209}{64}-\frac{1}{16}t-\frac{1}{8}t^2-\frac{1}{6}t^3$$

$$G(s)=\frac{1}{s^2(s-4)(s+5)}=\frac{\frac{1}{144}}{s-4}-\frac{\frac{1}{225}}{s+5}-\frac{\frac{1}{400}}{s}-\frac{\frac{1}{20}}{s^2}$$

$$g(t)=\frac{1}{144}e^{4t}-\frac{1}{225}e^{-5t}-\frac{1}{400}-\frac{1}{20}t$$

We can now get the solution to the differential equation. Starting with the transform we get,

$$Y(s)=F(s)+3e^{-6s}G(s) \qquad \Rightarrow \qquad y(t)=f(t)+3u_6(t)g(t-6)$$

where *f(t)* and *g(t)* are the functions shown above.

Okay, there is the one Laplace transform example with a differential equation with order greater than 2. As you can see the work in identical except for the fact that the partial fraction work (which we didn't show here) is liable to be messier and more complicated.

SYSTEMS OF DIFFERENTIAL EQUATIONS

In this section we want to take a brief look at systems of differential equations that are larger than 2 x 2. The problem here is that unlike the first few sections where we looked at n^{th} order differential equations we can't really come up with a set of formulas that will always work for every system. So, with that in mind we're going to look at all possible cases for a 3 x 3 system (leaving some details for you to verify at times) and then a couple of quick comments about 4 x 4 systems to illustrate how to extend things out to even larger systems and then we'll leave it to you to actually extend things out if you'd like to.

We will also not be doing any actual examples in this section. The point of this section is just to show how to extend out what we know about 2 x 2 systems to larger systems.

Initially the process is identical regardless of the size of the system. So, for a system of 3 differential equations with 3 unknown functions we first put the system into matrix form,

$$\vec{x}' = A\vec{x}$$

where the coefficient matrix, A, is a 3 x 3 matrix. We next need to determine the eigenvalues and eigenvectors for A and because A is a 3 x 3 matrix we know that there will be 3 eigenvalues (including repeated eigenvalues if there are any).

This is where the process from the 2 x 2 systems starts to vary. We will need a total of 3 linearly independent solutions to form the general solution. Some of what we know from the 2 x 2 systems can be brought forward to this point.

For instance, we know that solutions corresponding to simple eigenvalues (*i.e.* they only occur once in the list of eigenvalues) will be linearly independent. We know that solutions from a set of complex conjugate eigenvalues will be linearly independent. We also know how to get a set of linearly independent solutions from a double eigenvalue with a single eigenvector.

There are also a couple of facts about eigenvalues/eigenvectors that we need to review here as well. First, provided A has only real entries (which it always will here) all complex eigenvalues will occur in conjugate pairs (*i.e.* $\lambda = \alpha \pm \beta i$) and their associated eigenvectors will also be complex conjugates of each other. Next, if an eigenvalue has multiplicity $k \geq 2$(*i.e.* occurs at least twice in the list of eigenvalues) then there will be anywhere from 1 to k linearly independent eigenvectors for the eigenvalue.

With all these idea in mind let's start going through all the possible combinations of eigenvalues that we can possibly have for a 3 x 3 case. Let's also note that for a 3 x 3 system it is impossible to have only 2 real distinct eigenvalues. The only possibilities are to have 1 or 3 real distinct eigenvalues.

Here are all the possible cases.

3 REAL DISTINCT EIGENVALUES

In this case we'll have the real, distinct eigenvalues $\lambda_1 \neq \lambda_2 \neq \lambda_3$and their associated eigenvectors, $\vec{\eta}_1, \vec{\eta}_2$, $\vec{\eta}_3$ and are guaranteed to be linearly independent and so the three linearly independent solutions we get from this case are,

$$e^{\lambda_1 t}\,\vec{\eta}_1 \qquad e^{\lambda_2 t}\,\vec{\eta}_2 \qquad e^{\lambda_3 t}\,\vec{\eta}_3$$

1 REAL AND 2 COMPLEX EIGENVALUES

From the real eigenvalue/eigenvector pair, $\lambda_1$and, $\lambda_2$we get one solution,

$$e^{\lambda_1 t}\,\vec{\eta}_1$$

We get the other two solutions in the same manner that we did with the 2 x 2 case. If the eigenvalues are $\lambda_{2,3} = \alpha \pm \beta i$ with eigenvectors $\vec{\eta}_2$ and $\vec{\eta}_3$ w $\vec{\eta}_3 = \overline{(\vec{\eta}_2)}$ e can get two real-valued solution by using Euler's formula to expand,

$$e^{\lambda_2 t}\,\vec{\eta}_2 = e^{(\alpha+\beta i)t}\vec{\eta}_2 = e^{\alpha t}\left(\cos(\beta t) + i\sin(\beta t)\right)\vec{\eta}_2$$

into its real and imaginary parts, $\vec{u} + i\vec{v}$. The final two real valued solutions we need are then,

$$\vec{u} \qquad \vec{v}$$

1 REAL DISTINCT AND 1 DOUBLE EIGENVALUE WITH 1 EIGENVECTOR

From the real eigenvalue/eigenvector pair, λ_1 and $\vec{\eta}_1$, we get one solution,

$$e^{\lambda_1 t}\vec{\eta}_1$$

From our work in the 2 x 2 systems we know that from the double eigenvalue λ_2 with single eigenvector, $\vec{n}_2$, we get the following two solutions,

$$e^{\lambda_2 t}\vec{\eta}_2 \qquad te^{\lambda_2 t}\,\vec{\xi} + e^{\lambda_2 t}\vec{\rho}$$

where $\vec{\xi}$ and $\vec{\rho}$ must satisfy the following equations,

$$(A - \lambda_2 I)\vec{\xi} = \vec{0} \qquad (A - \lambda_2 I)\vec{\rho} = \vec{\xi}$$

Note that the first equation simply tells us that $\vec{\xi}$ must be the single eigenvector for this eigenvalue, $\vec{\eta}_2$, and we usually just say that the second solution we get from the double root case is,

$$te^{\lambda_2 t}\vec{\eta}_2 + e^{\lambda_2 t}\vec{\rho} \qquad \text{where } \vec{\rho} \text{ satisfies } (A - \lambda_2 I)\vec{\rho} = \vec{\eta}_2$$

1 REAL DISTINCT AND 1 DOUBLE EIGENVALUE WITH 2 LINEARLY INDEPENDENT EIGENVECTORS

We didn't look at this case back when we were examining the 2 × 2 systems but it is easy enough to deal with. In this case we'll have a single real distinct eigenvalue/eigenvector pair, $\lambda_1$and, $\vec{\eta}_1$ as well as a double eigenvalue $\lambda_2$and the double eigenvalue has two linearly independent eigenvectors, $\vec{\eta}_2$ and $\vec{\eta}_3$. In this case all three eigenvectors are linearly independent and so we get the following three linearly independent solutions,

$$e^{\lambda_1 t}\vec{\eta}_1 \qquad e^{\lambda_2 t}\vec{\eta}_2 \qquad e^{\lambda_2 t}\vec{\eta}_3$$

We are now out of the cases that compare to those that we did with 2 x 2 systems and we now need to move into the brand new case that we pick up for 3 x 3 systems. This new case involves eigenvalues with multiplicity of 3. As we noted above we can have 1, 2, or 3 linearly independent eigenvectors and so we actually have 3 sub cases to deal with here. So let's go through these final 3 cases for a 3 x 3 system.

1 TRIPLE EIGENVALUE WITH 1 EIGENVECTOR

The eigenvalue/eigenvector pair in this case are $\lambda$and $\vec{\eta}$.Because the eigenvalue is real we know that the first solution we need is,

$$e^{\lambda_1 t}\vec{\eta}$$

We can use the work from the double eigenvalue with one eigenvector to get that a second solution is,

$$te^{\lambda t}\vec{\eta} + e^{\lambda t}\vec{\rho} \qquad \text{where } \vec{\rho} \text{ satisfies } (A-\lambda I)\vec{\rho} = \vec{\eta}$$

For a third solution we can take a clue from how we dealt with n^{th} order differential equations with roots multiplicity 3. In those cases we multiplied the original solution by a t^2. However, just as with the double eigenvalue case that won't be enough to get us a solution. In this case the third solution will be,

$$\frac{1}{2}t^2 e^{\lambda t}\vec{\xi} + te^{\lambda t}\vec{\rho} + e^{\lambda t}\vec{\mu}$$

where, $\vec{\xi}$, $\vec{\rho}$ and $\vec{\mu}$ must satisfy,

$$(A-\lambda I)\vec{\xi} = \vec{0} \quad (A-\lambda I)\vec{\rho} = \vec{\xi} \quad (A-\lambda I)\vec{\mu} = \vec{\rho}$$

You can verify that this is a solution and the conditions by taking a derivative and plugging into the system.

Now, the first condition simply tells us that $\vec{\xi} = \vec{\eta}$ because we only have a single eigenvector here and so we can reduce this third solution to,

$$\frac{1}{2}t^2 e^{\lambda t}\vec{\eta} + te^{\lambda t}\vec{\rho} + e^{\lambda t}\vec{\mu}$$

where $\vec{\rho}$, and $\vec{\mu}$ must satisfy,

$$(A-\lambda I)\vec{\rho}=\vec{\eta} \quad (A-\lambda I)\vec{\mu}=\vec{\rho}$$

and finally notice that we would have solved the new first condition for in determining the second solution above and so all we really need to do here is solve the final condition.

As a final note in this case, the $\frac{1}{2}$ is in the solution solely to keep any extra constants from appearing in the conditions which in turn allows us to reuse previous results.

1 TRIPLE EIGENVALUE WITH 2 LINEARLY INDEPENDENT EIGENVECTORS

In this case we'll have the eigenvalue $\lambda$with the two linearly independent eigenvectors $\vec{\eta}_1$ and $\vec{\eta}_2$ so we get the following two linearly independent solutions,

$$e^{\lambda t}\vec{\eta}_1 \quad e^{\lambda t}\vec{\eta}_2$$

We now need a third solution. The third solution will be in the form,

$$te^{\lambda_2 t}\vec{\xi}+e^{\lambda_2 t}\vec{\xi}$$

where $\vec{\xi}$ and $\vec{\rho}$ must satisfy the following equations,

$$(A-\lambda_2 I)\vec{\xi}=\vec{0} \qquad (A-\lambda_2 I)\vec{\rho}=\vec{\xi}$$

We've already verified that this will be a solution with these conditions in the double eigenvalue case (that work only required a repeated eigenvalue, not necessarily a double one). However, unlike the previous times we've seen this we can't just say that $\vec{\xi}$ is an eigenvalue. In all the previous cases in which we've seen this condition we had a single eigenvalue and this time we have two linearly independent eigenvalues.

This means that the most general possible solution to the first condition is,

$$\vec{\xi}=c_1\vec{\eta}_1+c_2\vec{\eta}_2$$

This creates problems in solving the second condition. The second condition will not have solutions for every choice of $c_1$and $c_2$and the choice that we use will be dependent upon the eigenvectors. So upon solving the first condition we would need to plug the general solution into the second condition and then proceed to determine conditions on $c_1$and c_2 that would allow us to solve the second condition.

1 TRIPLE EIGENVALUE WITH 3 LINEARLY INDEPENDENT EIGENVECTORS

In this case we'll have the eigenvalue λ with the three linearly independent eigenvectors $\vec{\eta}_1, \vec{\eta}_2$ and $\vec{\eta}_3$ so we get the following three linearly independent solutions,

$$e^{\lambda t}\vec{\eta}_1 \quad e^{\lambda t}\vec{\eta}_2 \quad e^{\lambda t}\vec{\eta}_3 \cdot$$

4 × 4 Systems

We'll close this section out with a couple of comments about 4 × 4 systems. In these cases we will have 4 eigenvalues and will need 4 linearly independent solutions in order to get a general solution. The vast majority of the cases here are natural extensions of what 3 × 3 systems cases and in fact will use a vast majority of that work.

Here are a couple of new cases that we should comment briefly on however. With 4 × 4 systems it will now be possible to have two different sets of double eigenvalues and two different sets of complex conjugate eigenvalues. In either of these cases we can treat each one as a separate case and use our previous knowledge about double eigenvalues and complex eigenvalues to get the solutions we need.

It is also now possible to have a "double" complex eigenvalue. In other words we can have $\lambda = \alpha \pm \beta i$ each occur twice in the list of eigenvalues. The solutions f.or this case aren't too bad. We get two solutions in the normal way of dealing with complex eigenvalues. The remaining two solutions will come from the work we did for a double eigenvalue. The work we did in that case did not require that the eigenvalue/eigenvector pair to be real. Therefore if the eigenvector associated with $\lambda = \alpha \pm \beta i$ is $\vec{\eta}$ then the second solution will be,

$$te^{(\alpha+\beta i)t}\vec{\eta} + e^{(\alpha+\beta i)t}\vec{\rho} \qquad \text{where } \vec{\rho} \text{ satisfies} (A - \lambda I)\vec{\rho} = \vec{\eta}$$

and once we've determined $\vec{\rho}$ we can again split this up into its real and imaginary parts using Euler's formula to get two new real valued solutions.

Finally with 4 x 4 systems we can now have eigenvalues with multiplicity of 4. In these cases we can have 1, 2, 3, or 4 linearly independent eigenvectors and we can use our work with 3 x 3 systems to see how to generate solutions for these cases. The one issue that you'll need to pay attention to is the conditions for the 2 and 3 eigenvector cases will have the same complications that the 2 eigenvector case has in the 3 x 3 systems.

So, we've discussed some of the issues involved in systems larger than 2 x 2 and it is hopefully clear that when we move into larger systems the work can be become vastly more complicated.

SERIES SOLUTIONS

The purpose of this section is not to do anything new with a series solution problem. Instead it is here to illustrate that moving into a higher order differential equation does not really change the process outside of making it a little longer.

Back in the Series Solution chapter we only looked at 2nd order differential equations so we're going to do a quick example here involving a 3rd order differential equation so we can make sure and say that we've done at least one example with an order larger than 2.

Example: Find the series solution around $x_0 = 0$ for the following differential equation.

$$y''' + x^2 y' + xy = 0$$

Solution: Recall that we can only find a series solution about $x_0 = 0$ if this point is an ordinary point, or in other words, if the coefficient of the highest derivative term is not zero at $x_0 = 0$. That is clearly the case here so let's start with the form of the solutions as well as the derivatives that we'll need for this solution.

$$y(x) = \sum_{n=0}^{\infty} a_n x^n$$

$$y'(x) = \sum_{n=1}^{\infty} n a_n x^{n-1}$$

$$y'''(x) = \sum_{n=3}^{\infty} n(n-1)(n-2) a_n x^{n-3}$$

Plugging into the differential equation gives,

$$\sum_{n=3}^{\infty} n(n-1)(n-2)a_n x^{n-3} + x^2 \sum_{n=1}^{\infty} n a_n x^{n-1} + x \sum_{n=0}^{\infty} a_n x^n = 0$$

Now, move all the coefficients into the series and do appropriate shifts so that all the series are in terms of x_n.

$$\sum_{n=3}^{\infty} n(n-1)(n-2)a_n x^{n-3} + x^2 \sum_{n=1}^{\infty} n a_n x^{n+1} + x \sum_{n=0}^{\infty} a_n x^{n+1} = 0$$

$$\sum_{n=3}^{\infty} n(n+3)(n+2)(n+1)a_{n+3} x^n + \sum_{n=2}^{\infty} (n-1) a_{n-1} x^n + \sum_{n=1}^{\infty} a_{n-1} x^n = 0$$

Next, let's notice that we can start the second series at $n = 1$ since that term will be zero. So let's do that and then we can combine the second and third terms to get,

$$\sum_{n=0}^{\infty} (n+3)(n+2)(n+1)a_{n+3} x^n + \sum_{n=1}^{\infty} \left[(n-1)+1\right] a_{n-1} x^n = 0$$

$$\sum_{n=0}^{\infty} (n+3)(n+2)(n+1)a_{n+3} x^n + \sum_{n=1}^{\infty} n a_{n-1} x^n = 0$$

So, we got a nice simplification in the new series that will help with some further simplification. The new second series can now be started at $n = 0$and then combined with the first series to get,

$$\sum_{n=0}^{\infty} \left[(n+3)(n+2)(n+1)a_{n+3} + n a_{n-1}\right] + x^n = 0$$

We can now set the coefficients equal to get a fairly simply recurrence relation.

$$(n+3)(n+2)(n+1)a_{n+3}+na_{n-1}=0 \qquad n=0,1,2,\ldots$$

Solving the recurrence relation gives,

$$a_{n+3}=\frac{-na_{n-1}}{(n+1)(n+2)(n+3)} \qquad n=0,1,2,\ldots$$

Now we need to start plugging in values of n and this will be one of the main areas where we can see a somewhat significant increase in the amount of work required when moving into a higher order differential equation.

$$n=0: \quad a_3=0$$

$$n=1: \quad a_4=\frac{-a_0}{(2)(3)(4)}$$

$$n=2: \quad a_5=\frac{-2a_1}{(3)(4)(5)}$$

$$n=3: \quad a_6=\frac{-3a_2}{(4)(5)(6)}$$

$$n=5: \quad a_8=\frac{-5a_4}{(6)(7)(8)}=\frac{5a_0}{(2)(3)(4)(6)(7)(8)}$$

$$n=6: \quad a_9=\frac{-6a_5}{(7)(8)(9)}=\frac{(2)(6)a_1}{(3)(4)(5)(7)(8)(9)}$$

$$n=7: \quad a_{10}=\frac{-7a_6}{(8)(9)(10)}=\frac{(3)(7)a_2}{(4)(5)(6)(8)(9)(10)}$$

$$n=8: \quad a_{11}=\frac{-8a_7}{(9)(10)(11)}=0$$

$$n=9: \quad a_{12}=\frac{-9a_8}{(10)(11)(12)}=\frac{-(5)(9)a_0}{(2)(3)(4)(6)(7)(8)(10)(11)(12)}$$

$$n=10: \quad a_{13}=\frac{-10a_9}{(11)(13)(14)}=\frac{(2)(6)(10)a_1}{(3)(4)(5)(7)(8)(9)(11)(12)(13)}$$

$$n=11: \quad a_{14}=\frac{-11a_{10}}{(12)(13)(14)}=\frac{-(3)(7)(11)a_2}{(4)(5)(6)(8)(9)(10)(12)(13)(14)}$$

Okay, we can now break the coefficients down into 4 sub cases given by, $a_{4k}, a_{4k+1}, a_{4k+2}$, and a_{4k+3} for $k=0,1,2,3,\ldots$ We'll give a semi-detailed derivation for

a_{4k} and then leave the rest to you with only couple of comments as they are nearly identical derivations.

First notice that all the a_{4k} terms have a_0 in them and they will alternate in sign. Next notice that we can turn the denominator into a factorial, $(4k)!$ to be exact, if we multiply top and bottom by the numbers that are already in the numerator and so this will turn these numbers into squares. Next notice that the product in the top will start at 1 and increase by 4 until we reach $4k - 3$. So, taking all of this into account we get,

$$a_{4k} = \frac{(-1)^k (1)^2 (5)^2 \cdots (4k-3)^2 a_0}{(4k)!} \qquad k = 1,2,3,\ldots$$

and notice that this will only work starting with $k = 1$ as we won't get a_0 out of this formula as we should by plugging in $k = 0$.

Now, for a_{4k+1} the derivation is almost identical and so the formula is,

$$a_{4k+1} = \frac{(-1)^k (1)^2 (6)^2 \cdots (4k-2)^2 a_1}{(4k+1)!} \qquad k = 1,2,3,\ldots$$

and again notice that this won't work for $k = 0$

The formula for a_{4k+2} is again nearly identical except for this one note that we also need to multiply top and bottom by a 2 in order to get the factorial to appear in the denominator and so the formula here is,

$$a_{4k+2} = \frac{2(-1)^k (3)^2 (7)^2 \cdots (4k-1)^2 a_2}{(4k+2)!} \qquad k = 1,2,3,\ldots$$

noticing yet one more time that this won't work for $k = 0$.

Finally, we have $a_{4k+3} = 0$ for $k = 0,1,2,3,\ldots$

Now that we have all the coefficients let's get the solution,

$$\begin{aligned} y(x) &= a_0 + a_1 x + a_2 x^2 + a_3 x^3 + \cdots + a_{4k} x^{4k} + a_{4k+1} x^{4k+1} + a_{4k+3} x^{4k+3} + a_{4k+3} x^{4k+3} + \cdots \\ &= a_0 + a_1 x + a_2 x^2 \cdots + \frac{(-1)^k (1)^2 (5)^2 \cdots (4k-3)^2 a_0}{(4k)!} x^{4k} + \\ &\qquad \frac{(-1)^k (2)^2 (6)^2 \cdots (4k-2)^2 a_1}{(4k+1)!} x^{4k+1} + \\ &\qquad \frac{2(-1)^k (3)^2 (7)^2 \cdots (4k-1)^2 a_2}{(4k+2)!} x^{4k+2} + \cdots \end{aligned}$$

Collecting up the terms that contain the same coefficient (except for the first one in each case since the formula won't work for those) and writing everything as a set of series gives us our solution,

$$y(x) = a_0 \left\{ 1 + \sum_{k-1}^{\infty} \frac{(-1)^k (1)^2 (5)^2 \cdots (4k-3)^2 x^{4k}}{(4k)!} \right\} +$$

$$a_1 \left\{ x + \sum_{k=1}^{\infty} \frac{(-1)^k (1)^2 (6)^2 \cdots (4k-2)^2 x^{4k+1}}{(4k+1)!} \right\} +$$

$$a_2 \left\{ x^2 + \sum_{k=1}^{\infty} \frac{2(-1)^k (3)^2 (7)^2 \cdots (4k-1)^2 x^{4k+2}}{(4k+2)!} \right\}$$

So, there we have it. As we can see the work in getting formulas for the coefficients was a little messy because we had three formulas to get, but individually they were not as bad as even some of them could be when dealing with 2nd order differential equations.

Also note that while we got lucky with this problem and we were able to get general formulas for the terms the higher the order the less likely this will become.

Chapter 5

Step Functions

Before proceeding we should take a look at one more function. Without Laplace transforms it would be much more difficult to solve differential equations that involve this function in $g(t)$.

The function is the Heaviside function and is defined as,

$$u_c(t) = \begin{cases} 0 & \text{if } t < c \\ 1 & \text{if } t \geq c \end{cases}.$$

Here is a graph of the Heaviside function.

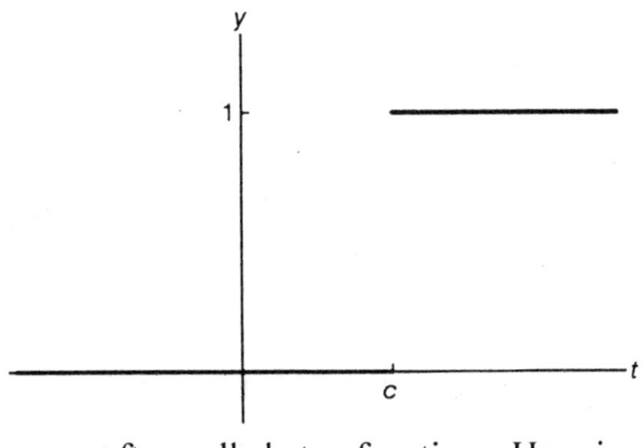

Heaviside functions are often called step functions. Here is some alternate notation for Heaviside functions.

$$u_c(t) = u(t-c) = H(t-c).$$

We can think of the Heaviside function as a switch that is off until $t = c$ at which point it turns on and takes a value of 1. So what if we want a switch that will turn on and takes some other value, say 4, or –7?

Heaviside functions can only take values of 0 or 1, but we can use them to get other kinds of switches. For instance $4u_c(t)$ is a switch that is off until $t = c$ and then turns on and takes a value of 4. Likewise, $-7u_c(t)$ will be a switch that will take a value of –7 when it turns on.

Now, suppose that we want a switch that is on (with a value of 1) and then turns off at $t = c$. We can use Heaviside functions to represent this as well. The following function will exhibit this kind of behavior.

$$1 - u_c(t) = \begin{cases} 1 - 0 = 1 & \text{if } t < c \\ 1 - 1 = 0 & \text{if } t \geq c \end{cases}.$$

Prior to $t = c$ the Heaviside is off and so has a value of zero. The function as whole then for $t < c$ has a value of 1. When we hit $t = c$ the Heaviside function will turn on and the function will now take a value of 0.

We can also modify this so that it has values other than 1 when it is one. For instance,

$$3 - 3u_c(t)$$

will be a switch that has a value of 3 until it turns off at $t = c$.

We can also use Heaviside functions to represent much more complicated switches.

Example: Write the following function (or switch) in terms of Heaviside functions.

$$f(t) = \begin{cases} -4 & \text{if } t < 6 \\ 25 & \text{if } 6 \leq t < 6 \\ 16 & \text{if } 8 \leq t < 30 \\ 10 & \text{if } t \geq 30 \end{cases}.$$

Solution: There are three sudden shifts in this function and so (hopefully) it's clear that we're going to need three Heaviside functions here, one for each shift in the function. Here's the function in terms of Heaviside functions.

$$f(t) = -4 + 29u_6(t) - 9u_8(t) - 6u_{30}(t).$$

It's fairly easy to verify this.

In the first interval, $t < 6$ all three Heaviside functions are off and the function has the value

$$f(t) = -4.$$

Notice that when we know that Heaviside functions are on or off we tend to not write them at all as we did in this case.

In the next interval, $6 \leq t < 8$the first Heaviside function is now on while the remaining two are still off. So, in this case the function has the value.

$$f(t) = -4. + 29 = 25.$$

In the third interval, $8 \leq t < 30$the first two Heaviside functions are one while the last remains off. Here the function has the value.

$$f(t) = -4 + 29 - 9 = 16.$$

In the last interval, $t \geq 30$all three Heaviside function are one and the function has the value.

$$f(t) = -4 + 29 - 9 - 6 = 10.$$

So, the function has the correct value in all the intervals.

All of this is fine, but if we continue the idea of using Heaviside function to represent switches, we really need to acknowledge that most switches will not turn on and take constant values. Most switches will turn on and vary continually with the value of t.

So, let's consider the following function.

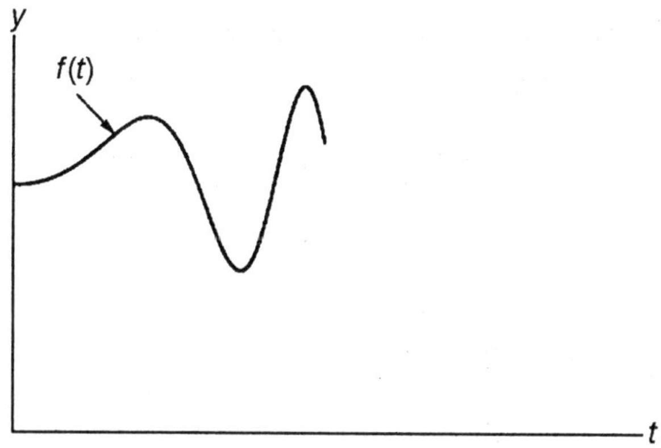

We would like a switch that is off until $t = c$ and then turns on and takes the values above. By this we mean that when $t = c$ we want the switch to turn on and take the value of $f(0)$ and when $t = c + 4$ we want the switch to turn on and take the value of $f(4)$, *etc.* In other words, we want the switch to look like the following,

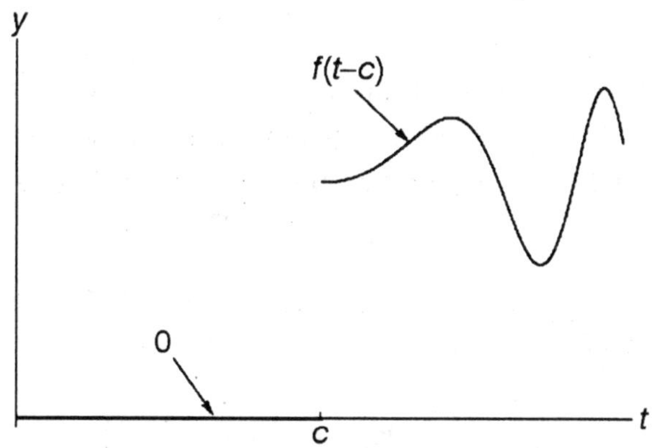

Notice that in order to take the values that we want the switch to take it needs to turn on and take the values of $f(t - c)$! We can use Heaviside functions to help us represent this switch as well. Using Heaviside functions this switch can be wrote as

$$g(t) = u_c(t) f(t - c).$$

Okay, we've talked a lot about Heaviside functions to this point, but we haven't even touched on Laplace transforms yet. So, let's start thinking about that. Let's determine the Laplace transform. This is actually easy enough to derive so let's do that. Plugging into the definition of the Laplace transform gives,

$$\mathfrak{L}\{u_c(t)f(t-c)\} = \int_0^{\infty} e^{-st}u_c(t)f(t-c)dt$$
$$= \int_0^{\infty} e^{-st}f(t-c)dt.$$

Notice that we took advantage of the fact that the Heaviside function will be zero if $t < c$ and 1 otherwise. This means that we can drop the Heaviside function and start the integral at c instead of 0. Now use the substitution $u = t - c$ and the integral becomes,

$$\mathfrak{L}\{u_c(t)f(t-c)\} = \int_0^{\infty} e^{-s(u+c)}f(u)du$$
$$= \int_0^{\infty} e^{-su}e^{-cs}(u)du.$$

The second exponential has no u's in it and so it can be factored out of the integral. Note as well that in the substitution process the lower limit of integration went back to 0.

$$\mathfrak{L}\{u_c(t)f(t-c)\} = e^{-cs}\int_0^{\infty} e^{-su}f(u)du.$$

Now, the integral left is nothing more than the integral that we would need to compute if we were going to find the Laplace transform of *f(t)*. Therefore, we get the following formula

$$\mathfrak{L}\{u_c(t)f(t-c)\} = e^{-cs}F(s).$$

In order to use the function *f(t)* must be shifted buy c, the same value that is used in the Heaviside function. Also note that we only take the transform of *f(t)* and not *f(t-c)*! We can also turn this around to get a useful formula for inverse Laplace transforms.-

$$\mathfrak{L}^{-1}\{e^{-cs}F(s)\} = u_c(t)f(t-c).$$

We can use to get the Laplace transform of a Heaviside function by itself. To do this we will consider the function to by $f(t) = 1$. Doing this gives us

$$\mathfrak{L}\{u_c(t)\} = \mathfrak{L}\{u_c(t)\cdot 1\} = e^{-cs}\mathfrak{L}\{1\} = \frac{1}{s}e^{-cs} = \frac{e^{-cs}}{s}.$$

Putting all of this together leads to the following two formulas.

$$\mathfrak{L}\{u_c(t)\} = \frac{e^{-cs}}{s} \qquad \mathfrak{L}^{-1}\left\{\frac{e^{-cs}}{s}\right\} = u_c(t).$$

Let' do some examples.

Example: Find the Laplace transform of each of the following.

1. $g(t)=10u_{12}(t)+2(t-6)^3u_6(t)-(7-e^{12-3t})u_4(t)$
2. $f(t)=-t^2u_3(t)+\cos(t))u_5(t)$
3. $h(t)=\begin{cases} t^4 & \text{if } t<5 \\ t^4+3\sin\left(\frac{t}{10}-\frac{1}{2}\right) & \text{if } t\geq 5\end{cases}$
4. $f(t)=\begin{cases} t & \text{if } t<6 \\ -8+(t-6)^2 & \text{if } t\geq 6\end{cases}$.

Solution: In all of these problems remember that the function MUST be in the form

$$u_c(t)f(t-c)$$

before we start taking transforms. If it isn't in that form we will have to put it into that form!

1. $g(t)=10u_{12}(t)+2(t-6)^3u_6(t)-\left(7-e^{12-3t}\right)u_4(t)$.

So there are three terms in this function. The first is simply a Heaviside function and so we can use on this term. The second and third terms however have functions with them and we need to identify the functions that are shifted for each of these. In the second term it the following function is shifted,

$$f(t)=2t^3 \quad\Rightarrow\quad f(t-6)=2(t-6)^3$$

and this has been shifted by the correct amount.

The third function uses,

$$f(t)=7-e^{-3t} \quad\Rightarrow\quad f(t-4)=7-e^{-3(t-4)}=7-e^{12-3t}$$

which has also been shifted by the correct amount.

With these functions identified we can now take the transform of the function.

$$G(s)=\frac{10e^{-12s}}{s}+e^{-6s}\frac{2(3!)}{s^{3+1}}-\left(\frac{7}{s}-\frac{1}{s+3}\right)e^{-4s}$$

$$=\frac{10e^{-12s}}{s}+\frac{12e^{-6s}}{s^{3+1}}-\left(\frac{7}{s}-\frac{1}{s+3}\right)e^{-4s}$$

2. $f(t)=-t^2u_3(t)+\cos(t)\,u_5(t)$.

This part is going to cause some problems. There are two terms and neither has been shifted by the proper amount. The first term needs to be shifted by 3 and the second needs to be shifted by 5. So, since they haven't been shifted, we will need to force the issue. We will need to add in the shifts, and then take them back out of course. Here they are.

$$f(t) = -(t-3+3)^2 u_3(t) + \cos(t-5+5)\, u_5(t).$$

Now we still have some potential problems here. The first function is still not really sifted correctly, so we'll need to use

$$(a+b)^2 a^2 + 2ab + b^2$$

to get this shifted correctly.

The second term can be dealt with in one of two ways. The first would be to use the formula

$$\cos(a+b) = \cos(a)\cos(b) - \sin(a)\sin(b)$$

to break it up into cosines and sines with arguments of t-5 which will be shifted as we expect. There is an easier way to do this one however. From out table of Laplace transforms we can see that if

$$g(t) = \cos(t+5) \quad \Rightarrow \quad g(t-5) = \cos(t-5+5).$$

This will make our life a little easier so we'll do it this way. Now, breaking up the first term and leaving the second term along gives us,

$$f(t) - ((t-3)^2 + 6(t-3) + 9)\, u_3(t) + \cos(t-5+5)\, u_5(t).$$

Okay, so it looks like the two functions that have been shifted here are

$$g(t) = t^2 + 6t + 9$$
$$g(t) = \cos(t+5).$$

Taking the transform then gives,

$$F(s) = -\left(\frac{2}{s^3} + \frac{6}{s^2} + \frac{9}{s}\right)e^{-3s} + \left(\frac{s\cos(5) - \sin(5)}{s^2+1}\right)e^{-5s}.$$

It's messy, especially the second term, but there it is. Also, do not get excited about the cos (5) and sin (5). They are just numbers.

3. $h(t) = \begin{cases} t^4 & \text{if } t < 5 \\ t^4 + 3\sin\left(\dfrac{t}{10} - \dfrac{1}{2}\right) & \text{if } t \geq 5 \end{cases}$

This one isn't as bad as it might look on the surface. The first thing that we need to do is write it in terms of Heaviside functions.

$$h(t) = t^4 + 3u_5(t)\sin\left(\frac{t}{10} - \frac{1}{2}\right)$$

$$= t^4 + 3u_5(t)\sin\left(\frac{t}{10}(t-5)\right).$$

Since the t^4 is in both terms there isn't anything to do when we add in the Heaviside function. The only thing that gets added in is the sine term. Notice as well that the sine has been shifted by the proper amount.

All we need to do now is to take the transform.

$$H(s) = \frac{4!}{s^5} + \frac{3\left(\frac{1}{10}\right)e^{-5s}}{s^2 + \left(\frac{1}{10}\right)^2} = \frac{24}{s^5} + \frac{\frac{3}{10}e^{-5s}}{s^2 + \frac{1}{100}}.$$

4. $f(t) = \begin{cases} t & \text{if } t < 6 \\ -8 + (t-6)^2 & \text{if } t \geq 6 \end{cases}.$

Again, the first thing that we need to do is write the function in terms of Heaviside functions.

$$f(t) = t + \left(-8 - t + (t-6)^2\right)u_6(t).$$

We had to add in a "-8" in the second term since that appears in the second part and we also had to subtract a t in the second term since the t in the first portion is no longer there. This subtraction of the t adds a problem because the second function is no longer correctly shifted. This is easier to fix than the previous example however.-

Here is the corrected function.

$$f(t) = t + \left(-8 - t + (t - 6 + 6) + (t-6)^2\right)u_6(t)$$

$$= t + \left(-8 - (t-6) - 6 + (t-6)^2\right)u_6(t)$$

$$= t + \left(-14 - (t-6) + (t-6)^2\right)u_6(t)$$

So, in the second term it looks like we are shifting

$$g(t) = t^2 - t - 14.$$

The transform is then,

$$F(s) = \frac{1}{s^2} + \left(\frac{2}{s^3} - \frac{1}{s^2} - \frac{14}{s}\right)e^{-6s}.$$

Without the Heaviside function taking Laplace transforms is not a terribly difficult process provided we have our trusty table of transforms. However, with the advent of Heaviside functions, taking transforms can become a fairly messy process on occasion.

So, let's do some inverse Laplace transforms to see how they are done.

Example: Find the inverse Laplace transform of each of the following.

1. $H(s)=\dfrac{se^{-4s}}{(3s+2)(s-2)}$

2. $G(s)=\dfrac{5e^{-6s}-3e^{-11s}}{(s+2)(s^2+9)}$

3. $F(s)=\dfrac{4s+e^{-s}}{(s-1)(s+2)}$

4. $G(s)=\dfrac{3s+8e^{-20s}-2se^{-3s}+6e^{-7s}}{s^2(s+3)}$

Solution: Notice that in order to use this formula the exponential doesn't really enter into the mix until the very end. The vast majority of the process is finding the inverse transform of the stuff without the exponential.

In these problems we are not going to go into detail on many of the inverse transforms. If you need a refresher on some of the basics of inverse transforms go back and take a look at the previous section.

1. $H(s)=\dfrac{se^{-4s}}{(3s+2)(s-2)}$

In light of the comments above let's first rewrite the transform in the following way.

$$H(s)=e^{-4s}\frac{s}{(3s+2)(s-2)}=e^{-4s}F(s).$$

Now, this problem really comes down to needing *f(t)*. So, let's do that. We'll need to partial fraction *F(s)* up. Here's the partial fraction decomposition.

$$F(s)=\frac{A}{3s+2}+\frac{B}{s-2}.$$

Setting numerators equal gives,

$$s=A(s-2)+B(3s+2).$$

We'll find the constants here by selecting values of s. Doing this gives,

$$s = 2 \qquad 2 = 8B \qquad \Rightarrow \qquad B = \frac{1}{4}$$

$$s = -\frac{2}{3} \qquad -\frac{2}{3} = -\frac{8}{3}A \qquad \Rightarrow \qquad B = \frac{1}{4}.$$

So, the partial fraction decomposition becomes,

$$F(s) = \frac{\frac{1}{4}}{3\left(s + \frac{2}{3}\right)} + \frac{\frac{1}{4}}{s - 2}.$$

Notice that we factored a 3 out of the denominator in order to actually do the inverse transform. The inverse transform of this is then,

$$f(t) = \frac{1}{12}e^{\frac{2t}{3}} + \frac{1}{4}e^{2t}.$$

Now, let's go back and do the actual problem. The original transform was,

$$H(s) = e^{-4s}F(s).$$

Note that we didn't bother to plug in *F(s)*. There really isn't a reason to plug it back in. Let's write down the inverse transform in terms of symbols. The inverse transform is,

$$h(t) = u_4(t)f(t - 4)$$

where, *f(t)* is,

$$f(t) = \frac{1}{12}e^{-\frac{2t}{3}} + \frac{1}{4}e^{2t}.$$

This is all the farther that we'll go with the answer. There really isn't any reason to plug in *f(t)* at this point. It would make the function longer and definitely messier. We will give almost all of our answers to these types of inverse transforms in this form.

2. $G(s) = \dfrac{5e^{-6s}}{(s+2)(s^2+9)}.$

This problem is not as difficult as it might at first appear to be. Because there are two exponentials we will need to deal with them separately eventually. Now, this might lead us to conclude that the best way to deal with this function is to split it up as follows,

$$G(s) = e^{-6s}\frac{5}{(s+2)(s^2+9)} - e^{-11s}\frac{3}{(s+2)(s^2+9)}.$$

Notice that we factored out the exponential, as we did in the last example, since we would need to do that eventually anyway. This is where a fairly common complication arises. Many people will call the first function *F(s)* and the second function *H(s)* and the partial fraction both of them.

However, if instead of just factoring out the exponential we would also factor out the coefficient we would get,

$$G(s) = 5e^{-6s}\frac{1}{(s+2)(s^2+9)} - 3e^{-11s}\frac{1}{(s+2)(s^2+9)}.$$

Upon doing this we can see that the two functions are in fact the same function. The only difference is the constant that was in the numerator. So, the way that we'll do these problems is to first notice that both of the exponentials have only constants as coefficients. Instead of breaking things up then, we will simply factor out the whole numerator and get,

$$G(s) = \left(5e^{-6s} - 3e^{-11s}\right)\frac{1}{(s+2)(s^2+9)}\left(5e^{-6s} - 3e^{-11s}\right)F(s)$$

and now we will just partial fraction *F(s)*.

Here is the partial fraction decomposition.

$$F(s) = \frac{A}{s+2} + \frac{Bs+C}{s^2+9}.$$

Setting numerators equal and combining gives us,

$$\begin{aligned} 1 &= A\left(s^2+9\right) + (s+2)(Bs+C) \\ &= (A+B)s^2 + (2B+C)s + 9A + 2C. \end{aligned}$$

Setting coefficient equal and solving gives,

$$\left.\begin{aligned} s^2 &: \quad A+B=0 \\ s^1 &: 2B+C=0 \\ s^0 &: 9A+2C=1 \end{aligned}\right\} \Rightarrow \quad A = \frac{1}{13}, \quad B = -\frac{1}{13}, C = \frac{2}{13}.$$

Substituting back into the transform gives and fixing up the numerators as needed gives,

$$\begin{aligned} F(s) &= \frac{1}{13}\left(\frac{1}{s+2} + \frac{-s+2}{s^2+9}\right) \\ &= \frac{1}{13}\left(\frac{1}{s+2} - \frac{s}{s^2+9} + \frac{2\frac{3}{3}}{s^2+9}\right). \end{aligned}$$

As we did in the previous section we factored out the common denominator to make our work a little simpler. Taking the inverse transform then gives,

$$f(t) = \frac{1}{13}\left(e^{-2t} - \cos(3t) + \frac{2}{3}\sin(3t)\right).$$

At this point we can go back and start thinking about the original problem.

$$G(s) = \left(5e^{-6s} - 3e^{-11s}\right)F(s)$$

$$= 5e^{-6s}F(s) - 3e^{-11s}F(s).$$

We'll also need to distribute the *F(s)* through as well in order to get the correct inverse transform. Recall that in order to take the inverse transform you must have a single exponential times a single transform. This means that we must multiply the *F(s)* through the parenthesis. We can now take the inverse transform,

$$g(t) = 5u_6(t)f(t-6) - 3u_u(t)f(t-11)$$

where,
$$f(t) = \frac{1}{13}\left(e^{-2t} - \cos(3t) + \frac{2}{3}\sin(3t)\right)$$

3. $F(s) = \dfrac{4s + e^{-s}}{(s-1)(s+2)}$.

In this case, unlike the previous part, we will need to break up the transform since one term has a constant in it and the other has an *s*. Note as well that we don't consider the exponential in this, only its coefficient. Breaking up the transform gives,

$$F(s) = \frac{4s}{(s-1)(s+2)} + e^{-s}\frac{1}{(s-1)(s+2)} = G(s) + e^{-s}H(s).$$

We will need to partial fraction both of these terms up. We'll start with *G(s)*.

$$G(s) = \frac{A}{s-1} + \frac{B}{s+2}.$$

Setting numerators equal gives,

$$4s = A(s+2) + B(s-1).$$

Now, pick values of *s* to find the constants.

$$s = -2 \qquad -8 = -3B \quad \Rightarrow \quad B = \frac{8}{3}$$

$$s = 1 \qquad -4 = 3A \quad \Rightarrow \quad A = \frac{4}{3}$$

So *G(s)* and its inverse transform is,

$$G(s) = \frac{\frac{4}{3}}{s-1} + \frac{\frac{8}{3}}{s+2}$$

$$g(t) = \frac{4}{3}e' + \frac{8}{3}e^{-2t}.$$

Now, repeat the process for *H(s)*.

$$H(s) = \frac{A}{s-1} + \frac{B}{s+2}.$$

Setting numerators equal gives,

$$1 = A(s+2) + B(s-1).$$

Now, pick values of s to find the constants.

$$s = -2 \qquad 1 = -3B \qquad \Rightarrow \qquad B = \frac{1}{3}$$

$$s = 1 \qquad 1 = 3A \qquad \Rightarrow \qquad A = \frac{1}{3}.$$

So *H(s)* and its inverse transform is,

$$H(s) = \frac{\frac{1}{3}}{s-1} - \frac{\frac{1}{3}}{s+2}$$

$$h(t) = \frac{1}{3}e' - \frac{1}{3}e^{-2t}.$$

Putting all of this together gives the following,

$$F(s) = G(s) + e^{-s}H(s)$$

$$f(t) = g(t) + u_1(t)h(t-1)$$

where,

$$g(t) = \frac{4}{3}e' + \frac{8}{3}e^{-2t} \quad \text{and} \quad h(t) = \frac{1}{3}e' - \frac{1}{3}e^{-2t}$$

4. $G(s) = \dfrac{3s + 8e^{-20s} - 2e^{-3s} + 6e^{-7s}}{s^2(s+3)}$

This one looks messier than it actually is. Let's first rearrange the numerator a little.

$$G(s) = \frac{s\left(3 - 2e^{-3s}\right) + \left(8e^{20s} + 6e^{-7s}\right)}{s^2(s+3)}.$$

In this form it looks like we can break this up into two pieces that will require partial fractions. When we break these up we should always try and break things up into as few pieces as possible for the partial fractioning. Doing this can save you a great deal of unnecessary work. Breaking up the transform as suggested above gives,

$$G(s) = \left(3 - 2e^{-3s}\right)\frac{1}{s(s+3)} + \left(8e^{-20s} + 6e^{-7s}\right)\frac{1}{s^2(s+3)}$$

$$= \left(3 - 2e^{-3s}\right)F(s) + \left(8e^{-20s} + 6e^{-7s}\right)H(s).$$

Note that we canceled an s in *F(s)*. You should always simplify as much a possible before doing the partial fractions.

Let's partial fraction up *F(s)* first.

$$F(s)=\frac{A}{s}+\frac{B}{s+3}.$$

Setting numerators equal gives, $1=A(s+3)+Bs$.

Now, pick values of s to find the constants.

$$s=-3 \qquad 1=-3B \quad \Rightarrow \quad B=\frac{1}{3}$$

$$s=0 \qquad 1=-3A \quad \Rightarrow \quad A=\frac{1}{3}.$$

So *F(s)* and its inverse transform is,

$$F(s)=\frac{\frac{1}{3}}{s}-\frac{\frac{1}{3}}{s+3}$$

$$f(t)=\frac{1}{s}-\frac{1}{3}e^{-3t}.$$

Now partial fraction *H(s)*.

$$H(s)=\frac{A}{s}-\frac{B}{s^2}+\frac{C}{s+3}.$$

Setting numerators equal gives,

$$1=As(s+3)+B(s+3)+Cs^2.$$

Pick values of s to find the constants.

$$s=-3 \qquad 1=9C \qquad \Rightarrow \quad C=\frac{1}{9}$$

$$s=0 \qquad 1=-3B \qquad \Rightarrow \quad B=\frac{1}{3}$$

$$s=1 \qquad 1=4A+4B+C=4A+\frac{13}{3} \quad \Rightarrow \quad B=\frac{1}{9}.$$

So *H(s)* and its inverse transform is,

$$H(s)=-\frac{\frac{1}{9}}{s}+\frac{\frac{1}{3}}{s^2}+\frac{\frac{1}{9}}{s+3}$$

$$h(t)=-\frac{1}{9}+\frac{1}{3}t+\frac{1}{9}e^{-3t}.$$

Now, let's go back to the original problem, remembering to multiply the transform through the parenthesis.

$$G(t) = 3F(s) - 2e^{-3s}F(s) + 8e^{-20s}H(s) + 6e^{-7s}H(s).$$

Taking the inverse transform gives,

$$g(t) = 3f(t) - 2u_3(t)f(t-3) + 8u_{20}(t)h(t-20) + 6u_7(t)h(t-7).$$

So, as this example has shown, these can be a somewhat messy. However, the mess is really only that of notation and amount of work. The actual partial fraction work was identical to the previous sections work. The main difference in this section is we had to do more of it. As far as the inverse transform process goes. Again, the vast majority of that was identical to the previous section as well.

So, don't let the apparent messiness of these problems get you to decide that you can't do them. Generally they aren't as bad as they seem initially.

SOLVING IVP'S WITH LAPLACE TRANSFORMS

It's now time to get back to differential equations. We've spent the last three sections learning how to take Laplace transforms and how to take inverse Laplace transforms. These are going to be invaluable skills for the next couple of sections so don't forget what we learned there. Before proceeding into differential equations we will need one more formula. We will need to know how to take the Laplace transform of a derivative. First recall that $f^{(n)}$ denotes the nth derivative of the function f. We now have the following fact.

FACT

Suppose that $f, f', f'', \ldots f^{(n-1)}$ are all continuous functions and $f^{(n)}$ is a piecewise continuous function. Then,-

$$\mathcal{L}\{f^{(n)}\} = s^n F(s) - s^{n-1}(0) - s^{n-2}f'(0) - \cdots - sf^{(n-2)}(0) - f^{(n-1)}(0).$$

Since we are going to be dealing with second order differential equations it will be convenient to have the Laplace transform of the first two derivatives.

Notice that the two function evaluations that appear in these formulas, $y(0)$and $y'(0)$, are often what we've been using for initial condition in out IVP's. So, this means that if we are to use these formulas to solve an IVP we will need initial conditions at $t = 0$.

While Laplace transforms are particularly useful for nonhomogeneous differential equations which have Heaviside functions in the forcing function we'll start off with a couple of fairly simple problems to illustrate how the process works.

Example: Solve the following IVP.

$$y' - 10y + 9y = 5t, \qquad y(0) = -1 \quad y'(0) = 2.$$

Solution:The first step in using Laplace transforms to solve an IVP is to take the transform of every term in the differential equation.

$$\mathcal{L}\{y'\} - 10\mathcal{L}\{y'\} + 9\mathcal{L}\{y\} = \mathcal{L}\{5t\}.$$

Using the appropriate formulas from our table of Laplace transforms gives us the following.

$$5^2 Y(s) - sy(0) - y'(0) - 10(sY(s) - y(0)) + 9Y(s) = \frac{5}{s^2}.$$

Plug in the initial conditions and collect all the terms that have a *Y(s)* in them.

$$\left(s^2 - 10s + 9\right)Y(s) + s - 12 = \frac{5}{s^2}.$$

Solve for *Y(s)*.

$$Y(s) = \frac{5}{s^2(s-9)(s-1)} + \frac{12 - s}{(s-9)(s-1)}.$$

At this point it's convenient to recall just what we're trying to do. We are trying to find the solution, *y(t)*, to an IVP. What we've managed to find at this point is not the solution, but its Laplace transform. So, in order to find the solution all that we need to do is to take the inverse transform.

Before doing that let's notice that in its present form we will have to do partial fractions twice. However, if we combine the two terms up we will only be doing partial fractions once. Not only that, but the denominator for the combined term will be identical to the denominator of the first term. This means that we are going to partial fraction up a term with that denominator no matter what so we might as well make the numerator slightly messier and then just partial fraction once.

This is one of those things where we are apparently making the problem messier, but in the process we are going to save ourselves a fair amount of work!

Combining the two terms gives,

$$Y(s) = \frac{5 + 12s^2 - s^2}{s^2(s-9)(s-1)}.$$

The partial fraction decomposition for this transform is,

$$Y(s) = \frac{A}{s} + \frac{B}{s^2} + \frac{C}{s-9} + \frac{D}{s-1}.$$

Setting numerators equal gives,

$$5 + 12s^2 - s^3 = As(s-9)(s-1) + B(s-9)(s-1) + Cs^2(s-1) + Ds^2(s-9).$$

Picking appropriate values of s and solving for the constants gives,

$$s=0 \qquad 5=9B \qquad \Rightarrow \qquad B=\frac{5}{9}$$

$$s=1 \qquad 16=-8D \qquad \Rightarrow \qquad D=-2$$

$$s=9 \qquad 248=648C \qquad \Rightarrow \qquad C=\frac{31}{81}$$

$$s=2 \qquad 45=-14A+\frac{4345}{81} \qquad \Rightarrow \qquad B=\frac{50}{81}$$

Plugging in the constants gives,

$$Y(s)\frac{\frac{50}{81}}{s}+\frac{\frac{5}{9}}{s^2}+\frac{\frac{31}{81}}{s-9}-\frac{2}{s-1}.$$

Finally taking the inverse transform gives us the solution to the IVP.

$$y(t)\frac{50}{81}+\frac{5}{9}t+\frac{31}{81}e^{5t}-2e^{t}.$$

That was a fair amount of work for a problem that probably could have been solved much quicker using the techniques for the previous chapter. The point of this problem however, was to show how we would use Laplace transforms to solve an IVP.

There are a couple of things to note here about using Laplace transforms to solve an IVP. First, using Laplace transforms reduces a differential equation down to an algebra problem. In the case of the last example the algebra was probably more complicated than the straight forward approach from the last chapter. However, in later problems this will be reversed. The algebra, while still very messy, will often be easier than a straight forward approach.

Second, unlike the approach in the last chapter, we did not need to first find a general solution, differentiate this, plug in the initial conditions and then solve for the constants to get the solution. With Laplace transforms, the initial conditions are applied during the first step and at the end we get the actual solution instead of a general solution. In many of the later problems Laplace transforms will make the problems significantly easier to work than if we had done the straight forward approach of the last chapter. Also, as we will see, there are some differential equations that simply can't be done using the techniques from the last chapter and so, in those cases, Laplace transforms will be our only solution.

Let's take a look at another fairly simple problem.

Example: Solve the following IVP.

$$2y'+3y'-2y=te^{-2t}, \qquad y(0)=0 \quad y'(0)=-2.$$

Solution: As with the first example, let's first take the Laplace transform of all the terms in the differential equation. We'll the plug in the initial conditions to get,

$$2(s^2Y(s)-sy(0)-y'(0))+3(sY(s)-y(0))-2Y(s) = \frac{1}{(s+2)^2}$$

$$(2s^2+3s-2)Y(s)+4 = \frac{1}{(s+2)^2}$$

Now solve for *Y(s)*.

$$Y(s)=\frac{1}{(2s-1)(s+2)^3}-\frac{4}{(2s-1)(s+2)}.$$

Now, as we did in the last example we'll go ahead and combine the two terms together as we will have to partial fraction up the first denominator anyway, so we may as well make the numerator a little more complex and just do a single partial fraction. This will give,

$$Y(s) = \frac{1-4(s+2)^2}{(2s-1)(s+2)}$$

$$= \frac{-4s^2-16s-15}{(2s-1)(s+2)^3}.$$

The partial fraction decomposition is then,

$$Y(s)=\frac{A}{2s-1}+\frac{B}{s+2}+\frac{C}{(s+2)^2}+\frac{D}{(s+2)}.$$

Setting numerator equal gives,

$$\begin{aligned}-4s^2-16s-15 &= A(s+2)^3+B(2s-1)(s+2)^2+C(2s-1)(s+2)+D(2s-1)\\ &= (A+2B)s^3+(6A+7B+2C)s^2+(12A+4B+2D)s\\ &\quad +8A-4B-2C-D.\end{aligned}$$

In this case it's probably easier to just set coefficients equal and solve the resulting system of equation rather than pick values of *s*. So, here is the system and its solution.

$$\left.\begin{aligned} s^3 &: & A+2B&=0\\ s^2 &: & 6A+7B+2C&=-4\\ s^1 &: & 12A+4B+3C+2D&=-16\\ s^0 &: & 8A-4B-2C-D&=-15\end{aligned}\right\} \Rightarrow \begin{array}{ll} A=-\dfrac{192}{125} & B=\dfrac{96}{125}\\ C=-\dfrac{2}{25} & B=\dfrac{1}{5}\end{array}.$$

We will get a common denominator of 125 on all these coefficients and factor that out when we go to plug them back into the transform. Doing this gives,

$$Y(s)=\frac{1}{125}\left(\frac{-192}{s\left(s-\frac{1}{2}\right)}+\frac{96}{s+2}-\frac{10}{s+2}-\frac{10}{(s+2)^2}-\frac{25\frac{2!}{2!}}{(s+2)^3}\right).$$

Notice that we also had to factor a 2 out of the denominator of the first term and fix up the numerator of the last term in order to get them to match up to the correct entries in our table of transforms.

Taking the inverse transform then gives,

$$Y(t)=\frac{1}{125}\left(-96e^{\frac{1}{2}}+96e^{-2t}-10te^{-2t}\frac{25}{2}t^2e^{-2t}\right).$$

Example: Solve the following IVP.

$$y'-6y'+15y=2\sin(3t),\qquad y(0)=-1\quad y'(0)=-4.$$

Solution: Take the Laplace transform of everything and plug in the initial conditions.

$$s^2y(s)-sy(0)-y'(0)-6(sY(s)-y(0))+15y(s) = 2\frac{3}{s^2+9}$$

$$\left(s^2-6s+15\right)Y(s)+s-2 = \frac{6}{s^2+9}.$$

Now solve for $Y(s)$ and combine into a single term as we did in the previous two examples.

$$Y(s)=\frac{-s^3+2s^2-9s+24}{(s^2+9)(s^2-6s+15)}.$$

Now, do the partial fractions on this. First let's get the partial fraction decomposition.

$$Y(s)=\frac{As+B}{s^2+9}+\frac{Cs+D}{s^2-6s+15}.$$

Now, setting numerators equal gives,

$$\begin{aligned}-s^3+2s^2-9s+24 &= (As+B)(s^2-6s+15)+(Cs+D)(s^2+9)\\ &= (A+C)s^2+(-6A+B+D)s^2+(15A-6B+9C)s+15B+9D.\end{aligned}$$

Setting coefficients equal and solving for the constants gives,

$$\left.\begin{aligned}s^3:&\quad A+C=-1\\ s^2:&\quad -6A+B+D=2\\ s^1:&\quad 15A-6B+9c=-9\\ s^0:&\quad 15B+9D=24\end{aligned}\right\}\Rightarrow\quad \begin{aligned}A&=\frac{1}{10} & B&=\frac{1}{10}\\ C&=\frac{11}{10} & D&=\frac{5}{2}\end{aligned}.$$

Now, plug these into the decomposition, complete the square on the denominator of the second term and then fix up the numerators for the inverse transform process.

$$Y(s) = \frac{1}{10}\left(\frac{s+1}{s^2+9} + \frac{-11s+25}{s^2-6s+15}\right)$$

$$= \frac{1}{10}\left(\frac{s+1}{s^2+9} + \frac{-11(s-3+3)+25}{(s-3)^2+6}\right)$$

$$= \frac{1}{10}\left(\frac{s}{s^2+9} + \frac{1\frac{2}{3}}{s^2+9} - \frac{11(s-3)}{(s-3)^2+6} - \frac{8\frac{\sqrt{6}}{\sqrt{6}}}{(s-3)^2+6}\right).$$

Finally, take the inverse transform.

$$y(t) = \frac{1}{10}\left(\cos(3t)\frac{1}{3} + \frac{1}{3}\sin(3t) - 11e^{3t}\cos\left(\sqrt{6t}\right) - \frac{8}{\sqrt{6}}e^{3t}\sin\left(\sqrt{6t}\right)\right).$$

To this point we've only looked at IVP's in which the initial values were at $t = 0$. This is because we need the initial values to be at this point in order to take the Laplace transform of the derivatives. The problem with all of this is that there are IVP's out there in the world that have initial values at places other than $t = 0$. Laplace transforms would not be as useful as it is if we couldn't use it on these types of IVP's. So, we need to take a look at an example in which the initial conditions are not at $t = 0$ in order to see how to handle these kinds of problems.

Example: Solve the following IVP.

$$y' + 4y' = \cos(t-3) + 4t, \qquad y(3) = 0 \quad y'(3) = 7.$$

Solution: The first thing that we will need to do here it to take care of the fact that initial conditions are not at $t = 0$. The only way that we can take the Laplace transform of the derivatives is to have the initial conditions at $t = 0$.

This means that we will need to formulate the IVP in such a way that the initial conditions are at $t = 0$. This is actually fairly simple to do, however we will need to do a change of variable to make it work. We are going to define

$$\eta = t - 3 \quad \Rightarrow \quad t = \eta + 3.$$

Let's start with the original differential equation.

$$y'(t) + 4y'(t) = \cos(t-3) + 4t.$$

Notice that we put in the *(t)* part on the derivatives to make sure that we get things correct here. We will next substitute in for *t*.

$$y'(\eta+3) + 4y'(\eta+3) = \cos(\eta) + 4(\eta+3).$$

Now, to simplify life a little let's define,

$$u(\eta) = y(\eta + 3).$$

Then, by the chain rule, we get that the following for the first derivative.

$$u'(\eta) = \frac{du}{d\eta} = \frac{dy}{dt}\frac{dt}{d\eta} = y'(\eta + 3).$$

By a similar argument we get the following for the second derivative.

$$u'(\eta) = y^{n}(\eta + 3).$$

The initial conditions for *u(ç)* are,

$$u(0) = y(0+3) = y(3) = 0$$

$$u'(0) = y'(0+3) = y'(3) = 7.$$

The IVP under these new variables is then,

$$u'' + 4u' = \cos(\eta) + 4\eta + 12, \quad u(0) = 0 \quad u'(0) = 7.$$

This is an IVP that we can use Laplace transforms on provided we replace all the *t*'s with ç's. So, taking the Laplace transform of this new differential equation and plugging in the new initial conditions gives,

$$\begin{aligned} s^2U(s) - su(0) - u'(0) + 4(sU(s) - u(0)) &= \frac{s}{s^2+1} + \frac{4}{s^2} + \frac{12}{s} \\ (s^2 + 4s)U(s) - 7 &= \frac{s}{s^2+1} + \frac{4+12s}{s^2}. \end{aligned}$$

Solving for *U(s)* gives,

$$\begin{aligned} (s^2 + 4s)\,U(s) &= \frac{s}{s^2+1} + \frac{4+12s+7s^2}{s^2} \\ U(s) &= \frac{1}{(s+4)(s^2+1)} + \frac{4+12s+7s^2}{s^2(s+4)}. \end{aligned}$$

Note that unlike the previous examples we did not completely combine all the terms this time. In all the previous examples we did this because the denominator of one of the terms was the common denominator for all the terms. Therefore, upon combining, all we did was make the numerator a little messier, and reduced the number of partial fractions required down from two to one. Note that all the terms in this transform that had only powers of *s* in denominator where combined for exactly this reason.

In this transform however, if we combined both of the remaining terms into a single term we would be left with a fairly involved partial fraction problem. Therefore, in this case, it would probably be easier to just do partial fractions twice.

We've done several partial fractions problems in this section and many partial fraction problems in the previous couple of sections so we're going to leave the details of the partial fractioning to you to check. Partial fractioning each of the terms in our transform gives us the following.

$$\frac{1}{(s+4)\left(s^2+1\right)} = \frac{\frac{1}{17}}{s+4} + \frac{1}{17}\left(\frac{-s+4}{s^2+1}\right)$$

$$\frac{4+12s+7s^2}{s^3(s+4)} = \frac{1}{s^3} + \frac{\frac{11}{4}}{s^3} - \frac{\frac{17}{16}}{s+4}.$$

Plugging these into our transform and combining like terms gives us

$$U(s) = \frac{1}{s^3} + \frac{\frac{11}{4}}{s^2} + \frac{\frac{17}{16}}{s} - \frac{\frac{273}{272}}{s+4} + \frac{1}{17}\left(\frac{-s+4}{s^2+1}\right)$$

$$= \frac{1\frac{2!}{2!}}{s^3} + \frac{\frac{11}{4}}{s^2} + \frac{\frac{17}{16}}{s} - \frac{\frac{273}{272}}{s+4} + \frac{1}{17}\left(\frac{-s}{s^2+1} + \frac{4}{s^2+1}\right).$$

Now, taking the inverse transform will give the solution to our new IVP. Don't forget to use ç's instead of *t*'s!

$$u(\eta) = \frac{1}{2}\eta^2 + \frac{11}{2}\eta + \frac{17}{16} - \frac{273}{272}e^{-4\eta} + \frac{1}{17}(4\sin(\eta) - \cos(\eta)).$$

This is not the solution that we are after of course. We are after *y(t)*. However, we can get this by noticing that

$$y(t) = y(\eta + 3) = u(\eta) = u(t-3).$$

So the solution to the original IVP is,

$$y(t) = \frac{1}{2}(t-3)^2 + \frac{11}{4}(t-3) + \frac{17}{16} - \frac{273}{272}e^{-4(t-3)} + \frac{1}{17}(4\sin(t-3) - \cos(t-3)).$$

$$y(t) = \frac{1}{2}t^2 - \frac{1}{4}t - \frac{43}{16} - \frac{273}{272}e^{-4(t-3)} + \frac{1}{17}(4\sin(t-3) - \cos(t-3)).$$

So, we can now do IVP's that don't have initial conditions that are at $t = 0$. We also saw in the last example that it isn't always the best to combine all the terms into a single partial fraction problem as we have been doing prior to this example.The examples worked in this section would have just as easy, if not easier, if we had used techniques from the previous chapter. They were worked here using Laplace transforms to illustrate the technique and method.

NONCONSTANT COEFFICIENT IVP'S

In this section we are going to see how Laplace transforms can be used to solve some differential equations that do not have constant coefficients. This is not always an easy thing to do. However, there are some simple cases that can be done.

To do this we will need a quick fact.

FACT

If *f(t)* is a piecewise continuous function on (0, ∞) of exponential order then,

$$\lim_{s\to\infty} F(s) = 0.$$

A function *f(t)* is said to be of exponential order áif there is exist positive constants T and M such that

$$|f(t)| \leq Me^{at} \quad \text{for all } t \geq T.$$

Put in other words, a function that is of exponential order will grow no faster than

$$Me^{at}$$

for some M and áand all sufficiently large t. One way to check whether a function is of exponential order or not is to compute the following limit.

$$\lim_{t\to\infty} \frac{|f(t)|}{e^{at}}.$$

If this limit is finite for some áthen the function will be of exponential order á. Likewise, if the limit is infinite for every áthen the function is not of exponential order. Almost all of the functions that you are liable to deal with in a first course in differential equations are of exponential order. A good example of a function that is not of exponential order is

$$f(t) = e^{t^3}.$$

We can check this by computing the above limit.

$$\lim_{t\to\infty} \frac{e^{t^3}}{e^{at}} = \lim_{t\to\infty} e^{t^3 - at} = \lim_{t\to\infty} e^{e(t^2-a)} = \infty.$$

This is true for any value of áand so the function is not of exponential order.

Do not worry too much about this exponential order stuff. This fact is occasionally needed in using Laplace transforms with non constant coefficients.

So, let's take a look at an example.

Example: Solve the following IVP.

$$y'' + 3ty' - 6y = 2, \qquad y(0) = 0 \quad y'(0) = 0.$$

Solution: Laplace transforms tells us that,

$$\begin{aligned}\mathfrak{L}\{ty'\} &= -\frac{d}{ds}(\mathfrak{L}\{y'\}) \\ &= \frac{d}{ds}(sY(s) - y(0)) \\ &= -sY'(s) - Y(s).\end{aligned}$$

So, upon taking the Laplace transforms of everything and plugging in the initial conditions we get,

$$s^2Y(s) - sy(0) - y'(0) + 3(-sy'(s) - Y(s)) - 6Y(s) = \frac{2}{s}$$

$$-3sY'(s) + (s^2 - 9)\,Y(s) = \frac{2}{s}$$

$$Y'(s) + \left(\frac{3}{s} - \frac{s}{3}\right)Y(s) = \frac{2}{3s^2}.$$

Unlike the examples in the previous section where we ended up with a transform for the solution, here we get a linear first order differential equation that must be solved in order to get a transform for the solution.

The integrating factor for this differential equation is,

$$\mu(t) = e^{\int\left(\frac{3}{s} - \frac{s}{3}\right)dt} = e^{\ln(s^3) - \frac{s^2}{6} = s^2 e^{\frac{s^2}{6}}}.$$

Multiplying through, integrating and solving for *Y(s)* gives,

$$\begin{aligned}\int\left(s^3 e^{\frac{s^2}{s}Y}(s)\right)' ds &= \int -\frac{2}{3}se^{-\frac{s^2}{6}ds} \\ s^3 e^{-\frac{s^2}{6}}Y(s) &= 2e^{-\frac{s^2}{6}} + c \\ Y(s) &= \frac{2}{s^3} + c\frac{e^{\frac{s^2}{6}}}{s^3}.\end{aligned}$$

Now, we have a transform for the solution. However that second term looks unlike anything we've seen to this point. This is where the fact about the transforms of exponential order functions comes into play. We are going to assume that whatever our solution is, it is of exponential order. This means that

$$\lim_{s\to\infty}\left(\frac{2}{s^3} + \frac{ce^{\frac{s^2}{6}}}{s^3}\right) = 0.$$

The first term does go to zero in the limit. The second term however, will only go to zero if $c = 0$. Therefore, we must have $c = 0$ in order for this to be the transform of our solution.

So, the transform of our solution, as well as the solution is,

$$Y(s) = \frac{2}{s^3} \qquad y(t) = t^2.$$

I'll leave it to you to verify that this is in fact a solution if you'd like to.

Now, not all nonconstant differential equations need to use. So, let's take a look at one more example.

Example: Solve the following IVP.

$$ty' - ty' + y = 2, \qquad y(0) = 2 \quad y'(0) = -4.$$

Solution: From the first example we have,

$$\mathfrak{L}\{ty'\} = -sY'(s) - Y(s).$$

We'll also need,

$$\begin{aligned} \mathfrak{L}\{ty''\} &= -\frac{d}{ds}(\mathfrak{L}\{ty''\}) \\ &= -\frac{d}{ds}(s^2Y(s) - sy(0) - y'(0)) \\ &= -s^2Y'(s) - 2sY(s) + y(0). \end{aligned}$$

Taking the Laplace transform of everything and plugging in the initial conditions gives,

$$-s^2Y'(s) - 2sY(s) + y(0) - (-sY')(s) - Y(s)) + Y(s) = \frac{2}{s}$$

$$(s - s^2)Y'(s) + (2 - 2s)Y(s) + 2 = \frac{2}{s}$$

$$s(1-s)Y'(s) + 2(1-s)Y(s) = \frac{2(1-s)}{s^2}$$

$$Y'(s) + \frac{2}{s}Y(s) = \frac{2}{s^2}.$$

Once again we have a linear first order differential equation that we must solve in order to get a transform for the solution. Notice as well that we never used the second initial condition in this work.

That is okay, we will use it eventually.

Since this linear differential equation is much easier to solve compared to the first one, we'll leave the details to you. Upon solving the differential equation we get,

$$Y'(s) = \frac{2}{s} + = \frac{c}{s^2}.$$

Now, this transform goes to zero for all values of c and we can take the inverse transform of the second term. Therefore, we won't need to get rid of the second term as did in the previous example.

Taking the inverse transform gives,

$$y(t) = 2 + ct.$$

Now, is where we will use the second initial condition. Upon differentiating and plugging in the second initial condition we can see that $c = -4$.

So, the solution to this IVP is,

$$y(t) = 2 - 4t.$$

So, we've seen how to use Laplace transforms to solve some nonconstant coefficient differential equations. Notice however that all we did was add in an occasional t to the coefficients. We couldn't get too complicated with the coefficients. If we had we would not have been able to easily use Laplace transforms to solve them.

Sometimes Laplace transforms can be used to solve nonconstant differential equations, however, in general, nonconstant differential equations are still very difficult to solve.

IVP'S WITH STEP FUNCTIONS

In this section we will use Laplace transforms to solve IVP's which contain Heaviside functions in the forcing function. This is where Laplace transform really starts to come into its own as a solution method.

To work these problems we'll just need to remember the following two formulas,

$$\mathfrak{L}\{u_c(t)f(t-c)\} = e^{-cs}F(s) \qquad \text{where} \quad F(s) = \mathfrak{L}\{f(t)\}$$

$$\mathfrak{L}\{e^{-cs}F(s)\} = u_c(t)f(t-c) \qquad \text{where} \quad f(s) = \mathfrak{L}^{-1}\{F(s)\}$$

In other words, we will always need to remember that in order to take the transform of a function that involves a Heaviside we've got to make sure the function has been properly shifted.

Let's work an example.

Example: Solve the following IVP.

$$y'' - y' + 5y = 4 + u_2(t)\,e^{4-2t}, \qquad y(0) = 2 \quad y'(0) = -1.$$

Solution: First let's rewrite the forcing function to make sure that it's being shifted correctly and to identify the function that is actually being shifted.

$$y''-y'+5y=4+u_2(t)\,e^{-2(t-2)}.$$

So, it is being shifted correctly and the function that is being shifted is e^{-2t}. Taking the Laplace transform of everything and plugging in the initial conditions gives,

$$s^2Y(s)-sy(0)-y'(0)-(sY(s)-y(0))+5Y(s)=\frac{4}{s}+\frac{e^{-2s}}{s+2}$$

$$(s^2-s+5)Y(s)-2s+3=\frac{4}{s}+\frac{e^{-2s}}{s+2}.$$

Now solve for *Y(s)*.

$$(s^2-s+5)Y(s) = \frac{4}{s}+\frac{e^{-2s}}{s+2}+2s-3$$

$$(s^2-s+5)Y(s) = \frac{2s^2-3s+4}{s}+\frac{e^{-2s}}{s+2}$$

$$Y(s)=\frac{2s^2-3s+4}{s(s^2-s+5)}+e^{-2s}\frac{1}{(s+2)(s^2-s+5)}$$

$$Y(s)= F(s)+e^{-2s}G(s).$$

Notice that we combined a couple of terms to simplify things a little. Now we need to partial fraction *F(s)* and *G(s)*. We'll leave it to you to check the details of the partial fractions.

$$F(s)=\frac{2s^2-3s+4}{s(s^2-s+5)}=\frac{1}{5}\left(\frac{4}{s}+\frac{6s-11}{s^2-s+5}\right)$$

$$G(s)=\frac{1}{(s+2)(s^2-s+5)}=\frac{1}{11}\left(\frac{1}{s+5}-\frac{s-3}{s+s+5}\right).$$

We now need to do the inverse transforms on each of these. We'll start with *F(s)*.

$$F(s) = \frac{1}{5}\left(\frac{4}{s}+\frac{6\left(s-\frac{1}{2}+\frac{1}{2}\right)-11}{\left(s-\frac{1}{2}\right)+\frac{19}{4}}\right)$$

$$=\frac{1}{5}\left(\frac{4}{s}+\frac{6\left(s-\frac{1}{2}\right)}{\left(s-\frac{1}{2}\right)^2+\frac{19}{4}}-\frac{8\frac{\sqrt{2}}{2}-\frac{2}{\sqrt{19}}}{\left(s-\frac{1}{2}\right)^2+\frac{19}{4}}\right)$$

$$f(t) = \frac{1}{5}\left(4 + 6e^{\frac{t}{2}}\cos\left(\frac{\sqrt{19}}{2}t\right) - \frac{16}{\sqrt{19}}e^{\frac{t}{2}} + \left(\frac{\sqrt{19}}{2}t\right)\right).$$

Now *G(s)*.

$$G(s) = \frac{1}{11}\left(\frac{1}{s+2} + \frac{s - \frac{1}{2} + \frac{1}{2} - 3}{\left(s - \frac{1}{2}\right)^2 + \frac{19}{4}}\right)$$

$$= \frac{1}{11}\left(\frac{1}{s+2} - \frac{s - \frac{1}{2}}{\left(s - \frac{1}{2}\right)^2 + \frac{19}{4}} + \frac{\frac{5}{2}\frac{\sqrt{19}}{\sqrt{19}}}{\left(s - \frac{1}{2}\right)^2 + \frac{19}{4}}\right)$$

$$g(t) = \frac{1}{11}\left(e^{-2t} - e^{\frac{t}{2}}\cos\left(\frac{\sqrt{19}}{2}t\right) + \frac{5}{\sqrt{19}}e^{\frac{t}{2}}\sin\left(\frac{\sqrt{19}}{2}t\right)\right).$$

Okay, we can now get the solution to the differential equation. Starting with the transform we get,

$$y(s) = F(s) + e^{-2s}G(s)$$

$$y(t) = f(t) + u_2(t)g(t-2)$$

where *f(t)* and *g(t)* are the functions shown above.

There is can be a fair amount of work involved in solving differential equations that involve Heaviside functions.

Let's take a look at another example or two.

Example: Solve the following IVP.

$$y'' - y' = \cos(2t) + \cos(2t - 12)u_6(t) \quad y(0) = -4,\ y'(0) = 0.$$

Solution: Let's rewrite the differential equation so we can identify the function that is actually being shifted.

$$y'' - y' = \cos(2t) + \cos(2(t-6))u_6(t).$$

So, the function that is being shifted is and it is being shifted correctly. Taking the Laplace transform of everything and plugging in the initial conditions gives,

$$s^2Y(s) - sy(0) - y'(0) - (sY(s) - y(0)) = \frac{s}{s^2+4} + \frac{se^{-6s}}{s^2+4}$$

$$(s^2 - s)Y(s) + 4s - 4 = \frac{s}{s^2+4} + \frac{se^{-6s}}{s^2+4}$$

Now solve for *Y(s)*.

$$(s^2 - s)Y(s) = \frac{s + se^{-6s}}{s^2 + 4} = 4s + 4$$

$$Y(s) = \frac{s\left(1 + e^{-6s}\right)}{(s-1)\left(s^2 + 4\right)} - 4\frac{s-1}{s(s-1)}$$

$$= \frac{1 + e^{-6s}}{(s-1)\left(s^2 + 4\right)} - \frac{4}{s}$$

$$Y(s) = \left(1 + e^{-6s}\right)F(s) + \frac{4}{s}.$$

Notice that we combined the first two terms to simplify things a little. Also there was some canceling going on in this one. Do not expect that to happen on a regular basis. We now need to partial fraction *F(s)*. We'll leave the details to you to check.

$$F(s) = \frac{1}{(s-1)\left(s^2 + 4\right)} = \frac{1}{5}\left(\frac{1}{s-1} - \frac{s+1}{s^2+1}\right)$$

$$f(t) = \frac{1}{5}\left(e^t - \cos(2t) - \frac{1}{2}\sin(2t)\right).$$

Okay, we can now get the solution to the differential equation. Starting with the transform we get,

$$Y(s) = F(s) + F(s)e^{-6s} + \frac{4}{s}$$

$$y(t) = f(t) + u_6(t)f(t-6) + 4.$$

where *f(t)* is given above.

Example: Solve the following IVP.

$$y'' - 5y' - 14y = 9 - u_3(t) + 4(t-1)u_1(t) \quad y(0) = 0,\ y'(0) = 10.$$

Solution: Let's take the Laplace transform of everything and note that in the third term we are shifting $4t$.

$$s^2Y(s) - sy(0) - y'(0) - 5(sY(s) - y(0)) - 14Y(s) = \frac{9}{s} + \frac{e^{-3s}}{s} + 4\frac{e^{-s}}{s^2}$$

$$(s^2 - 5s - 14)Y(s) - 10 = \frac{9 + e^{-3s}}{s} 4\frac{e^{-s}}{s^2}.$$

Now solve for *Y(s)*.

$$\left(s^2 - 5s - 14\right)Y(s) - 10 = \frac{9 + e^{-3s}}{s} 4\frac{e^{-s}}{s^2}$$

$$Y(s) = \frac{9+e^{-3s}}{s(s-7)(s+2)} + \frac{4e^{-s}}{s^2(s-7)(s+2)} + \frac{10}{(s-7)(s+2)}$$

$$Y(s) = \left(9+e^{-3s}\right)F(s) + 4e^{-s}G(s) + H(s)$$

So, we have three functions that we'll need to partial fraction for this problem. I'll leave it to you to check the details.

$$F(s) = \frac{1}{s(s-7)(s+2)} = -\frac{1}{14}\frac{1}{s} + \frac{1}{63}\frac{1}{s-7} + \frac{1}{18}\frac{1}{s+2}$$

$$f(t) = \frac{1}{14} + \frac{1}{63}e^{7t} + \frac{1}{18}e^{-2t}$$

$$G(s) = \frac{1}{s^2(s-7)(s+2)} = \frac{5}{196}\frac{1}{s} + \frac{1}{14}\frac{1}{s^2} + \frac{1}{441}\frac{1}{s-7} - \frac{1}{36}\frac{1}{s+2}$$

$$g(t) = \frac{5}{196}\frac{1}{14}t + \frac{1}{441}e^{7t} - \frac{1}{36}e^{-2t}$$

$$H(s) = \frac{10}{(s-7)(s+2)} = \frac{10}{9}\frac{1}{s-7} - \frac{10}{9}\frac{1}{s+2}$$

$$h(t) = \frac{10}{9}e^{7t} - \frac{10}{9}e^{-2t}.$$

Okay, we can now get the solution to the differential equation. Starting with the transform we get,

$$Y(s) = 9F(s) + e^{-3s}F(s) + 4e^{-s}G(s) + H(s)$$

$$y(t) = 9f(t) + u_3(t)f(t-3) + 4u_1(t)g(t-1) + h(t).$$

where *f(t), g(t)* and *h(t)* are given above.

Let's work one more example.

Example: Solve the following IVP.

$$y''(t) = 3y' + 2y = (t), \qquad y(0) = 0 \quad y'(0) = -2$$

where,

$$g(t) = \begin{cases} 2 & \text{t} < 6 \\ t & 6 \le t < 10. \\ 4 & t \ge 10 \end{cases}$$

Solution: The first step is to get *g(t)* written in terms of Heaviside functions so that we can take the transform.

$$g(t) = 2 + (t-2)u_6(t) + (4-t)u_{10}(t).$$

Now, while this is *g(t)* written in terms of Heaviside functions it is not yet in proper form for us to take the transform. Remember that each function must be shifted by a proper amount. So, getting things set up for the proper shifts gives us,

$$g(t) = 2 + (t-6+6-2)u_6(t) + (4-(t-10+10))u_{10}(t)$$

$$g(t) = 2 + (t-6+4)u_6(t) + (4-(t-10))u_{10}(t).$$

So, for the first Heaviside it looks like $f(t) = t + 4$is the function that is being shifted and for the second Heaviside it looks like $f(t) = -6 - t$is being shifted.

Now take the Laplace transform of everything and plug in the initial conditions.

$$s^2Y(s) - sy(0) - y'(0) + 3(sY(s) - y(0)) + 2Y(s) = \frac{2}{s} + e^{-6s}\left(\frac{1}{s^2} + \frac{4}{s}\right) - e^{-6s}\left(\frac{1}{s^2} + \frac{6}{s}\right)$$

$$(s^2 + 3s + 2)Y(s) + 2 = \frac{2}{s} + e^{-6s}\left(\frac{1}{s^2} + \frac{4}{s}\right) - e^{-6s}\left(\frac{1}{s^2} + \frac{6}{s}\right)$$

Solve for *Y(s)*.

$$\left(s^2 + 3s + 2\right)Y(s) = \frac{2}{s} + e^{-6s}\left(\frac{1}{s^2} + \frac{4}{s}\right) - e^{-10s}\left(\frac{1}{s^2} + \frac{6}{s}\right) - 2$$

$$\left(s^2 + 3s + 2\right)Y(s) = \frac{2 + 4e^{-6s} - 4e^{-10s}}{s} + \frac{e^{-6s} - e^{-10s}}{s^4} - 2$$

$$Y(s) = \frac{2 + 4e^{-6s} - 6e^{-10s}}{s(s+1)(s+2)} + \frac{e^{-6s} - e^{-10s}}{s^2(s+1)(s+2)} - \frac{2}{(s+1)(s+2)}$$

$$Y(s) = \left(2 + 4e^{-6s} - 6e^{-10s}\right)F(s) + \left(e^{-6s} - e^{-10s}\right)G(s) - H(s).$$

Now, in the solving process we simplified things into as few terms as possible. Even doing this, it looks like we'll still need to do three partial fractions.

I'll leave the details of the partial fractioning to you to verify. The partial fraction form and inverse transform of each of these are.

$$F(s) = \frac{1}{s(s+1)(s+2)} = \frac{\frac{1}{2}}{s} - \frac{1}{s+1} + \frac{\frac{1}{2}}{s+2}$$

$$f(t) = \frac{1}{2} - e^{-1} + \frac{1}{2}e^{-2t}.$$

$$G(s)=\frac{1}{s^2(s+1)(s+2)}=-\frac{\frac{3}{4}}{s}+\frac{\frac{1}{2}}{s^2}+\frac{\frac{1}{2}}{s^2}+\frac{\frac{1}{4}}{s+2}$$

$$g(t)=-\frac{3}{4}+\frac{1}{2}t+e^{-t}\frac{1}{4}e^{-2t}$$

$$H(s)=\frac{2}{(s+1)(s+2)}=\frac{2}{s+1}-\frac{2}{s+2}$$

$$h(t)=2e^{-t}-2e^{-2t}.$$

Putting this all back together is going to be a little messy. First rewrite the transform a little to make the inverse transform process possible.

$$Y(s)=2F(s)+e^{-6s}(4F(s)+G(s))-e^{-10s}(6F(s)+G(s))-H(s).$$

Now, taking the inverse transform of all the pieces gives us the final solution to the IVP.

$$y(t)=2f(t)-h(t)+u_6(t)(4f(t-6)+g(t-6))-u_{10}(t)(6f(t-10)+g(t-10))$$

where *f(t)*, *g(t)*, and *h(t)* are defined above.

So, the answer to this example is a little messy to write down, but overall the work here wasn't too terribly bad.

Before proceeding with the next section let's see how we would have had to solve this IVP if we hadn't had Laplace transforms. To solve this IVP we would have had to solve three separate IVP's. One for each portion of *g(t)*. Here is a list of the IVP's that we would have had to solve.

1. $0 < t < 6$

$$y''+3y+2y=2, \qquad y(0)=0 \quad y'(0)=-2.$$

The solution to this IVP, with some work, can be made to look like,

$$y_1(t)=2f(t)-h(t)$$

2. $6 \le t < 10$

$$y''+3y'+2=t, \qquad y(6)=y_1(6) \qquad y'(6)=y_1(6)$$

where, $y_1(t)$ is the solution to the first IVP. The solution to this IVP, with some work, can be made to look like,

$$y_2(t)=2f(t)-h(t)+4f(t-6)+g(t-6).$$

3. $t \ge 10$

$$y''+3y'+2y=4, \qquad y(10)=y_2(10) \qquad y'(10)=y_2(10)$$

where, $y_2(t)$ is the solution to the second IVP. The solution to this IVP, with some work, can be made to look like,

$$y_3(t) = 2f(t) - h(t) + 4f(t-6) + g(t-6) - f(t-10) - g(t-10).$$

There is a considerable amount of work required to solve all three of these and in each of these the forcing function is not that complicated. Using Laplace transforms saved us a fair amount of work.

DIRAC DELTA FUNCTION

When we first introduced Heaviside functions we noted that we could think of them as switches changing the forcing function, *g(t)*, at specified times. However, Heaviside functions are really not suited to forcing functions that exert a "large" force over a "small" time frame.

Examples of this kind of forcing function would be a hammer striking an object or a short in an electrical system. In both of these cases a large force (or voltage) would be exerted on the system over a very short time frame. The Dirac Delta function is used to deal with these kinds of forcing function.

DIRAC DELTA FUNCTION

There are many ways to actually define the Dirac Delta function. There three main properties of the Dirac Delta function that we need to be aware of. These are,

1. $\delta(t-a) = 0, t \neq a.$
2. $\int_{a=\varepsilon}^{a+c} \delta(t-a)dt = 1, \quad \varepsilon > 0.$
3. $\int_{a=\varepsilon}^{a+c} f(t)\,\delta\,(t=1)dt = f(a), \quad \varepsilon > 0.$

At the Dirac Delta function is sometimes thought of has having an "infinite" value. So, the Dirac Delta function is a function that is zero everywhere except one point and at that point it can be thought of as either undefined or as having an "infinite" value.

Note that the integrals in the second and third property are actually true for any interval containing $t = a$, provided it's not one of the endpoints. The limits given here are needed to prove the properties and so they are also given in the properties. We will however use the fact that they are true provided we are integrating over an interval containing $t = a$.

This is a very strange function. It is zero everywhere except one point and yet the integral of any interval containing that one point has a value of 1. The Dirac Delta function is not a real function as we think of them. It is instead an example of something called a generalized function or distribution.

Despite the strangeness of this "function" it does a very nice job of modeling sudden shocks or large forces to a system. Before solving an IVP we will need the transform of the Dirac Delta function. We can use the third property above to get this.

$$\mathfrak{L}\{\delta(t-a)\} = \int_0^\infty e^{-st}\delta(t-a)dt = e^{-as} \qquad \text{provided } a > 0.$$

Note that often the second and third properties are given with limits of infinity and negative infinity, but they are valid for any interval in which $t = a$is in the interior of the interval.

With this we can now solve an IVP that involves a Dirac Delta function.

Example: Solve the following IVP.

$$y'' + 2y' - 15y = 6\delta\,(t-9), \qquad y(0) = -5 \quad y'(0) = 7.$$

Solution:As with all previous problems we'll first take the Laplace transform of everything in the differential equation and apply the initial conditions.

$$s^2Y(s) - sy(0) - y'(0) + 2(sY(s) - y(0)) - 15Y(s) = 6e^{-9t}$$

$$(s^2 + 2s - 15)Y(s) + 5s + 3 = 6e^{-9s}.$$

Now solve for *Y(s)*.

$$\begin{aligned} Y(s) &= \frac{6e^{-9s}}{(s+5)(s-3)} - \frac{5s+3}{(s+5)(s-3)} \\ &= 6e^{-9s}F(s) - G(s). \end{aligned}$$

We'll leave it to you to verify the partial fractions and their inverse transforms are,

$$F(s) = \frac{1}{(s+5)(s-3)} = \frac{\frac{1}{8}}{s-3} - \frac{\frac{1}{8}}{s+5}$$

$$f(1) = \frac{1}{8}e^{3t} - \frac{1}{8}e^{-5t}$$

$$G(s) = \frac{5s+3}{(s+5)(s-3)} = \frac{\frac{9}{4}}{s-3} + \frac{\frac{11}{4}}{s+5}$$

$$g(t) = \frac{9}{4}e^{3t} + \frac{11}{4}e^{-5t}.$$

The solution is then,

$$Y(s) = 6e^{-9s}F(s) - G(s)$$

$$y(t) = 6u_9(t)f(t-9) - g(t).$$

where, *f(t)* and *g(t)* are defined above.

Example: Solve the following IVP.

$$2y''+10y=3u_{12}(t)-5\delta(t-4), \qquad y(0)=-1 \quad y'(0)=-2.$$

Solution: Take the Laplace transform of everything in the differential equation and apply the initial conditions.

$$2(s^2Y(s)-sy(0)-y'(0))+10Y(s)=\frac{3e^{-12s}}{s}-5e^{-4s}$$

$$(2(s^2+10)Y(s)+2s+4=\frac{3e^{-12s}}{s}-5e^{-4s}.$$

Now solve for *Y(s)*.

$$Y(s)=\frac{3e^{-12s}}{s(2s^2+10)}-\frac{5e^{-4s}}{2s^2+10}-\frac{2s+4}{2s^2+10}$$

$$=3e^{-12s}F(s)-5e^{-4s}G(s)-H(s).$$

We'll need to partial fraction the first function. The remaining two will just need a little work and they'll be ready. I'll leave the details to you to check.

$$F(s)=\frac{1}{s(2s^2+10)}=\frac{1}{10}\frac{1}{s}-\frac{1}{10}\frac{s}{s^2+5}$$

$$f(t)=\frac{1}{10}-\frac{1}{10}\cos\left(\sqrt{5}t\right)$$

$$g(t)=\frac{1}{2\sqrt{5}}\sin\left(\sqrt{5}t\right)$$

$$h(t)=\cos\left(\sqrt{5}t\right)+\frac{2}{\sqrt{5}}\sin\left(\sqrt{5}t\right).$$

The solution is then,

$$Y(s)=3e^{-12s}F(s)-5e^{-4s}G(s)-H(s)$$

$$y(t)=3u_{12}(t)f(t-12)-5u_4(t)g(t-4)-h(t).$$

where, *f(t)*,*g(t)* and *h(t)* are defined above.

So, with the exception of the new function these work the same way that all the problems that we've seen to this point work. Note as well that the exponential was introduced into the transform by the Dirac Delta function, but once in the transform it doesn't matter where it came from.

In other words, when we went to the inverse transforms it came back out as a Heaviside function.

Before proceeding to the next section let's take a quick side trip and note that we can relate the Heaviside function and the Dirac Delta function. Start with the following integral.

$$\int_{-\infty}^{1} \delta(u-a)du = \begin{cases} 0 & \text{if } t < a \\ 1 & \text{if } t > a \end{cases}.$$

However, this is precisely the definition of the Heaviside function. So,

$$\int_{-\infty}^{1} \delta(u-a)du = u_a(t).$$

Now, recalling the Fundamental Theorem of Calculus, we get,

$$u_a^t(t)\left(\int_{-\infty}^{2} \delta(u-a)du\right) = \delta(t-a).$$

CONVOLUTION INTEGRALS

On occasion we will run across transforms of the form,

$$H(s) = F(s)\,G(s)$$

that can't be dealt with easily using partial fractions. We would like a way to take the inverse transform of such a transform. We can use a convolution integral to do this.

CONVOLUTION INTEGRAL

If *f(t)* and *g(t)* are piecewise continuous function on $(0, \infty)$then the convolution integral of *f(t)* and *g(t)* is,

$$(f * g)(t) = \int_0^t f(t-\tau)g(\tau)d\tau.$$

A nice property of convolution integrals is.

$$(f * g)(t) = (g * f)(t).$$

Or,
$$\int_0^t f(t-\tau)g(\tau)d\tau = \int_0^t f(\tau)g(t-\tau)d\tau.$$

The following fact will allow us to take the inverse transforms of a product of transforms.

FACT

$$\mathfrak{L}\{f * g\} = F(s)G(s) \qquad \mathfrak{L}^{-1}\{F(s)\,G(s)\} = (f * g)(t).$$

Let's work a quick example to see how this can be used.

Example: Use a convolution integral to find the inverse transform of the following transform.

$$H(s) = \frac{1}{(s^2 + a^2)^2}.$$

Solution: Now, since we are going to use a convolution integral here we will need to write it as a product whose terms are easy to find the inverse transforms of. This is easy to do in this case.

$$H(s) = \left(\frac{1}{s^2+a^2}\right)\left(\frac{1}{s^2+a^2}\right).$$

So, in this case we have,

$$F(s) = G(s) = \frac{1}{s^2+a^2} \quad \Rightarrow \quad f(t) = g(t) = \frac{1}{a}\sin(at).$$

Using a convolution integral *h(t)* is,

$$\begin{aligned} h(t) &= (f * g)(t) \\ &= \frac{1}{a^2}\int_0^t \sin(at - a\tau)\sin(a\tau)\,d\tau \\ &= \frac{1}{2a^3}\left(\sin(at) - at\cos(at)\right). \end{aligned}$$

This is exactly what we would have gotten by using #11 from the table.

Convolution integrals are very useful in the following kinds of problems.

Example: Solve the following IVP

$$4y'' + y = h(t), \qquad y(0) = 3 \quad y'(0) = -7.$$

Solution: First, notice that the forcing function in this case has not been specified. Prior to this section we would not have been able to get a solution to this IVP. With convolution integrals we will be able to get a solution to this kind of IVP. The solution will be in terms of *g(t)* but it will be a solution.

Take the Laplace transform of all the terms and plug in the initial conditions.

$$4(s^2Y(s) - sy(0) - y'(0)) + Y(s) = G(s)$$

$$(4s^2+1)Y(s) - 12s + 28 = G(s).$$

Notice here that all we could do for the forcing function was to write down *G(s)* for its transform. Now, solve for *Y(s)*.

$$(4s^2+1)Y(s) = G(s) + 12s + 28$$

$$Y(s) = \frac{12s}{4\left(s^2+\frac{1}{4}\right)} + \frac{G(s)}{4\left(s^2+\frac{1}{4}\right)}.$$

We factored out a 4 from the denominator in preparation for the inverse transform process. To take inverse transforms we'll need to split up the first term and we'll also rewrite the second term a little.

$$Y(s) = \frac{12s-28}{4\left(s^2+\frac{1}{4}\right)} + \frac{G(s)}{4\left(s^2+\frac{1}{4}\right)}$$

$$= \frac{3s}{s^2+\frac{1}{4}} + \frac{7\frac{2}{2}}{s^2+\frac{1}{4}} + \frac{1}{4}G(s)\frac{\frac{2}{2}}{s^2+\frac{1}{4}}.$$

Now, the first two terms are easy to inverse transform. We'll need to use a convolution integral on the last term. The two functions that we will be using are,

$$g(t) \quad f(t) = 2\sin\left(\frac{t}{2}\right).$$

We can shift either of the two functions in the convolution integral. We'll shift *g(t)* in our solution. Taking the inverse transform gives us,

$$y(t) = 3\cos\left(\frac{t}{2}\right) + 14\sin\left(\frac{t}{2}\right) + \frac{1}{2}\int_0^t \sin\left(\frac{\tau}{2}\right) g(t-\tau)\,d\tau.$$

So, once we decide on a *g(t)* all we need to do is to an integral and we'll have the solution.

As this last example has shown, using convolution integrals will allow us to solve IVP's with general forcing functions. This could be very convenient in cases where we have a variety of possible forcing functions and don't which one we're going to use. With a convolution integral all that we need to do in these cases is solve the IVP once then go back and evaluate an integral for each possible *g(t)*. This will save us the work of having to solve the IVP for each and every *g(t)*.

Chapter 6

Systems of Differential Equations

To this point we've only looked as solving single differential equations. However, many "real life" situations are governed by a system of differential equations. Consider the population problems that we looked at back in the modeling section of the first order differential equations chapter. In these problems we looked only at a population of one species, yet the problem also contained some information about predators of the species. We assumed that any predation would be constant in these cases.

However, in most cases the level of predation would also be dependent upon the population of the predator. So, to be more realistic we should also have a second differential equation that would give the population of the predators. Also note that the population of the predator would be, in some way, dependent upon the population of the prey as well. In other words, we would need to know something about one population to find the other population. So to find the population of either the prey or the predator we would need to solve a system of at least two differential equations.

The next topic of discussion is then how to solve systems of differential equations. However, before doing this we will first need to do a quick review of Linear Algebra. Much of what we will be doing in this chapter will be dependent upon topics from linear algebra. This review is not intended to completely teach you the subject of linear algebra, as that is a topic for a complete class. The quick review is intended to get you familiar enough with some of the basic topics that you will be able to do the work required once we get around to solving systems of differential equations.

SYSTEMS OF EQUATIONS

Because we are going to be working almost exclusively with systems of equations in which the number of unknowns equals the number of equations we will restrict our review to these kinds of systems.

All of what we will be doing here can be easily extended to systems with more unknowns than equations or more equations than unknowns if need be.

Let's start with the following system of n equations with the n unknowns, $x_1, x_2, \ldots, x_n$.

$$\begin{aligned} a_{11}x_1 + a_{12}x_2 + \cdots + a_{1n}x_n &= b_1 \\ a_{21}x_1 + a_{22}x_2 + \cdots + a_{2n}x_n &= b_2 \\ &\vdots \\ a_{n1}x_1 + a_{n2}x_2 + \cdots + a_{nn}x_n &= b_n \end{aligned}$$

Note that in the subscripts on the coefficients in this system, a_{ij}, the i corresponds to the equation that the coefficient is in and the j corresponds to the unknown that is multiplied by the coefficient.

To use linear algebra to solve this system we will first write down the augmented matrix for this system. An augmented matrix is really just the all the coefficients of the system and the numbers for the right side of the system written in matrix form. Here is the augmented matrix for this system.

$$\begin{pmatrix} a_{11} & a_{12} & \cdots & a_{1n} & b_1 \\ a_{21} & a_{22} & \cdots & a_{2n} & b_2 \\ \vdots & \vdots & \ddots & \vdots & \vdots \\ a_{n1} & a_{n2} & \cdots & a_{nn} & b_n \end{pmatrix}.$$

To solve this system we will use elementary row operations (which we'll define these in a bit) to rewrite the augmented matrix in triangular form. The matrix will be in triangular form if all the entries below the main diagonal (the diagonal containing $a_{11}, a_{22}, \ldots, a_{nn}$) are zeroes.

Once this is done we can recall that each row in the augmented matrix corresponds to an equation. We will then convert our new augmented matrix back to equations and at this point solving the system will become very easy.

Before working an example let's first define the elementary row operations. There are three of them.

1. Interchange two rows. This is exactly what it says. We will interchange row i with row j. The notation that we'll use to denote this operation is: $R_i \leftrightarrow R_j$
2. Multiply row i by a constant, c. This means that every entry in row i will get multiplied by the constant c. The notation for this operation is: cR_i
3. Add a multiply of row i to row j. In our heads we will multiply row i by an appropriate constant and then add the results to row j and put the new row back into row j leaving row i in the matrix unchanged. The notation for this operation is: $cR_i + R_j$

It's always a little easier to understand these operations if we see them in action. So, let's solve a couple of systems.

Example: Solve the following system of equations.

$$-2x_1 + x_2 - x_3 = 4$$

$$x_1 + 2x_2 - 3x_3 = 13$$

$$3x_1 + x_3 = -1.$$

Solution: The first step is to write down the augmented matrix for this system. Don't forget that coefficients of terms that aren't present are zero.

$$\begin{pmatrix} -2 & 1 & -1 & 4 \\ 1 & 2 & 3 & 13 \\ 3 & 0 & 1 & -1 \end{pmatrix}.$$

Now, we want the entries below the main diagonal to be zero. The main diagonal has been colored red so we can keep track of it during this first example. For reasons that will be apparent eventually we would prefer to get the main diagonal entries to all be ones as well. We can get a one in the upper most spot by noticing that if we interchange the first and second row we will get a one in the uppermost spot for free. So let's do that.

$$\begin{pmatrix} -2 & 1 & -1 & 4 \\ 1 & 2 & 3 & 13 \\ 3 & 0 & 1 & -1 \end{pmatrix} \begin{matrix} R_1 \leftrightarrow R_2 \\ \rightarrow \end{matrix} \begin{pmatrix} 1 & 2 & 3 & 13 \\ -2 & 1 & -1 & 4 \\ 3 & 0 & 1 & -1 \end{pmatrix}.$$

Now we need to get the last two entries (the -2 and 3) in the first column to be zero. We can do this using the third row operation. Note that if we take 2 times the first row and add it to the second row we will get a zero in the second entry in the first column and if we take -3 times the first row to the third row we will get the 3 to be a zero. We can do both of these operations at the same time so let's do that.

$$\begin{pmatrix} 1 & 2 & 3 & 13 \\ -2 & 1 & -1 & 4 \\ 3 & 0 & 1 & -1 \end{pmatrix} \begin{matrix} 2R_1 + R_2 \\ -3R_1 + R_3 \\ \rightarrow \end{matrix} \begin{pmatrix} 1 & 2 & 3 & 13 \\ 0 & 5 & 5 & 30 \\ 0 & -6 & -8 & -40 \end{pmatrix}.$$

Before proceeding with the next step, let's make sure that you followed what we just did. Let's take a look at the first operation that we performed. This operation says to multiply an entry in row 1 by 2 and add this to the corresponding entry in row 2 then replace the old entry in row 2 with this new entry. The following are the four individual operations that we performed to do this.

$$2(1) + (-2) = 0$$

$$2(2) + (-1) = 5$$

$$2(3) + (-1) = 5$$

$$2(13) + 4 = 30.$$

Okay, the next step optional, but again is convenient to do. Technically, the 5 in the second column is okay to leave. However, it will make our life easier down the road if it is a 1. We can use the second row operation to take care of this. We can divide the whole row by 5. Doing this gives,

$$\begin{pmatrix} 1 & 2 & 3 & 13 \\ 0 & 5 & 5 & 30 \\ 0 & -6 & -8 & -40 \end{pmatrix} \overset{\frac{1}{5}R_2}{\rightarrow} \begin{pmatrix} 1 & 2 & 3 & 13 \\ 0 & 1 & 1 & 6 \\ 0 & -6 & -8 & -40 \end{pmatrix}.$$

The next step is to then use the third row operation to make the -6 in the second column into a zero.

$$\begin{pmatrix} 1 & 2 & 3 & 13 \\ 0 & 1 & 1 & 6 \\ 0 & -6 & -8 & -40 \end{pmatrix} \overset{6R_2 + R_3}{\rightarrow} \begin{pmatrix} 1 & 2 & 3 & 13 \\ 0 & 1 & 1 & 6 \\ 0 & 0 & -2 & -4 \end{pmatrix}.$$

Now, officially we are done, but again it's somewhat convenient to get all ones on the main diagonal so we'll do one last step.

$$\begin{pmatrix} 1 & 2 & 3 & 13 \\ 0 & 1 & 1 & 6 \\ 0 & 0 & -2 & -4 \end{pmatrix} \overset{-\frac{1}{2}R_3}{\rightarrow} \begin{pmatrix} 1 & 2 & 3 & 13 \\ 0 & 1 & 1 & 6 \\ 0 & 0 & 1 & -4 \end{pmatrix}.$$

We can now convert back to equations.

$$\begin{pmatrix} 1 & 2 & 3 & 13 \\ 0 & 1 & 1 & 6 \\ 0 & 0 & 1 & 2 \end{pmatrix} \Rightarrow \begin{aligned} x_1 + 2x_2 + 3x_3 &= 13 \\ x_2 + x_3 &= 6. \\ x_3 &= 2 \end{aligned}$$

At this point the solving is quite easy. We get x_3 for free and once we get that we can plug this into the second equation and get x_2. We can then use the first equation to get x_1. Note as well that having 1's along the main diagonal helped somewhat with this process.

The solution to this system of equation is

$$x_1 = 1 \qquad x_2 = 4 \qquad x_3 = 2.$$

The process used in this example is called Gaussian Elimination. Let's take a look at another example.

Example: Solve the following system of equations.

$$\begin{aligned} x_1 - 2x_3 + 3x_3 &= -2 \\ 2x_1 + x_2 + 2x_3 &= 3 \\ 2x_1 - x_2 + 3x_3 &= 1 \end{aligned}$$

Solution: First write down the augmented matrix.

$$\begin{pmatrix} 1 & -2 & 3 & -2 \\ -1 & 1 & -2 & 3 \\ 2 & -1 & 3 & 1 \end{pmatrix}.$$

We won't put down as many words in working this example. Here's the work for this augmented matrix.

$$\begin{pmatrix} 1 & -2 & 3 & -2 \\ -1 & 1 & -2 & 3 \\ 2 & -1 & 3 & 1 \end{pmatrix} \begin{matrix} R_1 + R_2 \\ -R_1 + R_3 \\ \to \end{matrix} \begin{pmatrix} 1 & -2 & 3 & -2 \\ 0 & -1 & 1 & 1 \\ 0 & 3 & -3 & 5 \end{pmatrix}$$

$$\begin{matrix} -R_2 \\ \to \end{matrix} \begin{pmatrix} 1 & -2 & 3 & -2 \\ 0 & 1 & -1 & -1 \\ 0 & 3 & -3 & 5 \end{pmatrix} \begin{matrix} -3R_1 + R_2 \\ \to \end{matrix} \begin{pmatrix} 1 & -2 & 3 & -2 \\ 0 & 1 & -1 & -1 \\ 0 & 0 & 0 & 8 \end{pmatrix}.$$

We won't go any farther in this example. Let's go back to equations to see why.

$$\begin{pmatrix} 1 & -2 & 3 & -2 \\ 0 & 1 & -1 & -1 \\ 0 & 0 & 0 & 8 \end{pmatrix} \Rightarrow \begin{aligned} x_1 - 2x_2 + 3x_3 &= -2 \\ x_2 - x_3 &= -1. \\ 0 &= 8 \end{aligned}$$

The last equation should cause some concern. There's one of three options here. First, we've somehow managed to prove that 0 equals 8 and we know that's not possible. Second, we've made a mistake, but after going back over our work it doesn't appear that we have made a mistake.

This leaves the third option. When we get something like the third equation that simply doesn't make sense we immediately know that there is no solution. In other words, there is no set of three numbers that will make all three of the equations true at the same time.

Let's work another example. We are going to get the system for this new example by making a very small change to the system from the previous example.

Example: Solve the following system of equations.

$$x_1 - 2x_2 + 3x_3 = -2$$

$$-x_1 + x_2 - 2x_3 = 3$$

$$2x_1 + x_2 - 3x_3 = -7$$

Solution: So, the only difference between this system and the system from the second example is we changed the 1 on the right side of the equal sign in the third equation to a –7.

Now write down the augmented matrix for this system.

$$\begin{pmatrix} 1 & -2 & 3 & -2 \\ -1 & 1 & -2 & 3 \\ 2 & -1 & 3 & -7 \end{pmatrix}.$$

The steps for this problem are identical to the steps for the second problem so we won't write them all down. Upon performing the same steps we arrive at the following matrix.

$$\begin{pmatrix} 1 & -2 & 3 & -2 \\ 0 & 1 & -1 & -1 \\ 0 & 0 & 0 & 0 \end{pmatrix}.$$

This time the last equation reduces to

$$0 = 0$$

and unlike the second example this is not a problem. Zero does in fact equal zero!

We could stop here and go back to equations to get a solution and there is a solution in this case. However, if we go one more step and get a zero above the one in the second column as well as below it our life will be a little simpler. Doing this gives,

$$\begin{pmatrix} 1 & -2 & 3 & -2 \\ 0 & 1 & -1 & -1 \\ 0 & 0 & 0 & 0 \end{pmatrix} \overset{2R_2 + R_1}{\Rightarrow} \begin{pmatrix} 1 & 0 & 1 & -4 \\ 0 & 1 & -1 & -1 \\ 0 & 0 & 0 & 0 \end{pmatrix}.$$

If we now go back to equation we get the following two equations.

$$\begin{pmatrix} 1 & 0 & 1 & -4 \\ 0 & 1 & -1 & -1 \\ 0 & 0 & 0 & 0 \end{pmatrix} \Rightarrow \begin{aligned} x_1 + x_3 &= -4 \\ x_2 - x_3 &= -1 \end{aligned}.$$

We have two equations and three unknowns. This means that we can solve for two of the variables in terms of the remaining variable. Since x_3 is in both equations we will solve in terms of that.

$$\begin{aligned} x_1 &= -x_3 - 4 \\ x_2 &= x_3 - 1 \end{aligned}.$$

What this solution means is that we can pick the value of x_3 to be anything that we'd like and then find values of x_1 and x_2. In these cases we typically write the solution as follows,

$$\begin{aligned} x_1 &= -t - 4 \\ x_2 &= t - 1 \qquad\qquad t = \text{any real number.} \\ x_3 &= t \end{aligned}$$

In this way we get an infinite number of solutions, one for each and every value of t.

These three examples lead us to a nice fact about systems of equations.

FACT

Given a system of equations, we will have one of the three possibilities for the number of solutions.

1. No solution.
2. Exactly one solution.
3. Infinitely many solutions.

Before moving on to the next section we need to take a look at one more situation. The system of equations is called a nonhomogeneous system if at least one of the b_i's is not zero. If however all of the b_i's are zero we call the system homogeneous and the system will be,

$$\begin{aligned} a_{11}x_1a_{12}x_2 + \cdots + a_{1n}x_n &= 0 \\ a_{21}x_1a_{22}x_2 + \cdots + a_{2n}x_n &= 0 \\ &\vdots \\ a_{n1}x_1a_{n2}x_2 + \cdots + a_{nn}x_n &= 0. \end{aligned}$$

Now, notice that in the homogeneous case we are guaranteed to have the following solution.

$$x_1 = x_2 = \cdots = x_n = 0.$$

This solution is often called the trivial solution.

For homogeneous systems the fact above can be modified to the following.

FACT

Given a homogeneous system of equations, we will have one of the two possibilities for the number of solutions.

1. Exactly one solution, the trivial solution
2. Infinitely many non-zero solutions in addition to the trivial solution.—

In the second possibility we can say non-zero solution because if there are going to be infinitely many solutions and we know that one of them is the trivial solution then all the rest must have at least one of the x_i's be non-zero and hence we get a non-zero solution.-

MATRICES AND VECTORS

This section is intended to be a catch all for many of the basic concepts that are used occasionally in working with systems of differential equations. There will not be a lot of details in this section, nor will we be working large numbers of examples. Also, in many cases we will not be looking at the general case since we won't need the general cases in our differential equations work.

Let's start with some of the basic notation for matrices. An n x m (this is often called the size or dimension of the matrix) matrix is a matrix with n rows and m columns and the entry in the ith row and j^{th} column is denoted by a_{ij}. A short hand method of writing a general n x m matrix is the following.

$$A = \begin{pmatrix} a_{11} & a_{12} & \cdots & a_{1m} \\ a_{21} & a_{22} & \cdots & a_{2m} \\ \vdots & \vdots & & \vdots \\ a_{n1} & a_{n2} & \cdots & a_{nm} \end{pmatrix}_{n \times m} = \left(a_1\right)_{n \times m}.$$

The size or dimension of a matrix is subscripted as shown if required. If it's not required or clear from the problem the subscripted size is often dropped from the matrix.

SPECIAL MATRICES

There are a few "special" matrices out there that we may use on occasion. The first special matrix is the square matrix. A square matrix is any matrix whose size (or dimension) is n x n. In other words it has the same number of rows as columns. In a square matrix the diagonal that starts in the upper left and ends in the lower right is often called the main diagonal.

The next two special matrices that we want to look at are the zero matrix and the identity matrix. The zero matrix, denoted $0_{n \times m}$, is a matrix all of whose entries are zeroes. The identity matrix is a square n x n matrix, denoted I_n, whose main diagonals are all 1's and all the other elements are zero. Here are the general zero and identity matrices.

$$0_{n \times m} = \begin{pmatrix} 0 & 0 & \cdots & 0 \\ \vdots & \vdots & & \vdots \\ 0 & 0 & \cdots & 0 \end{pmatrix}_{n \times m} \qquad I_{\mathrm{n}} = \begin{pmatrix} 1 & 0 & \cdots & 0 \\ 0 & 1 & \cdots & 0 \\ \vdots & \vdots & \ddots & \vdots \\ 0 & 0 & \cdots & 1 \end{pmatrix}_{n \times m}.$$

In matrix arithmetic these two matrices will act in matrix work like zero and one act in the real number system. The last two special matrices that we'll look at here are the column matrix and the row matrix. These are matrices that consist of a single column or a single row. In general they are,

$$x = \begin{pmatrix} x_1 \\ x_2 \\ x_n \end{pmatrix}_{n \times 1} \qquad y = (y_1 \quad y_2 \quad \cdots \quad y_m)_{1 \times m}.$$

We will often refer to these as vectors.

ARITHMETIC

We next need to take a look at arithmetic involving matrices. We'll start with addition and subtraction of two matrices. So, suppose that we have two *n* x *m* matrices, *A* and *B*. The sum (or difference) of these two matrices is then,

$$A_{n\times m} \pm B_{n\times m} = \left(a_{ij}\right)_{n\times m} \pm \left(b_{ij}\right)_{n\times m} = \left(a_{ij} \pm b_{ij}\right)_{n\times m}.$$

The sum or difference of two matrices of the same size is a new matrix of identical size whose entries are the sum or difference of the corresponding entries from the original two matrices. Note that we can't add or subtract entries with different sizes.

Next, let's look at scalar multiplication. In scalar multiplication we are going to multiply a matrix *A* by a constant (sometimes called a scalar) á. In this case we get a new matrix whose entries have all been multiplied by the constant, á.

$$\alpha\, A_{n\times m} = \alpha\left(a_{ij}\right)_{n\times m} = \left(\alpha\, a_{ij}\right)_{n\times m}.$$

Example: Given the following two matrices,

$$A = \begin{pmatrix} 3 & -2 \\ -9 & 1 \end{pmatrix} \qquad B = \begin{pmatrix} -4 & 1 \\ 0 & -5 \end{pmatrix}.$$

compute A-5B.

Solution: There isn't much to do here other than the work.

$$\begin{aligned} A - 5\mathrm{B} &= \begin{pmatrix} 3 & -2 \\ -9 & 1 \end{pmatrix} - 5\begin{pmatrix} -4 & 1 \\ 0 & -5 \end{pmatrix} \\ &= \begin{pmatrix} 3 & -2 \\ -9 & 1 \end{pmatrix} - \begin{pmatrix} 20 & 5 \\ 0 & -25 \end{pmatrix} \\ &= \begin{pmatrix} 23 & -7 \\ -9 & 26 \end{pmatrix}. \end{aligned}$$

We first multiplied all the entries of *B* by 5 then subtracted corresponding entries to get the entries in the new matrix.

The final matrix operation that we'll take a look at is matrix multiplication. Here we will start with two matrices, $A_{n\times p}$ and $B_{p\times m}$. Note that *A* must have the same number of columns as *B* has rows. If this isn't true then we can't perform the multiplication. If it is true then we can perform the following multiplication.

$$a_{n\times p}B_{p\times m} = \left(c_{ij}\right)_{n\times m}.$$

The new matrix will have size *n* x *m* and the entry in the ith row and j^{th} column, c_{ij}, is found by multiplying row i of matrix A by column j of matrix B. This doesn't always make sense in words so let's look at an example.

Example: Given

$$A=\begin{pmatrix} 2 & -1 & 0 \\ -3 & 6 & 1 \end{pmatrix}_{2\times 3} \qquad B=\begin{pmatrix} 1 & 0 & -1 & 2 \\ -4 & 3 & 1 & 0 \\ 0 & 3 & 0 & -2 \end{pmatrix}_{3\times 4}.$$

compute AB.

Solution: The new matrix will have size 2×4. The entry in row 1 and column 1 of the new matrix will be found by multiplying row 1 of A by column 1 of B. This means that we multiply corresponding entries from the row of A and the column of B and then add the results up. Here are a couple of the entries computed all the way out.

$$c_{11}=(2)(1)+(-1)(-4)+(0)(0)=6$$

$$c_{13}=(2)(-1)+(-1)(1)+(0)(0)=-3$$

$$c_{24}=(-3)(2)+(6)(0)+(1)(-2)=-9$$

Here's the complete solution.

$$C=\begin{pmatrix} 6 & -3 & -3 & 4 \\ -27 & 21 & 9 & -8 \end{pmatrix}.$$

In this last example notice that we could not have done the product BA since the number of columns of B does not match the number of row of A. It is important to note that just because we can compute AB doesn't mean that we can compute BA. Likewise, even if we can compute oth AB and BA they may or may not be the same matrix.

DETERMINANT

The next topic that we need to take a look at is the determinant of a matrix. The determinant is actually a function that takes a square matrix and converts it into a number. The actual formula for the function is somewhat complex and definitely beyond the scope of this review.

The main method for computing determinants of any square matrix is called the method of cofactors. Since we are going to be dealing almost exclusively with 2 x 2 matrices and the occasional 3 x 3 matrix we won't go into the method here. We can give simple formulas for each of these cases. The standard notation for the determinant of the matrix A is.

$$\det(A)=|A|.$$

Here are the formulas for the determinant of 2 x 2 and 3 x 3 matrices.

$$\begin{vmatrix} a & c \\ b & d \end{vmatrix} = ad - cd$$

$$\begin{vmatrix} a_{11} & a_{12} & a_{13} \\ a_{21} & a_{22} & a_{23} \\ a_{31} & a_{32} & a_{33} \end{vmatrix} = a_{11}\begin{vmatrix} a_{22} & c_{23} \\ b_{32} & d_{33} \end{vmatrix} - a_{12}\begin{vmatrix} a_{21} & c_{23} \\ b_{31} & d_{33} \end{vmatrix} + a_{13}\begin{vmatrix} a_{21} & c_{23} \\ b_{31} & d_{33} \end{vmatrix}.$$

Example: Find the determinant of each of the following matrices.

$$A = \begin{pmatrix} -9 & -18 \\ 2 & 4 \end{pmatrix} \qquad B = \begin{pmatrix} 2 & 3 & 1 \\ -1 & -6 & 7 \\ 4 & 5 & -1 \end{pmatrix}.$$

Solution: For the 2 x 2 there isn't much to do other than to plug it into the formula.

$$\det(A) = \begin{vmatrix} -9 & -18 \\ 2 & 4 \end{vmatrix} = (-9)(4) - (-18)(2) = 0.$$

For the 3 x 3 we could plug it into the formula, however unlike the 2 x 2 case this is not an easy formula to remember. There is an easier way to get the same result. A quicker way of getting the same result is to do the following. First write down the matrix and tack a copy of the first two columns onto the end as follows.

$$\det(B) = \left|\begin{matrix} 2 & 3 & 1 \\ -1 & -6 & 7 \\ 4 & 5 & -1 \end{matrix}\right| \begin{matrix} 2 & 3 \\ -1 & -6. \\ 4 & 5 \end{matrix}$$

Now, notice that there are three diagonals that run from left to right and three diagonals that run from right to left. What we do is multiply the entries on each diagonal up and the if the diagonal runs from left to right we add them up and if the diagonal runs from right to left we subtract them.

Here is the work for this matrix.

$$\det(B) = \left|\begin{matrix} 2 & 3 & 1 \\ -1 & -6 & 7 \\ 4 & 5 & -1 \end{matrix}\right| \begin{matrix} 2 & 3 \\ -1 & -6 \\ 4 & 5 \end{matrix}$$

$$= (2)(-6)(-1) + (3)(7)(4) + (1)(-1)(5) - (3)(-1)(-1) - (2)(7)(5) - (1)(-6)(4)$$

$$= 42$$

You can either use the formula or the short cut to get the determinant of a 3 × 3.

If the determinant of a matrix is zero we call that matrix singular and if the determinant of a matrix isn't zero we call the matrix nonsingular. The 2×2 matrix in the above example was singular while the 3×3 matrix is nonsingular.

MATRIX INVERSE

Next we need to take a look at the inverse of a matrix. Given a square matrix, A, of size n x n if we can find another matrix of the same size, B such that,

$$.AB = BA = I_n$$

then we call B the inverse of A and denote it by $B=A^{-1}$.

Computing the inverse of a matrix, A, is fairly simple. First we form a new matrix,

$$(A \;\; I_n).$$

and then use the row operations from the previous section and try to convert this matrix into the form,

$$(I_n \;\; B).$$

If we can then B is the inverse of A. If we can't then there is no inverse of the matrix A.

Example: Find the inverse of the following matrix, if it exists.

$$A = \begin{pmatrix} 2 & 1 & 1 \\ -5 & -3 & 0 \\ 1 & 1 & -1 \end{pmatrix}.$$

Solution: We first form the new matrix by tacking on the 3 x 3 identity matrix to this matrix. This is

$$\begin{pmatrix} 2 & 1 & 1 & 1 & 0 & 0 \\ -5 & -3 & 0 & 0 & 1 & 0 \\ 1 & 1 & -1 & 0 & 0 & 1 \end{pmatrix}.$$

We will now use row operations to try and convert the first three columns to the 3 x 3 identity. In other words we want a 1 on the diagonal that starts at the upper left corner and zeroes in all the other entries in the first three columns.

If you think about it, this process is very similar to the process we used in the last section to solve systems, it just goes a little farther. Here is the work for this problem.

$$\begin{pmatrix} 2 & 1 & 1 & 1 & 0 & 0 \\ -5 & -3 & 0 & 0 & 1 & 0 \\ 1 & 1 & -1 & 0 & 0 & 1 \end{pmatrix} \begin{matrix} R_1 \leftrightarrow R_3 \\ \Rightarrow \end{matrix} \begin{pmatrix} 1 & 1 & -1 & 0 & 0 & 1 \\ -5 & -3 & 0 & 0 & 1 & 0 \\ 2 & 1 & 1 & 1 & 0 & 0 \end{pmatrix} \begin{matrix} R_2 + 5R_1 \\ R_3 - 2R_1 \\ \Rightarrow \end{matrix}$$

$$\left(\begin{array}{cccccc} 1 & 1 & -1 & 0 & 0 & 1 \\ 0 & 2 & -5 & 0 & 1 & 5 \\ 0 & -1 & 3 & 1 & 0 & 5 \end{array}\right) \underset{\Rightarrow}{\tfrac{1}{2}R} \left(\begin{array}{cccccc} 1 & 1 & -1 & 0 & 0 & 1 \\ 0 & 1 & \frac{-5}{2} & 0 & \frac{1}{2} & \frac{5}{2} \\ 0 & -1 & 3 & 1 & 0 & -2 \end{array}\right) \underset{\Rightarrow}{R_3 + R_2}$$

$$\left(\begin{array}{cccccc} 1 & 1 & -1 & 0 & 0 & 1 \\ 0 & 1 & \frac{-5}{2} & 0 & \frac{1}{2} & \frac{5}{2} \\ 0 & 0 & \frac{1}{2} & 1 & \frac{1}{2} & \frac{1}{2} \end{array}\right) \underset{\Rightarrow}{2R_3} \left(\begin{array}{cccccc} 1 & 1 & -1 & 0 & 0 & 1 \\ 0 & 1 & \frac{-5}{2} & 0 & \frac{1}{2} & \frac{5}{2} \\ 0 & 0 & 1 & 2 & 1 & 1 \end{array}\right) \begin{array}{c} R_3 + \frac{5}{2}R_2 \\ R_1 + R_3 \\ \Rightarrow \end{array}$$

$$\left(\begin{array}{cccccc} 1 & 1 & 0 & 2 & 1 & 2 \\ 0 & 1 & 0 & 5 & 3 & 5 \\ 0 & 0 & 1 & 2 & 1 & 1 \end{array}\right) \underset{\Rightarrow}{R_1 - R_2} \left(\begin{array}{cccccc} 1 & 0 & 0 & -3 & -2 & -3 \\ 0 & 1 & 0 & 5 & 3 & 5 \\ 0 & 0 & 1 & 2 & 1 & 1 \end{array}\right)$$

So, we were able to convert the first three columns into the 3 x 3 identity matrix therefore the inverse exists and it is,

$$A^{-1} = \begin{pmatrix} -3 & -2 & -3 \\ 5 & 3 & 5 \\ 2 & 1 & 1 \end{pmatrix}.$$

So, there was an example in which the inverse did exist. Let's take a look at an example in which the inverse doesn't exist.

Example: Find the inverse of the following matrix, provided it exists.

$$B = \begin{pmatrix} 1 & -3 \\ -2 & 6 \end{pmatrix}$$

Solution: In this case we will tack on the 2×2 identity to get the new matrix and then try to convert the first two columns to the 2×2 identity matrix.

$$\left(\begin{array}{cccc} 1 & -3 & 1 & 0 \\ -2 & 6 & 0 & 1 \end{array}\right) \underset{\Rightarrow}{2R_1 + R_2} \left(\begin{array}{cccc} 1 & -3 & 1 & 0 \\ 0 & 0 & 0 & 1 \end{array}\right).$$

And we don't need to go any farther. In order for the 2×2 identity to be in the first two columns we must have a 1 in the second entry of the second column and a 0 in the second entry of the first column. However, there is no way to get a 1 in the second entry of the second column that will keep a 0 in the second entry in the first column. Therefore, we can't get the 2 x 2 identity in the first two columns and hence the inverse of B doesn't exist. We will leave off this discussion of inverses with the following fact.

FACT

Given a matrix A.

1. If A is nonsingular then A^{-1} will exist.

2. If A is singular then A^{-1} will NOT exist.

I'll leave it to you to verify this fact for the previous two examples. Systems of Equations Revisited. We need to do a quick revisit of systems of equations. Let's start with a general system of equations.

$$\begin{aligned} a_{11}x_1 + a_{12}x_2 + \cdots + a_{1n}x_n &= b_1 \\ a_{21}x_1 + a_{22}x_2 + \cdots + a_{2n}x_n &= b_2 \\ &\vdots \\ a_{n1}x_1 + a_{n2}x_2 + \cdots + a_{nn}x_n &= b_n. \end{aligned}$$

Now, covert each side into a vector to get,

$$\begin{pmatrix} a_{11}x_1 + a_{12}x_2 + \cdots + a_{1n}x_n \\ a_{21}x_1 + a_{22}x_2 + \cdots + a_{2n}x_{n2} \\ \vdots \\ a_{n1}x_1 + a_{n2}x_2 + \cdots + a_{nn}x_n \end{pmatrix} = \begin{pmatrix} b_1 \\ b_2 \\ \vdots \\ b_n \end{pmatrix}.$$

The left side of this equation can be thought of as a matrix multiplication.

$$\begin{pmatrix} a_{11} & a_{12} & \cdots & a_{1n} \\ a_{21} & a_{22} & \cdots & a_{2n} \\ \vdots & \vdots & \ddots & \vdots \\ a_{n1} & a_{n2} & \cdots & a_{nn} \end{pmatrix} \begin{pmatrix} x_1 \\ x_2 \\ \vdots \\ x_n \end{pmatrix} = \begin{pmatrix} b_1 \\ b_2 \\ \vdots \\ b_n \end{pmatrix}.$$

Simplifying up the notation a little gives,

$$A\vec{x} = \vec{h}.$$

where, is a vector whose components are the unknowns in the original system of equations.The solving process is identical. The augmented matrix for is

$$\left(A\,\vec{b}\right).$$

Once we have the augmented matrix we proceed as we did with a system that hasn't been wrote in matrix form.

FACT

Given the system of equation we have one of the following three possibilities for solutions.

1. There will be no solutions.
2. There will be exactly one solution.
3. There will be infinitely many solutions.

In fact we can go a little farther now. Since we are assuming that we've got the same number of equations as unknowns the matrix A is a square matrix and so we can compute its determinant. This gives the following fact.

FACT

Given the system of equations we have the following.

1. If A is nonsingular then there will be exactly one solution to the system.
2. If A is singular then there will either be no solution or infinitely many solutions to the system.

The matrix form of a homogeneous system is

$$A\vec{x} = \vec{0}.$$

where $\vec{0}$ is the vector of all zeroes. In the homogeneous system we are guaranteed to have a solution, $\vec{x} = \vec{0}$. The fact above for homogeneous systems is then,

FACT

Given the homogeneous system we have the following.

1. If A is nonsingular then the only solution will be $\vec{x} = \vec{0}$.
2. If A is singular then there will be infinitely many nonzero solutions to the system.

LINEAR INDEPENDENCE/LINEAR DEPENDENCE

This is not the first time that we've seen this topic. We also saw linear independence and linear dependence back when we were looking at second order differential equations. In that section we were dealing with functions, but the concept is essentially the same here. If we start with n vectors,

$$\vec{x}_1, \vec{x}_2, \ldots, \vec{x}_n.$$

If we can find constants, $c_1, c_2, \ldots, c_n$ with at least two nonzero such that

$$c_1\vec{x}_1 + c_2\vec{x}_2 + \ldots + c_n\vec{x}_n = \vec{0}.$$

then we call the vectors linearly dependent. If the only constants that work are $c_1 = 0$, $c_2 = 0, \ldots, c_n = 0$ then we call the vectors linearly independent.

If we further make the assumption that each of the n vectors has n components, *i.e.* each of the vectors look like,

$$\vec{x} = \begin{pmatrix} x_1 \\ x_2 \\ \vdots \\ x_n \end{pmatrix}.$$

We can get a very simple test for linear independence and linear dependence.

Note that this does not have to be the case, but in all of our work we will be working with n vectors each of which has n components.

FACT

Given the n vectors each with n components,

$$\vec{x}_1, \vec{x}_2, ..., \vec{x}_n.$$

form the matrix,

$$X = (\vec{x}_1 \quad \vec{x}_2 \quad ... \quad \vec{x}_n).$$

So, the matrix X is a matrix whose i^{th} column is the ith vector, $\vec{x}_i$. Then,

1. If X is nonsingular (*i.e.* det(X) is not zero) then the n vectors are linearly independent, and
2. If X is singular (*i.e.* det(X) = 0) then the n vectors are linearly dependent and the constants that make it true can be found by solving the system

$$X\vec{c} = \vec{0}.$$

where is a vector containing the constants.

Example: Determine if the following set of vectors are linearly independent or linearly dependent. If they are linearly dependent find the relationship between them.

$$\vec{x}^{(1)} = \begin{pmatrix} 1 \\ -3 \\ 5 \end{pmatrix}, \vec{x}^{(2)} = \begin{pmatrix} -2 \\ 1 \\ 4 \end{pmatrix}, \vec{x}^{(3)} = \begin{pmatrix} 6 \\ -2 \\ 1 \end{pmatrix}.$$

Solution: So, the first thing to do is to form X and compute its determinant.

$$X = \begin{pmatrix} 1 & -2 & 6 \\ -3 & 1 & -2 \\ 5 & 4 & 1 \end{pmatrix} \Rightarrow \det(X) = -79.$$

This matrix is non singular and so the vectors are linearly independent.

Example: Determine if the following set of vectors are linearly independent or linearly dependent. If they are linearly dependent find the relationship between them.

$$\vec{x}^{(1)} = \begin{pmatrix} 1 \\ -1 \\ 3 \end{pmatrix}, \vec{x}^{(2)} = \begin{pmatrix} -4 \\ 1 \\ -6 \end{pmatrix}, \vec{x}^{(3)} = \begin{pmatrix} 2 \\ -1 \\ 4 \end{pmatrix}.$$

Solution: As with the last example first form X and compute its determinant.

$$X = \begin{pmatrix} 1 & -4 & 2 \\ -1 & 1 & -1 \\ 3 & -6 & 4 \end{pmatrix}, \Rightarrow \quad \det(X) = 0.$$

So, these vectors are linearly dependent. We now need to find the relationship between the vectors. This means that we need to find constants that will make it true.

So we need to solve the system

$$X\vec{c} = \vec{0}.$$

Here is the augmented matrix and the solution work for this system.

$$\begin{pmatrix} 1 & -4 & 2 & 0 \\ -1 & 1 & -1 & 0 \\ 3 & -6 & 4 & 0 \end{pmatrix} \begin{matrix} R_2 + R_1 \\ R_3 - 3R_1 \\ \Rightarrow \end{matrix} \begin{pmatrix} 1 & -4 & 2 & 0 \\ 0 & -3 & 1 & 0 \\ 0 & -6 & -2 & 0 \end{pmatrix} \begin{matrix} R_3 - 3R_2 \\ \Rightarrow \end{matrix} \begin{pmatrix} 1 & -4 & 2 & 0 \\ 0 & -3 & 1 & 0 \\ 0 & 0 & 0 & 0 \end{pmatrix} \begin{matrix} -\frac{1}{3}R_2 \\ \Rightarrow \end{matrix}$$

$$\begin{pmatrix} 1 & -4 & 2 & 0 \\ 0 & 1 & -\frac{1}{3} & 0 \\ 0 & 0 & 0 & 0 \end{pmatrix} \begin{matrix} R_1 - 4R_2 \\ \Rightarrow \end{matrix} \begin{pmatrix} 1 & 0 & \frac{2}{3} & 0 \\ 0 & 1 & -\frac{1}{3} & 0 \\ 0 & 0 & 0 & 0 \end{pmatrix} \Rightarrow \begin{matrix} c_1 + \frac{2}{3}c_3 = 0 \\ c_1 + \frac{1}{3}c_3 = 0 \\ 0 = 0 \end{matrix} \Rightarrow \begin{matrix} c_1 = -\frac{2}{3}c_3 \\ c_2 = -\frac{1}{3}c_3 \end{matrix}.$$

Now, we would like actual values for the constants so, if use $c_3 = 3$we get the following solution $c_1 = -2$, $c_2 = 1$, and $c_3 = 3$. The relationship is then.

$$-2x^{-(1)} + \vec{x}^{(2)} + 3\vec{x}^{(3)} = \begin{pmatrix} 0 \\ 0 \\ 0 \end{pmatrix}.$$

CALCULUS WITH MATRICES

There really isn't a whole lot to this other than to just make sure that we can deal with calculus with matrices.

First, to this point we've only looked at matrices with numbers as entries, but the entries in a matrix can be functions as well. So we can look at matrices in the following form,

$$A(t) = \begin{pmatrix} a_{11}(t) & a_{12}(t) & \cdots & a_{1n}(t) \\ a_{21}(t) & a_{22}(t) & \cdots & a_{2n}(t) \\ \vdots & \vdots & & \vdots \\ a_{m1}(t) & a_{m2}(t) & \cdots & a_{mn}(t) \end{pmatrix}.$$

Now we can talk about differentiating and integrating a matrix of this form. To differentiate or integrate a matrix of this form all we do is differentiate or integrate the individual entries.

$$A(t) = \begin{pmatrix} a'_{11}(t) & a'_{12}(t) & \cdots & a'_{1n}(t) \\ a'_{21}(t) & a'_{22}(t) & \cdots & a'_{2n}(t) \\ \vdots & \vdots & & \vdots \\ a'_{m1}(t) & a'_{m2}(t) & \cdots & a'_{mn}(t) \end{pmatrix}$$

$$\int A(t)dt = \begin{pmatrix} \int a_{11}(t)dt & \int a_{12}(t)dt & \cdots & \int a_{1n}(t)dt \\ \int a_{21}(t)dt & \int a_{22}(t)dt & \cdots & \int a_{2n}(t)dt \\ \vdots & \vdots & & \vdots \\ \int a_{m1}(t)dt & \int a_{m2}(t)dt & \cdots & \int a_{mn}(t)dt \end{pmatrix}.$$

So when we run across this kind of thing don't get excited about it. Just differentiate or integrate as we normally would.

In this section we saw a very condensed set of topics from linear algebra. When we get back to differential equations many of these topics will show up occasionally and you will at least need to know what the words mean.

The main topic from linear algebra that you must know however if you are going to be able to solve systems of differential equations is the topic of the next section.

EIGENVALUES AND EIGENVECTORS

If you get nothing out of this quick review of linear algebra you must get this section. Without this section you will not be able to do any of the differential equations work that is in this chapter.

So let's start with the following. If we multiply an n x n matrix by an n x 1 vector we will get a new n x 1 vector back. In other words,

$$A\vec{\eta} = \vec{y}.$$

What we want to know is if it is possible for the following to happen. Instead of just getting a brand new vector out of the multiplication is it possible instead to get the following,

$$A\vec{\eta} = \lambda\vec{\eta}.$$

In other words is it possible, at least for certain ëand $\vec{\eta}$, to have matrix multiplication be the same as must multiplying the vector by a constant? Of course, we probably wouldn't be talking about this if the answer was no. So, it is possible for this to happen, however, it won't happen for just any value of ëor $\vec{\eta}$. If we do happen to have a ëand $\vec{\eta}$ for which this works (and they will always come in pairs) then we call ëan eigenvalue of A and $\vec{\eta}$ an eigenvector of A.

So, how do we go about find the eigenvalues and eigenvectors for a matrix? Well first notice that that if $\vec{\eta} \neq \vec{0}$. then it is going to be true for any value of ëand so we are going to make the assumption that. With that out of the way let's rewrite.

$$A\vec{\eta} - \lambda\vec{\eta} = \vec{0}$$

$$A\vec{\eta} - \lambda I_n\vec{\eta} = \vec{0}$$

$$(A - \lambda I_n)\,\vec{\eta} = \vec{0}$$

Notice that before we factored out the $\vec{\eta}$ we added in the appropriately sized identity matrix. This is equivalent to multiplying things by a one and so doesn't change the value of anything.

We needed to do this because without it we would have had the difference of a matrix, A, and a constant, ë, and this can't be done. We now have the difference of two matrices of the same size which can be done.

So, with this rewrite we see that

$$.(A - \lambda I_n)\,\vec{\eta} = \vec{0}.$$

is equivalent. In order to find the eigenvectors for a matrix we will need to solve a homogeneous system. We know that we will either have exactly one solution ($\vec{\eta} = \vec{0}$) or we will have infinitely many nonzero solutions. Since we've already said that don't want this means that we want the second case.

Knowing this will allow us to find the eigenvalues for a matrix. Recall from this fact that we will get the second case only if the matrix in the system is singular. Therefore we will need to determine the values of ëfor which we get,

$$\det (A - \lambda I) = 0.$$

Once we have the eigenvalues we can then go back and determine the eigenvectors for each eigenvalue. Let's take a look at a couple of quick facts about eigenvalues and eigenvectors.

FACT

If A is an n x n matrix then det$(A - \lambda I)$is an nth degree polynomial. This polynomial is called the characteristic polynomial.

To find eigenvalues of a matrix all we need to do is solve a polynomial. That's generally not too bad provided we keep n small. Likewise this fact also tells us that for an n x n matrix, A, we will have n eigenvalues if we include all repeated eigenvalues.

FACT

If λ_1, $\lambda_2,\ldots,\lambda_n$is the complete list of eigenvalues for A (including all repeated eigenvalues) then,

1. If λ occurs only once in the list then we call λ simple.
2. If ëoccurs $k>1$ times in the list then we say that ëhas multiplicity k.
3. If λ_1, $\lambda_2,\ldots,\lambda_n(k \le n)$ are the simple eigenvalues in the list with corresponding eigenvectors $\vec{\eta}^{(1)}, \vec{\eta}^{(2)}, \ldots, \vec{\eta}^{(k)}$ then the eigenvectors are all linearly independent.
4. If λ is an eigenvalue of $k > 1$ then ëwill have anywhere from 1 to k linearly independent eigenvectors.

The usefulness of these facts will become apparent when we get back into differential equations since in that work we will want linearly independent solutions.

Let's work a couple of examples now to see how we actually go about finding eigenvalues and eigenvectors.

Example: Find the eigenvalues and eigenvectors of the following matrix.

$$A = \begin{pmatrix} 2 & 7 \\ -1 & -6 \end{pmatrix}.$$

Solution: The first thing that we need to do is find the eigenvalues. That means we need the following matrix,

$$A - \lambda I = \begin{pmatrix} 2 & 7 \\ -1 & -6 \end{pmatrix} - \lambda \begin{pmatrix} 1 & 0 \\ 0 & 1 \end{pmatrix} = \begin{pmatrix} 2-\lambda & 7 \\ -1 & -6-\lambda \end{pmatrix}.$$

In particular we need to determine where the determinant of this matrix is zero.

$$\det\ (A - \lambda I) = (2-\lambda)(-6-\lambda) + 7 = \lambda^2 + 4\lambda - 5 = (\lambda - 1).$$

So, it looks like we will have two simple eigenvalues for this matrix, $\lambda_1 = -5$.and $\lambda_2 = 1$. We will now need to find the eigenvectors for each of these. Also note that according to the fact above, the two eigenvectors should be linearly independent.

To find the eigenvectors we simply plug in each eigenvalue and solve. So, let's do that.

$$\lambda_1 = -5.$$

In this case we need to solve the following system.

$$\begin{pmatrix} 7 & 7 \\ -1 & -1 \end{pmatrix} \vec{\eta} = \begin{pmatrix} 0 \\ 0 \end{pmatrix}.$$

Recall that officially to solve this system we use the following augmented matrix.

$$\begin{pmatrix} 7 & 7 & 0 \\ -1 & -1 & 0 \end{pmatrix} \overset{\frac{1}{7}R_1 + R_2}{\Rightarrow} \begin{pmatrix} 7 & 7 & 0 \\ 0 & 0 & 0 \end{pmatrix}.$$

Upon reducing down we see that we get a single equation

$$7\eta_1 + 7\eta_2 = 0 \quad \Rightarrow \quad \eta_1 = -\eta_2$$

that will yield an infinite number of solutions. This is expected behavior. Recall that we picked the eigenvalues so that the matrix would be singular and so we would get infinitely many solutions.

Notice as well that we could have identified this from the original system. This won't always be the case, but in the 2 × 2 case we can see from the system that one row will be a multiple of the other and so we will get infinite solutions. From this point on we won't be actually solving systems in these cases. We will just go straight to the equation and we can use either of the two rows for this equation. Now, let's get back to the eigenvector, since that is what we were after. In general then the eigenvector will be any vector that satisfies the following,

$$\vec{\eta} = \begin{pmatrix} \eta_1 \\ \eta_2 \end{pmatrix} = \begin{pmatrix} -\eta_2 \\ \eta_2 \end{pmatrix}, \ \eta_1 \neq 0.$$

To get this we used the solution to the equation that we found above.

We really don't want a general eigenvector however so we will pick a value for η_2 to get a specific eigenvector. We can choose anything (except $\eta_2 = 0$), so pick something that will make the eigenvector "nice". Note as well that since we've already assumed that the eigenvector is not zero we must choose a value that will not give us zero, which is why we want to avoid except in this case. Here's the eigenvector for this eigenvalue.

$$\vec{\eta}^{(1)} = \begin{pmatrix} -1 \\ 1 \end{pmatrix}, \quad \text{using } \eta_2 = 1.$$

Now we get to do this all over again for the second eigenvalue.

$$\lambda_2 = 1:$$

We'll do much less work with this part than we did with the previous part. We will need to solve the following system.

$$\begin{pmatrix} 1 & 7 \\ -1 & -7 \end{pmatrix} \vec{\eta} = \begin{pmatrix} 0 \\ 0 \end{pmatrix}.$$

Clearly both rows are multiples of each other and so we will get infinitely many solutions. We can choose to work with either row. We'll run with the first because to avoid having too many minus signs floating around. Doing this gives us,

$$\eta_1 + 7\eta_2 = 0 \qquad \eta_1 = -7\eta_2.$$

Note that we can solve this for either of the two variables. However, with an eye towards working with these later on let's try to avoid as many fractions as possible. The eigenvector is then,

$$\vec{\eta} = \begin{pmatrix} \eta_1 \\ \eta_2 \end{pmatrix} = \begin{pmatrix} -7\eta_2 \\ \eta_2 \end{pmatrix}, \quad \eta_2 \neq 0$$

$$\vec{\eta}^{(2)} = \begin{pmatrix} -7 \\ 1 \end{pmatrix}, \quad \eta_2 = 1.$$

Summarizing we have,

$$\lambda_1 = -5 \qquad \vec{\eta}^{(1)} = \begin{pmatrix} -1 \\ 1 \end{pmatrix}$$

$$\lambda_2 = 1 \qquad \vec{\eta}^{(1)} = \begin{pmatrix} -7 \\ 1 \end{pmatrix}.$$

Note that the two eigenvectors are linearly independent as predicted.

Example: Find the eigenvalues and eigenvectors of the following matrix.

$$A = \begin{pmatrix} 1 & -1 \\ \frac{A}{9} & -\frac{1}{3} \end{pmatrix}.$$

Solution: This matrix has fractions in it. That's life so don't get excited about it. First we need the eigenvalues.

$$\det(A - \lambda I) = \begin{vmatrix} 1-\lambda & -1 \\ \frac{4}{9} & -\frac{1}{3}-\lambda \end{vmatrix}$$

$$= (1-\lambda)\left(-\frac{1}{3}-\lambda\right)+\frac{4}{9} \;= \lambda^2 - \lambda\frac{2}{3}\lambda + \frac{1}{9}$$

$$= \left(\lambda - \frac{1}{3}\right)^2 \quad \Rightarrow \quad l_{1,2} = \frac{1}{3}.$$

So, it looks like we've got an eigenvalue of multiplicity 2 here. Remember that the power on the term will be the multiplicity. Now, let's find the eigenvector(s). This one is going to be a little different from the first example.

There is only one eigenvalue so let's do the work for that one. We will need to solve the following system,

$$\begin{pmatrix} \frac{2}{3} & -1 \\ \frac{4}{9} & -\frac{2}{3} \end{pmatrix}\begin{pmatrix} \eta_1 \\ \eta_2 \end{pmatrix} = \begin{pmatrix} 0 \\ 0 \end{pmatrix} \quad \Rightarrow \quad R_1 = \frac{3}{2}R_2.$$

So, the rows are multiples of each other. We'll work with the first equation in this example to find the eigenvector.

$$\frac{2}{3}\eta_1 - \eta_2 = 0 \qquad \eta_2 = \frac{2}{3}\eta_1.$$

Recall in the last example we decided that we wanted to make these as "nice" as possible and so should avoid fractions if we can. Sometimes, as in this case, we simply can't so we'll have to deal with it. In this case the eigenvector will be,

$$\vec{\eta} = \begin{pmatrix} \eta_1 \\ \eta_2 \end{pmatrix} = \begin{pmatrix} \eta_1 \\ \frac{2}{3}\eta_1 \end{pmatrix}, \qquad \eta_1 \neq 0$$

$$\vec{\eta}^{(1)} = \begin{pmatrix} 3 \\ 2 \end{pmatrix}, \qquad \eta_1 = 3.$$

Note that by careful choice of the variable in this case we were able to get rid of the fraction that we had. This is something that in general doesn't much matter if we do or not. However, when we get back to differential equations it will be easier on us if we don't have any fractions so we will usually try to eliminate them at this step.

Also in this case we are only going to get a single (linearly independent) eigenvector. We can get other eigenvectors, by choosing different values of η_1. However, each of these will be linearly dependent with the first eigenvector. If you're not convinced of this try it. Pick some values for η_1 and get a different vector and check to see if the two are linearly dependent. Recall from the fact above that an eigenvalue of multiplicity *k* will have anywhere from 1 to *k* linearly independent eigenvectors. In this case we got one. For most of the 2 x 2 matrices that we'll be working with this will be the case, although it doesn't have to be. We can, on occasion, get two.

Example: Find the eigenvalues and eigenvectors of the following matrix.

$$A = \begin{pmatrix} -4 & -17 \\ 2 & 2 \end{pmatrix}.$$

Solution: So, we'll start with the eigenvalues.

$$\det(A - \lambda I) = \begin{vmatrix} -4-\lambda & -17 \\ 2 & 2-\lambda \end{vmatrix}$$

$$= (-4-\lambda)(2-\lambda) + 34$$

$$= \lambda^2 + 2\lambda + 26.$$

This doesn't factor, so upon using the quadratic formula we arrive at,

$$\lambda_{1,2} = -1 \pm 5i.$$

In this case we get complex eigenvalues which are definitely a fact of life with eigenvalue/eigenvector problems so get used to them.

Finding eigenvectors for complex eigenvalues is identical to the previous two examples, but it will be somewhat messier. So, let's do that.

$$\lambda_1 = -1 + 5i.$$

The system that we need to solve this time is

$$\begin{pmatrix} -4-(-1+5i) & -17 \\ 2 & 2-(-1+5i) \end{pmatrix}\begin{pmatrix} \eta_1 \\ \eta_2 \end{pmatrix} = \begin{pmatrix} 0 \\ 0 \end{pmatrix}$$

$$\begin{pmatrix} -3-5i & -17 \\ 2 & 3-5i) \end{pmatrix}\begin{pmatrix} \eta_1 \\ \eta_2 \end{pmatrix} = \begin{pmatrix} 0 \\ 0 \end{pmatrix}.$$

Now, it's not super clear that the rows are multiples of each other, but they are. In this case we have,

$$R_1 = -\frac{1}{2}(3+5i)R_2.$$

This is not something that you need to worry about, we just wanted to make the point. For the work that we'll be doing later on with differential equations we will just assume that we've done everything correctly and we've got two rows are multiples of each other. Therefore, all that we need to do here is pick one of the rows and work with it.

We'll work with the second row this time.

$$2\eta_1 + (3-5i)\eta_2 = 0.$$

Now we can solve for either of the two variables. However, again looking forward to differential equations, we are going to need the "i" in the numerator so solve the equation in such a way as this will happen. Doing this gives,

$$2\eta_1 = -(3-5i)\eta_2$$

$$\eta_1 = -\frac{1}{2}(3-5i)\eta_2.$$

So, the eigenvector in this case is

$$\vec{\eta} = \begin{pmatrix} -\frac{1}{2}(3-5i)\eta_2 \end{pmatrix}, \quad \eta_2 \neq 0$$

$$\vec{\eta}^{(1)} = \begin{pmatrix} -3+5i \\ 2 \end{pmatrix}, \quad \eta_2 = 2.$$

As with the previous example we choose the value of the variable to clear out the fraction.

Now, the work for the second eigenvector is almost identical and so we'll not dwell on that too much.

$$\lambda_2 = -1 - 5i.$$

The system that we need to solve here is

$$\begin{pmatrix} -4-(-1-5i & -17 \\ 2 & 2-(-1-5i) \end{pmatrix}\begin{pmatrix} \eta_1 \\ \eta_2 \end{pmatrix} = \begin{pmatrix} 0 \\ 0 \end{pmatrix}$$

$$\begin{pmatrix} -3+5i & -17 \\ 2 & 3+5i) \end{pmatrix}\begin{pmatrix} \eta_1 \\ \eta_2 \end{pmatrix} = \begin{pmatrix} 0 \\ 0 \end{pmatrix}.$$

Working with the second row again gives,

$$2\eta_1 + (3+5i)\eta_2 = 0 \quad \Rightarrow \quad \eta_1 = -\frac{1}{2}(3+5i)\eta_2.$$

The eigenvector in this case is

$$\vec{\eta} = \begin{pmatrix} \eta_1 \\ \eta_2 \end{pmatrix} = \begin{pmatrix} -\frac{1}{2}(3+5i)\eta_2 \\ \eta_2 \end{pmatrix}, \qquad \eta_2 \neq 0$$

$$\vec{\eta}^{(2)} = \begin{pmatrix} -3-5i \\ 2 \end{pmatrix}, \qquad \eta_2 = 2.$$

Summarizing,

$$\lambda_1 = -1+5i \qquad \vec{\eta}^{(1)} = \begin{pmatrix} -3+5i \\ 2 \end{pmatrix}$$

$$\lambda_2 = -1-5i \qquad \vec{\eta}^{(2)} = \begin{pmatrix} -3+5i \\ 2 \end{pmatrix}.$$

There is a nice fact that we can use to simplify the work when we get complex eigenvalues. We need a bit of terminology first however. If we start with a complex number,

$$z = a + bi$$

then the complex conjugate of z is

$$\overline{z} = a - bi.$$

To compute the complex conjugate of a complex number we simply change the sign on the term that contains the "i". The complex conjugate of a vector is just the conjugate of each of the vectors components.

We now have the following fact about complex eigenvalues and eigenvectors.

FACT

If A is an n x n matrix with only real numbers and if $\lambda_1 = a + bi$ is an eigenvalue with eigenvector $\lambda_2 = \overline{\lambda}_1 = a - bi$. Then is also an eigenvalue and its eigenvector is the conjugate of $\vec{\eta}^{(0)}$.

This is a fact is something that you should feel free to use as you need to in our work.

Now, we need to work one final eigenvalue/eigenvector problem. To this point we've only worked with 2 × 2 matrices and we should work at least one that isn't 2 × 2. Also, we need to work one in which we get an eigenvalue of multiplicity greater than one that has more than one linearly independent eigenvector.

Example: Find the eigenvalues and eigenvectors of the following matrix.

$$A = \begin{pmatrix} 0 & 1 & 1 \\ 1 & 0 & 1 \\ 1 & 1 & 0 \end{pmatrix}.$$

Solution: Despite the fact that this is a 3 × 3 matrix, it still works the same as the 2 × 2 matrices that we've been working with. So, start with the eigenvalues

$$\det(A - \lambda 1) = \begin{vmatrix} -\lambda & 1 & 1 \\ 1 & -\lambda & 1 \\ 1 & 1 & -\lambda \end{vmatrix}$$

$$= -\lambda^3 + 3\lambda + 2$$

$$= (\lambda - 2)(\lambda + 1)^2 \qquad \lambda_1 = 2,\ \lambda_{2,3} = -1.$$

So, we've got a simple eigenvalue and an eigenvalue of multiplicity 2. Note that we used the same method of computing the determinant of a 3 × 3 matrix that we used in the previous section. We just didn't show the work.

Let's now get the eigenvectors. We'll start with the simple eigenvector.

$$\lambda_1 = 2.$$

Here we'll need to solve,

$$\begin{pmatrix} -2 & 1 & 1 \\ 1 & -2 & 1 \\ 1 & 1 & -2 \end{pmatrix} \begin{pmatrix} \eta_1 \\ \eta_2 \\ \eta_3 \end{pmatrix} = \begin{pmatrix} 0 \\ 0 \\ 0 \end{pmatrix}.$$

This time, unlike the 2 × 2 cases we worked earlier, we actually need to solve the system. So let's do that.

$$\begin{pmatrix} -2 & 1 & 1 & 0 \\ 1 & -2 & 1 & 0 \\ 1 & 1 & -2 & 0 \end{pmatrix} \overset{R_1 \leftrightarrow R_2}{\Rightarrow} \begin{pmatrix} 1 & -2 & 1 & 0 \\ -2 & 1 & 1 & 0 \\ 1 & 1 & -2 & 0 \end{pmatrix} \overset{\substack{R_2 + R_1 \\ R_3 - R_1}}{\Rightarrow} \begin{pmatrix} 1 & -2 & 1 & 0 \\ 0 & -3 & 3 & 0 \\ 0 & 3 & -3 & 0 \end{pmatrix}$$

$$\overset{-\frac{1}{3}}{\Rightarrow} \begin{pmatrix} 1 & -2 & 1 & 0 \\ 0 & 1 & -1 & 0 \\ 0 & 3 & -3 & 0 \end{pmatrix} \overset{\substack{R_3 - 3R_2 \\ R_1 + 2R_2}}{\Rightarrow} \begin{pmatrix} 1 & 0 & -1 & 0 \\ 0 & 1 & -1 & 0 \\ 0 & 0 & 0 & 0 \end{pmatrix} \overset{-\frac{1}{3}R_2}{\Rightarrow} \begin{pmatrix} 1 & -2 & 1 & 0 \\ 0 & 1 & -1 & 0 \\ 0 & 3 & -3 & 0 \end{pmatrix} \overset{\substack{R_3 - 3R_2 \\ R_1 + 2R_2}}{\Rightarrow} \begin{pmatrix} 1 & 0 & -1 & 0 \\ 0 & 1 & -1 & 0 \\ 0 & 0 & 0 & 0 \end{pmatrix}.$$

Going back to equations gives,

$$\eta_1 - \eta_3 = 0 \quad \Rightarrow \quad \eta_1 = \eta_3$$

$$\eta_2 - \eta_3 = 0 \quad \Rightarrow \quad \eta_2 = \eta_3.$$

So, again we get infinitely many solutions as we should for eigenvectors. The eigenvector is then,

$$\vec{\eta} = \begin{pmatrix} \eta_1 \\ \eta_2 \\ \eta_3 \end{pmatrix} = \begin{pmatrix} \eta_1 \\ \eta_2 \\ \eta_3 \end{pmatrix}, \qquad \eta_3 \neq 0$$

$$\vec{\eta}^{(1)} = \begin{pmatrix} 1 \\ 1 \\ 1 \end{pmatrix}, \qquad \eta_3 = 1.$$

Now, let's do the other eigenvalue.

$$\lambda_2 = -1.$$

Here we'll need to solve,

$$\begin{pmatrix} 1 & 1 & 1 \\ 1 & 1 & 1 \\ 1 & 1 & 1 \end{pmatrix} \begin{pmatrix} \eta_1 \\ \eta_2 \\ \eta_3 \end{pmatrix} = \begin{pmatrix} 0 \\ 0 \\ 0 \end{pmatrix}.$$

Okay, in this case is clear that all three rows are the same and so there isn't any reason to actually solve the system since we can clear out the bottom two rows to all zeroes in one step. The equation that we get then is,

$$\eta_1 + \eta_2 + \eta_3 = 0 \quad \Rightarrow \quad \eta_1 = -\eta_2 - \eta_3.$$

So, in this case we get to pick two of the values for free and will still get infinitely many solutions. Here is the general eigenvector for this case,

$$\vec{\eta} = \begin{pmatrix} \eta_1 \\ \eta_2 \\ \eta_3 \end{pmatrix} = \begin{pmatrix} -\eta_2 - \eta_3 \\ \eta_2 \\ \eta_3 \end{pmatrix}, \qquad \eta_2 \neq 0 \text{ and } \eta_3 \neq 0 \text{at the same time.}$$

Notice the restriction this time. Recall that we only require that the eigenvector not be the zero vector. This means that we can allow one or the other of the two variables to be zero, we just can't allow both of them to be zero at the same time!

What this means for us is that we are going to get two linearly independent eigenvectors this time. Here they are.

$$\vec{\eta}^{(2)} = \begin{pmatrix} -1 \\ 0 \\ 1 \end{pmatrix} \qquad \eta_2 = 0 \text{ and } \eta_3 = 1$$

$$\vec{\eta}^{(3)} = \begin{pmatrix} -1 \\ 1 \\ 0 \end{pmatrix} \qquad \eta_2 = 1 \text{ and } \eta_3 = 0.$$

Now when we talked about linear independent vectors in the last section we only looked at n vectors each with n components. We can still talk about linear independence in this case however. Recall back with we did linear independence for functions we saw at the time that if two functions were linearly dependent then they were multiples of each other. Well the same thing holds true for vectors. Two vectors will be linearly dependent if they are multiples of each other. In this case there is no way to get $\vec{\eta}^{(2)}$ by multiplying $\vec{\eta}^{(3)}$ by a constant. Therefore, these two vectors must be linearly independent.

So, summarizing up, here are the eigenvalues and eigenvectors for this matrix

$$\lambda_1 = 2 \qquad \vec{\eta}^{(1)} = \begin{pmatrix} 1 \\ 1 \\ 1 \end{pmatrix}$$

$$\lambda_2 = -1 \qquad \vec{\eta}^{(2)} = \begin{pmatrix} -1 \\ 0 \\ 1 \end{pmatrix}$$

$$\lambda_3 = -1 \qquad \vec{\eta}^{(3)} = \begin{pmatrix} -1 \\ 1 \\ 0 \end{pmatrix}$$

SYSTEMS OF DIFFERENTIAL EQUATIONS

In the introduction to this section we briefly discussed how a system of differential equations can arise from a population problem in which we keep track of the population of both the prey and the predator. It makes sense that the number of prey present will affect the number of the predator present. Likewise, the number of predator present will affect the number of prey present. Therefore the differential equation that governs the population of either the prey or the predator should in some way depend on the population of the other. This will lead to two differential equations that must be solved simultaneously in order to determine the population of the prey and the predator.

The whole point of this is to notice that systems of differential equations can arise quite easily from naturally occurring situations. Developing an effective predator-prey system of differential equations is not the subject of this chapter. However, systems can arise from n^{th} order linear differential equations as well. Before we get into this however, let's write down a system and get some terminology out of the way.

We are going to be looking at first order, linear systems of differential equations. These terms mean the same thing that they have meant up to this point. The largest derivative anywhere in the system will be a first derivative and all unknown functions and their derivatives will only occur to the first power and will not be multiplied by other unknown functions. Here is an example of a system of first order, linear differential equations.

$$x_1' = x_1 + 2x_2$$

$$x_2' = 3x_1 + 2x_2.$$

We call this kind of system a coupled system since knowledge of x_2 is required in order to find x_1 and likewise knowledge of x_1 is required to find x_2. We will worry about how to go about solving these later. At this point we are only interested in becoming familiar with some of the basics of systems. Now, as mentioned earlier, we can write an nth order linear differential equation as a system. Let's see how that can be done.

Example: Write the following 2nd order differential equations as a system of first order, linear differential equations.

$$2y' - 5y' + y = 0 \qquad y(3) = 6 \qquad y'(3) = -1.$$

Solution: We can write higher order differential equations as a system with a very simple change of variable. We'll start by defining the following two new functions.

$$x_1(t) = y(t)$$

$$x_2(t) = y'(t).$$

Now notice that if we differentiate both sides of these we get,

$$x_1' = y' = x_2$$

$$x_2' = y'' = -\frac{1}{2}y + \frac{5}{2}y' = -\frac{1}{2}x_1 + \frac{5}{2}x_2.$$

Note the use of the differential equation in the second equation. We can also convert the initial conditions over to the new functions.

$$x_1(3) = y(3) = 6$$

$$x_2(3) = y'(3) = -1.$$

Putting all of this together gives the following system of differential equations.

$$x_1' = x_2 \qquad x_1(3) = 6$$

$$x_1' = -\frac{1}{2}x_1 + \frac{5}{2}x_2 \quad x_2(3) = -1.$$

We will call the system in the above example an Initial Value Problem just as we did for differential equations with initial conditions.

Let's take a look at another example.

Example: Write the following 4th order differential equations as a system of first order, linear differential equations.

$$y^{(4)} + 3y' - \sin(t)y' + 8y = t^2 \qquad y(0) = 1 \quad y'(0) - 2 \quad y''(0) - 3\ y''(0) = 4.$$

Solution: Just as we did in the last example we'll need to define some new functions. This time we'll need 4 new functions.

$$x_1 = y \quad \Rightarrow \quad x_1' = y' = x_2$$

$$x_2 = y' \quad \Rightarrow \quad x_2' = y'' = x_3$$

$$x_3 = y'' \quad \Rightarrow \quad x_3' = y'' = x_4$$

$$x_4 = y' \quad \Rightarrow \quad x_4' = y^{(4)} = -8y + \sin(t)y' - 3y' + t^2 = -8x_1 + \sin(t)x_2 - 3x_3 + t^2.$$

The system along with the initial conditions is then,

$$x_1' = x_2 \qquad x_1(0) = 1$$

$$x_2' = x_3 \qquad x_2(0) = 2$$

$$x_3' = x_4 \qquad x_3(0) = 3$$

$$x_4' = -8x_1 + \sin(t)x_2 - 3x_3 + t^2 \quad x_4(0) = 4$$

Now, when we finally get around to solving these we will see that we generally don't solve systems in the form that we've given them in this section. Systems of differential equations can be converted to matrix form and this is the form that we usually use in solving systems.

Example: Convert the system the following system to matrix from.

$$x_1' = 4x_1 + 7x_2$$

$$x_2' = 2x_1 - 5x_2.$$

Solution: First write the system so that each side is a vector.

$$\begin{pmatrix} x_1' \\ x_2' \end{pmatrix} = \begin{pmatrix} 4x_1 + 7x_2 \\ -2x_1 - 5x_2 \end{pmatrix}.$$

Now the right side can be written as a matrix multiplication,

$$\begin{pmatrix} x_1' \\ x_2' \end{pmatrix} = \begin{pmatrix} 4 & 7 \\ -2 & -5 \end{pmatrix} \begin{pmatrix} x_1 \\ x_2 \end{pmatrix}.$$

Now, if we define,

$$\vec{x} = \begin{pmatrix} x_1 \\ x_2 \end{pmatrix}.$$

then,

$$\vec{x}' = \begin{pmatrix} x_1' \\ x_2' \end{pmatrix}.$$

The system can then be wrote in the matrix form,

$$\vec{x}' = \begin{pmatrix} 4 & 7 \\ -2 & 5 \end{pmatrix} \vec{x}.$$

Example: Convert the systems from Examples 1 and 2 into matrix form.

Solution: We'll start with the system from Example 1.

$$x_1' = x_2 \qquad x_1(3) = 6$$

$$x_2' = -\frac{1}{2}x_1 + \frac{5}{2}x_2 \quad x_2(3) = -1$$

First define,

$$\vec{x} = \begin{pmatrix} x_1 \\ x_2 \end{pmatrix}.$$

The system is then,

$$\vec{x}' = \begin{pmatrix} 0 & 1 \\ -\frac{1}{2} & \frac{5}{2} \end{pmatrix} \vec{x} \qquad \vec{x}(3) = \begin{pmatrix} x_1(3) \\ x_2(3) \end{pmatrix} = \begin{pmatrix} 6 \\ -1 \end{pmatrix}.$$

Now, let's do the system from Example 2.

$$x_1' = x_2 \qquad x_1(0) = 1$$

$$x_2' = x_3 \qquad x_2(0) = 2$$

$$x_3' = x_4 \qquad\qquad x_3(0) = 3$$

$$x_4' = -8x_1 + \sin(t)x_2 - 3x_3 + t^2 \qquad x_4(0) = 4$$

In this case we need to be careful with the t^2 in the last equation. We'll start by writing the system as a vector again and then break it up into two vectors, one vector that contains the unknown functions and the other that contains any known functions.

$$\begin{pmatrix} x_1' \\ x_2' \\ x_3' \\ x_4' \end{pmatrix} = \begin{pmatrix} x_2 \\ x_3 \\ x_4 \\ -8x_1 + \sin(t)x_2 - 3x_3 + t^2 \end{pmatrix} = \begin{pmatrix} x_2 \\ x_3 \\ x_4 \\ -8x_1 + \sin(t)\,x_2 - 3x_3 \end{pmatrix} + \begin{pmatrix} 0 \\ 0 \\ 0 \\ t^2 \end{pmatrix}.$$

Now, the first vector can now be written as a matrix multiplication and we'll leave the second vector alone.

$$\vec{x}\,' = \begin{pmatrix} 0 & 1 & 0 & 0 \\ 0 & 0 & 1 & 0 \\ 0 & 0 & 0 & 1 \\ -8 & \sin(t) & -3 & 0 \end{pmatrix}\vec{x} + \begin{pmatrix} 0 \\ 0 \\ 0 \\ t^2 \end{pmatrix} \qquad \vec{x}(0) = \begin{pmatrix} 1 \\ 2 \\ 3 \\ 4 \end{pmatrix}.$$

where,

$$\vec{x}(t) = \begin{pmatrix} x_1(t) \\ x_2(t) \\ x_3(t) \\ x_4(t) \end{pmatrix}.$$

Note that occasionally for "large" systems such as this we will one step farther and write the system as,

$$\vec{x}\,' = A\vec{x} + \vec{g}(t).$$

The last thing that we need to do in this section is get a bit of terminology out of the way. Starting with

$$\vec{x}\,' = A\vec{x} + \vec{g}(t).$$

we say that the system is homogeneous if $\vec{g}(t) = \vec{0}$ and we say the system is nonhomogeneous if $\vec{g}(t) \neq \vec{0}$.

NONHOMOGENEOUS SYSTEMS

We now need to address nonhomogeneous systems briefly. Both of the methods that we looked at back in the second order differential equations chapter can also be used here.

As we will see Undetermined Coefficients is almost identical when used on systems while Variation of Parameters will need to have a new formula derived, but will actually be slightly easier when applied to systems.

UNDETERMINED COEFFICIENTS

The method of Undetermined Coefficients for systems is pretty much identical to the second order differential equation case. The only difference is that the coefficients will need to be vectors now.

Example: Find the general solution to the following system.

$$\vec{x} = \begin{pmatrix} 1 & 2 \\ 3 & 2 \end{pmatrix} \vec{x} + t \begin{pmatrix} 2 \\ -4 \end{pmatrix}.$$

Solution: We already have the complimentary solution as we solved that part back in the real eigenvalue section. It is,

$$\vec{x}_c(t) = c_1 e \begin{pmatrix} -1 \\ 1 \end{pmatrix} + c_2 e^{4t} \begin{pmatrix} 2 \\ 3 \end{pmatrix}.$$

Guessing the form of the particular solution will work in exactly the same way it did back when we first looked at this method. We have a linear polynomial and so our guess will need to be a linear polynomial. The only difference is that the "coefficients" will need to be vectors instead of constants. The particular solution will have the form,

$$\vec{x}_p = t\vec{a} + \vec{b} = t \begin{pmatrix} a_1 \\ a_2 \end{pmatrix} + \begin{pmatrix} b_1 \\ b_2 \end{pmatrix}.$$

So, we need to differentiate the guess

$$\vec{x}'_p = \vec{a} = \begin{pmatrix} a_1 \\ a_2 \end{pmatrix}.$$

Before plugging into the system let's simplify the notation a little to help with our work. We'll write the system as,

$$\vec{x} = \begin{pmatrix} 1 & 2 \\ 3 & 2 \end{pmatrix} \vec{x} + t \begin{pmatrix} 2 \\ -4 \end{pmatrix} = A\vec{x} + t\vec{g}.$$

This will make the following work a little easier. Now, let's plug things into the system.

$$\vec{a} = A\left(t\vec{a} + \vec{b}\right) + t\vec{g}$$

$$\vec{a} = tA\vec{a} + A\vec{b} + t\vec{g}$$

$$\vec{0} = t\left(A\vec{a} + \vec{g}\right) + \left(A\vec{b} + \vec{a}\right)$$

Now we need to set the coefficients equal. Doing this gives,

$$t^1: \quad A\vec{a} + \vec{g} = \vec{0} \qquad A\vec{a} = -\vec{g}$$

$$t^0: \quad A\vec{b} + \vec{a} = \vec{0} \qquad A\vec{b} = \vec{a}.$$

Now only is unknown in the first equation so we can use Gaussian elimination to solve the system. We'll leave this work to you to check.

$$\begin{pmatrix} 1 & 2 \\ 3 & 2 \end{pmatrix}\begin{pmatrix} a_1 \\ a_2 \end{pmatrix} = -\begin{pmatrix} 2 \\ -4 \end{pmatrix} \quad \Rightarrow \quad \vec{a} = \begin{pmatrix} 3 \\ -\frac{5}{2} \end{pmatrix}.$$

Now that we know $\vec{a}$ we can solve the second equation for $\vec{b}$.

$$\begin{pmatrix} 1 & 2 \\ 3 & 2 \end{pmatrix}\begin{pmatrix} b_1 \\ b_2 \end{pmatrix} = \begin{pmatrix} 3 \\ -\frac{5}{2} \end{pmatrix} \quad \Rightarrow \quad \vec{b} = \begin{pmatrix} -\frac{11}{4} \\ -\frac{23}{8} \end{pmatrix}.$$

So, since we were able to solve both equations, the particular solution is then,

$$\vec{x}_p = t\begin{pmatrix} 3 \\ -\frac{5}{2} \end{pmatrix} + \begin{pmatrix} -\frac{11}{4} \\ -\frac{23}{8} \end{pmatrix}.$$

The general solution is then,

$$\vec{x}(t) = c_1 e^{-t}\begin{pmatrix} -1 \\ 1 \end{pmatrix} + c_2 e^{4t}\begin{pmatrix} 2 \\ 3 \end{pmatrix} + t\begin{pmatrix} 3 \\ -\frac{5}{2} \end{pmatrix} + \begin{pmatrix} -\frac{11}{4} \\ \frac{23}{8} \end{pmatrix}.$$

So, as you can see undetermined coefficients is nearly the same as the first time we saw it. The work in solving for the "constants" is a little messier however.

VARIATION OF PARAMETERS

In this case we will need to derive a new formula for variation of parameters for systems. The derivation this time will be much simpler than the when we first saw variation of parameters.

First let *X(t)* be a matrix whose i^{th} column is the ith linearly independent solution to the system,

$$\vec{x} = A\vec{x}.$$

Now it can be shown that *X(t)* will be a solution to the following differential equation.

$$X' = AX.$$

This is nothing more than the original system with the matrix in place of the original vector.

We are going to try and find a particular solution to

$$\vec{x}' = A\vec{x} + \vec{g}(t).$$

We will assume that we can find a solution of the form,

$$\vec{x}_p = X(t)\vec{v}(t).$$

where we will need to determine the vector $\vec{v}(t)$. To do this we will need to plug this into the nonhomogeneous system. Don't forget to product rule the particular solution when plugging the guess into the system.

$$X'\vec{v} + X\vec{v}' = A\,X\,\vec{v} + \vec{g}.$$

Note that we dropped the "*(t)*" part of things to simplify the notation a little. Now we can rewrite this a little.

$$X'\vec{v} + X\vec{v}' = X'\vec{v} + \vec{g}$$

$$X\vec{v}' = \vec{g}.$$

Because we formed X using linearly independent solutions we know that det(X) must be nonzero and this in turn means that we can find the inverse of X. So, multiply both sides by the inverse of X.

$$\vec{v}' = X^{-1}\vec{g}.$$

Now all that we need to do is integrate both sides to get $\vec{v}(t)$.

$$\vec{v}(t) = \int X^{-1}\vec{g}dt.$$

As with the second order differential equation case we can ignore any constants of integration. The particular solution is then,

$$\vec{x}_p = X\int X^{-1}\vec{g}dt.$$

Let's work a quick example using this.

Example: Find the general solution to the following system.

$$\vec{x}' = \begin{pmatrix} -5 & 1 \\ 4 & -2 \end{pmatrix}\vec{x} + e^{2t}\begin{pmatrix} 6 \\ -1 \end{pmatrix}.$$

Solution: We found the complimentary solution to this system in the real eigenvalue section. It is,

$$\vec{x}_c(t) = c_1 e^{-t}\begin{pmatrix} 1 \\ 4 \end{pmatrix} + c_2 e^{-6t}\begin{pmatrix} -1 \\ 1 \end{pmatrix}.$$

Now the matrix X is,

$$X = \begin{pmatrix} e^{-t} & -e^{-6t} \\ 4e^{-t} & e^{-6t} \end{pmatrix}.$$

Now, we need to find the inverse of this matrix. We saw how to find inverses of matrices back in the second linear algebra review section and the process is the same here even though we don't have constant entries. We'll leave the detail to you to check.

$$X^{-1} = \begin{pmatrix} \frac{1}{5}e^{t} & \frac{1}{5}e^{t} \\ -\frac{4}{5}e^{6t} & \frac{1}{5}e^{6t} \end{pmatrix}.$$

Now do the multiplication in the integral.

$$X^{-1}\vec{g} = \begin{pmatrix} \frac{1}{5}e^{t} & \frac{1}{5}e^{t} \\ -\frac{4}{5}e^{6t} & \frac{1}{5}e^{6t} \end{pmatrix}\begin{pmatrix} 6e^{2t} \\ -e^{2t} \end{pmatrix} = \begin{pmatrix} e^{3t} \\ -5e^{8t} \end{pmatrix}.$$

Now do the integral.

$$\int X^{-1}\vec{g}dt = \int\begin{pmatrix} e^{3t} \\ -5e^{8t} \end{pmatrix}dt = \begin{pmatrix} \int e^{3t}dt \\ \int -5e^{8t}dt \end{pmatrix} = \begin{pmatrix} \frac{1}{3}e^{3t} \\ -\frac{5}{8}e^{8t} \end{pmatrix}.$$

Remember that to integrate a matrix or vector you just integrate the individual entries.

We can now get the particular solution.

$$\begin{aligned} \vec{x}_p &= X\int X^{-1}\vec{g}\,dt \\ &= \begin{pmatrix} e^{-t} & -e^{-6t} \\ 4e^{-t} & e^{-6t} \end{pmatrix}\begin{pmatrix} \frac{1}{3}e^{3t} \\ -\frac{5}{8}e^{8t} \end{pmatrix} \\ &= \begin{pmatrix} \frac{23}{24}e^{2t} \\ \frac{17}{24}e^{2t} \end{pmatrix} \\ &= e^{2t}\begin{pmatrix} \frac{23}{24} \\ \frac{17}{24} \end{pmatrix}. \end{aligned}$$

The general solution is then,

$$\vec{x}(t) = c_1e^{-t}\begin{pmatrix} 1 \\ 4 \end{pmatrix} + c_1e^{-6t}\begin{pmatrix} -1 \\ 1 \end{pmatrix} + e^{2t}\begin{pmatrix} \frac{23}{24} \\ \frac{17}{24} \end{pmatrix}.$$

So, some of the work can be a little messy, but overall not too bad.

We looked at two methods of solving nonhomogeneous differential equations here and while the work can be a little messy they aren't too bad. Of course we also kept the nonhomogeneous part fairly simple here. More complicated problems will have significant amounts of work involved.

LAPLACE TRANSFORMS

There's not too much to this section. We're just going to work an example to illustrate how Laplace transforms can be used to solve systems of differential equations.

Example: Solve the following system.

$$x_1' = 3x_1 - 3x_2 + 2 \qquad x_1(0) = 1$$

$$x_2' = -6x_1 - t \qquad x_1(0) = -1$$

Solution: First notice that the system is not given in matrix form. This is because the system won't be solved in matrix form. Also note that the system is nonhomogeneous.

We start just as we did when we used Laplace transforms to solve single differential equations. We take the transform of both differential equations.

$$sX_1(s) - x_1(0) = 3X_1(s) - 3X_2(s) + \frac{2}{s}$$

$$sX_2(s) - x_2(0) = -6X_1(s) - (s) - \frac{1}{s^2}.$$

Now plug in the initial condition and simplify things a little.

$$(s-3)X_1(s) + 3X_2(s) = \frac{2}{s} + 1 = \frac{2+s}{s}$$

$$6X_1(s) + sX_2(s) = -\frac{1}{s^2} - 1 = -\frac{s^2+1}{s^2}.$$

Now we need to solve this for one of the transforms. We'll do this by multiplying the top equation by s and the bottom by –3 and then adding. This gives,

$$\left(s^2 - 3s - 18\right)X_1(s) = 2 + 5 + \frac{3s^2+3}{s^2}.$$

Solving for X_1 gives,

$$X_1(s) = \frac{s^2 + 5s^2 + 3}{s^2(s+3)(s-6)}.$$

Partial fractioning gives,

$$X_1(s) = \frac{1}{108}\left(\frac{133}{s-6} - \frac{28}{s+3} + \frac{3}{s} - \frac{18}{s^2}\right).$$

Taking the inverse transform gives us the first solution,

$$X_1(t) = \frac{1}{108}(133e^{6t} - 28e^{-3t} + 3 - 18t).$$

Now to find the second solution we could go back up and eliminate X_1 to find the transform for X_2 and sometimes we would need to do that. However, in this case notice that the second differential equation is,

$$x_2' = -6x_1 - t \quad \Rightarrow \quad x_2 = \int -6x_1 - t dt.$$

So, plugging the first solution in and integrating gives,

$$\begin{aligned} x_2(t) &= -\frac{1}{18}\int 133e^{6t} - 28e^{-3t} + 3dt \\ &= \frac{1}{108}(133e^{6t} + 56e^{-3t} + 18t) + c. \end{aligned}$$

Now, reapplying the second initial condition to get the constant of integration gives

$$-1 = -\frac{1}{108}(133 + 56) + c \quad \Rightarrow \quad c = \frac{3}{4}.$$

The second solution is then,

$$x_2(t) = -\frac{1}{108}(133e^{6t} + 56e^{-3t} + 18t - 81).$$

So, putting all this together gives the solution to the system as,

$$x_1(t) = \frac{1}{108}(133e^{6t} - 28e^{-3t} + 3 - 81t)$$

$$x_2(t) = -\frac{1}{108}(133e^{6t} + 56e^{-3t} + 18 - 81).$$

Compared to the last section the work here wasn't too bad. That won't always be the case of course, but you can see that using Laplace transforms to solve systems isn't too bad in at least some cases.

MODELING

In this section we're going to go back and revisit the idea of modeling only this time we're going to look at it in light of the fact that we now know how to solve systems of differential equations.

We're not actually going to be solving any differential equations in this section. Instead we'll just be setting up a couple of problems that are extensions of some of the work that we've done in earlier modeling sections whether it is the first order modeling or the vibrations work we did in the second order chapter.

Almost all of the systems that we'll be setting up here will be nonhomogeneous systems (which we only briefly looked at), will be nonlinear (which we didn't look at) and/or will involve systems with more than two differential equations (which we didn't look at, although *most* of what we do know will still be true).

MIXING PROBLEMS

Let's start things by looking at a mixing problem. The last time we saw these was back in the first order chapter. In those problems we had a tank of liquid with some type of contaminate dissolved in it. Liquid, possibly with more contaminate dissolved in it, entered the tank and liquid left the tank. In this situation we want to extend things out to the following situation.

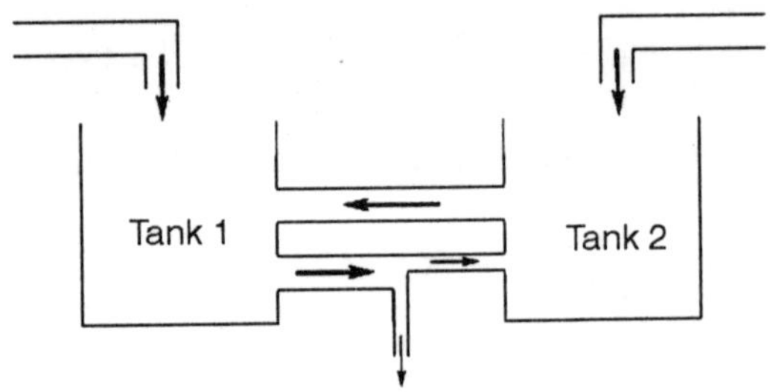

We'll now have two tanks that are interconnected with liquid potentially entering both and with an exit for some of the liquid if we need it (as illustrated by the lower connection). For this situation we're going to make the following assumptions.

1. The inflow and outflow from each tank are equal, or in other words the volume in each tank is constant. When we worked with a single tank we didn't need to worry about this, but here if we don't well end up with a system with nonconstant coefficients and those can be quite difficult to solve.
2. The concentration of the contaminate in each tank is the same at each point in the tank. In reality we know that this won't be true but without this assumption we'd need to deal with partial differential equations.
3. The concentration of contaminate in the outflow from tank 1 (the lower connection in the figure above) is the same as the concentration in tank 1. Likewise, the concentration of contaminate in the outflow from tank 2 (the upper connection) is the same as the concentration in tank 2.
4. The outflow from tank 1 is split and only some of the liquid exiting tank 1 actually reaches tank 2. The remainder exits the system completely. Note that if we don't want any liquid to completely exit the system we can think of the exit as having a value that is turned off. Also note that we could just as easily done the same thing for the outflow from tank 2 if we'd wanted to.

Let's take a look at a quick example.

Example: Two 1000 liter tanks are with salt water. Tank 1 contains 800 liters of water initially containing 20 grams of salt dissolved in it and tank 2 contains 1000 liters of water and initially has 80 grams of salt dissolved in it. Salt water with a concentration of gram/ liter of salt enters tank 1 at a rate of 4 liters/hour.

Fresh water enters tank 2 at a rate of 7 liters/hour. Through a connecting pipe water flows from tank 2 into tank 1 at a rate of 10 liters/hour. Through a different connecting pipe 14 liters/hour flows out of tank 1 and 11 liters/hour are drained out of the pipe (and hence out of the system completely) and only 3 liters/hour flows back into tank 2. Set up the system that will give the amount of salt in each tank at any given time.

Solution: Okay, let $Q_1(t)$and $Q_2(t)$be the amount of salt in tank 1 and tank 2 at any time t respectively. Now all we need to do is set up a differential equation for each tank just as we did back when we had a single tank. The only difference is that we now need to deal with the fact that we've got a second inflow to each tank and the concentration of the second inflow will be the concentration of the other tank.

Recall that the basic differential equation is the rate of change of salt (Q') equals the rate at which salt enters minus the rate at salt leaves. Each entering/leaving rate is found by multiplying the flow rate times the concentration.

Here is the differential equation for tank 1.

$$\begin{aligned} Q_1' &= (4)\left(\frac{1}{2}\right)+(10)\left(\frac{Q_2}{1000}\right)-(14)\left(\frac{Q_1}{800}\right) \qquad Q_1(0)=20 \\ &= 2+\frac{Q_2}{100}-\frac{7Q_1}{400}. \end{aligned}$$

In this differential equation the first pair of numbers is the salt entering from the external inflow. The second set of numbers is the salt that entering into the tank from the water flowing in from tank 2. The third set is the salt leaving tank as water flows out.

Here's the second differential equation.

$$\begin{aligned} Q_2' &= (7)(0)+(3)\left(\frac{Q_1}{800}\right)-(10)\left(\frac{Q_2}{1000}\right) \qquad Q_2(0)=80 \\ &= \frac{3Q_1}{800}-\frac{Q_2}{100}. \end{aligned}$$

Note that because the external inflow into tank 2 is fresh water the concentration of salt in this is zero.

In summary here is the system we'd need to solve,

$$\begin{aligned} Q_1' &= 2+\frac{Q_2}{100}-\frac{7Q_1}{400} \qquad Q_1(0)=20 \\ Q_2' &= \frac{3Q_1}{800}-\frac{Q_2}{100} \qquad Q_2(0)=80. \end{aligned}$$

This is a nonhomogeneous system because of the first term in the first differential equation. If we had fresh water flowing into both of these we would in fact have a homogeneous system.

POPULATION

The next type of problem to look at is the population problem. Back in the first order modeling section we looked at some population problems. In those problems we looked at a single population and often included some form of predation. The problem in that section was we assumed that the amount of predation would be constant. This however clearly won't be the case in most situations. The amount of predation will depend upon the population of the predators and the population of the predators will depend, as least partially, upon the population of the prey.

So, in order to more accurately (well at least more accurate than what we originally did) we really need to set up a model that will cover both populations, both the predator and the prey. These types of problems are usually called predator-prey problems. Here are the assumptions that we'll make when we build up this model.

1. The prey will grow at a rate that is proportional to its current population if there are no predators.
2. The population of predators will decrease at a rate proportional to its current population if there is no prey.
3. The number of encounters between predator and prey will be proportional to the product of the populations.
4. Each encounter between the predator and prey will increase the population of the predator and decrease the population of the prey.

So, given these assumptions let's write down the system for this case.

Example: Write down the system of differential equations for the population of predators and prey using the assumptions above.

Solution: We'll start off by letting x represent the population of the predators and y represent the population of the prey.

Now, the first assumption tells us that, in the absence of predators, the prey will grow at a rate of $ay$where $a > 0$. Likewise the second assumption tells us that, in the absence of prey, the predators will decrease at a rate of $-bx$ where $a > 0$.

Next, the third and fourth assumptions tell us how the population is affected by encounters between predators and prey. So, with each encounter the population of the predators will increase at a rate of and the population of the prey will decrease at a rate of $-\beta xy$where $\alpha > 0$and $\beta > 0$.

Putting all of this together we arrive at the following system.

$$x' = -bx + \alpha\ xy = x(\alpha y - b)$$
$$y' = -ay - \beta xy = y(a - \beta x)$$

Note that this is a nonlinear system and we've not (nor will we here) discuss how to solve this kind of system. We simply wanted to give a "better" model for some population problems and to point out that not all systems will be nice and simple linear systems.

MECHANICAL VIBRATIONS

When we first looked at mechanical vibrations we looked at a single mass hanging on a spring with the possibility of both a damper and/or an external force acting on the mass. Here we want to look at the following situation.

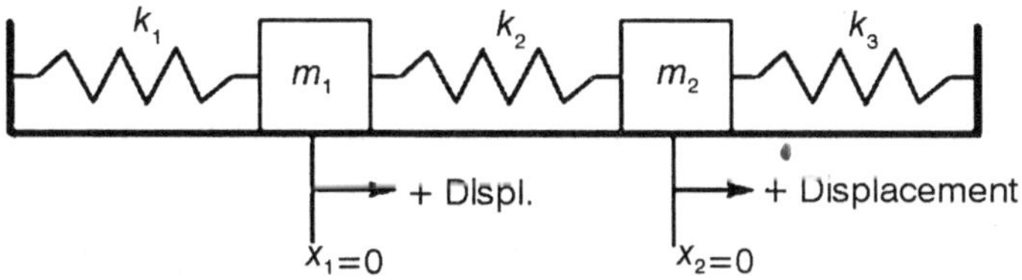

In the figure above we are assuming that the system is at rest. In other words all three springs are currently at their natural lengths and are not exerting any forces on either of the two masses and that there are no currently any external forces acting on either mass.

We will use the following assumptions about this situation once we start the system in motion.

1. x_1 will measure the displacement of mass m_1 from its equilibrium (*i.e.* resting) position and m_2 will measure the displacement of mass from its equilibrium position.
2. As noted in the figure above all displacement will be assumed to be positive if it is to the right of equilibrium position and negative if to the left of the equilibrium position.
3. All forces acting to the right are positive forces and all forces acting to the left are negative forces.
4. The spring constants, k_1, k_2, and k_3, are all positive and may or may not be the same value.
5. The surface that the system is sitting on is frictionless and so the mass of each of the objects will not affect the system in any way.

Before writing down the system for this case recall that the force exerted by the spring on the each mass is the spring constant times the amount that the spring has been compressed or stretched and we'll need to be careful with signs to make sure that the force is acting in the correct direction.

Example: Write down the system of differential equations for the spring and mass system above.

Solution: To help us out let's first take a quick look at a situation in which both of the masses have been moved. This is shown below.

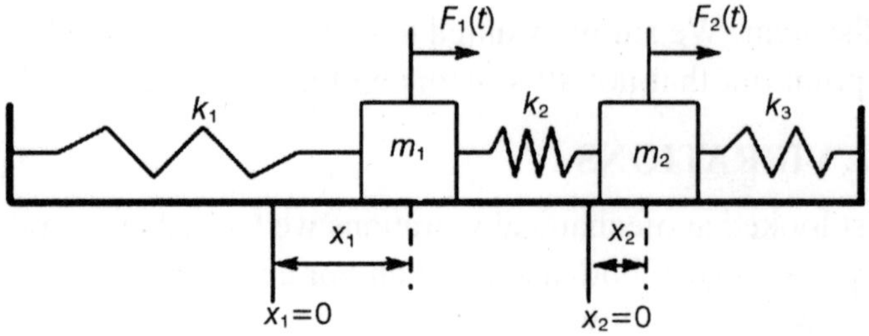

Before proceeding let's note that this is only a representation of a typical case, but most definitely not all possible cases.

In this case we're assuming that both $x_1$and $x_2$are positive and that $x_2 - x_1 < 0$, or in other words, both masses have been moved to the right of their respective equilibrium points and that $m_1$has been moved farther than m_2. So, under these assumption on $x_1$and $x_2$we know that the spring on the left (with spring constant k_1) has been stretched past it's natural length while the middle spring (spring constant k_2) and the right spring (spring constant k_3) are both under compression.

Also, we've shown the external forces, $F_1(t)$and F_2 (t), as present and acting in the positive direction. They do not, in practice, need to be present in every situation in which case we will assume that $F_1(t) = 0$and/or $F_1(t) = 0$. Likewise, if the forces are in fact acting in the negative direction we will then assume that $F_1(t) < 0$and/or $F_2(t) < 0$.

Before proceeding we need to talk a little bit about how the middle spring will behave as the masses move. Here are all the possibilities that we can have and the affect each will have on $x_2 - x_1$. Note that in each case the amount of compression/stretch in the spring is given by $|x_2 - x_1|$although we won't be using the absolute value bars when we set up the differential equations.

1. If both mass move the same amount in the same direction then the middle spring will not have changed length and we'll have $x_2 - x_1 = 0$.
2. If both masses move in the positive direction then the sign of $x_2 - x_1$will tell us which has moved more. If moves more than then the spring will be in compression and $x_2 - x_1 < 0$. Likewise, if $m_2$moves more than $m_1$then the spring will have been stretched and $x_2 - x_1 < 0$.
3. If both masses move in the negative direction we'll have pretty much the opposite behavior. If $m_1$moves more than $m_2$then the spring will have been stretched and $x_2 – x_1 < 0$ Likewise, if $m_2$moves more than $m_1$then the spring will be in compression and $x_2 – x_1 < 0$.

4. If m_1 moves in the positive direction and m_2 moves in the negative direction then the spring will be in compression and $x_2 - x_1 < 0$.
5. Finally, if m_1 moves in the negative direction and m_2 moves in the positive direction then the spring will have been stretched and $x_2 - x_1 < 0$.

Now, we'll use the figure above to help us develop the differential equations (the figure corresponds to case 2 above...) and then make sure that they will also hold for the other cases as well.

Let's start off by getting the differential equation for the forces acting on m_1. Here is a quick sketch of the forces acting on m_1 for the figure above.

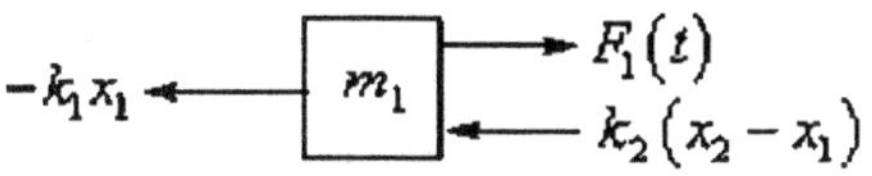

In this case $x_1 > 0$ and so the first spring has been stretched and so will exert a negative (*i.e.* to the left) force on the mass. The force from the first spring is then $-k_1x_1$ and the "-" is needed because the force is negative but both k_1 and x_1 are positive.

Next, because we're assuming that m_1 has moved more than m_2 and both have moved in the positive direction we also know that $x_2 - x_1 < 0$. Because m_1 has moved more than m_2 we know that the second spring will be under compression and so the force should be acting in the negative direction on m_1 and so the force will be $k_2(x_2 - x_1)$. Note that because k_2 is positive and $x_2 - x_1$ is negative this force will have the correct sign (*i.e.* negative).

The differential equation for m_1 is then,

$$m_1 x_1' = -k_1 x_1 + k_2(x_2 - x_1) + F_1(t).$$

Note that this will also hold for all the other cases. If m_1 has been moved in the negative direction the force form the spring on the right that acts on the mass will be positive and $-k_1x_1$ will be a positive quantity in this case. Next, if the middle is has been stretched (*i.e.* $x_2 - x_1 > 0$) then the force from this spring on m_1 will be in the positive direction and $k_2(x_2 - x_1)$ will be a positive quantity in this case. Therefore, this differential equation holds for all cases not just the one we illustrated at the start of this problem.

Let's now write down the differential equation for all the forces that are acting on m_2. Here is a sketch of the forces acting on this mass for the situation sketched out in the figure above.

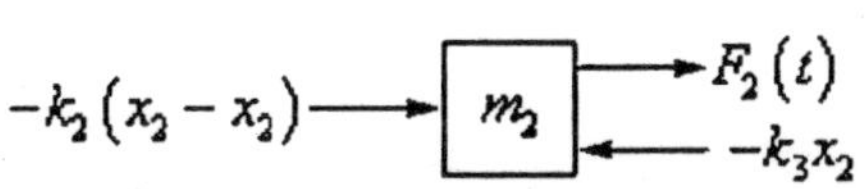

In this case x_2 is positive and so the spring on the right is under compression and will exert a negative force on m_2 and so this force should be $-k_3x_2$, where the "-" is required because both and are positive. Also, the middle spring is still under compression but the force that it exerts on this mass is now a positive force, unlike in the case of m_1, and so is given by $-k_2(x_2-x_1)$. The "-" on this force is required because x_2-x_1 is negative and the force must be positive.—

The differential equation for m_2 is then,

$$m_2 x_2' = -k_3 x_2 - k_2(x_2 - x_1) + F_2(t).$$

We'll leave it to you to verify that this differential equation does in fact hold for all the other cases.

Putting all of this together and doing a little rewriting will then give the following system of differential equations for this situation.

$$m_1 x_1' = -(k_1 - k_2)x_1 + k_2 x_2 F_1(t)$$

$$m_2 x_2' = -k_2 x_1 - (k_2 - k_3)x_2 + F_2(t).$$

This is a system to two linear second order differential equations that may or may not be nonhomogeneous depending whether there are any external forces, $F_1(t)$ and $F_2(t)$, acting on the masses.

We have not talked about how to solve systems of second order differential equations. However, it can be converted to a system of first order differential equations as the next example shows and in many cases we could solve that.

Example: Convert the system from the previous example to a system of 1st order differential equations.

Solution: This isn't too hard to do. Recall that we did this for single higher order differential equations earlier in the chapter when we first started to look at systems. To convert this to a system of first order differential equations we can make the following definitions.

$$u_1 = x_1 \qquad u_2 = x_1' \qquad u_3 = x_2 \qquad u_4 = x_2'.$$

We can then convert each of the differential equations as we did earlier in the chapter.

$$u_1' = x_1' = u_2$$

$$u_2' = x_1^n = \frac{1}{m_1}\left(-(k_1 + k_2)x_1 + k_2 x_2 + F_1(t)\right) = \frac{1}{m_1}(-(k_1 + k_2)u_1 + k_2 u_3 + F_1(t))$$

$$u_3' = x_2' = u_4$$

$$u_4' = x_2^n = \frac{1}{m_2}\left(k_2 x_1 - (k_2 + k_3)x_2 + F_2(t)\right) = \frac{1}{m_2}(k_2 u_1 - (k_2 + k_3)u_3 + F_2(t))$$

Eliminating the "middle" step we get the following system of first order differential equations.

$$u_1' = u_2$$

$$u_2' = \frac{1}{m_1}\left(-(k_1 + k_2)u_1 + k_2 u_3 + F_1(t)\right)$$

$$u_2' = u_4$$

$$u_4' = \frac{1}{m_2}\left(k_2 u_1 - (k_2 + k_3)u_3 + F_2(t)\right).$$

The matrix form of this system would be,

$$\vec{u}' \begin{bmatrix} 0 & 1 & 0 & 0 \\ \frac{-(k_1 + k_2)}{m_1} & 0 & \frac{k_2}{m_1} & 0 \\ 0 & 0 & 0 & 1 \\ \frac{k_2}{m_2} & 0 & \frac{-(k_1 + k_3)}{m_2} & 0 \end{bmatrix} \vec{u} + \begin{pmatrix} 0 \\ \frac{F_1(0)}{m_1} \\ 0 \\ \frac{F_2(0)}{m_2} \end{pmatrix} \qquad \text{where, } \vec{u} = \begin{pmatrix} u_1 \\ u_2 \\ u_3 \\ u_4 \end{pmatrix}.$$

While we never discussed how to solve systems of more than two linear first order differential equations we know most of what we need to solve this.

In an earlier section we discussed briefly solving nonhomogeneous systems and all of that information is still valid here.

For the homogenous system, that we'd still need to solve for the general solution to the nonhomogeneous system, we know most of what we need to know in order to solve this. The only issues that we haven't dealt with are what to do with repeated complex eigenvalues (which are now a possibility) and what to do with eigenvalues of multiplicity greater than 2 (which are again now a possibility).

Chapter 7

Periodic and Orthogonal Functions

This is going to be a short section. We just need to have a brief discussion about a couple of ideas that we'll be dealing with on occasion as we move into the next topic of this chapter.

PERIODIC FUNCTION

The first topic we need to discuss is that of a periodic function. A function is said to be periodic with period T if the following is true,

$$f(x+T) = f(x) \qquad \text{for all } x.$$

The following is a nice little fact about periodic functions.

FACT 1

If f and g are both periodic functions with period T then so is and fg.

This is easy enough to prove so let's do that.

$$(f+g)(x+T) = f(x+T) + g(x+T) = f(x) + g(x) = (f+g)(x)$$

$$(fg)(x+T) = f(x+T)g(x+T) = f(x)g(x) = (fg)(x).$$

The two periodic functions that most of us are familiar are sine and cosine and in fact we'll be using these two functions regularly in the remaining sections of this chapter. So, having said that let's close off this discussion of periodic functions with the following fact,

FACT 2

sin (ω x) and cos (ω x)are periodic functions with period $T = \dfrac{2\pi}{\omega}$.

EVEN AND ODD FUNCTIONS

The next quick idea that we need to discuss is that of even and odd functions.

Recall that a function is said to be even if,

$$f(-x) = f(x)$$

and a function is said to be odd if,

$$f(-x) = -f(x).$$

The standard examples of even functions are $f(x) = x^2$and $g(x) = \cos(x)$while the standard examples of odd functions are $f(x) = x^3$and $g(x) = \sin(x)$. The following fact about certain integrals of even/odd functions will be useful in some of our work.

FACT 3

1. If $f(x)$ is an even function then,

$$\int_{-L}^{L} f(x)dx = 2\int_{0}^{L} f(x)dx.$$

2. If is an odd function then,

$$\int_{-L}^{L} f(x)dx = 0.$$

Note that this fact is only valid on a "symmetric" interval, *i.e.* an interval in the form $[-L, L]$. If we aren't integrating on a "symmetric" interval then the fact may or may not be true.

ORTHOGONAL FUNCTIONS

The final topic that we need to discuss here is that of orthogonal functions. This idea will be integral to what we'll be doing in the remainder of this chapter and in the next chapter as we discuss one of the basic solution methods for partial differential equations.

Let's first get the definition of orthogonal functions out of the way.

1. Two non-zero functions, $f(x)$and $g(x)$, are said to be orthogonal on $a \le x \le b$if,

$$\int_{a}^{b} f(x)g(x)dx = 0.$$

2. A set of non-zero functions, $\{f_i(x)\}$, is said to be mutually orthogonal on $a \le x \le b$(or just an orthogonal set if we're being lazy) if $f_i(x)$and $f_j(x)$are orthogonal for every $i \ne j$. In other words,-

$$\int_{a}^{b} f_i(x)f_j(x)dx = \begin{cases} 0 & i \ne j \\ c > 0 & i = j \end{cases}.$$

Note that in the case of $i = j$for the second definition we know that we'll get a positive value from the integral because,

$$\int_{a}^{b} f_i(x)f_i(x)dx = \int_{a}^{b} [f_i(x)]^2 dx > 0.$$

Recall that when we integrate a positive function we know the result will be positive as well.

Also note that the non-zero requirement is important because otherwise the integral would be trivially zero regardless of the other function we were using.

Before we work some examples there are a nice set of trig formulas that we'll need to help us with some of the integrals.

$$\sin\alpha\cos\beta = \frac{1}{2}[\sin(\alpha-\beta)+\sin(\alpha+\beta)]$$

$$\sin\alpha\cos\beta = \frac{1}{2}[\sin(\alpha-\beta)-\sin(\alpha+\beta)]$$

$$\cos\alpha\cos\beta = \frac{1}{2}[\cos(\alpha-\beta)+\cos(\alpha+\beta)].$$

Now let's work some examples that we'll need over the course of the next couple of sections.

Example: Show that $\left\{\cos\left(\frac{n\pi x}{L}\right)\right\}_{n=0}^{\infty}$ is mutually orthogonal on $-L \le x \le L$.

Solution: This is not too difficult to do. All we really need to do is evaluate the following integral.

$$\int_{-L}^{L} \cos\left(\frac{n\pi x}{L}\right)\cos\left(\frac{m\pi x}{L}\right)dx.$$

Before we start evaluating this integral let's notice that the integrand is the product of two even functions and so must also be even. This means that we can use Fact 3 above to write the integral as,

$$\int_{-L}^{L} \cos\left(\frac{n\pi x}{L}\right)\cos\left(\frac{m\pi x}{L}\right)dx = 2\int_{0}^{L} \cos\left(\frac{n\pi x}{L}\right)\cos\left(\frac{m\pi x}{L}\right)dx.$$

There are two reasons for doing this. First having a limit of zero will often make the evaluation step a little easier and that will be the case here. We'll discuss the second reason after we're done with the example.

Now, in order to do this integral we'll actually need to consider three cases.

$\boldsymbol{n = m = 0}$

In this case the integral is very easy and is,

$$\int_{-L}^{L} dx = 2\int_{0}^{L} dx = 2L.$$

$n = m \neq 0$

This integral is a little harder than the first case, but not by much (provided we recall a simple trig formula). The integral for this case is,

$$\int_{-L}^{L} \cos^2\left(\frac{n\pi x}{L}\right)dx = 2\int_{0}^{L} \cos^2\left(\frac{n\pi x}{L}\right)dx = \int_{0}^{L} 1+\cos\left(\frac{n\pi x}{L}\right)dx$$

$$= \left(x + \frac{L}{2n\pi}\sin\left(\frac{2n\pi x}{L}\right)\right)\Bigg|_0^L = L + \frac{L}{2n\pi}\sin(2n\pi).$$

Now, at the point we need to recall that n is an integer and so $\sin(2n\pi) = 0$and our final value for the is,

$$\int_{-L}^{L} \cos^2\left(\frac{n\pi x}{L}\right)dx = 2\int_{0}^{L} \cos^2\left(\frac{n\pi x}{L}\right)dx = L.$$

The first two cases are really just showing that if $n = m$the integral is not zero (as it shouldn't be) and depending upon the value of n (and hence m) we get different values of the integral. Now we need to do the third case and this, in some ways, is the important case since we must get zero out of this integral in order to know that the set is an orthogonal set. So, let's take care of the final case.

$n \neq m$

This integral is the "messiest" of the three that we've had to do here. Let's just start off by writing the integral down.

$$\int_{-L}^{L} \cos\left(\frac{n\pi x}{L}\right)\cos\left(\frac{m\pi x}{L}\right)dx = 2\int_{0}^{L} \cos\left(\frac{n\pi x}{L}\right)\cos\left(\frac{m\pi x}{L}\right)dx.$$

In this case we can't combine/simplify as we did in the previous two cases. We can however, acknowledge that we've got a product of two cosines with different arguments and so we can use one of the trig formulas above to break up the product as follows,

$$\int_{-L}^{L} \cos\left(\frac{n\pi x}{L}\right)\cos\left(\frac{m\pi x}{L}\right)dx = 2\int_{0}^{L} \cos\left(\frac{n\pi x}{L}\right)\cos\left(\frac{m\pi x}{L}\right)dx$$

$$= \int_{0}^{L} \cos\left(\frac{(n-m)\pi x}{L}\right) + \cos\left(\frac{(n+m)\pi x}{L}\right)dx$$

$$= \left[\frac{L}{(n-m)\pi}\sin\left(\frac{(n-m)\pi x}{L}\right) + \frac{L}{(n+m)\pi}\sin\left(\frac{(n+m)\pi x}{L}\right)\right]_0^L$$

$$= \frac{L}{(n-m)\pi}\sin((n-m)\pi) + \frac{L}{(n+m)\pi}\sin((n+m)\pi).$$

Now, we know that n and m are both integers and so $n - m$and $n + m$are also integers and so both of the sines above must be zero and all together we get,

$$\int_{-L}^{L} \cos\left(\frac{n\pi x}{L}\right)\cos\left(\frac{m\pi x}{L}\right)dx = 2\int_{0}^{L} \cos\left(\frac{n\pi x}{L}\right)\cos\left(\frac{m\pi x}{L}\right)dx = 0.$$

So, we've shown that if $n \neq m$the integral is zero and if $n = m$the value of the integral is a positive constant and so the set is mutually orthogonal.

In all of the work above we kept both forms of the integral at every step. Let's discuss why we did this a little bit. By keeping both forms of the integral around we were able to show that not only is $\left\{\cos\left(\frac{n\pi x}{L}\right)\right\}_{n=0}^{\infty}$ mutually orthogonal on $-L \leq x \leq L$but it is also mutually orthogonal on. The only difference is the value of the integral when and we can get those values from the work above.

Let's take a look at another example.

Example: Show that $\left\{\sin\left(\frac{n\pi x}{L}\right)\right\}_{n=1}^{\infty}$ is mutually orthogonal on $-L \leq x \leq L$and on $0 \leq x \leq L.$

Solution: First we'll acknowledge from the start this time that we'll be showing orthogonality on both of the intervals. Second, we need to start this set at $n = 1$because if we used $n = 0$the first function would be zero and we don't want the zero function to show up on our list.

As with the first example all we really need to do is evaluate the following integral.

$$\int_{-L}^{L} \sin\left(\frac{n\pi x}{L}\right)\sin\left(\frac{m\pi x}{L}\right)dx.$$

In this case integrand is the product of two odd functions and so must be even. This means that we can again use Fact 3 above to write the integral as,

$$\int_{-L}^{L} \sin\left(\frac{n\pi x}{L}\right)\sin\left(\frac{m\pi x}{L}\right)dx = 2\int_{-L}^{L} \sin\left(\frac{n\pi x}{L}\right)\sin\left(\frac{m\pi x}{L}\right)dx.$$

We only have two cases to do for the integral here.

n* = *m

Not much to this integral. It's pretty similar to the previous examples second case.

$$\int_{-L}^{L} \sin^2\left(\frac{n\pi x}{L}\right)dx = 2\int_{0}^{L} \sin^2\left(\frac{m\pi x}{L}\right)dx \int_{0}^{L} 1 - \cos\left(\frac{2n\pi x}{L}\right)dx$$

$$= \left(x - \frac{L}{2n\pi}\sin\left(\frac{2n\pi x}{L}\right)\right)\Bigg|_{0}^{L} = L - \frac{L}{2n\pi}\sin(2n\pi) = L.$$

Summarizing up we get,

$$\int_{-L}^{L} \sin^2\left(\frac{n\pi x}{L}\right)dx = 2\int_0^L \sin^2\left(\frac{n\pi x}{L}\right)dx = L.$$

n ≠ m

As with the previous example this can be a little messier but it is also nearly identical to the third case from the previous example so we'll not show a lot of the work.

$$\begin{aligned}\int_{-L}^{L} \sin\left(\frac{n\pi x}{L}\right)\sin\left(\frac{m\pi x}{L}\right)dx &= 2\int_0^L \sin\left(\frac{n\pi x}{L}\right)\sin\left(\frac{m\pi x}{L}\right)dx \\ &= \int_0^L \cos\left(\frac{(n-m)\pi x}{L}\right)-\cos\left(\frac{(n+m)\pi x}{L}\right)dx \\ &= \left[\frac{L}{(n-m)\pi}\sin\left(\frac{(n-m)\pi x}{L}\right)-\frac{L}{(n-m)\pi}\sin\left(\frac{(n+m)\pi x}{L}\right)\right]_0^L \\ &= \frac{L}{(n-m)\pi}\sin((n-m)\pi)-\frac{L}{(n-m)\pi}\sin((n+m)\pi).\end{aligned}$$

As with the previous example we know that n and m are both integers a and so both of the sines above must be zero and all together we get,

$$\int_{-L}^{L} \sin\left(\frac{n\pi x}{L}\right)\sin\left(\frac{m\pi x}{L}\right)dx = 2\int_0^L \sin\left(\frac{n\pi x}{L}\right)\sin\left(\frac{m\pi x}{L}\right)dx = 0.$$

So, we've shown that if $n \neq 0$the integral is zero and if $n = m$the value of the integral is a positive constant and so the set is mutually orthogonal.

We've now shown that $\left\{\sin\left(\frac{n\pi x}{L}\right)\right\}_{n=1}^{\infty}$ mutually orthogonal on $-L \leq x \leq L$and on $0 \leq x \leq L$.

We need to work one more example in this section.

Example: Show that $\left\{\sin\left(\frac{n\pi x}{L}\right)\right\}_{n=1}^{\infty}$ and $\left\{\cos\left(\frac{n\pi x}{L}\right)\right\}_{n=0}^{\infty}$ are mutually orthogonal on $-L \leq x \leq L$.

Solution: This example is a little different from the previous two examples. Here we want to show that together both sets are mutually orthogonal on $-L \leq x \leq L$. To show this we need to show three things. First (and second actually) we need to show that individually each set is mutually orthogonal and we've already done that in the previous two examples.

The third (and only) thing we need to show here is that if we take one function from one set and another function from the other set and we integrate them we'll get zero.

Also, note that this time we really do only want to do the one interval as the two sets, taken together, are not mutually orthogonal on $0 \le x \le L$.

You might want to do the integral on this interval to verify that it won't always be zero.

So, let's take care of the one integral that we need to do here and there isn't a lot to do. Here is the integral.

$$\int_{-L}^{L} \sin\left(\frac{n\pi x}{L}\right)\cos\left(\frac{n\pi x}{L}\right)dx.$$

The integrand in this case is the product of an odd function (the sine) and an even function (the cosine) and so the integrand is an odd function. Therefore, since the integral is on a symmetric interval, *i.e.* $-L \le x \le L$, and so by Fact 3 above we know the integral must be zero or,

$$\int_{-L}^{L} \sin\left(\frac{n\pi x}{L}\right)\cos\left(\frac{m\pi x}{L}\right)dx = 0.$$

So, in previous examples we've shown that on the interval $-L \le x \le L$the two sets are mutually orthogonal individually and here we've shown that integrating a product of a sine and a cosine gives zero.

Therefore, as a combined set they are also mutually orthogonal.

We've now worked three examples here dealing with orthogonality and we should note that these were not just pulled out of the air as random examples to work. In the following sections (and following chapter) we'll need the results from these examples. So, let's summarize those results up here.

1. $\left\{\cos\left(\frac{n\pi x}{L}\right)\right\}_{n=0}^{\infty}$ and $\left\{\sin\left(\frac{n\pi x}{L}\right)\right\}_{n=1}^{\infty}$ are mutually orthogonal on $-L \le x \le L$as individual sets and as a combined set.

2. $\left\{\cos\left(\frac{n\pi x}{L}\right)\right\}_{n=0}^{\infty}$ is mutually orthogonal on $0 \le x \le L$.

3. $\left\{\sin\left(\frac{n\pi x}{L}\right)\right\}_{n=1}^{\infty}$ is mutually orthogonal on $0 \le x \le L$.

We will also be needing the results of the integrals themselves, both on $-L \le x \le L$and on $0 \le x \le L$so let's also summarize those up here as well so we can refer to them when we need to.

1. $$\int_{-L}^{L} \cos\left(\frac{n\pi x}{L}\right)\cos\left(\frac{m\pi x}{L}\right)dx = \begin{cases} 2L & \text{if } n = m = 0 \\ L & \text{if } n = m \neq 0 \\ 0 & \text{if } n \neq m \end{cases}$$

2. $$\int_{0}^{L} \cos\left(\frac{n\pi x}{L}\right)\cos\left(\frac{m\pi x}{L}\right)dx = \begin{cases} L & \text{if } n = m = 0 \\ \frac{L}{2} & \text{if } n = m \neq 0 \\ 0 & \text{if } n \neq m \end{cases}$$

3. $$\int_{-L}^{L} \sin\left(\frac{n\pi x}{L}\right)\sin\left(\frac{m\pi x}{L}\right)dx = \begin{cases} L & \text{if } n = m \\ 0 & \text{if } n \neq m \end{cases}$$

4. $$\int_{0}^{L} \sin\left(\frac{n\pi x}{L}\right)\sin\left(\frac{m\pi x}{L}\right)dx = \begin{cases} \frac{L}{2} & \text{if } n = m \\ 0 & \text{if } n \neq m \end{cases}$$

5. $$\int_{-L}^{L} \sin\left(\frac{n\pi x}{L}\right)\cos\left(\frac{m\pi x}{L}\right)dx = 0.$$

With this summary we'll leave this section and move off into the second major topic of this chapter: Fourier Series.

FOURIER SINE SERIES

In this section we are going to start taking a look at Fourier series. We should point out that this is a subject that can span a whole class and what we'll be doing in this section (as well as the next couple of sections) is intended to be nothing more than a very brief look at the subject. The point here is to do just enough to allow us to do some basic solutions to partial differential equations in the next chapter. There are many topics in the study of Fourier series that we'll not even touch upon here.

So, with that out of the way let's get started, although we're not going to start off with Fourier series. Let's instead think back to our Calculus class where we looked at Taylor Series. With Taylor Series we wrote a series representation of a function, $f(x)$, as a series whose terms were powers of $x - a$for some $x = a$. With some conditions we were able to show that,

$$f(x) = \sum_{n=0}^{\infty} \frac{f^{(n)}(a)}{n!}(x-a)^n$$

and that the series will converge to $f(x)$on $|x-a| < R$for some R that will be dependent upon the function itself.

There is nothing wrong with this, but it does require that derivatives of all orders exist at $x = a\,x = ax = a$. Or in other words $f^{(n)}(a)$exists for $n = 0,1,2,3,...$Also for some functions the value of R may end up being quite small.

These two issues (along with a couple of others) mean that this is not always the best way or writing a series representation for a function. In many cases it works fine and there will be no reason to need a different kind of series. There are times however where another type of series is either preferable or required.

We're going to build up an alternative series representation for a function over the course of the next couple of sections. The ultimate goal for the rest of this chapter will be to write down a series representation for a function in terms of sines and cosines.

We'll start things off by assuming that the function, $f(x)$, we want to write a series representation for is an odd function (*i.e.* $f(-x) = -f(x)$). Because $f(x)$is odd it makes some sense that should be able to write a series representation for this in terms of sines only (since they are also odd functions).

What we'll try to do here is write $f(x)$as the following series representation, called a Fourier sine series, on $-L \le x \le L$.

$$\sum_{n=1}^{\infty} B_n \sin\left(\frac{n\pi x}{L}\right).$$

There are a couple of issues to note here. First, at this point, we are going to assume that the series representation will converge to $f(x)$on $-L \le x \le L$. We will be looking into whether or not it will actually converge in a later section. However, assuming that the series does converge to $f(x)$it is interesting to note that, unlike Taylor Series, this representation will always converge on the same interval and that the interval does not depend upon the function.

Second, the series representation will not involve powers of sine (again contrasting this with Taylor Series) but instead will involve sines with different arguments.

Finally, the argument of the sines, $\frac{n\pi x}{L}$, may seem like an odd choice that was arbitrarily chosen and in some ways it was. For Fourier sine series the argument doesn't have to necessarily be this but there are several reasons for the choice here. First, this is the argument that will naturally arise in the next chapter when we use Fourier series (in general and not necessarily Fourier sine series) to help us solve some basic partial differential equations.

The next reason for using this argument is that fact that the set of functions that we chose to work with, $\left\{\sin\left(\frac{n\pi x}{L}\right)\right\}_{n=1}^{\infty}$ in this case, need to be orthogonal on the given interval, $-L \le x \le L$in this case, and note that in the last section we showed that in fact they are. In other words, the choice of functions we're going to be working with and the interval we're working on will be tied together in some way. We can use a different argument, but will need to also choose an interval on which we can prove that the sines (with the different argument) are orthogonal.

So, let's start off by assuming that given an odd function, $f(x)$, we can in fact find a Fourier sine series, of the form given above, to represent the function on $-L \le x \le L$. This means we will have,

$$f(x)=\sum_{n=1}^{\infty} B_n \sin\left(\frac{n\pi x}{L}\right).$$

As noted above we'll discuss whether or not this even can be done and if the series representation does in fact converge to the function in later section. At this point we're simply going to assume that it can be done. The question now is how to determine the coefficients, B_n, in the series.

Let's start with the series above and multiply both sides by $\sin\left(\frac{m\pi x}{L}\right)$ where m is a fixed integer in the range $\{1,2,3,...\}$. In other words we multiply both sides by any of the sines in the set of sines that we're working with here. Doing this gives,

$$f(x)\sin\left(\frac{m\pi x}{L}\right)=\sum_{n=1}^{\infty} B_n \sin\left(\frac{n\pi x}{L}\right)\sin\left(\frac{m\pi x}{L}\right).$$

Now, let's integrate both sides of this from $x = -L$ to $x = L$.

$$\int_{-L}^{L} f(x)\sin\left(\frac{m\pi x}{L}\right)dx=\int_{-L}^{L}\sum_{n=1}^{\infty} B_n \sin\left(\frac{n\pi x}{L}\right)\sin\left(\frac{m\pi x}{L}\right)dx.$$

At this point we've got a small issue to deal with. We know from Calculus that an integral of a finite series (more commonly called a finite sum....) is nothing more than the (finite) sum of the integrals of the pieces. In other words for finite series we can interchange an integral and a series. For infinite series however, we cannot always do this. For some integrals of infinite series we cannot interchange an integral and a series. Luckily enough for us we actually can interchange the integral and the series in this case. Doing this gives and factoring the constant, B_n B_nB_n, out of the integral gives,

$$\begin{aligned}\int_{-L}^{L} f(x)\sin\left(\frac{m\pi x}{L}\right)dx &= \sum_{n=1}^{\infty}\int_{-L}^{L} B_n \sin\left(\frac{n\pi x}{L}\right)\sin\left(\frac{m\pi x}{L}\right)dx \\ &= \sum_{n=1}^{\infty} B_n\int_{-L}^{L} \sin\left(\frac{n\pi x}{L}\right)\sin\left(\frac{m\pi x}{L}\right)dx.\end{aligned}$$

Now, recall from the last section we proved that $\left\{\sin\left(\frac{n\pi x}{L}\right)\right\}_{n=1}^{\infty}$ is orthogonal on $-L \le x \le L$ and that,

$$\int_{-L}^{L} \sin\left(\frac{n\pi x}{L}\right)\sin\left(\frac{n\pi x}{L}\right)dx = \begin{cases} L & \text{if } n = m \\ 0 & \text{if } n \neq m \end{cases}.$$

So, what does this mean for us. As we work through the various values of n in the series and compute the value of the integrals all but one of the integrals will be zero. The only non-zero integral will come when we have $n = m$, in which case the integral has the value of L. Therefore, the only non-zero term in the series will come when we have $n = m$and our equation becomes,

$$\int_{-L}^{L} f(x)\sin\left(\frac{m\pi x}{L}\right)dx = B_m L.$$

Finally all we need to do is divide by L and we know have an equation for each of the coefficients.

$$B_m = \frac{1}{L}\int_{-L}^{L} f(x)\sin\left(\frac{m\pi x}{L}\right)dx \qquad m = 1,2,3,...$$

Next, note that because we're integrating two odd functions the integrand of this integral is even and so we also know that,

$$B_m = \frac{2}{L}\int_{-L}^{L} f(x)\sin\left(\frac{m\pi x}{L}\right)dx \qquad m = 1,2,3,...$$

Summarizing all this work up the Fourier sine series of an odd function $f(x)$on $-L \leq x \leq L$is given by,

$$f(x) = \sum_{n=1}^{\infty} B_n \sin\left(\frac{n\pi x}{L}\right) \qquad B_n = \frac{1}{L}\int_{-L}^{L} f(x)\sin\left(\frac{n\pi x}{L}\right)dx \qquad m = 1,2,3,...$$

$$= \frac{2}{L}\int_{-L}^{L} f(x)\sin\left(\frac{n\pi x}{L}\right)dx \qquad m = 1,2,3,...$$

Let's take a quick look at an example.

Example: Find the Fourier sine series for $f(x) = x$on $-L \leq x \leq L$.

Solution: First note that the function we're working with is in fact an odd function and so this is something we can do. There really isn't much to do here other than to compute the coefficients for $f(x) = x$.

Here is that work and note that we're going to leave the integration by parts details to you to verify. Don't forget that n, L, and $\pi$are constants!

$$B_n = \frac{2}{L}\int_{0}^{L} x\sin\left(\frac{n\pi x}{L}\right)dx = \frac{2}{L}\left(\frac{L}{n^2\pi^2}\right)\left(L\sin\left(\frac{n\pi x}{L}\right) - n\pi x\cos\left(\frac{n\pi x}{L}\right)\right)\Bigg|_0^L$$

$$= \frac{2}{n^2\pi^2}(L\sin(n\pi) - n\pi L\cos(n\pi)).$$

These integrals can, on occasion, be somewhat messy especially when we use a general L for the endpoints of the interval instead of a specific number.

Now, taking advantage of the fact that n is an integer we know that $\sin(n\pi) = 0$and that $\cos(n\pi) = (-1)^n$. We therefore have,

$$B_n = \frac{2}{n^2\pi^2}\left(-n\pi L(-1)^n\right) = \frac{(-1)^{n+1}}{n\pi} \qquad n = 1, 2, 3, \ldots$$

The Fourier sine series is then,

$$x = \sum_{n=1}^{\infty} \frac{(-1)^{n+1}}{n\pi} \sin\left(\frac{n\pi x}{L}\right) = \frac{2L}{\pi}\sum_{n=1}^{\infty} \frac{(-1)^{n+1}}{n} \sin\left(\frac{n\pi x}{L}\right).$$

At this point we should probably point out that we'll be doing most, if not all, of our work here on a general interval ($-L \le x \le L$ or $0 \le x \le L$) instead of intervals with specific numbers for the endpoints.

There are a couple of reasons for this. First, it gives a much more general formula that will work for any interval of that form which is always nice. Secondly, when we run into this kind of work in the next chapter it will also be on general intervals so we may as well get used to them now.

Now, finding the Fourier sine series of an odd function is fine and good but what if, for some reason, we wanted to find the Fourier sine series for a function that is not odd? To see how to do this we're going to have to make a change. The above work was done on the interval. In the case of a function that is not odd we'll be working on the interval. The reason for this will be made apparent in a bit.

So, we are now going to do is to try to find a series representation for $f(x)$on the interval $0 \le x \le L$that is in the form,

$$\sum_{n=1}^{\infty} B_n \sin\left(\frac{n\pi x}{L}\right)$$

or in other words,

$$f(x) = \sum_{n=1}^{\infty} B_n \sin\left(\frac{n\pi x}{L}\right).$$

As we did with the Fourier sine series on $-L \le x \le L$we are going to assume that the series will in fact converge to $f(x)$and we'll hold off discussing the convergence of the series for a later section. There are two methods of generating formulas for the coefficients, B_n, although we'll see in a bit that they really the same way, just looked at from different perspectives.

The first method is to just ignore the fact that $f(x)$is odd and proceed in the same manner that we did above only this time we'll take advantage of the fact that we proved in the previous section that $\left\{\sin\left(\frac{n\pi x}{L}\right)\right\}_{n=1}^{\infty}$ also forms an orthogonal set on $0 \le x \le L$and that,

$$\int_0^L \sin\left(\frac{n\pi x}{L}\right)\sin\left(\frac{m\pi x}{L}\right)dx = \begin{cases} \frac{L}{2} & \text{if } n=m \\ 0 & \text{if } n \neq m \end{cases}.$$

So, if we do this then all we need to do is multiply both sides of our series by$\left(\frac{m\pi x}{L}\right)$, integral from 0 to L and interchange the integral and series to get,

$$\int_0^L f(x)\sin\left(\frac{m\pi x}{L}\right)dx = \sum_{n=1}^{\infty} B_n \int_0^L \sin\left(\frac{n\pi x}{L}\right)\sin\left(\frac{m\pi x}{L}\right)dx.$$

Now, plugging in for the integral we arrive at,

$$\int_0^L f(x)\sin\left(\frac{n\pi x}{L}\right)dx = B_n\left(\frac{1}{2}\right).$$

Upon solving for the coefficient we arrive at,

$$B_m = \frac{2}{L}\int_0^L f(x)\sin\left(\frac{n\pi x}{L}\right)dx \quad m = 1.2.3,...$$

Note that this is identical to the second form of the coefficients that we arrived at above by assuming $f(x)$was odd and working on the interval $-L \le x \le L$. The fact that we arrived at essentially the same coefficients is not actually all the surprising as we'll see once we've looked the second method of generating the coefficients.

Before we look at the second method of generating the coefficients we need to take a brief look at another concept. Given a function, $f(x)$, we define the odd extension of $f(x)$to be the new function,

$$g(x) = \begin{cases} f(x) & \text{if } 0 \le x \le L \\ -f(x) & \text{if} - L \le x \le 0 \end{cases}.$$

It's pretty easy to see that this is an odd function.

$$g(-x) = -f(-x)) = -f(x) = -g(x) \qquad \text{for } 0 < x < L$$

and we can also know that on $0\le x \le L$we have that $g(x) = f(x)$. Also note that if $f(x)$is already an odd function then we in fact get $g(x) = f(x)$on $-L \le x \le L$.

Let's take a quick look at a couple of odd extensions before we proceed any further.

Example: Sketch the odd extension of each of the given functions.

1. $f(x) = L - x \quad 0 \le x \le L$ on

2. $f(x)=1+x^2$ on $0 \le x \le L$

3 $f(x)=\begin{cases} \frac{1}{2} & \text{if} 0 \le x \le \frac{L}{2} \\ x-\frac{1}{2} & \text{if} \frac{L}{2} \le x \le L \end{cases}$.

Solution: Not much to do with these other than to define the odd extension and then sketch it.

1. $f(x)=L-x$ on $0 \le x \le L$.

Here is the odd extension of this function.

$$g(x) = \begin{cases} f(x) & \text{if } 0 \le x \le L \\ -f(-x) & \text{if } -L \le x \le 0 \end{cases}$$

$$= \begin{cases} L-x & \text{if } 0 \le x \le L \\ -L-x & \text{if } -L \le x \le 0 \end{cases}.$$

Below is the graph of both the function and its odd extension. Note that we've put the "extension" in with a dashed line to make it clear the portion of the function that is being added to allow us to get the odd extension.

2. $f(x)=1+x^2$ on $0 \le x \le L$.

First note that this is clearly an even function. That does not however mean that we can't define the odd extension for it. The odd extension for this function is,

$$g(x) = \begin{cases} f(x) & \text{if } 0 \le x \le L \\ -f(-x) & \text{if} -L \le x \le 0 \end{cases}$$

$$= \begin{cases} 1+x^2 & \text{if } 0 \le x \le L \\ -1-x^2 & \text{if} -L \le x \le 0 \end{cases}.$$

The sketch of the original function and its odd extension are,

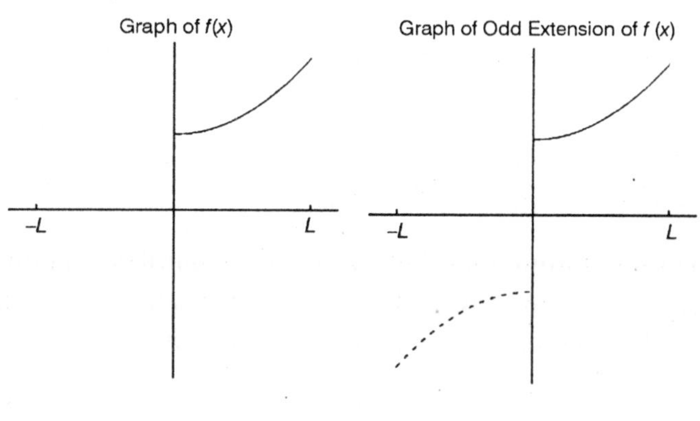

3. $f(x)=\begin{cases} \frac{L}{2} & \text{if } 0 \le x \le \frac{L}{2} \\ x-\frac{L}{2} & \text{if } \le \frac{L}{2} x \le L \end{cases}.$

Let's first write down the odd extension for this function.

$$g(x)=\begin{cases} f(x) & \text{if } 0 \le x \le L \\ -f(-x) & \text{if} -L \le x \le 0 \end{cases} = \begin{cases} x-\frac{L}{2} & \text{if } \frac{L}{2} \le x \le L \\ \frac{L}{2} & \text{if } 0 \le x \le \frac{L}{2} \\ -\frac{L}{2} & \text{if } -\frac{L}{2} \le x \le 0 \\ x+\frac{L}{2} & \text{if } -L \le x \le -\frac{L}{2} \end{cases}.$$

The sketch of the original function and its odd extension are,

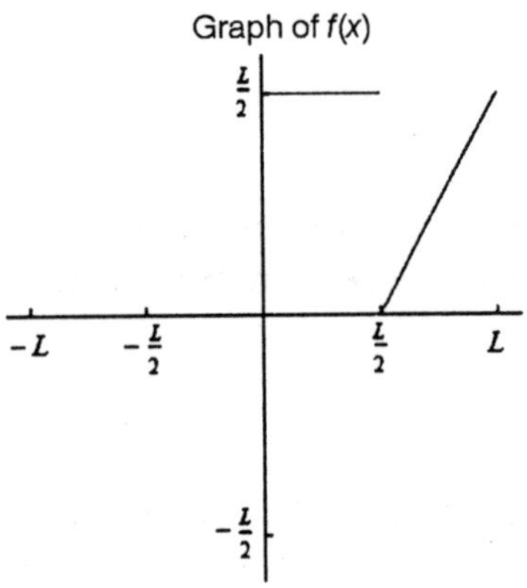

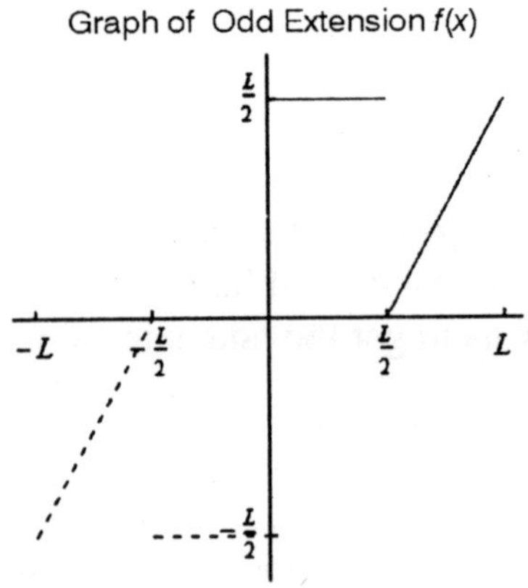

With the definition of the odd extension (and a couple of examples) out of the way we can now take a look at the second method for getting formulas for the coefficients of the Fourier sine series for a function $f(x)$on $0 \le x \le L$. First, given such a function define its odd extension as above. At this point, because $g(x)$is an odd function, we know that on $-L \le x \le L$the Fourier sine series for $g(x)$(and NOT $f(x)$yet) is,

$$g(x)=\sum_{n=1}^{\infty} B_n \sin\left(\frac{n\pi x}{L}\right) \qquad B_n=\frac{2}{L}\int_0^L g(x)\sin\left(\frac{n\pi x}{L}\right)dx \quad n=1,2,3,...$$

However, because we know that $g(x) = f(x)$on $0 \le x \le L$we can also see that as long as we are on we have,

$$f(x)=\sum_{n=1}^{\infty} B_n \sin\left(\frac{n\pi x}{L}\right) \qquad B_n=\frac{2}{L}\int_0^L f(x)\sin\left(\frac{n\pi x}{L}\right)dx \quad n=1,2,3,...$$

So, exactly the same formula for the coefficients regardless of how we arrived at the formula and the second method justifies why they are the same here as they were when we derived them for the Fourier sine series for an odd function.

Now, let's find the Fourier sine series for each of the functions that we looked at in Example.

Note that again we are working on general intervals here instead of specific numbers for the right endpoint to get a more general formula for any interval of this form and because again this is the kind of work we'll be doing in the next chapter.

Also, we'll again be leaving the actually integration details up to you to verify. In most cases it will involve some fairly simple integration by parts complicated by all the constants (n, L, π, etc.) that show up in the integral.

Example: Find the Fourier sine series for $f(x) = L - x$on $0 \le x \le L$.

Solution: There really isn't much to do here other than computing the coefficients so here they are,

$$B_n = \frac{2}{L}\int_0^L f(x)\sin\left(\frac{n\pi x}{L}\right)dx = \frac{2}{L}\int_0^L (L-x)\sin\left(\frac{n\pi x}{L}\right)dx$$

$$= \frac{2}{L}\left(-\frac{L}{n^2\pi^2}\right)\left[L\sin\left(\frac{n\pi x}{L}\right) - n\pi(x-L)\cos\left(\frac{n\pi x}{L}\right)\right]_0^l$$

$$= \frac{2}{L}\left[\frac{L^2}{n^2\pi^2}(n\pi - \sin(n\pi))\right] = \frac{2L}{n\pi}.$$

In the simplification process don't forget that n is an integer.

So, with the coefficients we get the following Fourier sine series for this function.

$$f(x) = \sum_{n=1}^{\infty}\frac{2L}{n\pi}\sin\left(\frac{n\pi x}{L}\right).$$

In the next example it is interesting to note that while we started out this section looking only at odd functions we're now going to be finding the Fourier sine series of an even function on $0 \le x \le L$. Recall however that we're really finding the Fourier sine series of the odd extension of this function and so we're okay.

Example: Find the Fourier sine series for $f(x) = 1 + x^2$on $0 \le x \le L$.

Solution: In this case the coefficients are liable to be somewhat messy given the fact that the integrals will involve integration by parts twice. Here is the work for the coefficients.

$$B_n = \frac{2}{L}\int_0^L f(x)\sin\left(\frac{n\pi x}{L}\right)dx = \frac{2}{L}\int_0^L \left(1+x^2\right)\sin\left(\frac{n\pi x}{L}\right)dx$$

$$= \frac{2}{L}\left(\frac{L}{n^3\pi^3}\right)\left[\left(2L^2 - n^2\left(1+x^2\right)\right)\cos\left(\frac{n\pi x}{L}\right) + 2Ln\pi x\sin\left(\frac{n\pi x}{L}\right)\right]_0^L$$

$$= \frac{2}{L}\left(\frac{L}{n^3\pi^3}\right)\left[\left(2L^2 - n^2\pi^2\left(1+x^2\right)\right)\cos(n\pi) + 2L^2 n\pi x\sin(n\pi) - (2L^2 - n^2\pi^2)\right]$$

$$= \frac{2}{n^3\pi^3}\left[\left(2L^2 - n^2\pi^2\left(1+x^2\right)\right)(-1)^n - 2L^2 + n^2\pi^2\right].$$

As noted above the coefficients are not the most pleasant ones, but there they are. The Fourier sine series for this function is then,

$$f(x) = \sum_{n=1}^{\infty}\frac{2}{n^3\pi^3}\left[\left(2L^2 - n^2\pi^2\left(1+L^2\right)\right)(-1)^n - 2L^2 + n^2\pi^2\right]\sin\left(\frac{n\pi x}{L}\right).$$

In the last example of this section we'll be finding the Fourier sine series of a piecewise function and can definitely complicate the integrals a little but they do show up on occasion and so we need to be able to deal with them.

Example: Find the Fourier sine series for $f(x) = \begin{cases} \frac{L}{2} & \text{if } 0 \le x \le \frac{1}{2} \\ x - \frac{L}{2} & \text{if } \frac{1}{2} \le x \le L \end{cases}$ on $0 \le x \le L$.

Solution: Here is the integral for the coefficients.

$$B_n = \frac{2}{L}\int_0^L f(x)\sin\left(\frac{n\pi x}{L}\right)dx = \frac{2}{L}\left[\int_0^{\frac{L}{2}} f(x)\sin\left(\frac{n\pi x}{L}\right)dx + \int_{\frac{1}{2}}^{L} f(x)\sin\left(\frac{n\pi x}{L}\right)dx\right]$$

$$= \frac{2}{L}\left[\int_0^{\frac{L}{2}} \sin\left(\frac{n\pi x}{L}\right)dx + \int_{\frac{1}{2}}^{L}\left(x - \frac{L}{2}\right)\sin\left(\frac{n\pi x}{L}\right)dx\right].$$

Note that we need to split the integral up because of the piecewise nature of the original function. Let's do the two integrals separately

$$\int_0^{\frac{L}{2}} \sin\left(\frac{n\pi x}{L}\right)dx = -\left(\frac{L}{2}\right)\left(\frac{L}{n\pi}\right)\cos\left(\frac{n\pi x}{L}\right)\Bigg|_0^{\frac{L}{2}} = \frac{L^2}{2n\pi}\left(1 - \cos\left(\frac{2\pi}{L}\right)\right)$$

$$\int_0^{\frac{L}{2}}\left(x - \frac{L}{2}\right)\sin\left(\frac{n\pi x}{L}\right)dx = \frac{L}{n^2\pi^2}\left[L\sin\left(\frac{n\pi L}{2}\right) - n\pi\left(x - \frac{L}{2}\right)\cos\left(\frac{n\pi x}{L}\right)\right]\Bigg|_{\frac{L}{2}}^{L}$$

$$= \frac{L^2}{n^2\pi^2}\left[L\sin(n\pi) - \frac{n\pi L}{2}\cos(n\pi) - L\sin\left(\frac{n\pi}{2}\right)\right]$$

$$= \frac{L}{n^2\pi^2}\left[\frac{n\pi(-1)^n}{2} + \sin\left(\frac{n\pi}{2}\right)\right].$$

Putting all of this together gives,

$$B_n = \frac{2}{L}\int_0^L f(x)\sin\left(\frac{n\pi x}{L}\right)dx = \frac{2}{L}\left(\frac{L^2}{2n\pi}\right)\left[1+(-1)^{n+1} - \cos\left(\frac{n\pi}{2}\right) + \frac{2}{n\pi}\sin\left(\frac{n\pi}{2}\right)\right]$$

$$= \frac{L}{n\pi}\left[1+(-1)^{n+1} - \cos\left(\frac{n\pi}{2}\right) + \frac{2}{n\pi}\sin\left(\frac{n\pi}{2}\right)\right].$$

So, the Fourier sine series for this function is,

$$f(x) = \sum_{n=1}^{\infty}\frac{L}{n\pi}\left[1+(-1)^{n+1} - \cos\left(\frac{n\pi}{2}\right) + \frac{2}{n\pi}\sin\left(\frac{n\pi}{2}\right)\right]\sin\left(\frac{n\pi x}{L}\right).$$

As the previous two examples has shown the coefficients for these can be quite messy but that will often be the case and so we shouldn't let that get us too excited.

FOURIER COSINE SERIES

In this section we're going to take a look at Fourier cosine series. We'll start off much as we did in the previous section where we looked at Fourier sine series. Let's start by assuming that the function, $f(x)$, we'll be working with initially is an even function (*i.e.* $f(-x) = f(x)$) and that we want to write a series representation for this function on $-L \leq x \leq L$in terms of cosines (which are also even). In other words we are going to look for the following,

$$f(x) = \sum_{n=0}^{\infty} A_n \cos\left(\frac{n\pi x}{L}\right).$$

This series is called a Fourier cosine series and note that in this case (unlike with Fourier sine series) we're able to start the series representation at since that term will not be zero as it was with sines. Also, as with Fourier Sine series, the argument of $\frac{n\pi x}{L}$ in the cosines is being used only because it is the argument that we'll be running into in the next chapter. The only real requirement here is that the given set of functions we're using be orthogonal on the interval we're working on.

Note as well that we're assuming that the series will in fact converge to $f(x)$on $-L \leq x \leq L$at this point. In a later section we'll be looking into the convergence of this series in more detail.

So, to determine a formula for the coefficients, A_n, we'll use the fact that $\left\{\cos\left(\frac{n\pi x}{L}\right)\right\}_{n=0}^{\infty}$ do form an orthogonal set on the interval $-L \leq x \leq L$as we showed in a previous section. In that section we also derived the following formula that we'll need in a bit.

$$\int_{-L}^{L}\cos\left(\frac{n\pi x}{L}\right)\cos\left(\frac{m\pi x}{L}\right)dx = \begin{cases} 2L & \text{if } n=m=0 \\ L & \text{if } n=m\neq 0. \\ 0 & \text{if } n\neq m \end{cases}$$

We'll get a formula for the coefficients in almost exactly the same fashion that we did in the previous section.

We'll start with the representation above and multiply both sides by $\cos\left(\frac{m\pi x}{L}\right)$ where m is a fixed integer in the range $\{0,1,2,3,...\}$. Doing this gives,

$$f(x)\cos\left(\frac{m\pi x}{L}\right)=\sum_{n=0}^{\infty}A_n\cos\left(\frac{n\pi x}{L}\right)\cos\left(\frac{n\pi x}{L}\right).$$

Next, we integrate both sides from $x = -L$ to $x = L$ and as we were able to do with the Fourier Sine series we can again interchange the integral and the series.

$$\begin{aligned}\int_{-L}^{L}f(x)\cos\left(\frac{m\pi x}{L}\right)dx &= \int_{-L}^{L}\sum_{n=0}^{\infty}A_n\cos\left(\frac{n\pi x}{L}\right)\cos\left(\frac{m\pi x}{L}\right)dx\\ &= \sum_{n=0}^{\infty}A_n\int_{-L}^{L}\cos\left(\frac{n\pi x}{L}\right)\cos\left(\frac{m\pi x}{L}\right)dx.\end{aligned}$$

We now know that the all of the integrals on the right side will be zero except when because the set of cosines form an orthogonal set on the interval $-L \le x \le L$.

However, we need to be careful about the value of m (or n depending on the letter you want to use). So, after evaluating all of the integrals we arrive at the following set of formulas for the coefficients.

***m* = 0**

$$\int_{-L}^{L}f(x)dx=A_0(2L)\ \Rightarrow\ A_0=\frac{1}{2L}\int_{-L}^{L}f(x)dx.$$

***m* ≠ 0**

$$\int_{-L}^{L}f(x)\cos\left(\frac{m\pi x}{L}\right)dx=A_m(L)\quad\Rightarrow\quad A_m=\frac{1}{L}\int_{-L}^{L}f(x)\cos\left(\frac{m\pi x}{L}\right)dx.$$

Summarizing everything up then, the Fourier cosine series of an even function, $f(x)$ on $-L \le x \le L$ is given by,

$$f(x)=\sum_{n=0}^{\infty}A_n\cos\left(\frac{n\pi x}{L}\right)\qquad A_n=\begin{cases}\frac{1}{2L}\int_{-L}^{L}f(x)dx & n=0\\ \frac{1}{L}\int_{-L}^{L}f(x)\cos\left(\frac{n\pi x}{L}\right)dx & n\neq 0\end{cases}.$$

Finally, before we work an example, let's notice that because both $f(x)$ and the cosines are even the integrand in both of the integrals above is even and so we can write the formulas for the A_n's as follows,

$$A_n = \begin{cases} \frac{1}{2}\int_0^L f(x)dx & n=0 \\ \frac{1}{L}\int_0^L f(x)\cos\left(\frac{n\pi x}{L}\right)dx & n \neq 0 \end{cases}.$$

Now let's take a look at an example.

Example: Find the Fourier cosine series for $f(x) = x^3$on $-L \le x \le L$.

Solution: We clearly have an even function here and so all we really need to do is compute the coefficients and they are liable to be a little messy because we'll need to do integration by parts twice. We'll leave most of the actual integration details to you to verify.

$$A_0 = \frac{1}{L}\int_0^L f(x)dx = \frac{1}{L}\int_0^L x^2 dx = \frac{1}{L}\left(\frac{L^3}{3}\right) = \frac{L^2}{3}.$$

$$A_0 = \frac{2}{L}\int_0^L f(x)\cos\left(\frac{n\pi x}{L}\right)dx = \frac{2}{L}\int_0^L x^2\cos\left(\frac{n\pi x}{L}\right)dx$$

$$= \frac{2}{L}\left(\frac{L}{n^3\pi^3}\right)\left(2Ln\pi x\cos\left(\frac{n\pi x}{L}\right) + (n^2\pi^2 n^2 - 2L^2)\sin\left(\frac{n\pi x}{L}\right)\right)\Bigg|_0^L$$

$$= \frac{2}{n^2\pi^2}\left(2Ln\pi x\cos(n\pi) + (n^2\pi^2 n^2 - 2L^2)\sin(n\pi)\right)$$

$$= \frac{4L^2(-1)^n}{n^2\pi^2} \qquad n = 1, 2, 3, \ldots$$

The coefficients are then,

$$A_0 = \frac{L^2}{3} \qquad A_0 = \frac{4L^2(-1)^n}{n^2\pi^2}, \quad n = 1, 2, 3, \ldots$$

The Fourier cosine series is then,

$$x^2 = \sum_{n=0}^{\infty} A_n\cos\left(\frac{n\pi x}{L}\right) = A_0 + \sum_{n=1}^{\infty} A_n\cos\left(\frac{n\pi x}{L}\right) = \frac{L^2}{3} + \sum_{n=1}^{\infty}\frac{4L^2(-1)^2}{n^2\pi^2}\cos\left(\frac{n\pi x}{L}\right).$$

Note that we'll often strip out the $n = 0$from the series as we've done here because it will almost always be different from the other coefficients and it allows us to actually plug the coefficients into the series.

Now, just as we did in the previous section let's ask what we need to do in order to find the Fourier cosine series of a function that is not odd. As with Fourier sine series when we make this change we'll need to move onto the interval $0 \le x \le L$now instead of

$-L \le x \le L$and again we'll assume that the series will converge to $f(x)$at this point and leave the discussion of the convergence of this series to a later section.

We could go through the work to find the coefficients here twice as we did with Fourier sine series, however there's no real reason to.

So, while we could redo all the work above to get formulas for the coefficients let's instead go straight to the second method of finding the coefficients.

In this case, before we actually proceed with this we'll need to define the even extension of a function, $f(x)$on $-L \le x \le L$. So, given a function $f(x)$we'll define the even extension of the function as,

$$g(x) = \begin{cases} f(x) & \text{if } 0 \le x \le L \\ f(-x) & \text{if } -L \le x \le 0 \end{cases}.$$

Showing that this is an even function is simple enough.

$$g(-x) = g(-(-x)) = f(x) = g(x) \qquad \text{for } 0 < x < L$$

and we can see that $g(x) = f(x)$on $0 \le x \le L$and if $f(x)$is already an even function we get $g(x) = f(x)$on $-L \le x \le L$.

Let's take a look at some functions and sketch the even extensions for the functions.

Example: Sketch the odd extension of each of the given functions.

1. $f(x) = L - x$ on $0 \le x \le L$
2. $f(x) = x^3$on $0 \le x \le L$
3. $f(x) = \begin{cases} \frac{L}{2} & \text{if } 0 \le x \le \frac{L}{2} \\ x - \frac{L}{2} & \text{if } \frac{L}{2} \le x \le L \end{cases}.$

Solution:

1. $f(x) = L - x$ on $0 \le x \le L$.

Here is the even extension of this function.

$$g(x) = \begin{cases} f(x) & \text{if } 0 \le x \le L \\ f(-x) & \text{if} -L \le x \le 0 \end{cases}$$

$$= \begin{cases} L - x & \text{if } 0 \le x \le L \\ L + x & \text{if} -L \le x \le 0 \end{cases}.$$

Here is the graph of both the original function and its even extension. Note that we've put the "extension" in with a dashed line to make it clear the portion of the function that is being added to allow us to get the even extension

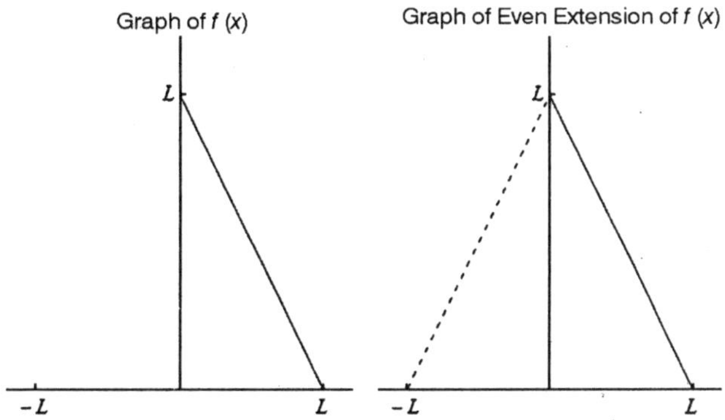

2. $f(x) = x^3$ on $0 \le x \le L$

The even extension of this function is,

$$g(x) = \begin{cases} f(x) & \text{if } 0 \le x \le L \\ f(-x) & \text{if } -L \le x \le 0 \end{cases}$$

$$= \begin{cases} x^3 & \text{if } 0 \le x \le L \\ -x^3 & \text{if } -L \le x \le 0 \end{cases}.$$

The sketch of the function and the even extension is,

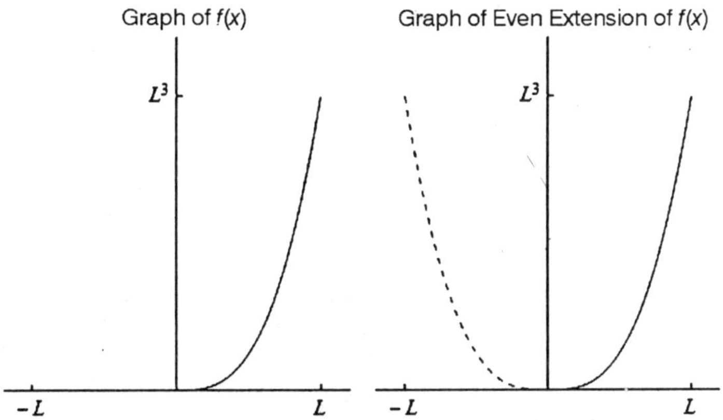

3. $f(x) = \begin{cases} \frac{L}{2} & \text{if } 0 \le x \le \frac{L}{2} \\ x - \frac{L}{2} & \text{if } \frac{L}{2} \le x \le L \end{cases}.$

Here is the even extension of this function,

$$g(x) = \begin{cases} f(x) & \text{if } 0 \le x \le L \\ f(-x) & \text{if} - L \le x \le 0 \end{cases} = \begin{cases} x - \frac{L}{2} & \text{if } 0 \le x \le L \\ \frac{L}{2} & \text{if} - L \le x \le \frac{L}{2} \\ \frac{L}{2} & \text{if } \frac{1}{2} \le x \le 0 \\ -x - \frac{L}{2} & \text{if} - L \le x \le -\frac{L}{2} \end{cases}.$$

The sketch of the function and the even extension is,

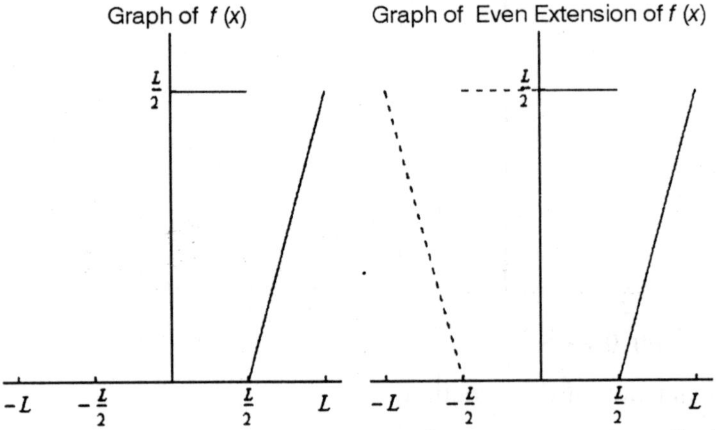

Okay, let's now think about how we can use the even extension of a function to find the Fourier cosine series of any function $f(x)$on $0 \le x \le L$.

So, given a function $f(x)$we'll let $g(x)$be the even extension as defined above. Now, $g(x)$is an even function on $-L \le x \le L$and so we can write down its Fourier cosine series. This is, and note that we'll use the second form of the integrals to compute the constants.

Now, because we know that on $0 \le x \le L$we have $f(x) = g(x)$and so the Fourier cosine series of $f(x)$on $0 \le x \le L$is also given by,

$$g(x) = \sum_{n=0}^{\infty} A_n \cos\left(\frac{n\pi x}{L}\right) \qquad A_n = \begin{cases} \frac{1}{L}\int_0^L f(x)dx & n = 0 \\ \frac{2}{L}\int_0^L f(x)\cos\left(\frac{n\pi x}{L}\right) & n \neq 0 \end{cases}$$

Let's take a look at a couple of examples.

Example: Find the Fourier cosine series for $f(x) = L - x$on $0 \le x \le L$.

Solution: All we need to do is compute the coefficients so here is the work for that,

$$A_0 = \frac{1}{L}\int_0^L f(x)dx = \frac{1}{L}\int_0^L L - xdx = \frac{L}{2}.$$

$$A_n = \frac{2}{L}\int_0^L f(x)\cos\left(\frac{n\pi x}{L}\right)dx = \frac{2}{L}\int_0^L (L-x)\cos\left(\frac{n\pi x}{L}\right)dx$$

$$= \frac{2}{L}\left(\frac{L}{n^2\pi^2}\right)\left(n\pi(L-x)\sin\left(\frac{n\pi x}{L}\right) - L\cos\left(\frac{n\pi x}{L}\right)\right)\Bigg|_0^L$$

$$= \frac{2}{L}\left(\frac{L}{n^2\pi^2}\right)(-L\cos(n\pi) + L) = \frac{2L}{n^2\pi^2}\left(1 + (-1)^{n+1}\right) \qquad n = 1,2,3,\ldots$$

The Fourier cosine series is then,

$$f(x) = \frac{L}{2} + \sum_{n=1}^{\infty} \frac{2L}{n^2\pi^2}\left(1 + (-1)^{n+1}\right)\cos\left(\frac{n\pi x}{L}\right).$$

Note that as we did with the first example in this section we stripped out the term before we plugged in the coefficients.

Next, let's find the Fourier cosine series of an odd function.A_0 Note that this is doable because we are really finding the Fourier cosine series of the even extension of the function.

Example: Find the Fourier cosine series for $f(x) = x^3$on $0 \le x \le L$.

Solution: The integral for $A_0$is simple enough but the integral for the rest will be fairly messy as it will require three integration by parts.

We'll leave most of the details of the actual integration to you to verify. Here's the work,

$$A_0 = \frac{1}{L}\int_0^L f(x)dx = \frac{1}{L}\int_0^L x^3 dx = \frac{L^3}{4}$$

$$A_n = \frac{2}{L}\int_0^L f(x)\cos\left(\frac{n\pi x}{L}\right)dx = \frac{2}{L}\int_0^L x^3\cos\left(\frac{n\pi x}{L}\right)dx$$

$$= \frac{2}{L}\left(\frac{L}{n^4\pi^4}\right)\left(n\pi x(n^2\pi^2x^2 - 6L^2)\sin\left(\frac{n\pi x}{L}\right) + (3Ln^2\pi^2x^2 - 6L^3)\cos\left(\frac{n\pi x}{L}\right)\right)\Bigg|_0^L$$

$$= \frac{2}{L}\left(\frac{L}{n^4\pi^4}\right)\left(n\pi L\left(n^2\pi^2L^2 - 6L^2\right)\sin(n\pi) + \left(3L^3n^3\pi^2 - 6L^3\right)\cos(n\pi) + 6L^3\right)$$

$$= \frac{2}{L}\left(\frac{3L^4}{n^4\pi^4}\right)\left(2 + \left(n^2\pi^2\right)(-1)^n\right) = \frac{6L^3}{n^4\pi^4}\left(2 + \left(n^2\pi^2 - 2\right)(-1)^n\right) \quad n = 1,2,3,\ldots$$

The Fourier cosine series for this function is then,

$$f(x) = \frac{L^3}{4} + \sum_{n=1}^{\infty} \frac{6L^3}{n^4\pi^4}\left(2 + \left(n^2\pi^2\right)(-1)^n\right)\cos\left(\frac{n\pi x}{L}\right).$$

Finally, let's take a quick look at a piecewise function.

Example: Find the Fourier cosine series for $f(x) = \begin{cases} \frac{L}{2} & \text{if } 0 \le x \le \frac{L}{2} \\ x - \frac{L}{2} & \text{if } \frac{L}{2} \le x \le L \end{cases}$ on $0 \le x \le L$.

Solution: We'll need to split up the integrals for each of the coefficients here. Here are the coefficients.

$$A_0 = \frac{1}{L}\int_0^L f(x)dx = \frac{1}{L}\left[\int_0^{\frac{L}{2}} f(x)dx + \int_{\frac{L}{2}}^L f(x)dx\right]$$

$$= \frac{1}{L}\left[\int_0^{\frac{L}{2}} dx + \int_{\frac{L}{2}}^L x\frac{L}{2}dx\right] = \frac{1}{L}\left[\frac{L^2}{4} + \frac{L^2}{8}\right] = \frac{1}{L}\left[\frac{3L^2}{8}\right] = \frac{3L}{8}.$$

For the rest of the coefficients here is the integral we'll need to do.

$$A_n = \frac{2}{L}\int_0^L f(x)\cos\left(\frac{n\pi x}{L}\right)dx = \frac{2}{L}\left[\int_0^{\frac{L}{2}} f(x)\cos\left(\frac{n\pi x}{L}\right)dx + \int_{\frac{L}{2}}^L f(x)\cos\left(\frac{n\pi x}{L}\right)dx\right]$$

$$= \frac{2}{L}\left[\int_0^{\frac{L}{2}} \frac{L}{2}\cos\left(\frac{n\pi x}{L}\right)dx + \int_{\frac{L}{2}}^L \left(x - \frac{L}{2}\right)\cos\left(\frac{n\pi x}{L}\right)dx\right].$$

To make life a little easier let's do each of these separately.

$$\int_{\frac{L}{2}}^L \left(x - \frac{L}{2}\right)dx = \frac{L}{2}\left(\frac{L}{n\pi}\right)\sin\left(\frac{n\pi x}{L}\right)\Bigg|_{\frac{L}{2}}^L = \frac{L}{2}\left(\frac{L}{n\pi}\right)\sin\left(\frac{n\pi}{2}\right) = \frac{L^2}{2n\pi}\sin\left(\frac{n\pi}{2}\right)$$

$$\int_0^{\frac{L}{2}} \cos\left(x - \frac{L}{2}\right)\cos\left(\frac{n\pi x}{L}\right)dx = \frac{L}{n\pi}\left(\frac{L}{n\pi}\cos\left(\frac{n\pi x}{L}\right) + \left(x - \frac{L}{2}\right)\sin\left(\frac{n\pi x}{L}\right)\right)\Bigg|_{\frac{L}{2}}^L$$

$$= \frac{L}{n\pi}\left(\frac{L}{n\pi}\cos(n\pi) + \frac{L}{2}\sin(n\pi) - \frac{L}{n\pi}\cos\left(\frac{n\pi}{2}\right)\right)$$

$$= \frac{L^2}{n^2\pi^2}\left((-1)^n - \cos\left(\frac{n\pi}{2}\right)\right).$$

Putting these together gives,

$$A_n = \frac{2}{L}\left[\frac{L^2}{2n\pi}\sin\left(\frac{n\pi}{2}\right) + \frac{L^2}{n^2\pi^2}(-1)^n - \cos\left(\frac{n\pi}{2}\right)\right]$$

$$= \frac{2L}{n\pi}\left[(-1)^n - \cos\left(\frac{n\pi}{2}\right) + \frac{n\pi}{2}\sin\left(\frac{n\pi}{2}\right)\right].$$

So, after all that work the Fourier cosine series is then,

$$f(x) = \frac{3L}{8} + \sum_{n=1}^{\infty}\frac{2L}{n\pi}\left[(-1)^n - \cos\left(\frac{n\pi}{2}\right) + \frac{n\pi}{2}\sin\left(\frac{n\pi}{2}\right)\right]\cos\left(\frac{n\pi x}{L}\right).$$

Note that much as we saw with the Fourier sine series many of the coefficients will quite messy to deal with.

FOURIER SERIES

Okay, in the previous two sections we've looked at Fourier sine and Fourier cosine series. It is now time to look at a Fourier series. With a Fourier series we are going to try to write a series representation for $f(x)$ on $-L \le x \le L$on in the form,

$$f(x) = \sum_{n=0}^{\infty} A_n \cos\left(\frac{n\pi x}{L}\right) + \sum_{n=0}^{\infty} B_n \sin\left(\frac{n\pi x}{L}\right).$$

So, a Fourier series is, in some way a combination of the Fourier sine and Fourier cosine series.

Also, like the Fourier sine/cosine series we'll not worry about whether or not the series will actually converge to $f(x)$or not at this point. For now we'll just assume that it will converge and we'll discuss the convergence of the Fourier series in a later section.

Determining formulas for the coefficients, $A_n$and B_n, will be done in exactly the same manner as we did in the previous two sections. We will take advantage of the fact that $\left\{\cos\left(\frac{n\pi x}{L}\right)\right\}_{n=0}^{\infty}$ and $\left\{\sin\left(\frac{n\pi x}{L}\right)\right\}_{n=1}^{\infty}$ are mutually orthogonal on $-L \le x \le L$as we proved earlier. We'll also need the following formulas that we derived when we proved the two sets were mutually orthogonal.

$$\int_{-L}^{L} \cos\left(\frac{n\pi x}{L}\right) \cos\left(\frac{m\pi x}{L}\right) dx = \begin{cases} 2L & \text{if } n = m = 0 \\ L & \text{if } n = m \ne 0 \\ 0 & \text{if } n \ne m \end{cases}$$

$$\int_{-L}^{L} \sin\left(\frac{n\pi x}{L}\right) \sin\left(\frac{m\pi x}{L}\right) dx = \begin{cases} L & \text{if } n = m \\ 0 & \text{if } n \ne m \end{cases}$$

$$\int_{-L}^{L} \sin\left(\frac{n\pi x}{L}\right) \cos\left(\frac{m\pi x}{L}\right) dx = 0.$$

So, let's start off by multiplying both sides of the series above by $\cos\left(\frac{m\pi x}{L}\right)$and integrating from L to L. Doing this gives,

$$\int_{-L}^{L} f(x)\cos\left(\frac{m\pi x}{L}\right) dx = \int_{-L}^{L} \sum_{n=0}^{\infty} A_n \cos\left(\frac{n\pi x}{L}\right) \cos\left(\frac{m\pi x}{L}\right) dx + \int_{-L}^{L} \sum_{n=1}^{\infty} B_n \cos\left(\frac{n\pi x}{L}\right) \cos\left(\frac{m\pi x}{L}\right) dx.$$

Now, just as we've been able to do in the last two sections we can interchange the integral and the summation. Doing this gives,

$$\int_{-L}^{L} f(x)\cos\left(\frac{m\pi x}{L}\right) dx = \sum_{n=0}^{\infty} A_n \int_{-L}^{L} \cos\left(\frac{n\pi x}{L}\right) \cos\left(\frac{m\pi x}{L}\right) dx + \sum_{n=1}^{\infty} B_n \int_{-L}^{L} \sin\left(\frac{n\pi x}{L}\right) \cos\left(\frac{m\pi x}{L}\right) dx.$$

We can now take advantage of the fact that the sines and cosines are mutually orthogonal. The integral in the second series will always be zero and in the first series the integral will be zero if $n \neq m$and so this reduces to,

$$\int_{-L}^{L} f(x)\cos\left(\frac{m\pi x}{L}\right)dx = \begin{cases} A_m(2L) & \text{if } n = m = 0 \\ A_m(L) & \text{if } n = m \neq 0 \end{cases}.$$

Solving for gives,

$$A_0 \frac{1}{2L}\int_{-L}^{L} f(x)dx$$

$$A_m \frac{1}{L}\int_{-L}^{L} f(x)\cos\left(\frac{m\pi x}{L}\right)dx \qquad m = 1,2,3,....$$

Now, do it all over again only this time multiply both sides by $\sin\left(\frac{m\pi x}{L}\right)$, integrate both sides from L to L and interchange the integral and summation to get,

$$\int_{-L}^{L} f(x)\sin\left(\frac{m\pi x}{L}\right)dx = \sum_{n=0}^{\infty} A_n \int_{-L}^{L} \cos\left(\frac{m\pi x}{L}\right)\sin\left(\frac{m\pi x}{L}\right)dx + \sum_{n=1}^{\infty} B_n \int_{-L}^{L} \sin\left(\frac{m\pi x}{L}\right)\sin\left(\frac{m\pi x}{L}\right)dx.$$

In this case the integral in the first series will always be zero and the second will be zero if and so we get,

$$B_m \int_{-L}^{L} f(x)\sin\left(\frac{m\pi x}{L}\right)dx = B_m(L).$$

Finally, solving for gives,

$$B_m \frac{1}{L}\int_{-L}^{L} f(x)\sin\left(\frac{m\pi x}{L}\right)dx \quad m = 1,2,3,...$$

In the previous two sections we also took advantage of the fact that the integrand was even to give a second form of the coefficients in terms of an integral from 0 to L. However, in this case we don't know anything about whether $f(x)$will be even, odd, or more likely neither even nor odd. Therefore, this is the only form of the coefficients for the Fourier series.

Before we start examples let's remind ourselves of a couple of formulas that we'll make heavy use of here in this section, as we've done in the previous two sections as well. Provided n in an integer then,

$$\cos(n\pi) = (-1)^n \qquad \sin(n\pi) = 0.$$

In all of the work that we'll be doing here n will be an integer and so we'll use these without comment in the problems so be prepared for them. Also don't forget that sine is an odd function, *i.e.* $\sin(-x) = -\sin(x)$and that cosine is an even function, *i.e.* $\cos(-x) = \cos$

(*x*). We'll also be making heave use of these ideas without comment in many of the integral evaluations so be ready for these as well.

Now let's take a look at an example.

Exasmple: Find the Fourier series for $f(x) = L - x$ on $-L \leq x \leq L$.

Solution: So, let's go ahead and just run through formulas for the coefficients.

$$A_0 = \frac{1}{2L}\int_{-L}^{L} f(x)dx = \frac{1}{2L}\int_{-L}^{L} L - xdx = L$$

$$A_n = \frac{1}{L}\int_{-L}^{L} f(x)\cos\left(\frac{n\pi x}{L}\right)dx = \frac{1}{L}\int_{-L}^{L}(L-x)\cos\left(\frac{n\pi x}{L}\right)dx$$

$$= \frac{1}{L}\left(\frac{L^2}{n^2\pi^2}\right)\left(n\pi(L-x)\sin\left(\frac{n\pi x}{L}\right) - L\cos\left(\frac{n\pi x}{L}\right)\right)\Bigg|_{-L}^{L}$$

$$= \frac{1}{L}\left(\frac{L^2}{n^2\pi^2}\right)(-2n\pi L\sin(-n\pi)) = 0 \qquad n = 1,2,3,....$$

$$B_n = \frac{1}{L}\int_{-L}^{L} f(x)\sin\left(\frac{n\pi x}{L}\right)dx = \frac{1}{L}\int_{-L}^{L}(L-x)\sin\left(\frac{n\pi x}{L}\right)dx$$

$$= \frac{1}{L}\left(-\frac{L}{n^2\pi^2}\right)\left[L\sin\left(\frac{n\pi x}{L}\right) - n\pi(x-L)\cos\left(\frac{n\pi x}{L}\right)\right]\Bigg|_{-L}^{L}$$

$$= \frac{1}{L}\left[\frac{L}{n^2\pi^2}(2n\pi\cos(n\pi) - 2\sin(n\pi))\right] = \frac{2L(-1)}{n\pi} \qquad n = 1,2,3,...$$

Note that in this case we had $A_0 \neq 0$ and $A_n = 0$, $n = 1,2,3,...$ This will happen on occasion so don't get excited about this kind of thing when it happens.

The Fourier series is then,

$$f(x) = \sum_{n=0}^{\infty} A_n \cos\left(\frac{n\pi x}{L}\right) + \sum_{n=0}^{\infty} B_n \sin\left(\frac{n\pi x}{L}\right)$$

$$= A_0 + \sum_{n=0}^{\infty} A_n \cos\left(\frac{n\pi x}{L}\right) + \sum_{n=0}^{\infty} B_n \sin\left(\frac{n\pi x}{L}\right) = L + \sum_{n=1}^{\infty}\frac{2L(-1)}{n\pi}\sin\left(\frac{n\pi x}{L}\right).$$

As we saw in the previous example sometimes we'll get $A_0 \neq 0$ and $A_n = 0$, $n = 1,2,3,...$ Whether or not this will happen will depend upon the function and often won't happen, but when it does don't get excited about it.

Let's take a look at another problem.

Example: Find the Fourier series for $f(x)=\begin{cases} L & \text{if } -L \le x \le 0 \\ 2x & \text{if } 0 \le x \le L \end{cases}$ on $-L \le x \le L$.

Solution: Because of the piece-wise nature of the function the work for the coefficients is going to be a little unpleasant but let's get on with it.-

$$A_0 = \frac{1}{2L}\int_{-L}^{L} f(x)dx = \frac{1}{2L}\left[\int_{-L}^{0} f(x)dx + \int_{0}^{L} f(x)dx\right]$$

$$= \frac{1}{2L}\left[\int_{-L}^{0} Ldx + \int_{0}^{L} 2xdx\right] = \frac{1}{2L}\left[L^2 + L^2\right] = L$$

$$A_n = \frac{1}{L}\int_{0}^{L} f(x)\cos\left(\frac{n\pi x}{L}\right)dx = \frac{1}{L}\left[\int_{-L}^{0} f(x)\cos\left(\frac{n\pi x}{L}\right)dx + \int_{0}^{L} f(x)\cos\left(\frac{n\pi x}{L}\right)dx\right]$$

$$= \frac{1}{L}\left[\int_{-L}^{0} L\cos\left(\frac{n\pi x}{L}\right)dx + \int_{0}^{L} 2x\cos\left(\frac{n\pi x}{L}\right)dx\right].$$

At this point it will probably be easier to do each of these individually.

$$\int_{-L}^{0} L\cos\left(\frac{n\pi x}{L}\right)dx = \left(\frac{L^2}{n\pi}\sin\left(\frac{n\pi x}{L}\right)\right)\Bigg|_{-L}^{0} = \frac{L^2}{n\pi}\sin(n\pi) = 0$$

$$\int_{-L}^{0} 2x\cos\left(\frac{n\pi x}{L}\right)dx = \left(\frac{2L}{n^2\pi^2}\right)\left(L\cos\left(\frac{n\pi x}{L}\right) + n\pi x\sin\left(\frac{n\pi x}{L}\right)\right)\Bigg|_{-L}^{0}$$

$$= \left(\frac{2L}{n^2\pi^2}\right)\left(L\cos(n\pi) + n\pi L\sin(n\pi) - L\cos(0)\right)$$

$$= \left(\frac{2L}{n^2\pi^2}\right)\left((-1)^n - 1\right).$$

So, if we put all of this together we have,

$$A_n = \frac{1}{L}\int_{0}^{L} f(x)\cos\left(\frac{n\pi x}{L}\right)dx = \frac{1}{L}\left[0 + \left(\frac{2L^2}{n^2\pi^2}\right)\left((-1)^n\right)\right]$$

$$= \frac{2L^2}{n^2\pi^2}\left((-1)^n - 1\right), \qquad n = 1, 2, 3, \ldots$$

So, we've gotten the coefficients for the cosines taken care of and now we need to take care of the coefficients for the sines.

$$Bn = \frac{1}{L}\int_{0}^{L} f(x)\sin\left(\frac{n\pi x}{L}\right)dx = \frac{1}{L}\left[\int_{-L}^{0} f(x)\sin\left(\frac{n\pi x}{L}\right)dx + \int_{0}^{L} f(x)\sin\left(\frac{n\pi x}{L}\right)dx\right]$$

$$= \frac{1}{L}\left[\int_{-L}^{0} L\sin\left(\frac{n\pi x}{L}\right)dx + \int_{0}^{L} 2x\sin\left(\frac{n\pi x}{L}\right)dx\right].$$

As with the coefficients for the cosines will probably be easier to do each of these individually.

$$\int_{-L}^{0} L\sin\left(\frac{n\pi x}{L}\right)dx = \left(-\frac{L^2}{n\pi}\cos\left(\frac{n\pi x}{L}\right)\right)\Bigg|_{-L}^{0} = \frac{L^2}{n\pi}(-1+\cos(n\pi)) = \frac{L^2}{n\pi}\left((-1)^n - 1\right)$$

$$\int_{0}^{L} 2x\sin\left(\frac{n\pi x}{L}\right)dx = \left(\frac{2L}{n^2\pi^2}\right)\left(L\sin\left(\frac{n\pi x}{L}\right) - n\pi x\cos\left(\frac{n\pi x}{L}\right)\right)\Bigg|_{0}^{L}$$

$$= \left(\frac{2L}{n^2\pi^2}\right)(L\sin(n\pi) - n\pi L\cos(n\pi))$$

$$= \left(\frac{2L}{n^2\pi^2}\right)\left(-n\pi(-1)^n\right) = -\frac{2L^n}{n\pi}(-1)^n.$$

So, if we put all of this together we have,

$$B_n = \frac{1}{L}\int_{0}^{L} f(x)\sin\left(\frac{n\pi x}{L}\right)dx = \frac{1}{L}\left[\frac{L^2}{n\pi}\left((-1)^n - 1\right) - \frac{2L^2}{n\pi}(-1)^n\right]$$

$$= \frac{L}{n\pi}\left[-1-(-1)^2\right] = -\frac{L}{n\pi}\left(1+(-1)^n\right) \qquad n = 1,2,3,....$$

So, after all that work the Fourier series is,

$$f(x) = \sum_{n=0}^{\infty} A_n\cos\left(\frac{n\pi x}{L}\right) + \sum_{n=0}^{\infty} B_n\cos\left(\frac{n\pi x}{L}\right)$$

$$= A_0 + \sum_{n=1}^{\infty} A_n\cos\left(\frac{n\pi x}{L}\right) + \sum_{n=1}^{\infty} B_n\sin\left(\frac{n\pi x}{L}\right)$$

$$= L + \sum_{n=1}^{\infty}\frac{2L}{n^2\pi^2}\left((-1)^n - 1\right)\cos\left(\frac{n\pi x}{L}\right) - \sum_{n=1}^{\infty}\frac{L}{n\pi}\left(1+(-1)^n\right)\sin\left(\frac{n\pi x}{L}\right).$$

As we saw in the previous example there is often quite a bit of work involved in computing the integrals involved here.

The next couple of examples are here so we can make a nice observation about some Fourier series and their relation to Fourier sine/cosine series

Example: Find the Fourier series for $f(x) = x$on $-L \le x \le L$.

Solution: Let's start with the integrals for A_n.

$$A_0 = \frac{1}{2L}\int_{-L}^{L} f(x)dx = \frac{1}{2L}\int_{-L}^{L} xdx = 0$$

$$A_n = \frac{1}{L}\int_{-L}^{L} f(x)\,dx = \frac{1}{L}\int_{-L}^{L} x\cos\left(\frac{n\pi x}{L}\right)dx = 0.$$

In both cases note that we are integrating an odd function (x is odd and cosine is even so the product is odd) over the interval $[-L,L]$and so we know that both of these integrals will be zero.

Next here is the integral for B_n

$$B_n = \frac{1}{L}\int_{-L}^{L} f(x)\sin\left(\frac{n\pi x}{L}\right)dx = \frac{1}{L}\int_{-L}^{L} x\sin\left(\frac{n\pi x}{L}\right)dx = \frac{2}{L}\int_{0}^{L} x\sin\left(\frac{n\pi x}{L}\right)dx.$$

In this case we're integrating an even function (x and sine are both odd so the product is even) on the interval and so we can "simplify" the integral as shown above. The reason for doing this here is not actually to simplify the integral however. It is instead done so that we can note that we did this integral back in the Fourier sine series section and so don't need to redo it in this section. Using the previous result we get,

$$B_n = \frac{(-1)^{n-1}2L}{n\pi} \qquad n = 1,2,3,....$$

In this case the Fourier series is,

$$f(x) = \sum_{n=0}^{\infty} A_n \cos\left(\frac{n\pi x}{L}\right) + \sum_{n=0}^{\infty} B_n \sin\left(\frac{n\pi x}{L}\right) = \sum_{n=0}^{\infty} \frac{(-1)^{n+1}2L}{n\pi}\sin\left(\frac{n\pi x}{L}\right).$$

If you go back and take a look at Example in the Fourier sine series section, the same example we used to get the integral out of, you will see that in that example we were finding the Fourier sine series for$f(x) = x$on $-L \le x \le L$. The important thing to note here is that the answer that we got in that example is identical to the answer we got here.

If you think about it however, this should not be too surprising. In both cases we were using an odd function on $-L \le x \le L$and because we know that we had an odd function the coefficients of the cosines in the Fourier series, A_n, will involve integrating and odd function over a symmetric interval, $-L \le x \le L$, and so will be zero. So, in these cases the Fourier sine series of an odd function on $-L \le x \le L$is really just a special case of a Fourier series.

Note however that when we moved over to doing the Fourier sine series of any function $0 \le x \le L$ on we should no longer expect to get the same results. You can see this by comparing Example 1 above with Example in the Fourier sine series section. In both examples we are finding the series for $f(x) = x - L$and yet got very different answers.

So, why did we get different answers in this case? Recall that when we find the Fourier sine series of a function on $0 \le x \le L$we are really finding the Fourier sine series of the odd extension of the function on $-L \le x \le L$and then just restricting the result down to $0 \le x \le L$. For a Fourier series we are actually using the whole function on $-L \le x \le L$instead of its odd extension. We should therefore not expect to get the same results since we are really using different functions (at least on part of the interval) in each case.

So, if the Fourier sine series of an odd function is just a special case of a Fourier series it makes some sense that the Fourier cosine series of an even function should also be a special case of a Fourier series. Let's do a quick example to verify this.

Example: Find the Fourier series for $f(x) = x^2$on $-L \le x \le L$.

Solution: Here are the integrals for the $A_n$and in this case because both the function and cosine are even we'll be integrating an even function and so can "simplify" the integral.

$$A_0 = \frac{1}{2L}\int_{-L}^{L} f(x)dx = \frac{1}{2L}\int_{-L}^{L} x^2 dx = \frac{1}{L}\int_{0}^{L} x^2 dx$$

$$A_n = \frac{1}{L}\int_{-L}^{L} f(x)\cos\left(\frac{n\pi x}{L}\right)dx = \frac{1}{L}\int_{-L}^{L} x^2\cos\left(\frac{n\pi x}{L}\right)dx = \frac{2}{L}\int_{0}^{L} x^2\cos\left(\frac{n\pi x}{L}\right)dx.$$

As with the previous example both of these integrals were done in Example in the Fourier cosine series section and so we'll not bother redoing them here. The coefficients are,

$$A_0 = \frac{L^2}{3} \qquad A_0 = \frac{4L^2(-1)^n}{n^2\pi^2}, \qquad n = 1, 2, 3, \ldots.$$

Next here is the integral for the B_n

$$B_n = \frac{1}{L}\int_{-L}^{L} f(x)\sin\left(\frac{n\pi x}{L}\right)dx = \frac{1}{L}\int_{-L}^{L} x^2\sin\left(\frac{n\pi x}{L}\right)dx = 0.$$

In this case the function is even and sine is odd so the product is odd and we're integrating over $-L \le x \le L$and so the integral is zero.

The Fourier series is then,

$$f(x) = \sum_{n=0}^{\infty} A_n\cos\left(\frac{n\pi x}{L}\right) + \sum_{n=0}^{\infty} B_n\sin\left(\frac{n\pi x}{L}\right) = \frac{L^2}{3} + \sum_{n=0}^{\infty}\frac{4L^2(-1)^n}{n^2\pi^2}\cos\left(\frac{n\pi x}{L}\right).$$

As suggested before we started this example the result here is identical to the result from Example in the Fourier cosine series section and so we can see that the Fourier cosine series of an even function is just a special case a Fourier series.

CONVERGENCE OF FOURIER SERIES

Over the last few sections we've spent a fair amount of time to computing Fourier series, but we've avoided discussing the topic of convergence of the series. In other words, will the Fourier series converge to the function on the given interval?

In this section we're going to address this issue as well as a couple of other issues about Fourier series. We'll be giving a fair number of theorems in this section but are not going to be proving any of them. We'll also not be doing a whole lot of in the way of examples in this section.

Before we get into the topic of convergence we need to first define a couple of terms that we'll run into in the rest of the section. First, we say that $f(x)$has a jump discontinuity at $x = a$if the limit of the function from the left, denoted $f(a^-)$, and the limit of the function from the right, denoted $f(a^+)$, both exist and $f(a^+) \neq f(a^+)$.

Next, we say that $f(x)$is piecewise smooth if the function can be broken into distinct pieces and on each piece both the function and its derivative, $f'(x)$, are continuous. A piecewise smooth function may not be continuous everywhere however the only discontinuities that are allowed are a finite number of jump discontinuities.

Let's consider the function,

$$f(x) = \begin{cases} L & \text{if } -L \le x \le 0 \\ 2x & \text{if } 0 \le x \le L \end{cases}.$$

We found the Fourier series for this function in Example of the previous section. Here is a sketch of this function on the interval on which it is defined, *i.e.* $-L \le x \le L$.

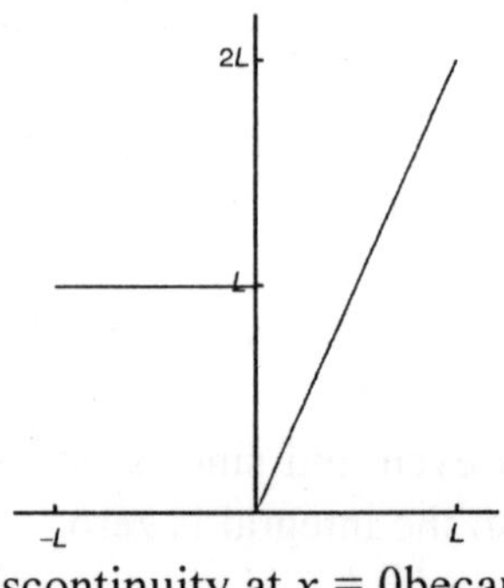

This function has a jump discontinuity at $x = 0$because $f(0^-) = L \neq 0 = f(0^+)$and note that on the intervals $-L \le x \le 0$and $0 \le x \le L$both the function and its derivative are continuous. This is therefore an example of a piecewise smooth function. Note that the function itself is not continuous at $x = 0$but because this point of discontinuity is a jump discontinuity the function is still piecewise smooth.

The last term we need to define is that of periodic extension.Given a function, $f(x)$, defined on some interval, we'll be using $-L \le x \le L$exclusively here, the periodic extension of this function is the new function we get by taking the graph of the function on the given interval and then repeating that graph to the right and left of the graph of the original function on the given interval. It is probably best to see an example of a periodic extension at this point to help make the words above a little clearer. Here is a sketch of the period extension of the function we looked at above,

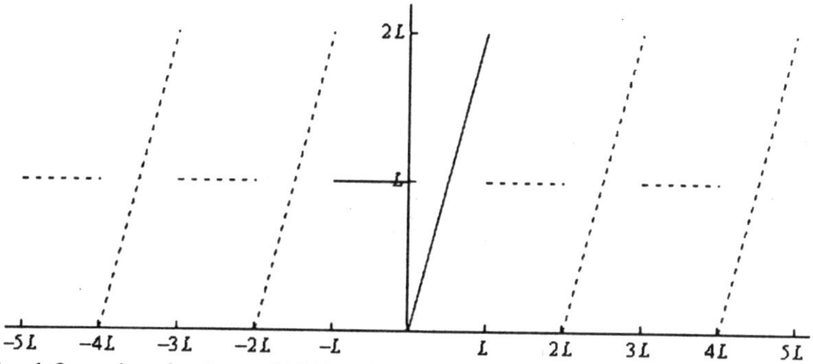

The original function is the solid line in the range $-L \le x \le L$. We then got the periodic extension of this by picking this piece up and copying it every interval of length $2L$ to the right and left of the original graph. This is shown with the two sets of dashed lines to either side of the original graph.

Note that the resulting function that we get from defining the periodic extension is in fact a new periodic function that is equal to the original function on $-L \le x \le L$.

With these definitions out of the way we can now proceed to talk a little bit about the convergence of Fourier series.

We will start off with the convergence of a Fourier series and once we have that taken care of the convergence of Fourier Sine/Cosine series will follow as a direct consequence. Here then is the theorem giving the convergence of a Fourier series.

CONVERGENCE OF FOURIER SERIES

Suppose $f(x)$is a piecewise smooth on the interval $-L \le x \le L$. The Fourier series of $f(x)$will then converge to,

1. The periodic extension of $f(x)$if the periodic extension is continuous.
2. The average of the two one-sided limits, $\frac{1}{2}\left[f(a^-)+f(a^+)\right]$, if the periodic extension has a jump discontinuity at $x = a$.

The first thing to note about this is that on the interval $-L \le x \le L$both the function and the periodic extension are equal and so where the function is continuous on $-L \le x \le L$the periodic extension will also be continuous and hence at these points the Fourier series will in fact converge to the function. The only points in the interval $-L \le x \le L$where the Fourier series will not converge to the function is where the function has a jump discontinuity.

Let's again consider Example of the previous section. In that section we found that the Fourier series of,

$$f(x) = \begin{cases} L & \text{if } -L \le x \le 0 \\ 2x & \text{if } 0 \le x \le L \end{cases}.$$

on to be,

$$f(x) = L + \sum_{n=1}^{\infty} \frac{2L}{n^2\pi^2}\left((-1)^2 - 1\right)\cos\left(\frac{n\pi x}{L}\right) - \sum_{n=1}^{\infty} \frac{L}{n\pi}\left(1 + (-1)^n\right)\sin\left(\frac{n\pi x}{L}\right).$$

We now know that in the intervals $-L \le x \le 0$and $0 < x < L$the function and hence the periodic extension are both continuous and so on these two intervals the Fourier series will converge to the periodic extension and hence will converge to the function itself.

At the point $x = 0$the function has a jump discontinuity and so the periodic extension will also have a jump discontinuity at this point. That means that at $x = 0$the Fourier series will converge to,

$$\frac{1}{2}\left[f\left(0^-\right) + f\left(0^+\right)\right] = \frac{1}{2}[L + 0] = \frac{L}{2}.$$

At the two endpoints of the interval, $x = -L$and $x = L$, we can see from the sketch of the periodic extension above that the periodic extension has a jump discontinuity here and so the Fourier series will not converge to the function there but instead the averages of the limits.

So, at $x = -L$the Fourier series will converge to,

$$\frac{1}{2}\left[f\left(-L^-\right) + f\left(-L^+\right)\right] = \frac{1}{2}[2L + L] = \frac{3L}{2}$$

and at $x = L$the Fourier series will converge to,

$$\frac{1}{2}\left[f\left(L^-\right) + f\left(L^+\right)\right] = \frac{1}{2}[2L + L] = \frac{3L}{2}.$$

Now that we have addressed the convergence of a Fourier series we can briefly turn our attention to the convergence of Fourier sine/cosine series. First, as noted in the previous section the Fourier sine series of an odd function on $-L \le x \le L$and the Fourier cosine series of an even function on $-L \le x \le L$are both just special cases of a Fourier series we now know that both of these will have the same convergence as a Fourier series.

Next, if we look at the Fourier sine series of any function, $g(x)$, on $0 \le x \le L$then we know that this is just the Fourier series of the odd extension of $g(x)$restricted down to the interval $0 \le x \le L$. Therefore we know that the Fourier series will converge to the odd extension on $-L \le x \le L$where it is continuous and the average of the limits where the odd extension has a jump discontinuity. However, on $0 \le x \le L$we know that $g(x)$and the odd extension are equal and so we can again see that the Fourier sine series will have the same convergence as the Fourier series.

Likewise, we can go through a similar argument for the Fourier cosine series using even extensions to see that Fourier cosine series for a function on $0 \le x \le L$will also have the same convergence as a Fourier series.

The next topic that we want to briefly discuss here is when will a Fourier series be continuous. From the theorem on the convergence of Fourier series we know that where the function is continuous the Fourier series will converge to the function and hence be continuous at these points. The only places where the Fourier series may not be continuous is if there is a jump discontinuity on the interval $-L \le x \le L$and potentially at the endpoints as we saw that the periodic extension may introduce a jump discontinuity there.

So, if we're going to want the Fourier series to be continuous everywhere we'll need to make sure that the function does not have any discontinuities in $-L \le x \le L$. Also, in order to avoid having the periodic extension introduce a jump discontinuity we'll need to require that $f(-L) = f(L)$. By doing this the two ends of the graph will match up when we form the periodic extension and hence we will avoid a jump discontinuity at the end points.

Here is a summary of these ideas for a Fourier series.

Suppose $f(x)$is a piecewise smooth on the interval $-L \le x \le L$. The Fourier series of $f(x)$will be continuous and will converge to $f(x)$on $-L \le x \le L$provided $f(x)$is continuous on $-L \le x \le L$and $f(-L) = f(L)$.

Now, how can we use this to get similar statements about Fourier sine/cosine series on $0 \le x \le L$? Let's start with a Fourier cosine series. The first thing that we do is form the even extension of $f(x)$on $-L \le x \le L$. For the purposes of this discussion let's call the even extension $g(x)$. As we saw when we sketched several even extensions in the Fourier cosine series section that in order for the sketch to be the even extension of the function we must have both,

$$g(0-) = g(0^+) \qquad g(-L) = g(L).$$

If one or both of these aren't true then $g(x)$will not be an even extension of $g(x)$.

So, in forming the even extension we do not introduce any jump discontinuities at $x = 0$and we get for free that $g(-L) = g(L)$. If we now apply the above theorem to the even extension we see that the Fourier series of the even extension is continuous on $-L \le x \le L$. However, because the even extension and the function itself are the same on $0 \le x \le L$then the Fourier cosine series of $f(x)$must also be continuous on $0 \le x \le L$.

Here is a summary of this discussion for the Fourier cosine series.

Suppose $f(x)$is a piecewise smooth on the interval $0 \le x \le L$. The Fourier cosine series of $f(x)$will be continuous and will converge to $f(x)$on $0 \le x \le L$provided is continuous on $0 \le x \le L$.

Note that we don't need any requirements on the end points here because they are trivially satisfied when we convert over to the even extension.

For a Fourier sine series we need to be a little more careful. Again, the first thing that we need to do is form the odd extension on $-L \le x \le L$and let's call it $g(x)$. We know that in order for it to be the odd extension then we know that at all points in $-L \le x \le L$ it must satisfy $g(-x) = -g(x)$ and that is what can lead to problems.

As we saw in the Fourier sine series section it is very easy to introduce a jump discontinuity at $x = 0$ when we form the odd extension. In fact, the only way to avoid forming a jump discontinuity at this point is to require that $f(0) = 0$.

Next, the requirement that at the endpoints we must have $g(-L) = -g(L)$ will practically guarantee that we'll introduce a jump discontinuity here as well when we form the odd extension. Again, the only way to avoid doing this is to require $f(L) = 0$.

So, with these two requirements we will get an odd extension that is continuous and so we know that the Fourier series of the odd extension on $-L \le x \le L$ will be continuous and hence the Fourier sine series will be continuous on $0 \le x \le L$.

Here is a summary of all this for the Fourier sine series.

Suppose $f(x)$ is a piecewise smooth on the interval $0 \le x \le L$. The Fourier sine series of $f(x)$ will be continuous and will converge to $f(x)$ on $0 \le x \le L$ provided $f(x)$ is continuous on $0 \le x \le L$, $f(0) = 0$ and $f(L) = 0$.

The next topic of discussion here is differentiation and integration of Fourier series. In particular we want to know if we can differentiate a Fourier series term by term and have the result be the Fourier series of the derivative of the function. Likewise we want to know if we can integrate a Fourier series term by term and arrive at the Fourier series of the integral of the function.

Note that we'll not be doing much discussion of the details here. All we're really going to be doing is giving the theorems that govern the ideas here so that we can say we've given them.

Let's start off with the theorem for term by term differentiation of a Fourier series.

Given a function $f(x)$ if the derivative, $f'(x)$, is piecewise smooth and the Fourier series of $f(x)$ is continuous then the Fourier series can be differentiated term by term. The result of the differentiation is the Fourier series of the derivative, $f'(x)$.

One of the main condition of this theorem is that the Fourier series be continuous and from above we also know the conditions on the function that will give this. So, if we add this into the theorem to get this form of the theorem,

Suppose $f(x)$ is a continuous function, its derivative $f'(x)$ is piecewise smooth and $f(-L) = f(L)$ then the Fourier series of the function can be differentiated term by term and the result is the Fourier series of the derivative. For Fourier cosine/sine series the basic theorem is the same as for Fourier series. All that's required is that the Fourier cosine/sine series be continuous and then you can differentiate term by term. The theorems that we'll give here will merge the conditions for the Fourier cosine/sine series to be continuous into the theorem.

Let's start with the Fourier cosine series.

Suppose $f(x)$is a continuous function and its derivative $f'(x)$is piecewise smooth then the Fourier cosine series of the function can be differentiated term by term and the result is the Fourier sine series of the derivative.

Next the theorem for Fourier sine series.

Suppose $f(x)$is a continuous function, its derivative $f'(x)$is piecewise smooth, $f(0) = 0$and $f(L) = 0$ then the Fourier sine series of the function can be differentiated term by term and the result is the Fourier cosine series of the derivative.

The theorem for integration of Fourier series term by term is simple so there it is.

Suppose $f(x)$is piecewise smooth then the Fourier sine series of the function can be integrated term by term and the result is a convergent infinite series that will converge to the integral of $f(x)$.

Note however that the new series that results from term by term integration may not be the Fourier series for the integral of the function.

Chapter 8

Laplace Transforms

There are many kinds of transforms out there in the world. Laplace transforms and Fourier transforms are probably the main two kinds of transforms that are usedo solve differential equations. As we will see in later sections we can use Laplace transforms to reduce a differential equation to an algebra problem. The algebra can be messy on occasion, but it will be simpler than actually solving the differential equation directly in many cases. Laplace transforms can also be used to solve IVP's that we can't use any previous method on. For "simple" differential equations such as those in the first few sections of the last chapter Laplace transforms will be more complicated than we need.

In fact, for most homogeneous differential equations such as those in the last chapter Laplace transforms is significantly longer and not so useful. Also, many of the "simple" nonhomogeneous differential equations that we saw in the Undetermined Coefficients and Variation of Parameters are still simpler (or at the least no more difficult than Laplace transforms) to do as we did them there. However, at this point, the amount of work required for Laplace transforms is starting to equal the amount of work we did in those sections.

Laplace transforms comes into its own when the forcing function in the differential equation starts getting more complicated. In the previous chapter we looked only at nonhomogeneous differential equations in which $g(t)$ was a fairly simple continuous function. In this chapter we will start looking at $g(t)$'s that are not continuous. It is these problems were the reasons for using Laplace transforms starts to become clear. We will also see that, for some of the more complicated nonhomogeneous differential equations from the last chapter, Laplace transforms are actually easier on those problems as well.

Laplace transforms (or just transforms) can seem scary when we first start looking at them. However, as we will see, they aren't as bad as they may appear at first. Before we start with the definition of the Laplace transform we need to get another definition out of the way.

A function is called piecewise continuous on an interval if the interval can be broken into a finite number of subintervals on which the function is continuous on each open subinterval (*i.e.* the subinterval without it's endpoints) and has a finite limit at the endpoints of each subinterval. In other words, a piecewise continuous function is a function that has a finite number of breaks in it and doesn't blow up to infinity anywhere.

Definition: Suppose that *f(t)* is a piecewise continuous function. The Laplace transform of *f(t)* is denoted $\mathfrak{L}\{f(t)\}$ and defined as

$$\mathfrak{L}\{f(t)\} = \int_0^\infty e^{-st} f(t)dt.$$

There is an alternate notation for Laplace transforms. For the sake of convenience we will often denote Laplace transforms as,

$$\mathfrak{L}\{f(t)\} = F(s).$$

With this alternate notation, note that the transform is really a function of a new variable, *s*, and that all the *t*'s will drop out in the integration process. Now, the integral in the definition of the transform is called an improper integral and it would probably be best to recall how these kinds of integrals work before we actually jump into computing some transforms.

Example: If c $\neq$ 0, evaluate the following integral.

$$\int_0^\infty e^{ct} dt.$$

Solution: Remember that you need to convert improper integrals to limits as follows,

$$\int_0^\infty e^{ct} dt \lim_{n\to\infty} \int_0^\infty e^{ct} dt.$$

Now, do the integral, then evaluate the limit.

$$\begin{aligned}\int_0^\infty e^{ct} dt &= \lim_{n\to\infty} \int_0^\infty e^{ct} dt \\ &= \lim_{n\to\infty} \left(\frac{1}{c} e^{ct}\right)\Bigg|_0^n \\ &= \lim_{n\to\infty} \left(\frac{1}{c} e^{cn} - \frac{1}{c}\right).\end{aligned}$$

Now, at this point, we've got to be careful. The value of *c* will affect our answer. We've already assumed that *c* was non-zero, now we need to worry about the sign of *c*. If *c* is positive the exponential will go to infinity. On the other hand, if *c* is negative the exponential will go to zero. So, the integral is only convergent (*i.e.* the limit exists and is finite) provided $c<0$. In this case we get,

$$\int_0^\infty e^{ct} dt = -\frac{1}{c} \text{ provided } c < 0.$$

Now that we remember how to do these, let's compute some Laplace transforms. We'll start off with probably the simplest Laplace transform to compute.

Example: Compute $\mathfrak{L}\{1\}$.

Solution: There's not really a whole lot do here other than plug the function $f(t) = 1$

$$\mathfrak{L}\{1\} = \int_0^\infty e^{-s} dt.$$

Now, at this point notice that this is nothing more than the integral in the previous example with $c = -s$. Therefore, all we need to do the appropriate substitution. Doing this gives,

$$\mathfrak{L}\{1\} = \int_0^\infty e^{-s} dt = -\frac{1}{-s} \quad \text{provided } -s < 0$$

Or, with some simplification we have,

$$\mathfrak{L}\{1\} = \frac{1}{s} \qquad \text{provided } -\mathrm{s} < 0.$$

Notice that we had to put a restriction on s in order to actually compute the transform. All Laplace transforms will have restrictions on s.

At this stage of the game, this restriction is something that we tend to ignore, but we really shouldn't ever forget that it's there.

Let's do another example.

Example: Compute $\mathfrak{L}\{e^{at}\}$

Solution: Plug the function into the definition of the transform and do a little simplification.

$$\mathfrak{L}\{e^{at}\} = \int_0^\infty e^{-st} e^{at} dt = \int_0^\infty e^{(a-s)t} dt.$$

Once again, notice that we can provide $c = a - s$. So let's do this.

$$\begin{aligned} \mathfrak{L}\{e^{at}\} &= \int_0^\infty e^{(a-s)t} dt \\ &= \frac{1}{a-s} \qquad \text{provided } a - s < 0 \\ &= \frac{1}{s-a} \qquad \text{provided } s > a. \end{aligned}$$

Let's do one more example that doesn't come down to an application.

Example : Compute $\mathfrak{L}\{\sin(at)\}$.

Solution: Note that we're going to leave it to you to check most of the integration here. Plug the function into the definition. This time let's also use the alternate notation.

$$\mathcal{L}\{\sin(at)\} = F(s)$$

$$= \int_0^{\infty} e^{-st} \sin(at) dt$$

$$= \lim_{n\to\infty} \int_0^{\infty} e^{-st} \sin(at)\, dt.$$

Now, if we integrate by parts we will arrive at,

$$F(s) = \lim_{n\to\infty}\left(-\left(\frac{1}{a}e^{-st}\cos(at)\right)\Bigg|_0^n - \frac{s}{a}\int_0^n e^{-st}\cos(at)\, dt \right).$$

Now, evaluate the first term to simplify it a little and integrate by parts again on the integral. Doing this arrives at,

$$F(s) = \lim_{n\to\infty}\left(\frac{1}{a}\left(1 - e^{-in}\cos(at)\right) - \frac{s}{a}\left(\left(\frac{1}{a}e^{-st}\sin(at)\right)\Bigg|_0^n + \frac{s}{a}\int_0^n e^{-st}\cos(at)\, dt\right)\right).$$

Now, evaluate the second term, take the limit and simplify.

$$F(s) = \lim_{n\to\infty}\left(\frac{1}{a}(1 - e^{-n}\cos(an)) - \frac{s}{a}\left(\frac{1}{a}e^{-sn}\sin(an) + \frac{s}{a}\int_0^n e^{-st}\sin(at)\, dt.\right)\right)$$

$$= \frac{1}{a} - \frac{s}{a}\left(\frac{s}{a}\int_0^{\infty} e^{-st}\sin(at)\, dt\right) = \frac{1}{a} - \frac{s^2}{a^2}\int_0^{\infty} e^{-st}\sin(at)\, dt.$$

Now, notice that in the limits we had to assume that $s>0$ in order to do the following two limits.

$$\lim_{n\to\infty} e^{-sn}\cos(an) = 0$$

$$\lim_{n\to\infty} e^{-sn}\sin(an) = 0.$$

Without this assumption, we get a divergent integral again. Also, note that when we got back to the integral we just converted the upper limit back to infinity.

The reason for this is that, if you think about it, this integral is nothing more than the integral that we started with. Therefore, we now get,

$$F(s) = \frac{1}{a} - \frac{s^2}{a^2}F(s).$$

Now, simply solve for *F(s)* to get,

$$\mathfrak{L}\{\sin(at)\} = F(s) = \frac{a}{s^2 + a} \qquad \text{provided s > 0.}$$

As this example shows, computing Laplace transforms is often messy.

Before moving on to the next section, we need to do a little side note. On occasion you will see the following as the definition of the Laplace transform.

$$\mathfrak{L}\{f(t)\} = \int_0^{\infty} e^{-st} f(t)dt.$$

Note the change in the lower limit from zero to negative infinity. In these cases there is almost always the assumption that the function *f(t)* is in fact defined as follows,

$$f(t) = \begin{cases} 0 & \text{if } t < 0 \\ f(t) & \text{if } t < 0 \end{cases}.$$

In other words, it is assumed that the function is zero if $t<0$. In this case the first half of the integral will drop out since the function is zero. A Heaviside function is usually used to make the function zero for $t<0$. We will be looking at these in a later section

LAPLACE TRANSFORMS

As we saw in the last section computing Laplace transforms directly can be fairly complicated. Usually we just use a table of transforms when actually computing Laplace transforms. The table that is provided here is not an inclusive table, but does include most of the commonly used Laplace transforms and most of the commonly needed formulas pertaining to Laplace transforms.

Before doing a couple of examples to illustrate the use of the table let's get a quick fact out of the way.

FACT

Given *f(t)* and *g(t)* then,

$$\mathfrak{L}\{af(t) + bg(t)\} = aF(s) + bG(s).$$

for any constants *a* and *b*.

In other words, we don't worry about constants and we don't worry about sums or differences of functions in taking Laplace transforms. All that we need to do is take the transform of the individual functions, then put any constants back in and add or subtract the results back up.

So, let's do a couple of quick examples.

Example: Find the Laplace transforms of the given functions.

1. $f(t) = 6e^{-5t} + e^{3t} + 5t^3 - 9$
2. $g(t) = 4\cos(4t) - 9\sin(4t) + 2\cos(10t)$
3. $h(t) = 3\sin(2t) + 3\sin(2t)$
4. $g(t) = e^{st} + \cos(6t) - e^{3t}\cos(6t)$.

Solution: We'll do these examples in a little more detail than is typically used since this is the first time we're using the tables.

1. $f(t) = 6e^{-5t} + e^{3t} + 5t^3 - 9$

$$F(s) = 6\frac{1}{s-(-5)} + \frac{1}{s-3} + 5\frac{3!}{s^{3+1}} - 9\frac{1}{s}$$

$$= \frac{6}{s+5} + \frac{1}{s-3} + \frac{30}{s^4} - \frac{9}{s}$$

2. $g(t) = 4\cos(4t) - 9\sin(4t) + 2\cos(10t)$

$$G(s) = 4\frac{s}{s^2+(4)^2} - 9\frac{4}{s^2+(4)^2} + 2\frac{s}{s^2+(10)^2}$$

$$= \frac{4s}{s^2+16} - \frac{36}{s^2+16} + \frac{2s}{s^2+100}$$

3. $h(t) = 3\sin h(2t) + 3\sin(2t)$

$$H(s) = 3\frac{2}{s^2-(2)^2} + 3\frac{2}{s^2+(2)^2}$$

$$= \frac{6}{s^2-4} + \frac{6}{s^2+4}$$

4. $g(t) = e^{3t} + \cos(6t) - e^{3t}\cos(6t)$

$$G(s) = \frac{1}{s-3} + \frac{s}{s^2+(6)^2} - \frac{s-3}{(s-3)^2+(6)^2}$$

$$= \frac{1}{s-3} + \frac{s}{s^2+36} - \frac{s-3}{(s-3)^2+36}.$$

Make sure that you pay attention to the difference between a "normal" trig function and hyperbolic functions. The only difference between them is the "+ a^2" for the "normal" trig functions becomes a "- a^2" in the hyperbolic function! It's very easy to get in a hurry and not pay attention and grab the wrong formula.

Let's do one final set of examples.

Example: Find the transform of each of the following functions.

1. $f(t) = t\cosh(3t)$
2. $h(t) = t^2 \sin(2t)$
3. $g(t) = t^{\frac{3}{2}}$
4. $f(t) = (10t)^{\frac{3}{2}}$
5. $f(t) = tg'(t)$

Solution:

1. $f(t) = t\cosh(3t)$

This function is not in the table of Laplace transforms. If we take *n=1*.

$$F(s) = \mathfrak{L}\{tg(t)\} = -G'(s), \quad \text{where } g(t) = \cosh(3t).$$

So, we then have,

$$G(s) = \frac{s}{s^2 - 9} \qquad G'(s) = -\frac{s^2 + 9}{(s^2 - 9)^2}$$

we then have,

$$F(s) = \frac{s^2 = 9}{\left(s^2 - 9\right)^2}$$

2. $h(t) = t^2 \sin(2t)$

We could use it with $n = 1$ n = 1$n = 1$.

$$H(s) = \mathfrak{L}\{tf(t)\} = -F'(s), \quad \text{where } f(t) = t\sin(2t).$$

Or we could use it with $n = 2$.

$$H(s) = \mathfrak{L}\{t^2 f(t)\} = F''(s), \quad \text{where } f(t) = t\sin(2t).$$

Since it's less work to do one derivative, let's do it the first way. So we have,

$$F(s) = \frac{4s}{\left(s^2 + 4\right)^2} \qquad F'(s) = -\frac{12s^2 - 16}{\left(s^2 + 4\right)^2}.$$

The transform is then,

$$H(s) = \frac{12s^2 - 16}{\left(s^2 + 4\right)}$$

3. $g(t) = t^{\frac{3}{2}}$

This part can be with $n = 2$. We need to notice that

$$\int_0^t \sqrt{v}\,dv = \frac{2}{3} t^{\frac{2}{3}} \qquad \Rightarrow \qquad t^{\frac{2}{3}} = \frac{2}{3}\int_0^t \sqrt{v}\,dv.$$

Now,

$$f(t) = \sqrt{t} \qquad\qquad F(s) = \frac{\sqrt{\pi}}{2s^{\frac{3}{2}}}.$$

we get the following.

$$G(s) = \frac{3}{2}\left(\frac{\sqrt{\pi}}{2s^{\frac{3}{2}}}\right)\left(\frac{1}{s}\right) = \frac{3\sqrt{\pi}}{4s^{\frac{3}{2}}}$$

4. $f(t) = (10t)^{\frac{3}{2}}$

To see this note that if

$$g(t) = t^{\frac{3}{2}}$$

then

$$f(t) = g(10t).$$

Therefore, the transform is.

$$F(s) = \frac{1}{10} G\left(\frac{s}{10}\right)$$

$$= \frac{1}{10}\left(\frac{3\sqrt{\pi}}{4\left(\frac{s}{10}\right)^{\frac{5}{2}}}\right) = 10^{\frac{3}{2}} \frac{3\sqrt{\pi}}{4s^{\frac{5}{2}}}$$

5. $f(t) = tg'(t)$

$$\mathfrak{L}\{tg'(t)\} = -\frac{d}{ds}\mathfrak{L}\{g'\} = -\frac{d}{ds}\{sG(s) - g(0)\}$$

$$= \left(G(s) + sG'(s) - 0\right)$$

$$= G(s) - sG'(s).$$

Remember that *g(0)* is just a constant so when we differentiate it we will get zero!

As this set of examples has show us we can't forget to use some of the general formulas to derive new Laplace transforms for functions that aren't explicitly listed in the table!

INVERSE LAPLACE TRANSFORMS

Finding the Laplace transform of a function is not terribly difficult if we've got a table of transforms in front of us to use as we saw in the last section. What we would like to do now is go the other way.

We are going to be given a transform, *F(s)*, and ask what function (or functions) did we have originally. As you will see this can be a more complicated and lengthy process than taking transforms. In these cases we say that we are finding the Inverse Laplace Transform of *F(s)* and use the following notation.

$$f(t) = \mathfrak{L}^{-1}\{F(s)\}.$$

As with Laplace transforms, we've got the following fact to help us take the inverse transform.

FACT

Given the two Laplace transforms *F(s)* and *G(s)* then

$$\mathfrak{L}^{-1}\{aF(s) + bG(s)\} = a\mathfrak{L}^{-1}\{F(s)\} + b\mathfrak{L}^{-1}\{G(s)\}$$

for any constants *a* and *b*.

So, we take the inverse transform of the individual transforms, put any constants back in and then add or subtract the results back up.

Let's take a look at a couple of fairly simple inverse transforms.

Example: Find the inverse transform of each of the following.

1. $F(s) = \frac{6}{s} - \frac{1}{s-8} + \frac{4}{s-3}$
2. $H(s) = \frac{19}{s+2} - \frac{1}{3s-5} + \frac{7}{s^5}$
3. $F(s) = \frac{6s}{s^2+25} + \frac{3}{s^2+25}.$
4. $G(s) = \frac{8}{3s^2+12} + \frac{3}{s^2+49}$

Solution: I've always felt that the key to doing inverse transforms is to look at the denominator and try to identify what you've got based on that. If there is only one entry in the table that has that particular denominator, the next step is to make sure the numerator is correctly set up for the inverse transform process. If it isn't, correct it (this is always easy to do) and then take the inverse transform.

If there is more than one entry in the table has a particular denominator, then the numerators of each will be different, so go up to the numerator and see which one you've got. If you need to correct the numerator to get it into the correct form and then take the inverse transform.

So, with this advice in mind let's see if we can take some inverse transforms.

1. $F(s)=\frac{6}{s}-\frac{1}{s-8}+\frac{4}{s+3}$.

From the denominator of the first term it looks like the first term is just a constant. The correct numerator for this term is a "1" so we'll just factor the 6 out before taking the inverse transform. The second term appears to be an exponential with $a = 8$ and the numerator is exactly what it needs to be. The third term also appears to be an exponential, only this time $a =$ 3 and we'll need to factor the 4 out before taking the inverse transforms.

So, with a little more detail than we'll usually put into these,

$$F(s) = 6\frac{1}{s}-\frac{1}{s-8}+4\frac{1}{s-3}$$

$$f(t) = 6(1)-e^{8t}+4\left(e^{3t}\right)$$

$$= 6-e^{8t}+4e^{3t}$$

2. $H(s)=\frac{19}{s+2}-\frac{1}{3s-5}+\frac{7}{s^5}$.

The first term in this case looks like an exponential with $a = -2$ and we'll need to factor out the 19. Be careful with negative signs in these problems, it's very easy to lose track of them.

The second term almost looks like an exponential, except that it's got a *3s* instead of just an *s* in the denominator. It is an exponential, but in this case we'll need to factor a 3 out of the denominator before taking the inverse transform.

The denominator of the third term appears with $n = 4$. The numerator however, is not correct for this. There is currently a 7 in the numerator and we need a $4! = 24$ in the numerator. This is very easy to fix.

Whenever a numerator is off by a multiplicative constant, as in this case, all we need to do is put the constant that we need in the numerator. We will just need to remember to take it back out by dividing by the same constant.

So, let's first rewrite the transform.

$$H(s) = \frac{19}{s-(-2)} - \frac{1}{3\left(s-\frac{5}{3}\right)} + \frac{7}{s^{4+1}}$$

$$= 19\frac{1}{s-(-2)} - \frac{1}{3}\frac{1}{s-\frac{5}{3}} + \frac{7}{4!s^{4+1}}.$$

So, what did we do here? We factored the 19 out of the first term. We factored the 3 out of the denominator of the second term since it can't be there for the inverse transform and in the third term we factored everything out of the numerator except the 4! since that is the portion that we need in the numerator for the inverse transform process.

Let's now take the inverse transform.

$$h(t) = 19e^{-2t} - \frac{1}{3}e^{\frac{5t}{3}} + \frac{7}{24}t^4$$

3. $F(s) = \dfrac{6s}{s^2+25} + \dfrac{2}{s^2+25}.$

The numerators will tell us which we've actually got. The first one has an s in the numerator and so this means that the first term must be #8 and we'll need to factor the 6 out of the numerator in this case. The second term has only a constant in the numerator and so this term must be #7, however, in order for this to be exactly #8 we'll need to multiply/divide a 5 in the numerator to get it correct for the table.

The transform becomes,

$$F(s) = 6\frac{s}{s^2+(5)^2} + \frac{3\frac{5}{5}}{s^2+(5)^2}$$

$$= 6\frac{s}{s^2+(5)^2} + \frac{3}{5}\frac{5}{s^2+(5)^2}.$$

Taking the inverse transform gives,

$$f(t)\, 6\cos(5t) + \frac{3}{5}\sin(5t)$$

4. $G(s) = \dfrac{8}{3s^2+12} + \dfrac{2}{s^2-49}.$

In this case the first term will be a sine once we factor a 3 out of the denominator, while the second term appears to be a hyperbolic sine. Again, be careful with the difference between these two. Both of the terms will also need to have their numerators fixed up. Here is the transform once we're done rewriting it.

$$G(s) = \frac{1}{3}\frac{8}{s^2+4} + \frac{3}{s^2-49} = \frac{1}{3}\frac{(4)(2)}{s^2+(2)^2} + \frac{3\frac{7}{7}}{s^2-(7)^2}.$$

Notice that in the first term we took advantage of the fact that we could get the 2 in the numerator that we needed by factoring the 8.The inverse transform is then,

$$g(t) = \frac{4}{3}\sin(2t) + \frac{3}{7}\sin h(7t).$$

So, probably the best way to identify the transform is by looking at the denominator. If there is more than one possibility use the numerator to identify the correct one. Fix up the numerator if needed to get it into the form needed for the inverse transform process. Finally, take the inverse transform.

Let's do some slightly harder problems. These are a little more involved than the first set.

Example: Find the inverse transform of each of the following.

1. $F(s) = \dfrac{6s - 5}{s^2 + 7}$
2. $F(s) = \dfrac{1 - 3s}{s^2 + 8s + 21}$
3. $G(s) = \dfrac{3s - 2}{2s^2 - 6s - 2}$
4. $H(s) = \dfrac{s + 7}{s^2 - 3s - 10}$

Solution:

1. $F(s) = \dfrac{6s - 5}{s^2 + 7}$.

From the denominator of this one it appears that it is either a sine or a cosine. However, the numerator doesn't match up to either of these in the table. A cosine wants just an s in the numerator with at most a multiplicative constant, while a sine wants only a constant and no s in the numerator. We've got both in the numerator. This is easy to fix however. We will just split up the transform into two terms and then do inverse transforms.

$$F(s) = \frac{6s - 5}{s^2 + 7} - \frac{5\frac{\sqrt{7}}{\sqrt{7}}}{s^2 + 7}$$

$$f(t) = f(t) = 6\cos\left(\sqrt{7}t\right) - \frac{5}{\sqrt{7}}\sin\left(\sqrt{7}t\right).$$

Do not get too used to always getting the perfect squares in sines and cosines that we saw in the first set of examples. More often than not (at least in my class) they won't be perfect squares!

2. $F(s)=\dfrac{1-3s}{s^2+8s+21}$

In this case there are no denominators in our table that look like this. We can however make the denominator look like one of the denominators in the table by completing the square on the denominator. So, let's do that first.

$$\begin{aligned} s^2+8s+21 &= s^2+8s+16-16+21 \\ &= s^2+8s+16+5 \\ &= (s+4)^2+5 \end{aligned}$$

Recall that in completing the square you take half the coefficient of the s, square this, and then add and subtract the result to the polynomial. After doing this the first three terms should factor as a perfect square.

So, the transform can be written as the following.

$$F(s)=\frac{1-3s}{(s+4)^2+5}.$$

However, note that in order for it to be a #19 we want just a constant in the numerator and in order to be a #20 we need an s __ a in the numerator. We've got neither of these so we'll have to correct the numerator to get it into proper form.

In correcting the numerator always get the s __ a first. This is the important part. We will also need to be careful of the 3 that sits in front of the s. One way to take care of this is to break the term into two pieces, factor the 3 out of the second and then fix up the numerator of this term. This will work, however it will put three terms into our answer and there are really only two terms.

So, we will leave the transform as a single term and correct it as follows,

$$\begin{aligned} F(s) &= \frac{1-3(s+4-4)}{(s+4)^2+5} \\ &= \frac{1-3(s+4)+12}{(s+4)^2+5} \\ &= \frac{-3(s+4)+13}{(s+4)^2+5}. \end{aligned}$$

We needed an $s + 4$ in the numerator, so we put that in. We just needed to make sure and take the 4 back out by subtracting it back out. Also, because of the 3 multiplying the s we needed to do all this inside a set of parenthesis. Then we partially multiplied the 3 through the second term and combined the constants. With the transform in this form, we can break it up into two transforms each of which are in the tables and so we can do inverse transforms on them,

$$F(s) = -3\frac{s+4}{(s+4)^2+5} + \frac{13\frac{\sqrt{5}}{\sqrt{5}}}{(s+4)^2+5}$$

$$f(t) = -3e^{-4t}\cos\left(\sqrt{5}t\right) + \frac{13}{\sqrt{5}}e^{-4t}\sin\left(\sqrt{5}t\right)$$

3. $G(s) = \dfrac{3s-2}{2s^2-6x-2}$.

This one is similar to the last one. We just need to be careful with the completing the square however. The first thing that we should do is factor a 2 out of the denominator, then complete the square. Remember that when completing the square a coefficient of 1 on the s^2 term is needed! So, here's the work for this transform.

$$G(s) = \frac{3s-2}{2\left(s^2-3s-1\right)}$$

$$= \frac{1}{2}\frac{3s-2}{s^2-3s+\frac{9}{4}-\frac{9}{4}-1}$$

$$= \frac{1}{2}\frac{3s-2}{\left(s^2-\frac{3}{2}\right)-\frac{13}{4}}.$$

Here's the work for that and the inverse transform.

$$G(s) = \frac{1}{2}\frac{\left(s-\frac{3}{2}+\frac{3}{2}\right)-2}{\left(s-\frac{3}{2}\right)^2-\frac{13}{4}}$$

$$= \frac{1}{2}\frac{3\left(s-\frac{3}{2}\right)+\frac{5}{2}}{\left(s-\frac{3}{2}\right)^2-\frac{13}{4}}$$

$$= \frac{1}{2}\left(\frac{3\left(s-\frac{3}{2}\right)}{\left(s-\frac{3}{2}\right)^2-\frac{13}{4}} + \frac{\frac{5}{2}\frac{\sqrt{13}}{\sqrt{13}}}{\left(s-\frac{3}{2}\right)^2-\frac{13}{4}}\right)$$

$$g(t) = \frac{1}{2}\left(3e^{\frac{3t}{2}}\cosh\left(\frac{\sqrt{13}}{2}t\right) + \frac{5}{\sqrt{13}}e^{\frac{3t}{2}}\sin h\left(\frac{\sqrt{13}}{2}t\right)\right).$$

In correcting the numerator of the second term, notice that I only put in the square root since we already had the "over 2" part of the fraction that we needed in the numerator.

4. $H(s) = \frac{s+7}{s^2 - 3s - 10}$.

This one appears to be similar to the previous two, but it actually isn't. The denominators in the previous two couldn't be easily factored. In this case the denominator does factor and so we need to deal with it differently. Here is the transform with the factored denominator.

$$H(s) = \frac{s+7}{(s+2)(s-5)}.$$

The denominator of this transform seems to suggest that we've got a couple of exponentials, however in order to be exponentials there can only be a single term in the denominator and no s's in the numerator.

To fix this we will need to do partial fractions on this transform. In this case the partial fraction decomposition will be

$$H(s) = \frac{A}{s+2} + \frac{B}{a-5}.$$

Don't remember how to do partial fractions? In this example we'll show you one way of getting the values of the constants and after this example we'll review how to get the correct form of the partial fraction decomposition.

Okay, so let's get the constants. There is a method for finding the constants that will always work, however it can lead to more work than is sometimes required.

Eventually, we will need that method, however in this case there is an easier way to find the constants.

Regardless of the method used, the first step is to actually add the two terms back up. This gives the following.

$$\frac{s+7}{(s+2)(s-5)} = \frac{A(s-5)+B(s+2)}{(s+2)(s-5)}.$$

Now, this needs to be true for any s that we should choose to put in. So, since the denominators are the same we just need to get the numerators equal. Therefore, set the numerators equal.

$$s+7 = A(s-5)+B(s+2).$$

Again, this must be true for ANY value of s that we want to put in. So, let's take advantage of that. If it must be true for any value of s then it must be true for $s = -2$, to pick a value at random. In this case we get,

$$s = A(-7) + B(0) \qquad \Rightarrow \quad A = -\frac{5}{7}.$$

We found A by appropriately picking s. We can B in the same way if we chose $s = 5$.

$$12 = A(0) + B(7) \qquad \Rightarrow \quad B = \frac{12}{7}.$$

This will not always work, but when it does it will usually simplify the work considerably.

So, with these constants the transform becomes,

$$H(s) = \frac{-\frac{5}{7}}{s+2} + \frac{\frac{12}{7}}{s-5}.$$

We can now easily do the inverse transform to get,

$$h(t) = -\frac{5}{7}e^{-2t} + \frac{12}{7}e^{5t}.$$

The last part of this example needed partial fractions to get the inverse transform. When we finally get back to differential equations and we start using Laplace transforms to solve them, you will quickly come to understand that partial fractions are a fact of life in these problems. Almost every problem will require partial fractions to one degree or another.

Note that we could have done the last part of this example as we had done the previous two parts. If we had we would have gotten hyperbolic functions. However, recalling the definition of the hyperbolic functions we could have written the result in the form we got from the way we worked our problem.

However, most students have a better feel for exponentials than they do for hyperbolic functions and so it's usually best to just use partial fractions and get the answer in terms of exponentials. It may be a little more work, but it will give a nicer (and easier to work with) form of the answer.

Be warned that in my class I've got a rule that if the denominator can be factored with integer coefficients then it must be.

So, let's remind you how to get the correct partial fraction decomposition. The first step is to factor the denominator as much as possible. Then for each term in the denominator we will use the following table to get a term or terms for our partial fraction decomposition.

Factor in denominator	*Term in partial fraction decomposition*
$ax+b$	$\dfrac{A}{ax+b}$
$(ax+b)^k$	$\dfrac{A_1}{ax+b}+\dfrac{A_2}{(ax+b)^2}+\cdots+\dfrac{A_k}{(ax+b)^k}$
ax^2+bx+c	$\dfrac{Ax+B}{ax^2+bx+c}$
$(ax^2+bx+c)^k$	$\dfrac{A_1x+B_1}{ax^2+bx+c}+\dfrac{A_2x+B_2}{(ax^2+bx+c)}+\cdots+\dfrac{A_kx+B_k}{(ax^2+bx+c)^k}.$

Notice that the first and third cases are really special cases of the second and fourth cases respectively. So, let's do a couple more examples to remind you how to do partial fractions.

Example: Find the inverse transform of each of the following.

1. $G(s)=\dfrac{86s-78}{(s+3)(s-4)(5s-1)}$

2. $F(s)=\dfrac{2-5s}{(s-6)(s^2+11)}$

3. $G(s)=\dfrac{25}{s^3(s^2+4s+5)}$

Solution:

1. $G(s)=\dfrac{86s-78}{(s+3)(s-4)(5s-1)}$

Here's the partial fraction decomposition for this part.

$$G(s)=\frac{A}{(s+3)}+\frac{A}{(s-4)}+\frac{A}{(5s-1)}.$$

Now, this time we won't go into quite the detail as we did in the last example.

We are after the numerator of the partial fraction decomposition and this is usually easy enough to do in our heads.

Therefore, we will go straight to setting numerators equal.

$$86s-78=A(s-4)(5s-1)+B(s+3)(5s-1)+C(s+3)(s-4).$$

As with the last example, we can easily get the constants by correctly picking values of s.

$$s=-3 \qquad -336=A(-7)(-16) \qquad \Rightarrow \qquad A=-3$$

$$s=\frac{1}{5} \qquad -\frac{304}{5}=C\left(\frac{16}{5}\right)\left(-\frac{19}{5}\right) \qquad \Rightarrow \qquad C=5.$$

$$s=4 \qquad 266=B(7)(19) \qquad \Rightarrow \qquad B=2$$

So, the partial fraction decomposition for this transform is,

$$G(s)=-\frac{3}{s+3}+\frac{2}{s-4}+\frac{5}{5s-1}.$$

Now, in order to actually take the inverse transform we will need to factor a 5 out of the denominator of the last term. The corrected transform as well as its inverse transform is.

$$G(s)=-\frac{3}{s+3}+\frac{2}{s-4}+\frac{5}{s-\frac{1}{5}}$$

$$g(t)=-3e^{-3t}+2e^{4t}+e^{\frac{t}{5}}$$

2. $F(s)=\dfrac{2-5s}{(s-6)(s^2+11)}$.

So, for the first time we've got a quadratic in the denominator. Here's the decomposition for this part.

$$F(s)=\frac{A}{s-6}+\frac{Bs+C}{s^2+11}.$$

Setting numerators equal gives,

$$2-5s=A(s^2+11)+(Bs+C)(s-6).$$

Okay, in this case we could use $s=6$to quickly find A, but that's all it would give. In this case we will need to go the "long" way around to getting the constants.

Note that this way will always work, but is sometimes more work than is required.

The "long" way is to completely multiply out the right side and collect like terms.

$$\begin{aligned} 2-5s &= A(s^2+11)+(Bs+C)(s-6) \\ &= As^2+11A+Bs^2+6B+Cs-6C \\ &= (A+B)s^2+(-6B+C)s+11A-6C. \end{aligned}$$

In order for these two to be equal the coefficients of the s^2, s and the constants must all be equal. So, setting coefficients equal gives the following system of equations that can be solved.

$$\left.\begin{aligned} s^2 &: \quad A+B=0 \\ s^1 &: -6B+C=-5 \\ s^0 &: 11A-6C=2 \end{aligned}\right\} \Rightarrow \quad \text{A}=-\frac{28}{47}, \quad B=\frac{28}{47}, \quad C=-\frac{67}{47}.$$

Notice that I used $s0$ to denote the constants. This is habit on my part and isn't really required, it's just what I'm used to doing.

Also, the coefficients are fairly messy fractions in this case. Get used to that. They will often be like this when we get back into solving differential equations.

There is a way to make our life a little easier as well with this. Since all of the fractions have a denominator of 47 we'll factor that out as we plug them back into the decomposition. This will make dealing with them much easier.

The partial fraction decomposition is then,

$$F(s)=\frac{1}{47}\left(\frac{28}{s-6}+\frac{28s-67}{s^2+11}\right)$$

$$=\frac{1}{47}\left(\frac{28}{s-6}+\frac{28s}{s^2+11}-\frac{67\frac{\sqrt{11}}{\sqrt{11}}}{s^2+11}\right).$$

The inverse transform is then.

$$f(t)=\frac{1}{47}\left(-28e^{6t}+28\cos\left(\sqrt{11}t\right)-\frac{67}{\sqrt{11}}\sin\left(\sqrt{11}t\right)\right).$$

3. $G(s)=\dfrac{25}{s^3(s^2+4s+5)}$

With this last part do not get excited about the s^3. We can think of this term as

$$s^3=(s-0)^3$$

and it becomes a linear term to a power. So, the partial fraction decomposition is

$$G(s)=\frac{A}{s}+\frac{B}{s^2}+\frac{C}{s^3}+\frac{Ds+E}{s^2+4s+5}.$$

Setting numerators equal and multiplying out gives.

$$25=As^2(s^2+4s+5)+Bs(s^2+4s+5)+C(s^2+4s+5)+(Ds+E)s^3$$
$$=(A+D)s^4+(4A+B+E)s^3+(5A+4B+C)s^2(5B+4C)s+5C.$$

Setting coefficients equal gives the following system.

$$\left.\begin{aligned} s^4: &\quad A+D=0\\ s^3: &\quad 4A+B+E=0\\ s^2: &\quad 5A+4B+C=0\\ s^1: &\quad 5B+4C=0\\ s^0: &\quad 5C=25 \end{aligned}\right\} \Rightarrow A=\frac{11}{5},\ B=-4,\ C=5, D=-\frac{11}{5}, E=-\frac{24}{5}.$$

This system looks messy, but it's easier to solve than it might look. First we get C for free from the last equation. We can then use the fourth equation to find B. The third equation will then give A, *etc.*

When plugging into the decomposition we'll get everything with a denominator of 5, then factor that out as we did in the previous part in order to make things easier to deal with.

$$G(s)=\frac{1}{5}\left(\frac{11}{s}-\frac{20}{s^2}+\frac{25}{s^3}-\frac{11s+24}{s^2+4s+5}\right).$$

Notc that wc also factored a minus sign out of the last two terms. To complete this part we'll need to complete the square on the later term and fix up a couple of numerators. Here's that work.

$$\begin{aligned} G(s) &= \frac{1}{5}\left(\frac{11}{s}-\frac{20}{s^2}+\frac{25}{s^3}-\frac{11s+24}{s^2+4s+5}\right)\\ &= \frac{1}{5}\left(\frac{11}{s}-\frac{20}{s^2}+\frac{25}{s^3}-\frac{11(s+2-2)+24}{(s+2)^2+1}\right)\\ &= \frac{1}{5}\left(\frac{11}{s}-\frac{20}{s^2}+\frac{25\frac{2!}{2!}}{s^3}-\frac{11(s+2)}{(s+2)^2+1}-\frac{2}{(s+2)^2+1}\right). \end{aligned}$$

The inverse transform is then.

$$g(t)=\frac{1}{5}\left(11-20t+\frac{25}{2}t^2-11e^{-2t}\cos(t)-2e^{-2t}\sin(t)\right).$$

So, one final time. Partial fractions are a fact of life when using Laplace transforms to solve differential equations. Make sure that you can deal with them.

APPLICATION FOR LAPLACE TRANSFORM

In this Section we discuss some theorem on Laplace transform. Also, we discuss the application of Laplace transform.

THEOREMS OF THE LAPLACE TRANSFORM

The Laplace transform of the function f(t) is denoted by

$$L(f(t)) = \int_0^\infty e^{-st} f(t)\,dt$$

(a) *Superposition theorem:* For arbitrary constants a_1 and a_2 it follows that

$$\{a_1 f_1(t) + a_2 f_2(t)\} = a_1 F_1(s) + a_2 F_2(s).$$

The Laplace transformation is a linear integral transformation.

(b) *Similarity theorem:* For an arbitrary constant a > 0

$$\{f(at)\} = \frac{1}{a} F\left(\frac{s}{a}\right)$$

is valid. This follows from Equation by the substitution of $\tau = at$.

(c) *Real Shifting theorem:* For an arbitrary constant a > 0

$$\{f(t-a)\} = e^{-as} F(s)$$

is valid. This follows directly from Equation by the substitution of $\tau = t - a$.

(d) *Complex Shifting theorem:* For an arbitrary constant a > 0

$$\{e^{-at} f(t)\} = F(s+a)$$

is valid.

(e) *Derivative theorem:* For a causal function of time, $f(t)$, for which the derivative for $t > 0$ exists.

$$\left\{\frac{\mathrm{d}f(t)}{\mathrm{d}t}\right\} = s\ F(s) - f(0+),$$

and in the case of multiple differentiation

$$\left\{\frac{\mathrm{d}^n f(t)}{\mathrm{d}t^n}\right\} = s^n\ F(s) - \sum_{i=1}^{n} s^{n-i} \left.\frac{\mathrm{d}^{(i-1)} f(t)}{\mathrm{d}t^{(i-1)}}\right|_{t=0+}$$

(f) *Complex differentiation theorem:* This theorem shows that a differentiation of the mapped function F(s) corresponds to a multiplication with the time t in the time domain:

$$\{t^k = f(t)\} = (-1)^k \frac{d^k F(s)}{ds^k}.$$

(g) *Integral theorem:* The integral of a function is mapped by

$$\left\{\int_0^t = f(\tau)\mathrm{d}\tau\right\} = \frac{1}{s} F(s).$$

(h) *Convolution in the time domain:* The convolution of two functions of time $f_1(t)$ and $f_2(t)$, presented by the symbolic notation $f_1(t) * f_2(t)$, is defined as

$$f_1(t) * f_2(t) = \int_0^t f_1(\tau)\, f_2(t-\tau)\,d\tau .$$

In the previous section it is shown that the convolution of the two original functions corresponds to the multiplication of the related mapped functions, that is

$$\{f_1(t) * f_2(t)\} = F_1(s)\, F_2(s)$$

(i) *Convolution in the frequency domain:* Whereas in (h) the convolution of two functions of time was given, a similar result for the convolution of two functions in the frequency domain exists and is given by

$$\{f_1(t)\, f_2(t)\} = \frac{1}{2\pi j} \int_{c-j\infty}^{c+j\infty} F_1(p)\, F_2(s-p)\,dp$$

Here $F_1(s) \bullet\!\!-\!\!\circ\, f_1(t)$ and $F_2(s) \bullet\!\!-\!\!\circ\, f_2(t)$ is valid. Furthermore, p is the complex variable of integration. According to this theorem the Laplace transform of the product of two functions of time is equal to the convolution of $F_1(s)$ and $F_2(s)$ in the mapped domain.

(j) *Initial and final value theorems:* The *theorem of the initial condition* allows the direct calculation of the function value $f(0+)$ of a causal function of time $f(t)$ from the Laplace transform $F(s)$. If the Laplace transform of $f(t)$ and $f(t)$ exist, then

$$f(0+) = \lim_{t \to 0+} f(t) = \lim_{s \to \infty} s\, F(s)$$

is valid if the $\lim_{t \to 0} f(t)$ exists.

Using the *theorem of the final value* the value of $f(t)$ for $t \to \infty$ can be determined from $F(s)$, if the Laplace transform of $f(t)$ and $f(t)$ exist and the limit $\lim_{t \to \infty} f(t)$ also exists. Then it follows from previous section that

$$f(\infty) = \lim_{t \to \infty} f(t) = \lim_{t \to 0} s\, F(s)$$

One has to observe that

$$\lim_{t \to \infty} f(t) \text{ or } \lim_{t \to 0} f(t)$$

can be calculated only from the corresponding Laplace transform $\{f(t)\}$ by application of the theorems of the initial or final value, if the existence of the related limit in the time domain is a priori assured. The following two examples should explain this:

Example: $f(t) = e^{\alpha t}\,(\alpha > 0) \bullet\!\!-\!\!\circ\, F(s) = \dfrac{1}{s-\alpha}$

The limit $\lim_{t\to\infty} e^{\alpha t}$ does not exist so that the final value theorem may not be applied.

Example: $f(t) = \cos w_0 t \bullet\!\!-\!\!\circ\ F(s) = \dfrac{s}{s^2 + w_0^2}$

The limit $\lim_{t\to\infty} \cos w_0 t$ does not exist and therefore the final value theorem may not be applied. It can be concluded from the last two examples that the following general statement is valid: If the Laplace transform $F(s)$ has, apart from a single pole at the origin $s = 0$, poles on the imaginary axis or in the right-half plane, then the initial or final value theorems cannot be applied.

THE INVERSE LAPLACE TRANSFORM

As already mentioned in previous section in many cases a direct evaluation of the complex inverse integral is not necessary. A complicated function, $F(s)$, must be decomposed into a sum of simple functions of s that is

$$F(s) = F_1(s) + F_2(s) + ... + F_n(s),$$

which have a known inverse Laplace transform:

$$L^{-1}\{F(s)\} = L^{-1}\{F_1(s)\} + L^{-1}\{F_2(s)\} + ... + L^{-1}\{F_3(s)\}$$
$$= f_1(t) + f_2(t)_n(t) = f(t).$$

For many problems in control the function $F(s)$ is a ratio of polynomials in s, known as *rational fraction,* that is

$$\text{F}(s) = \frac{n_0 + n_1 s + ...n_m s^m}{d_0 + d_1 s + ... + s^n} = \frac{N(s)}{D(s)},$$

where $N(s)$ and $D(s)$ are the numerator and the denominator, respectively.

If $m > n$, then $N(s)$ is divided by $D(s)$, where a polynomial in s and a ratio of polynomials are obtained. The numerator of the fraction $N_1(s)$ has a lower order than n. E.g, if $m = n + 2$, then

$$\frac{N(s)}{D(s)} = k_2 s^2 + k_1 s + k_0 + \frac{N_1(s)}{D(s)},$$

where by degree $N_1(s) < n$ and k_0, k_1 and k_2 are constants.

A rational fraction $F(s)$ given in Equation can be decomposed into more simple functions by application of *partial fraction decomposition*, as shown in Equation. In order to perform this decomposition the denominator polynomial $D(s)$ must be factorised into the form

$$F(s) = \frac{N(s)}{(s - s - 1)\,(s - s_2)...(s - s_n)}.$$

For a denominator polynomial of n-th order one obtains roots or zeros $s = s_1, s_2,..., s_n$. The zeros of $D(s)$ are also known as the *poles* of $F(s)$, since they define where $F(^s)$ is infinite. The partial fraction decomposition for different types of poles is shown in the following.

***Case 1**: $F(s)$ has only single poles.*

Here $F(s)$ can be expanded into the form

$$F(s) = \sum_{k=1}^{n} \frac{c_k}{s - s_k},$$

where the *residuals* c_k are real or complex constants. Using the table of correspondences one immediately can obtain the corresponding function of time

$$F(t) = \sum_{k=1}^{n} c_k e^{s_k t} \quad \text{for} \qquad t > 0.$$

The values c_k can be determined either by comparing the coefficients or by using the theorem of residuals from the theory of functions according to

$$c_k = \frac{N(s_k)}{D'(s_k)} = (s - s_k)\frac{N(s)}{D(s)}\bigg|_{s=s_k}$$

for $k = 1, 2,..., n$ with $D'(s_k) = \mathrm{d}\,D/\mathrm{d}s\,|_{s=s_k}$

***Case 2**: F(s) has multiple poles.*

For multiple poles of $F(s)$ each with multiplicity $r_k(k = 1, 2,..., l)$ the corresponding partial fraction decomposition is

$$F(s) = \sum_{k=1}^{l} \sum_{v=1}^{rk} \frac{c_{kv}}{(s - s_k)^v} \quad \text{with} \quad n = \sum_{k=1}^{l} r_k$$

The back transformation of Equation into the time domain is

$$f(t) = \sum_{k=1}^{l} e^{s_k t} \sum_{v=1}^{rk} \frac{c_{kv} t^{v-1}}{(v-1)!} \quad \text{for } t > 0$$

The real or complex coefficients $c_k v$ for v = 1, 2,...,r_k determined by the theorem of residuals are

$$c_{kv} = \frac{1}{(r_k - v)!}\left\{\frac{\mathrm{d}(r_k - v)}{\mathrm{d}s(r_k - v)!}[F(s)\,(s - s_k)^{r_k}]\right\}_{s=s_k}$$

This general relation also contains the case of single poles of $F(s)$. The poles may be real or complex.

Case 3: $F(s)$ has also *conjugate complex poles.*

As both, the numerator $N(s)$ and the denominator $D(s)$ of the function $F(s)$ are rational algebraic functions, complex factors always arise as conjugate complex pairs. If $F(s)$ has a conjugate complex pair of poles $s_1, 2 = \sigma_1 \pm jw_1$, then for the function $F_1,2(s)$ in the partial fraction decomposition of

$$F(s) = \frac{N(s)}{D(s)} = F_{1,2}(s) + F_3(s) + \ldots + F_n(s)$$

Equation can be applied to give

$$F_{1,2}(s) = \frac{c_1}{s - (\sigma + jw_1)} + \frac{c_2}{s - (\sigma_1 - jw_1)},$$

where the residuals

$$c_{1,2} = \delta_1 \pm j\varepsilon_1$$

are also a conjugate complex pair. Therefore, both fractions of $F_{1,2}(s)$ can be combined, and one obtains

$$F_{1,2}(s) = \frac{\beta_0 + \beta_1 s}{\alpha_0 + \alpha_1 s + s^2}$$

with the real coefficients

$$\left.\begin{array}{ll} \alpha_0 = \alpha_1^2 + w_1^2 & ; \alpha_1 = -2\sigma_1 \\ \beta_0 = -2(\sigma_2\beta_1 + w_1\varepsilon_1) & ; \beta_1 = 2\delta_1 \end{array}\right\}.$$

The determination of the coefficients β_0 and β_1 is performed again using the theorem of residuals by

$$(\beta_0 + \beta_1 s)\big|_{s=s_1} = (s - s_1)(s - s_2)\frac{N(s)}{D(s)}\bigg|_{s=s_1}$$

As s_1 is complex, both sides of this equation are complex. Comparing the real and imaginary parts of both sides one gets two equations for the calculation of β_0 and β_1. This procedure is demonstrated now using the following example.

Example: Find the inverse Laplace transform $f(t)$ of

$$F(s) = \frac{1}{(s^2 + 2s + 2)(s + 2)}$$

The partial fraction decomposition of $F(s)$ is

$$F(s) = F_{1,2}(s) + F_3(s) = \frac{\beta_0 + \beta_1 s}{s^2 + 2s + 2} + \frac{c_3}{s+2},$$

where the function $F_{1,2}(s)$ contains the conjugate pair of poles

$$s_{1,2} = -1 \pm j.$$

In addition the third pole of $F(s)$ is

$$s_3 = -2$$

For the coefficients β_0 and β_1 it follows from Equation

$$(\beta_0 + \beta_1 s)|_{s=s_1} = \left.\frac{1}{s+2}\right|_{s=s_1}$$

$$(\beta_0 + \beta_1) + j\beta_1 = \frac{1}{-1+j+2} = \frac{1}{2} - j\frac{1}{2}.$$

Comparing the real and imaginary parts on both sides one obtains

$$\beta_0 - \beta_1 = \frac{1}{2} \text{ and } \beta_1 - -\frac{1}{2}$$

and from this finally $\beta_0 = 0$.

Using Equation the residual is

$$c_3 = (s+2)\left.\frac{1}{(s^2 + 2s + 2)}\right|_{s=s_3} = \frac{1}{2}.$$

The partial fraction decomposition of $F(s)$ is thus

$$F(s) = -\frac{1}{2}\left[\frac{s}{s^2 + 2s + 2}\right] + \frac{1}{2}\frac{1}{s+2},$$

which can be rearranged in the form

$$F(s) = \frac{1}{2}\left[\frac{s+1}{(s+1)^2 + 1} - \frac{1}{(s+1)^2 + 1} - \frac{1}{s+2}\right]$$

which can be directly applied to find the inverse transformation.

$$f(t) = -\frac{1}{2}[e^{-t}\cos t - e^{-t}\sin t - e^{-2t}] \text{ for } t > 0,$$

which can be rearranged as

$$f(t) = \frac{1}{2}e^{-t}[e^{-t} + \sin t - \cos t] \text{ for } t > 0.$$

The graphical representation of $f(t)$ is shown in Figure. Figure shows the corresponding poles, marked by a x, for this $F(s)$ in the complex s plane.

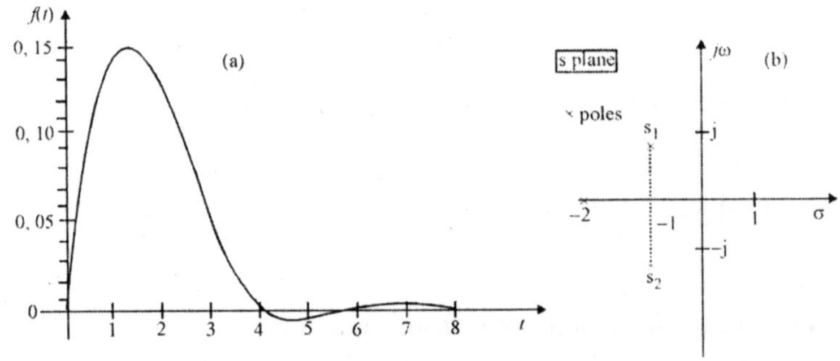

Fig. (a) Graph of the Original Function $f(t)$ (Function in the time Domain) and (b) Position of the Poles of $F(s)$ in the s Plane

It can be seen from this example that the position of the poles s_1, s_2 and s_3 affects the shape of the graph of $f(t)$. In this case all poles of $F(s)$ have negative real parts, therefore the graph of $f(t)$ shows a damped behaviour, i.e. it decreases to zero for $t \to \infty$. If the real part of one pole be positive, then the graph of $f(t)$ would be infinitely large for $t \to \infty$.

Since in control problems the original function $f(t)$ always represents the time behaviour of a system variable, the behaviour of this system variable $f(t)$ can be judged to a large extent by investigation of the positions of the poles of the corresponding mapped function $F(s)$.

SOLVING LINEAR DIFFERENTIAL EQUATIONS USING THE LAPLACE TRANSFORM

The Laplace transform, the basics of which have been introduced in the sections above, is an elegant way for fast and schematic solving of linear differential equations with constant coefficients. In the following the importance of this approach is demonstrated. Instead of solving the differential equation with the initial conditions directly in the original domain, the detour via a mapping into the frequency domain is taken, where only an algebraic equation has to be solved. Thus solving differential equations is performed according to Figure in the following three steps: Transformation of the differential equation into the mapped space, Solving the algebraic equation in the mapped space, Back transformation of the solution into the original space.

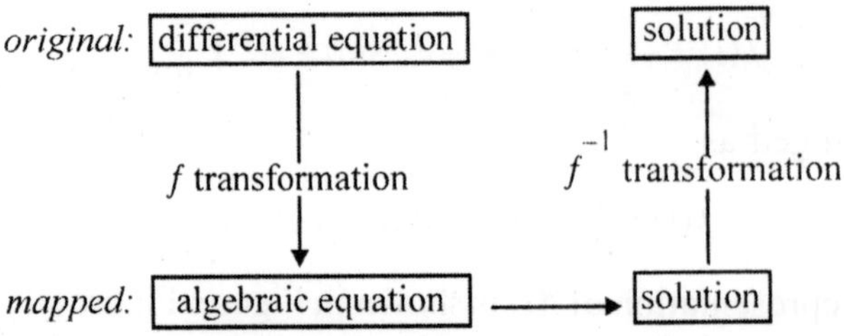

Fig. Schema for Solving Differential Equations using the Laplace Transformation

The procedure will be demonstrated by the following two examples.

Example: Consider the differential equation

$$\ddot{f}(t)+3\dot{f}(t)+2f(t)=e^{-t}$$

with the initial conditions $f(0+)=\dot{f}(0+)=0$

Solution: Proceeding using the steps given above one has

Step 1: $s^2F(s)+3sF(s)+2F(s)=\dfrac{1}{s+1}$

Step 2: $F(s)=\dfrac{1}{s+1}\,\dfrac{1}{s^2+3s+2}$

Step 3: The complex function $F(s)$ must be decomposed into partial fractions in order to use the tables of correspondences. This gives

$$F(s)=\frac{1}{s+1}-\frac{1}{s+1}+\frac{1}{(s+1)^2}.$$

By using the inverse Laplace transformation that the solution of the given differential equation is

$$f(t)=e^{-2t}-e^{-t}+te^{-t}.$$

Example: Given the differential equation

$$\ddot{x}+a_1\dot{x}+a_0x=0,$$

where a_0 and a_1 are constants and the initial conditions $\dot{x}(0+)$ and $x(0+)$ are known. Then

Solution:

Step 1: $s^2X(s)-s\;x(0+)-\dot{x}(0+)+a_2[s\;X(s)-x(0+)]+a_0\;X(s)=0$

Step 2: $X(s)=\dfrac{s+a_1}{s^2+a_1s+a_0}x(0+)+\dfrac{1}{s^2+a_1s+a_0}\dot{x}(0+),$

$$X(s)=L_0(s)\;x(0+)+L(s)\;\dot{x}(0+)$$

with the abbreviation

$$L_0(s)=\frac{N_0(s)}{D(s)}=\frac{s+a_1}{s^2+a_1s+a_0}\quad\text{and}\quad L(s)=\frac{N(s)}{D(s)}=\frac{1}{s^2+a_1s+a_0}.$$

Step 3: ***Case (a):*** two *single real zeros of the denominator:*

This means

$$D(s)=s^2+a_1s+a_0=(s-\alpha_1)(s-\alpha_2).$$

For both rational expressions $L_0(s)$ and $L(s)$ it follows by partial fraction decomposition that

$$L_0(s) = \frac{A_1}{s-\alpha_1} + \frac{A_2}{s-\alpha_2} \text{ and } L(s) = \frac{B_1}{s-\alpha_1} + \frac{B_2}{s-\alpha_2}.$$

The coefficients A_i and B_i can now be determined by comparing coefficients or by applying Equation.

$$A_i = \frac{N_0(\alpha_i)}{D'(\alpha_i)};\ B_i = \frac{N(\alpha_i)}{D'(\alpha_i)} \text{ for } i = 1, 2.$$

Thus for Equation follows

$$X(s) = \left[\frac{A_1}{s-\alpha_1} + \frac{A_2}{s-\alpha_2}\right] x(0+) + \left[\frac{B_1}{s-\alpha_1} + \frac{B_2}{s-\alpha_2}\right] \dot{x}(0+),$$

and by applying the formula the solution of the differential equation is

$$x(t) = [A_1 e^{\alpha_1 t} + A_2 e^{\alpha_2 t}]\, x(0+) + [B_1 e^{\alpha_1 t} + B_2 e^{\alpha_2 t}]\, \dot{x}(0+)$$

$$= [A_1 x(0+) + B_i \dot{x}(0+)] e^{\alpha_1 t} + [A_2\, x(0+) + B_2 \dot{x}(0+)] e^{\alpha_2 t}.$$

Case (b): One *double real zero of the denominator:* Here is $D(s) = (s-\alpha)^2$

For the two rational expressions $L0(s)$ and $L(s)$ of Equation the partial fraction decomposition is

$$L_0(s) = \frac{A_1}{s-\alpha} + \frac{A_2}{(s-\alpha)^2} \text{ and } L(s) = \frac{B_1}{s-\alpha} + \frac{B_2}{(s-\alpha)^2}.$$

The coefficients ^{Ai} and ^{Bi} are determined by comparing both sides or by evaluation of Equation:

$$A_1 = \left\{\frac{d}{ds}\left[\frac{N_0(s)}{D(s)}(s-\alpha)^2\right]\right\}_{s=\alpha} = 1,\ A_2 = \left[\frac{N_0(s)}{D(s)}(s-\alpha)^2\right]_{s=\alpha} = \alpha + \alpha_1$$

and
$$B_1 = \left\{\frac{d}{ds}\left[\frac{N(s)}{D(s)}(s-\alpha)^2\right]\right\}_{s=\alpha} = 0,\ B_2 = \left[\frac{N(s)}{D(s)}(s-\alpha)^2\right]_{s=\alpha} = 1.$$

From these results one obtains the solution

$$X(s) = \frac{x(0+)}{s-\alpha} + \frac{(\alpha + a_1)\, x(0+) + \dot{x}(0+)}{(s-\alpha)^2}$$

in the mapped space. By applying the inverse Laplace transformation the required solution of the differential equation is

$$x(t) = x(0+)e^{\alpha t} + [(\alpha + \alpha_1)\, x(0+) + \dot{x}(0+)]\, te^{\alpha t}.$$

Case (c): Two *conjugate complex zeroes of the denominator:* Here

$$D(s) = (s - \alpha_1)(s - \alpha_2) \text{ with } \alpha_{1,2} = \sigma_1 \pm jw_1 .$$

Introducing the values of α_1 and α_2 and after multiplication of this expression one obtains

$$D(s) = (s - \sigma_1)^2 + w_1^2 .$$

Comparison with the denominator of the original relation, Equation, gives according to Equation

$$a_0 = \sigma_1^2 + w_1^2 \text{ and } a_1 = -2\sigma_1 .$$

With these coefficients Equation is in the form

$$X(s) = \frac{s - 2\sigma_1}{(s - \sigma_1)^2 + w_1^2} x(0+) + \frac{1}{(s - \sigma_1)^2 + w_1^2} \dot{x}(0+)$$

$$= \left[\frac{s - \sigma_1}{(s - \sigma_1)^2 + w_1^2} - \frac{\sigma_1}{w_1} \frac{w_1}{(s - \sigma_1)^2 + w_1^2} \right] x(0+) + \frac{1}{w_1} \frac{w_1}{(s - \sigma_1)^2 + w_1^2} \dot{x}(0+) ,$$

and from this one gets $X(s)$ the corresponding time function

$$x(t) = e^{\sigma_1 t} \left[\cos w_1 t - \frac{\sigma_1}{w_1} \sin w_1 t \right] x(0+) + \frac{1}{w_1} e^{\sigma_1 t} \dot{x}(0+) \sin w_1 t \quad \text{or rearranged}$$

$$x(t) = e^{\sigma_1 t} \left\{ x(0+) \cos w_1 t + \left[\frac{1}{w_1} \dot{x}(0+) - \frac{\sigma_1}{w_1} x(0+) \right] \sin w_1 t \right\} .$$

Also from Equation of this example the importance of the position of the zeros of $D(s)$, the poles of $X(s)$, on the solution is clear. For all three cases the solution of the differential equation according to Equations are mainly influenced by the position of the poles of $X(s)$. These poles of $X(s)$ are - as one can see from the two examples - only depend on the left side of the corresponding differential equation, i.e. the homogeneous part of it. As is generally known the solution of the homogeneous differential equation describes the *modes* of the system, that is the behaviour, which depends only on the initial conditions. Therefore, consider for the general case only the homogeneous part of an nth-order ordinary homogeneous linear differential equation with constant coefficients that is

$$\sum_{i=0}^{n} a_i \frac{d^i x_a(t)}{dt^i} = 0$$

with all n initial conditions $(d^i x_a(t)/dt^i)|_{t=0+}$ for $i = 0, 1, \ldots, n-1$.

One obtains by Laplace transformation

$$X_a(s)\sum_{i=0}^{n} a_i s^i - \left[\sum_{i=1}^{n} a_i \sum_{v=1}^{i} s^{i-v} \frac{d^{v-1}}{dt^{v-1}} x_a(t)|_{t=0+}\right] = 0$$

and a form according to Equation

$$X_a(s) = \frac{\sum_{i=1}^{n} a_i \sum_{v=1}^{i} s^{i-v} \frac{d^{v-1}}{dt^{v-1}} x_a(t)|_{t=0+}}{\sum_{i=0}^{n} a_i s^i} = \frac{N(s)}{D(s)},$$

where $N(s)$ and $D(s)$ are polynomials in s and the initial conditions are only in the numerator polynomial $N(s)$. The poles $s_k(k = 1, 2,..., n)$ of $X_a(s)$ can be determined directly from the solution of the equation

$$\sum_{i=1}^{n} a_i s^i = 0$$

After factorisation of this equation one obtains

$$a_n(s - s_1)(s - s_2)... (s - s_n) = 0.$$

The poles s_k of $X_a(s)$ make it possible to perform a partial fraction decomposition of $X_a(s)$, e.g. for the case of single poles according to Equation. For this case one obtains following Equation the solution of the homogeneous differential equation, in the form

$$x_a(t) = \sum_{k=1}^{n} c_k e^{s_k t} \text{ for } t > 0.$$

From this one can realise that the position of the poles s_k of $X_a(s)$ in the plane completely characterises the modes or inherent behaviour of the system described by Equation.

Thus one obtains for Re $s_k < 0$ (left-half s plane) a decreasing and for Re $s_k > 0$ (right-half s plane) an increasing behaviour of $x_a(t)$, while for pairs of poles with Re $s_k = 0$ permanent oscillations occur. Therefore, Equation is called the *characteristic equation* and the poles s_k of $X_a(s)$ are often called *eigenvalues* of the equation.

Therefore investigation of the characteristic equation provides the most important information about the oscillating behaviour of a system. Laplace transform of the impulse $\delta(t)$ and step $\sigma(t)$ The impulse function $\delta(t)$ is *not a function* in the sense of classical analysis, but a distribution (pseudo-function). Therefore without entering the theory of distributions the integral

$$\{\delta(t)\} = \int_0^\infty \delta(t)e^{-st}\,dt$$

is not defined. The singularity exactly matches with the lower integration limit. The impulse function can be approximately described by the limit

$$\delta(t) = \lim_{e \to 0} r_e(t)$$

with the rectangular impulse function

$$r_e = \begin{cases} 1/\varepsilon & \text{for } 0 \le t \le \varepsilon \\ 0 & \text{otherwise.} \end{cases}$$

Strictly speaking this representation of $\delta(t)$ is not a distribution, as $r_e(t)$ for $0 \le t \le \infty$ is not arbitrarily often differentiable. Because of the simple description compared with other functions (e.g. Gaussian functions) this approach is preferred here. From Equation it follows that

$$\{\delta(t)\} = \left[\int_0^{\infty} \lim_{e \to 0} r_e(t) \right] e^{-st} \, dt.$$

As Equation can also be represented in the form

$$r_e(t) = \frac{1}{\varepsilon}[\sigma(t) - \sigma(t - \varepsilon)],$$

where $\sigma(t)$ is the unit step. Since the integration is independent of ε, the limit and integration can be permuted so that

$$\{\delta(t)\} = \lim_{e \to 0} \left\{ \frac{1}{\varepsilon} \int_0^{\infty} [\sigma(t) - \sigma(t - \varepsilon)] \, e^{-st} \, dt \right\}$$

$$\{\delta(t)\} = \lim_{e \to 0} \left\{ \frac{1}{\varepsilon} \frac{1}{s} (1 - e^{-\varepsilon s}) \right\}.$$

By applying l' Hospital's rule one obtains

$$\{\delta(t)\} = \lim_{e \to 0} \frac{s e^{-\varepsilon s}}{s} = 1.$$

As the impulse $\delta(t)$ has an area of unity it is also called unit impulse.

The solution can be determined in the following three steps:

Step 1: The Laplace transform of the given differential equation is:

$$sY(s) - y(0-) = 1 \text{ with } y(0-) = 0.$$

Step 2: The solution of the algebraic equation is:

$$Y(s) = \frac{1}{s}$$

Step 3: From the back transformation the solution follows as

$$y(t) = \sigma(t)$$

where $\sigma(t)$ is the unit step function.

TRANSFER FUNCTIONS

Transfer functions simplify working with linear differential equations, which describe the transition behavior of linear dynamical systems. As transfer functions contain at least a rational function of a complex variable, the physical interpretation of poles and zeros is discussed. Finally, the most popular operations of combining systems described by transfer functions are presented. The properties of transfer functions are shown by examples and interactive questions allow the readers to test their knowledge of transfer functions.

Definition: Linear, continuous-time time-invariant systems with lumped parameters - as initially a dead time is not being taken into account - will be described by the ordinary differential equation

$$\sum_{i=0}^{n} a_i \frac{d^i x_{3.}(t)}{dt^i} = \sum_{j=0}^{m} bj \frac{djx_e(t)}{dt^j}.$$

If all *initial conditions* are set to *zero* and the Laplace transformation is applied to both sides of this equation, one obtains

$$X_{3.}(s)\sum_{i=0}^{n} a_i s^i = X_e(s)\sum_{j=0}^{m} b_j s^j,$$

or reordering

$$\frac{X_{3.}(s)}{X_e(s)} = \frac{b_0 + b_1 s + \ldots + b_m s^m}{a_0 + a_1 a + \ldots + a_n s^n} = G(s) = \frac{N(s)}{D(s)},$$

where $N(s)$ and $D(s)$ describe the numerator and denominator polynomials, respectively. The quotient of the Laplace-transformed output and input of such a type of system is a rational fraction. The coefficients of this fraction depend only on the structure and parameters of the system. Such a type of function $G(s)$, which describes completely the transfer behaviour of a system, is called the *transfer function* of the system. With such a transfer function the output

$$X_{3.}(s) = G(s)X_e(s)$$

INTERPRETATION OF THE TRANSFER FUNCTION

Comparing Equation with the convolution theorem, it then follows for the representation of Equation in the time domain that

$$x_{3.}(t) = \int_0^t g(t-r)x_e(\tau)\,d\tau,$$

where obviously the inverse Laplace transform of $G(s)$ is the function $g(t)$. This function is generally known as the *weighting function* of the system. In other words, the transfer function is the Laplace-transformed weighting function according to

$$G(s) = \{g(t)\}.$$

If the unit impulse $\delta(t)$ is taken as the input signal $x_{3.}(t)$ for a system described by the transfer function $G(s)$, one obtains according to Equation

$$X_{3.}(s) = G(s)\,\{\delta(t)\}$$

and after using the Laplace transform of the unit impulse $\delta(t)$ from Equation

$$X_{3.}(s) = G(s)$$

or from Equation

$$x_{3.}(t) = g(t).$$

This shows that the response to an unit impulse $\delta(t)$ is the weighting function. Therefore the weighting function is also called the *impulse response*.

Another interpretation is when the system is excited by the input signal

$$x_e(t) = \bar{x}_e e^{\sigma t}\sin(wt).$$

For the steady-state case one obtains for the output signal

$$x_{35}(t) = |G(\sigma + jw)|\,\bar{x}_e e^{\sigma t}\sin(wt + \phi(\sigma + jw)).$$

This shows that the modulus $|G(\sigma + jw)|$ of the transfer function describes the gain, and that $\phi(\sigma + jw) = \arg|G(\sigma + jw)$ describes the phase shift of a sinusoidal function with the frequency w and with increasing or decreasing amplitude according to $e^{\sigma t}$ can be immediately calculated for a known input signal $x_e(t)$, and therefore $X_e(s)$

REALISABILITY AND PROPERNESS OF TRANSFER FUNCTIONS

It must be mentioned that a transfer function with $m > n$ is physically not realisable. The transfer function of an ideal differentiator is described by $G(s) = s$ according to Equation. Any transfer function with $m > n$ can be decomposed into

$$G(s) = \frac{N(s)}{D(s)} = \frac{N_1(s)}{D(s)} + k_0 + k_1 s + \ldots + k_{m-m} s^{n-n},$$

where degree $N_1(s) = n - 1$ and terms in s with positive powers also occur. Such derivative elements would deliver for input signals of arbitrary high frequency corresponding output signals of arbitrary high amplitude, which are physically not realisable.

The *condition of realisability* of the transfer function according to Equation is

$$\text{degree } N(s) \leq \text{ degree } D(s) \text{ or } m \leq n.$$

This condition is also called as the *condition of properness*. If a transfer function does not follow this condition, it is called an *improper* transfer function. It has the property that $G(s) \to \infty$ as $s \to \infty$. A realisable transfer function is called as *proper* and it always follows $G(s) \to k_0$ as $s \to \infty$. A transfer function with $k_0 = 0$ is called as *strictly proper*.

Transfer functions with dead time. If a time delay or *dead time* T_t is introduced in the input signal $x_e(t)$, one obtains instead of Equation the differential equation

$$\sum_{i=0}^{n} a_i \frac{d^i x_3(t)}{dt^i} = \sum_{j=0}^{m} bj \frac{djx_e(t-T_t)}{dt^j}.$$

In this case taking the Laplace transformation gives the *transcendental* transfer function

$$G(s) = \frac{N(s)}{D(s)} e^{-sT_t}.$$

Poles and zeros of the transfer function

In some cases (e.g. stability analysis) it is expedient to represent the rational transfer function $G(s)$ according to Equation in the factorised form

$$G(s) = \frac{N(s)}{D(s)} = k_0 \frac{(s - s_{z_1})(s - s_{z_2})...(s - s_{z_m})}{(s - s_{p_1})(s - s_{p_2})...(s - s_{p_n})}.$$

For physical reasons only real coefficients a_i, b_j occur. Therefore the poles sp_i and the *zeros* s_{z_j} of $G(s)$, respectively, can be *real or complex conjugate* pairs. The terms zeros and poles are chosen, because the transfer function is zero at s_{z_j} and infinite at s_{p_i}. Zeros and poles can be graphically represented in the complex plane as shown in Figure. A linear time-invariant system *without* dead time is described completely by the distribution of its poles and zeros and the gain factor k_0.

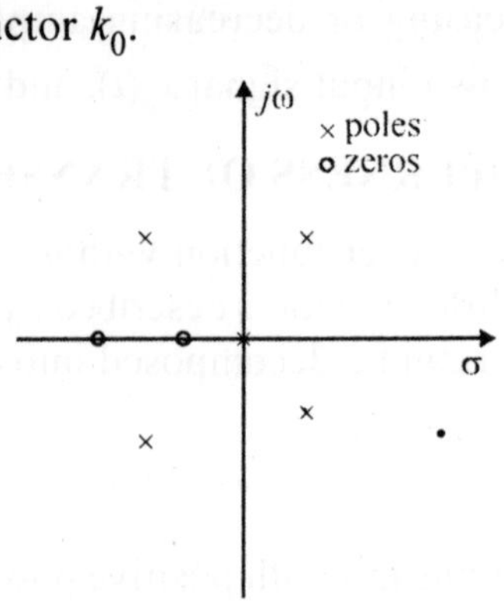

Fig. Example of the Pole and zero Distribution of a Rational Transfer Function in the Complex Plane

Moreover, the poles and zeros of a transfer function have a further significance. Observing a system without input ($x_e(t) \equiv 0$) according to Equation and determining the time response $x_3(t)$ for the given *n* initial conditions, one has to solve the associated homogeneous differential equation

$$\sum_{i=0}^{n} a_i \frac{d^i x_3(t)}{dt^i} = 0,$$

which corresponds exactly to Equation. For the approach $x_3(t) = e^{st}$ of Equation one obtains for the solution in s the *characteristic equation*

$$P(s) = \sum_{i=0}^{n} a_i s^i = 0$$

which was already mentioned in Equation. This relation can be directly determined by setting the denominator of $G(s)$ to zero ($D(s) = 0$), as long as $D(s)$ and $N(s)$ have no common factor. The zeros s_k of the characteristic equation are the poles sp_i of the transfer function. As already shown in previous section the modes (i.e. $x_e(t) \equiv 0$) are described by the characteristic equation, so that the poles s_{p_i} of a transfer function contain all of this information.

The zeros of a transfer function are those values $s = s_{z_j}$ for which $|G(s_{z_j})| = 0$. This means that the output signal $X_3(s)$ does not contain any components which depend on s_{z_j}. In order to explain this in more detail a stable system with a transfer function according to Equation is excited by the input signal

$$x_e(t) = e^{s_{z_j} t}$$

First for simplification the zero $s_{z_j} = \sigma_{z_j}$ is assumed to be real. For this case the input signal is $x_e(t) = e^{\sigma_{z_j} t}$. Because $|G(s_{z_j})| = 0$ one obtains from Equation the steady-state output signal as

$$I_{35}(t) = 0$$

In the case of complex conjugate pairs of zeros $s_{z_j}, s_{z_j+1} = s^*_{z_j}$ both zeros have to be taken into consideration in the input signal

$$\begin{aligned} x_e(t) &= 2e^{sz_i t} + e^{s z_j^* t} \\ &= 2e^{\sigma z_i t} \cos \omega t \\ &= 2e^{\sigma z_i t} \sin\left(\omega t + \frac{\pi}{2}\right). \end{aligned}$$

Equation leads also to the result $x_{as}(t) = 0$. This shows that a zero sz_i of a system blocks the transmission of the input signal $e^{\sigma_{z_i} t}$.

Example: The mass-spring-damper mechanical system in Figure with the mechanical constants $c_1 = 1$, $c_2 = 2$, $d = 1.5$, $m_1 = 1$ and $m_2 = 4$ is excited by the force x_e. The transfer function between the force and the position can be shown to be

$$G(s) = \frac{s^2 + 1}{s^1 + 0.5s^3 + 1.75s^2 + 0.5s + 0.5}$$

which has the zeros $sz_{1,2} = \pm j$. If this system is excited by the sinusoidal input signal

$$X_e(s) = \frac{1}{(s - sz_1)(s - sz_2)} = \frac{1}{s^2 + 1}$$

$$^{-1}\{X_e(s)\} = \sin t,$$

which is derived from this pair of zeros, the output signal $x_a(t)$ decays to zero as shown in Figure even though the input signal is a sinusoidal signal and the mass m_1 shows an undamped oscillation. The system does not pass this oscillation to the mass when the frequency matches the zeros.

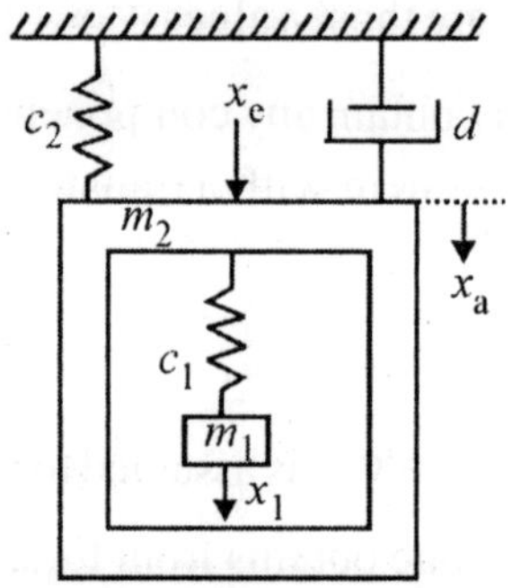

Fig.. Mass-Spring-Damper Mechanical System used for the Interpretation of zeros

Using transfer functions for calculations.

For combinations of transfer functions simple rules for determining the resulting transfer function can be derived. The combinations are of the type that transfer function blocks are connected. In making any transfer function block connection it is assumed that the connection does not load the block to which the connection is being made.

1. *Series connection:* From the diagram in Figure it follows that

$$Y(s) = G_2(s)X_{e2}(s)$$

$$X_{e2}(s) = X_{a1}(s) = G_1(s)\,U(s)$$

$$Y(s) = G_2(s)G_1(s)\,U(s).$$

The total transfer function of this series connection is

$$G(s) = \frac{Y(s)}{U(s)} = G_1(s)G_2(s)$$

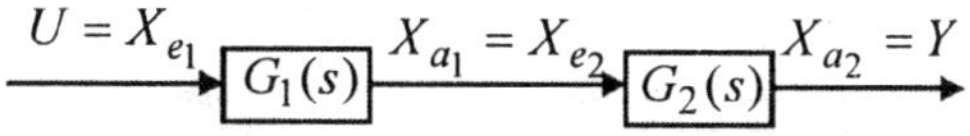

Fig. Series Connection of two Transfer Functions

2. *Parallel connection:* For the output of both functions it follows from Figure that

$$X_{a1}(s) = G_1(s)\, U(s)$$
$$X_{a2}(s) = G_2(s)\, U(s).$$

The output of the total system is

$$Y(s) = X_a(s) = X_{a_2}(s) + X_{a_2}(s) = [G_1(s) + G_2(s)]U(s),$$

and from this the transfer function of a parallel connection is

$$G(s) = \frac{Y(s)}{U(s)} = G_1(s) + G_2(s)$$

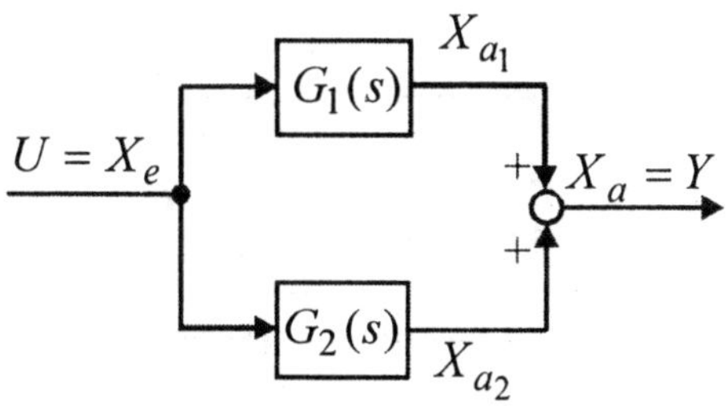

Fig. Parallel Connection of two Transfer Functions

3. *Feedback loop:* From Figure the output is

$$Y(s) = X_a(s) = [U(s) \underset{(+)}{-} X_{a_2}(s) + G_1(s)$$

Since

$$X_{a_2} = G_2(s)Y(s)$$

one obtains

$$Y(s) = [U(s) \underset{(+)}{-} G_2(s)Y(s)]G_1(s),$$

and from this

$$Y(s) = \frac{G_1(s)}{1 \underset{(-)}{+} G_1(s)G_2(s)} U(s)$$

The total transfer function of the feedback loop is

$$G(s) = \frac{Y(s)}{U(s)} = \frac{G_1(s)}{1 \underset{(-)}{+} G_1(s)G_2(s)}$$

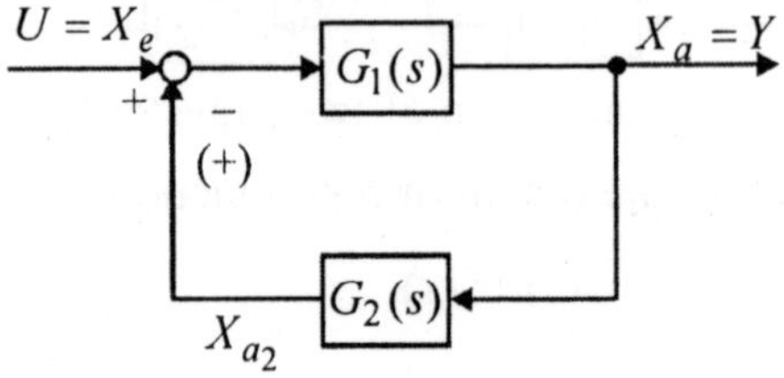

Fig. Feedback using two Elements

As the output of $G_1(s)$ is fed back via $G_2(s)$ to the input, this is called *feedback*. One has to distinguish between a positive feedback for positive adding of $X_{a_2}(s)$ and a negative feedback for negative adding of $X_{a_2}(s)$.

Example: For the special case of $G_1(s)$ being a pure amplifier with a high gain $K \to \infty$, one obtains for negative feedback

$$G(s) = \frac{K}{1 + KG_2(s)} = \frac{1}{\frac{1}{K} + G_2(s)} \approx \frac{1}{G_2(s)}.$$

The entire technique of operational amplifiers is based on this principle. In a feedback loop for which $G_1(s)$ is an amplifier with $K \to \infty$ then an element $G_2(s)$ can be used to realise any transfer function within certain limits.

Chapter 9

Infinite Series

An infinite series is simply a way of representing a limit, where each successive approximation is found from the preceding one by addition of one more term. In principle, the expression of a number as a decimal is merely the representation of a number a in the form of an infinite series $a = a_1+a_2+a_3+...$ where, if $0 < a < 1$, the term a_n is put equal to $a_n\ 10^{-n}$ and a_n is a whole number between 0 and 9, inclusively.

Since every limiting value can be written in the form of an infinite series, it may seem that a special study of series is superfluous. However, in many cases, it happens that limiting values occur naturally in the form of infinite series, which often exhibit particularly simple laws of formation.

Of course, it is not true that every series has an easily recognizable law of formation. For example, the number p can certainly be represented as a decimal, yet we know no simple law which enables us to state the value of an arbitrary digit, say its 7000-th decimal. However, if we set aside the representation of π by a decimal and consider instead Gregory's series, we have an expression with a perfectly clear general law of formation.

Infinite series, in which approximations to the limit are found by repeated addition of new terms, are analogous to infinite products, in which the approximations to the limit arise from repeated multiplication by new factors. However, we shall not go deeply into the theory of infinite products.

The Concepts of Convergence and Divergence

The Fundamental Ideas: We consider an infinite series the general term of which we denote * by a_n; the series then has the form $a_1+a_2+... = \sum_{\nu=1}^{\infty} a_\nu$.

The symbol with the summation sign on the right hand side is merely an abbreviated way of writing the expression on the left hand side. *For formal reasons, we include the possibility that certain of the numbers a_n may be zero. If all a_n vanish onwards from a number N (i.e., when $n > N$), we speak of a terminating series.

If, as n increases, the n-th partial sum $s_n = a_1 + a_2 + ... + a_n = \sum_{\nu=1}^{n} a_\nu$ approaches a limit

$$S = \lim_{n\to\infty} s_n$$

we say that the series converges, and otherwise that it diverges. In the first case, we call S the sum of the series. We have already met with many examples of convergent series; for instance, the $1 + q + q^2 + ...$, geometric series which converges to the sum $1/(1-q)$ for $q < 1$, Gregory's series, the series for log 2, the series for e, and others converge. In the language of infinite series, Cauchy's convergence test is expressed as follows:

A necessary and sufficient condition for the convergence of a series is that the number $|s_m - s_n| = |a_{n+1} + a_{n+2} + ... + a_m|$ becomes arbitrarily small if m and n are chosen sufficiently large ($m > n$).

In other words: A series converges if and only if the following condition is met: Given a positive number ε, however small, it is possible to choose an index $N = N(\varepsilon)$, which, in general, increases beyond all bounds as $\varepsilon \to 0$, in such a way that the $|s_m - s_n|$ is less than e, provided only that $m > N$ and $n > N$.

We can make the meaning of the convergence test clearer by considering the geometric series with $q = 1/2$. If we choose $\varepsilon = 1/10$, we need only take $N = 4$, because

$$|s_m - s_n| = \frac{1}{2^n} + ... + \frac{1}{2^{m-1}}$$

$$= \frac{1}{2^{n-1}}\left(\frac{1}{2} + \frac{1}{2^2} + ... + \frac{1}{2^{m-n}}\right) < \frac{1}{2^{n-1}}$$

and $\frac{1}{2^{n-1}} < \frac{1}{10}, if\ n > 4$.

If we choose ε equal to 1/100, it is sufficient to take 7 as the corresponding value of N, as is easily verified.

Obviously, it is a necessary condition for the convergence of a series that $\lim_{n\to\infty} a_n = 0$.

Otherwise, undoubtedly, the convergence criterion cannot be fulfilled. However, this necessary condition is by no means sufficient for convergence; on the contrary, it is easy to find infinite series the general term a_n of which approaches 0 as n increases, yet their sum does not exist, because the partial sum s_n increases without limit as n increases.

An example of this is the series $1 + \frac{1}{\sqrt{2}} + \frac{1}{\sqrt{3}} + ... + \frac{1}{\sqrt{n}} + ...$ with the general term $\frac{1}{\sqrt{n}}$. We immediately see that

$$s_n > \frac{1}{\sqrt{n}} + ... + \frac{1}{\sqrt{n}} = \frac{n}{\sqrt{n}} = \sqrt{n}$$

The n-th partial sum increases beyond all bounds as n increases, whence the the series diverges

The same is true for the classical example of the harmonic series $1+\frac{1}{2}+\frac{1}{3}+\frac{1}{4}+...$, for which

$$a_{n+1}+...+a_{2n} = \frac{1}{n+1}+...+\frac{1}{2n} > \frac{1}{2n}+...+\frac{1}{2n} = \frac{1}{2}$$

Since n and $m = 2n$ can be taken as large as we please, the series diverges, because Cauchy's test is not fulfilled; in fact, the n-th partial sum tends obviously to infinity, since all the terms are positive. On the other hand, the series, formed from the same numbers with alternating signs, $1-\frac{1}{2}+\frac{1}{3}-\frac{1}{4}+\frac{1}{5}-+..+\frac{(-1)^{n-1}}{n}+...$ converges and has the sum log 2.

It is by no means true that in every divergent series s_n tends to $+1$ or -1. Thus, in the case of the series $1 - 1 + 1 - 1 + 1 + -...$, we see that the partial sum s_n has alternately the values 1 and 0, and, on account of this backwards / forwards oscillation neither approaches a definite limit nor increases numerically beyond all bounds.

As the convergence and divergence of an infinite series is concerned, the following fact which, though self-evident, is very important should be noted: The convergence or divergence of a series is not changed by insertion or removal of finite numbers of terms. As far as convergence or divergence is concerned, it does not matter in the least whether we begin the series at the term a_0, or a_1, or a_5, or at any other arbitrarily chosen term.

ABSOLUTE CONVERGENCE AND CONDITIONAL CONVERGENCE

The series $1+\frac{1}{2}+\frac{1}{3}+\frac{1}{4}...$ diverges; but if we change the sign of every second term, the resulting series converges. On the other hand, the geometric series 1-*q*+*q*ýÿ-*q*ýÿÿýÿÿýÿ converges and has the sum $1/(1 + q)$, provided that $0 < q < 1$; on converting all the signs to +, we obtain the series $1 + q + q^2 + q^3 +...$, which also converges and has the sum $1/(1 - q)$.

There appears here a distinction which we must examine more closely. With a series all terms of which are positive, there are only two possible cases; either it converges or the partial sum increases beyond all bounds with n. In fact, the partial sums, being a monotonic increasing sequence, must converge, if they remain bounded. Convergence occurs, if the terms approach zero sufficiently rapidly as n increases; on the other hand, divergence occurs, if the terms do not approach zero at all or if they approach zero too slowly. However, in series, where some terms are positive and others negative, the changes of sign may bring about convergence, since a too great increase in the partial sums, due to the positive terms, is compensated by negative terms, as a final result of which a definite limit is approached.

In order to grasp this fact better with a series $\sum_{\nu=1}^{\infty} a_\nu$, which has positive and negative terms, we compare it with the series which has the same terms all with positive signs, that is,

$$|a_1| + |a_2| + \dots = \sum_{\nu=1}^{\infty} |a_\nu|$$

If this series converges, then for sufficiently large values of n and $m > n$, the expression $|a_1| + |a_{n+2}| + \dots + |a_m|$ will certainly be as small as we please; on account of the relation $|a_{n+1} + \dots + a_m| \leq |a_{n+1}| + \dots + |a_m|$, the expression on the left hand side is also arbitrarily small, whence the original series $\sum_{\nu=1}^{\infty} a_\nu$ converges. In this case, the original series is said to be absolutely convergent. Its convergence is due to the numerical smallness of its terms and does not depend on the change of the signs. On the other hand, if the series with all the terms taken positively diverges and the original series still converges, we say that the original series is conditionally convergent. Conditional convergence results from the terms of opposite signs compensating one another.

For conditional convergence, Leibnitz's convergence test is frequently of use: If the terms of a series have alternating sign and, moreover, their absolute values $|a_n|$ tend monotonically to 0 (so that $|a_{n+1}| < |a_n|$), the series $\sum_{\nu=1}^{\infty} a_\nu$ converges.

In the proof, we assume that $a_1 > 0$, which does not essentially limit the generality of the argument, and write our series in the form $b_1 - b_2 + b_3 - + \dots$, where all the terms b_n are now positive, b_n tends to 0 and the condition $b_{n+1} < b_n$ is satisfied. If we bracket the terms together in the two ways $b_1 - (b_2 - b_3) - (b_4 - b_5) - \dots$ and $(b_1 - b_2) + (b_3 + b_4) + (b_5 + b_6) + \dots$ we see at once that the partial sums satisfy the two relations:

$$s_1 > s_2 > s_3 > \dots > s_{2m+1} > \dots$$

$$s_2 < s_4 < s_6 < \dots < s_{2m} < \dots$$

On the other hand, $s_{2n} < s_{2n+1} < s_1$ and $s_{2n+1} > s_{2n} > s_2$, whence the odd partial sums s_1, $s_3, \dots$ form a monotonic decreasing sequence, which in no case drops below the value s_2 and this sequence has a limit L. The even partial sums $s_2, s_4, \dots$ likewise form a monotonic increasing sequence the terms of which in no case exceed the fixed number s_1, whence this sequence must have a limiting value L'.

Since the numbers s_{2n} and s_{2n+1} differ only by the number b_{2n+1}, which approaches 0 as n increases, the limiting values L and L' are equal to each other. In other words, the even and the odd partial sums approach the same limit, which we will now denote by S. However, this implies that our series is convergent, as has been asserted; its sum is S.

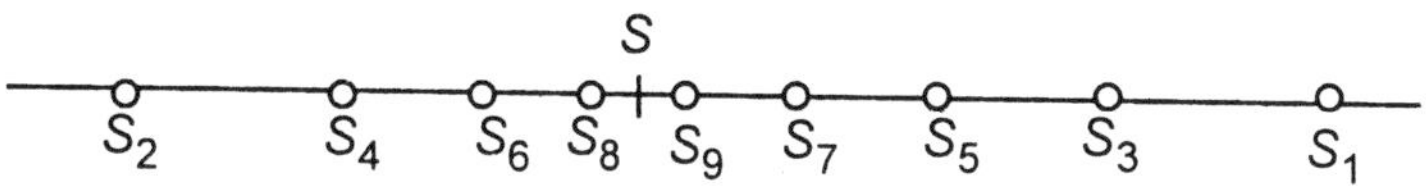

Fig. Convergence of An Alternating Series

In conclusion, we make another general remark about the fundamental difference between absolute convergence and conditional convergence. We consider a convergent series $\sum_{\nu=1}^{\infty} a_\nu$ and denote the positive terms of the series by $p_1, p_2, p_3,...$ and the negative terms by $-q_1, -q_2, -q_3,...$ If we form the n-th partial sum $s_n = \sum_{\nu=1}^{n} a_\nu$ of the given series, a certain nmnber, say, n' positive terms and a certain number, say n'' negative terms must appear, where $n' + n''=n$. Moreover, if the number of positive terms as well as that of negative terms in the series is infinite, then the two numbers n'' and n'' will increase beyond allboundsw ith n. We see immediately that the partial sum s_n is simply equal to the partial sum $\sum_{\nu=1}^{n'} p_\nu$ of the positive terms of the series plus the partial sum $-\sum_{\nu=1}^{n''} q_\nu$ of the negative terms. If the given series converges absolutely, then both the series of positive terms $\sum_{\nu=1}^{\infty} p_\nu$ and the series of the absolute values of the negative terms $\sum_{\nu=1}^{\infty} q_\nu$ certainly converge. In fact, as m increases, the partial sums $\sum_{\nu=1}^{m} p_\nu$ *and* $\sum_{\nu=1}^{m} q_\nu$ are monotonic non-decreasing with the upper bound $\sum_{\nu=1}^{\infty} |a_\nu|$.

The sum of an absolutely convergent series is then simply equal to the sum of the series consisting only of the positive terms plus the sum of the series consisting of the negative terms, or, in other words, it is equal to the difference of the two series with positive terms.

However, $\sum_{\nu=1}^{n} a_\nu = \sum_{\nu=1}^{n'} p_\nu - \sum_{\nu=1}^{n''} q_\nu$; as n increases, n' and n'' must also increase beyond all bounds, and the limit of the left hand side must therefore be equal to the difference of the two sums on the right hand side. If the series contains only a finite number of terms of one particular sign, the facts are correspondingly simplified. On the other hand, if the series does not converge absolutely, but does converge conditionally, then both the series $\sum_{\nu=1}^{\infty} p_\nu$ and $\sum_{\nu=1}^{\infty} q_\nu$ must be divergent, because, if both were convergent, the series would converge absolutely, contrary to our hypothesis. If only one diverges, say $\sum_{\nu=1}^{\infty} p_\nu$, and the other converges, then separation into positive and negative parts, $s_n = \sum_{\nu=1}^{n'} p_\nu - \sum_{\nu=1}^{n''} q_\nu$, shows that the series could not converge; in fact, as n increases, n' and $\sum_{\nu=1}^{n'} p_\nu$ would increase beyond all bounds, while the term $\sum_{\nu=1}^{n''} q_\nu$ would approach a definite limit, so that the partial sum s_n would increase beyond all bounds.

Hence, we see that a conditionally convergent series cannot be thought of as the difference of two convergent series, one consisting of its positive terms and the other consisting of the absolute values of its negative terms.

Closely connected with this fact, there arises another difference between absolutely and conditionally convergent series which we shall now mention briefly.

Rearrangement of Terms: It is a property of finite sums that we may change the order of the terms or, as we say, rearrange the terms at will without changing their values. There arises the question: What is the exact meaning of a change of the order of terms in an infinite series and does such a rearrangement leave the value of the sum unchanged? While in the case of finite sums there arises no difficulty, for example, in adding the terms in reverse order, in the case of infinite series such a possibility does not exist; there is no last term with which to begin.

Now a change of the order of terms in an infinite series can only mean this: A series $a_1 + a_2 + a_3 + ...$ is rearranged into a series $b_1 + b_2 + b_3 + ...$, provided that every term a_n of the first series occurs exactly once in the second series and conversely. For example, the amount by which a_n is displaced may increase beyond all bounds as n does; the only point is that it must appear somewhere in the new series. If some of the terms are moved to later positions in the series, other terms must, of course, be moved to earlier positions. For example, the series $1 + q + q^2 + q^4 + q^3 + q^8 + q^7 + q^6 + q^5 + q^{16} + ...$ is a rearrangement of the geometric series $1 + q + q^2 + ...$

As regards the change of order, there is a fundamental distinction between absolutely convergent and conditionally convergent series. In absolutely convergent series, rearrangement of terms does not affect the convergence and the value of the sum of the series is not changed, exactly as in the case of finite sums. On the other hand, in conditionally convergent series, the value of the sum of the series can be changed at will by suitable rearrangement of the series, and the series can even be made to diverge, if so desired.

The first of these facts, referring to absolutely convergent series, is easily established. Let us assume, to begin with, that our series has only positive terms and consider the n-th partial sum $s_n = \sum_{\nu=1}^{n} a_\nu$. All the terms of this partial sum occur in the m-th partial sum $t_m = \sum_{\nu=1}^{m} b_\nu$ of the rearranged series, provided only that m is chosen large enough. Hence $t_m \leq s_n$. On the other hand, we can determine an index n' so large that the partial sum $s_{n'} = \sum_{\nu=1}^{n'} a_\nu$ of the first series contains all the terms b_1, b_2, b_m. It then follows that $t_m \leq s_{n'} \leq A$, where A is the sum of the first series. Thus, for all sufficiently large values of m, we have $s_n \leq t_m \leq A$; since s_n can be made to differ from A by an arbitrarily small amount, it follows that the rearranged series also converges, in fact, to the same limit A as the original series.

If the absolutely convergent series has both positive and negative terms, we may regard it as the difference of two series each of which has only positive terms. Since in the rearrangement of the original series each of these two series merely undergoes rearrangement and therefore converges to the same value as before, the same is true of the original series when rearranged. In fact, by the case just considered, the new series is absolutely convergent and is therefore the difference of the two rearranged series of positive terms.

The fact just proved may seem to the beginner to be a triviality. The fact that it really does require a proof and that in this proof absolute convergence is essential can be demonstrated by an example of the opposite behaviour of conditionally convergent series. Consider the familiar series $1 - \frac{1}{2} + \frac{1}{3} - \frac{1}{4} + \frac{1}{5} - \frac{1}{6} + \frac{1}{7} - \frac{1}{8} + - \ldots = \log 2$.

Write below it the result of multiplication by the factor 1/2 $\frac{1}{2} - \frac{1}{4} + \frac{1}{6} - \frac{1}{8} + - \ldots = \frac{1}{2}\log 2$ and add the two series, combining the terms in vertical columns. We thus obtain $1 - \frac{1}{3} - \frac{1}{2} + \frac{1}{5} + \frac{1}{7} - \frac{1}{4} + \frac{1}{9} + \frac{1}{11} + + \ldots = \frac{3}{2}\log 2$.

This last series can obviously be obtained by rearranging the original series and yet the value of the sum of the series has been multiplied by the factor 3/2. It is easy to imagine the effect of the discovery of this apparent paradox on the mathematicians of the Eighteenth Century, who were accustomed to operate with infinite series without regard to their convergence. We shall give the proof of the theorem stated above concerning the change in the sum of a conditionally convergent series which arises from a change of order of its terms, although we shall have no occasion to make use of the result.

Let $p_1, p_2, \ldots$ be the positive terms and $-q_1, -q_2, \ldots$ the negative terms of the series. Since the absolute value $|a_n|$ tends to 0 as n increases, the numbers p_n and q_n must also tend to 0 with n. Moreover, as we have already seen, the sum $\sum_1^\infty p_\nu$ must diverge and the same is true for $\sum_1^\infty q_\nu$.

This abbreviated notation for $\sum_1^\infty p_v$ and analogous expressions for other series will be used often in what follows.

We can now find easily a rearrangement of the original series which has an arbitrary number a as its limit. In order to be specific, let a be positive. We then add the first n_1 positive terms, just enough to ensure that the sum $\sum_1^{n_1} p_v$ is less than a. Since the sum $\sum_1^{n_1} p_v$ increases with n_1 beyond all bounds, it is always possible, by using enough terms, to make the partial sum larger than a. The sum will then differ from the exact value a by at most p_n. We now add just enough negative terms $-\sum_1^{m_1} p_v$ in order to ensure that the sum $-\sum_1^{m_1} p_v$ $-\sum_1^{m_1} q_v$ is less than a. This is also possible, as follows from the divergence of the series from the divergence of the series $\sum_1^\infty q_v$. The difference between this sum and a is now at most q_{m1}. We now add just enough other positive terms $\sum_{n_1+1}^{n_2} p_v$ in order to make the partial sum again larger than a, as is again possible since the series of positive terms diverges. The difference between the partial sum and a is now at most p_{ms}. We again add just just enough negative terms $\sum_{m_1+1}^{m_2} q_v$, which start next after the last one previously used, to make the sum once more less than a and continue in the same way. The values of the sums thus obtained will oscillate about the number a and, when the process is carried out far enough, the oscillation will only take place between arbitrarily narrow bounds; in fact, since the terms p_n and q_n themselves tend to 0 when n is sufficiently large, the length of the interval in which the oscillation occurs will also tend to 0, and the theorem is proved.

In the same manner, we can rearrange s series in such a way as to make it diverge; we merely have to choose such largc numbers of positive terms as compared with the negative ones that there occurs no longer compensation.

Operations with Infinite Series: It is clear that two convergent infinite series $a_1 + a_2 +... = S$ and $b_1 + b_2 +... = T$ can be added term by term, i.e., that the series formed from the terms $c_n = a_n + b_n$ converges and has the sum $S + T$, because

$$\sum_{v=1}^{n} c_v = \sum_{v=1}^{n} a_v + \sum_{v=1}^{n} b_v \to S+T$$

This theorem is really nothing more than another statement of the fact that the limit of the sum of two terms is the sum of their limits.

It is also clear that, if we multiply each term of a convergent infinite series by the same factor, the series remains convergent and its sum is multiplied by the same factor.

In the cases just mentioned, it is immaterial whether the convergence is absolute or conditional. On the other hand, a further study, which is not necessary for us here, shows that, if two infinite series are multiplied by each other by the method used in multiplying finite sums, the product series will not usually converge or have the product of the two sums as its sum unless at least one of the two series is absolutely convergent.

Tests for Convergence and Divergence

We have already encountered a test of a general nature for the convergence of series, which applies to series with terms of alternating signs and decreasing absolute values and which asserts that such series are at least conditionally convergent. In what follows, we shall only consider criteria referring to absolute convergence.

The Comparison Test

All such considerations of convergence depend on the comparison of the series in question with a second series; this second series is chosen in such a way that its convergence can readily be tested. The general comparison test may be stated as follows: If all the numbers b_1, b_2... are positive and the series $\sum_{\nu=1}^{\infty} b_\nu$ converges and if $|a_n| \leqq b_n$ for all values of n, then the series $\sum_{\nu=1}^{\infty} a_n$ is absolutely convergent. If we apply Cauchy's test, the proof becomes almost trivial, because, if m n, we have $|a_n + ... + a_m| \leqq |a_n| + ... + |a_m| \leqq b_n + ... + b_m$.

Since the series $\sum_{\nu=1}^{\infty} b_\nu$ converges, the right hand side is arbitrarily small, provided that n and m are sufficiently large, whence for such values of n and m the left hand side is also arbitrarily small, so that, by Cauchy's test, the given series converges. The convergence is absolute, since our argument applies equally well to the convergence of series of absolute values $|a_n|$.

The analogous proof for the following fact will be left to the reader:

If $|a_n| \geqq b_n > 0$ and the series $\sum_{\nu=1}^{\infty} b_\nu$ diverges, then the series $\sum_{n=1}^{\infty} a_\nu$ is certainly not absolutely convergent.

Comparison with the Geometric Series

In applications of the test, the most frequently used comparison series is the geometric series. We obtain immdediately the theorem: The series $\sum_{n=1}^{\infty} a_n$ is absolutely convergent, if from a certain term onwards there holds a relation of the form

$$|a_n| < c q^n,$$

where c is a positive number independent of n and q is any fixed positive number less than 1.

This test is usually expressed in one of the following, weaker forms: The series $\sum_{n=1}^{\infty} a_n$ converges absolutely, if there holds from a certain term onwards a relation of the form

$$\left|\frac{a_{n+1}}{a_n}\right| < q,$$

where q is again a positive number less than 1 and independent of n, or, if from a certain term onward there holds a relation of the form

$$\sqrt[n]{|a_n|} < q,$$

where q is again a positive number less than 1. In particular, the conditions of these tests are satisfied if there applies a relation of the form

$$\lim_{n\to\infty}\left|\frac{a_{n+1}}{a_n}\right| = < 1$$

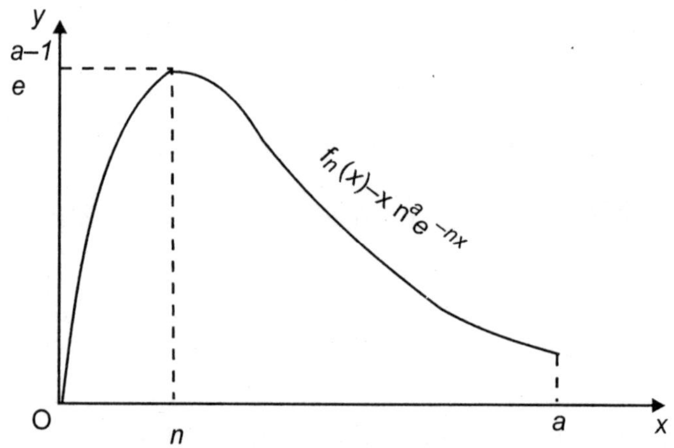

These statements are easily established in the following manner: Let the criterion II*a*, the ratio test, be satisfied from the suffix n_0 onward, i.e., when $n > n_0$. For the sake of brevity, let $a_{n_0+m+1} = b_m$ and find that $|b_1| < q|b_0|$, $|b_2| < q|b_1| < q^2|b_0|$, $|b_3| < q\ |b_2| < q^3|b_0|$, etc.; hence $|b_m| < q^m|b_0|$, which establishes our statement. For the Criterion II*b* - the root test - we have at once $|a_n| < q_n$, and our statement follows immediately.

Finally, in order to prove III, we consider an arbitrary number q such that $k < q < 1$. Then, from a certain n_0 onwards, i.e.,, when $n > n_0$, it is certain that $\left.\frac{a_{n+1}}{a_n}\right| < q$ or $\sqrt[n]{|a_n|} < q$, as the case may be, since from a certain term onwards the values of $\left.\frac{a_{n+1}}{a_n}\right| < q$ *of* $\sqrt[n]{|a_n|}$ differ from k by less than $(g - k)$. The statement is then established by reference to the results already proved.

We stress the point that the four tests derived from the original criterion $|a_n|<cq^n$ are not equivalent to each other or to the original one, i.e., that they cannot be derived from one another in both directions. We shall soon see from examples that, if a series satisfies one of these conditions, it need not by any means satisfy all the other ones.

* More exactly: If III*a* is fulfilled, then II*a* ie fulfilled: if III*b*, then II*b*; if III*a*, then III*b*; if II*a*, then II*b*, and if any of the four is satisfied, then so is I. None of these statements can he reversed.

For the sake of completeness, it may be pointed out that a series certainly diverges if from a certain term onwards $|a_n| > c$ for a properly chosen positive number c or if from a certain term onwards $\sqrt[n]{|a_n|} > 1$, or if

$$\lim_{n\to\infty}\left|\frac{a_{n+1}}{a_n}\right| = k, \qquad \lim_{n\to\infty}\left|\sqrt[n]{a_n}\right| = k,$$

where $k > 1$. In fact, as we immediately recognize, in such a series the terms cannot tend to zero as n increases, whence the series must diverge. (Under these conditions, the series cannot even be conditionally convergent.)

Our tests furnish sufficient conditions for the absolute convergence of a series, i.e.. if they are satisfied, we can conclude that the series converges absolutely. However, they are definitely not necessary conditions, i.e., absolutely convergent series can be found which do not satisfy these conditions.

For example, the knowledge that

$$\lim_{n\to\infty}\left|\frac{a_{n+1}}{a_n}\right| = 1, \qquad \lim_{n\to\infty}\left|\sqrt[n]{a_n}\right| = 1,$$

does not entitle us to make any statement about the convergence of the series. Such a series may converge or diverge. For example, the series $\sum_{n=1}^{\infty}\frac{1}{n}$, for which $\lim_{n\to\infty}\sqrt[n]{|a_n|} = 1$ *and* $\lim_{n\to\infty}\sqrt[n]{\left|\frac{a_{n+1}}{a_n}\right|} = 1$ is divergent, as we have seen before. On the other hand, we shall soon see that the series $\sum_{n=1}^{\infty}\frac{1}{n^2}$, which satisfies the same relations, converges. As an example of the application of our tests, we first consider the series $q + 2q^2 + 3q^3 + \ldots + nq^n + \ldots$

For this series,

$$\lim_{n\to\infty}\sqrt[n]{|a_n|} = |q|\cdot\lim_{n\to\infty}\sqrt[n]{n} = |q|,$$

$$\lim_{n\to\infty}\left|\frac{a_{n+1}}{a_n}\right| = |q|\cdot\lim_{n\to\infty}\frac{n+1}{n} = |q|.$$

It follows from the ratio test as well as from the root test even in the weaker form III that the series converges, if $|q| < 1$.

On the other hand, if we consider the series $1 + 2q + q^2 + 2q^3 + ... + q^{2n} + 2q^{2n+1} + ...$,

we can no longer prove convergence by the ratio test when }ÿ } $|q| < 1$, because then $|2q^{2n+1}/q^{2n}| = 2|q|$ } 1. But the root test yields immediately $\lim \sqrt[n]{|a_n|} = |q|$ and shows that the series converges provided $|q| < 1$, which, of course, we could have observed directly.

Comparison with an Integral

We now proceed to a discussion of convergence which is independent of the preceding considerations. We shall carry it out for the particularly simple and important case of the series

$$\sum_{n-1}^{\infty} \frac{1}{n^{\alpha}} = 1 + \frac{1}{2^{\alpha}} + \frac{1}{3^{\alpha}} + ...,$$

where the general term a_n is $1/n^{\alpha}$, $\alpha > 0$. In order to investigate the properties of this series, we consider the graph of the function $y = 1/x^{\alpha}$ and mark off on the x-axis the integral values $x = 1, x = 2$. We first construct the rectangle of height $1/n^{\alpha}$ over the interval n-1 ýÿ x ýÿ n of the x-axis ($n > 1$) and compare it with the area of the region bounded by the same interval of the x -axis, the ordinates at the ends and the curve $y=1/x^{\alpha}$.

Secondly, we construct the rectangle of height $1/n^{\alpha}$ lying above the interval n}x}n+1 and similarly compare it with the area of the region lying above the same interval and below the curve.

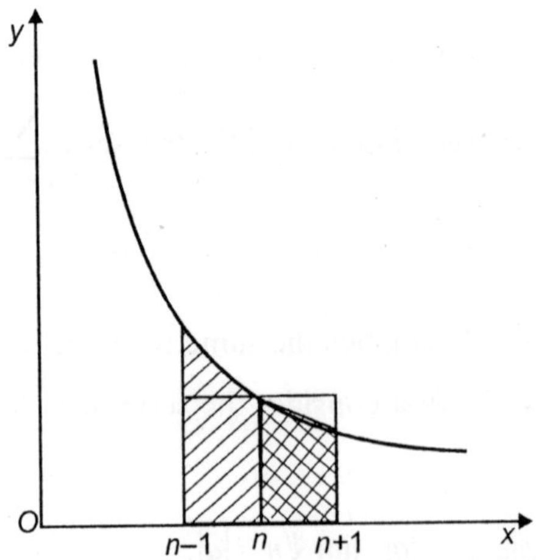

Fig. Comparison of a Series with an Integral

In the first case, the area under the curve is obviously larger than the area of the rectangle; in the second case, it is less than the area of the rectangle.

In other words, $\int_n^{n+1}\frac{dx}{x^\alpha}<\frac{1}{n^\alpha}<\int_{n-1}^{n}\frac{dx}{x^\alpha}$, as we may also prove directly from the integral itself. Writing down this inequality for $n = 2, n = 3, \ldots, n = m$ and summing, we obtain the estimate * for the m-th partial sum $s_m=\sum_{n=1}^{m}\frac{1}{n^\alpha}$

$$1+\int_2^{m+1}\frac{dx}{x^\alpha}<s_m<1+\int_1^{m}\frac{dx}{x^\alpha}$$

* From this relation follows at once that the sequence of numbers $C_n = 1+\frac{1}{2}+\frac{1}{3}+\ldots+\frac{1}{n}-\log n$ is bounded below. Since we see from the inequality $\frac{1}{n+1}<\int_n^{n+1}\frac{dx}{x}=\log(n+1)$ that the sequence is monotonic decreasing, it must approach a limit

$$\lim_{n\to\infty} C_n=\lim_{n\to\infty}\left(1+\frac{1}{2}+\frac{1}{3}+\ldots+\frac{1}{n}-\log n\right)$$

The number C with the value 0.577,2 is called *Euler's constant*. In contrast to the other important special numbers of analysis, such as p and e, no other expression with a simple law of formation has been found for Euler's constant.

Now, as m increases, the integral $\int_1^m\frac{1}{x^\alpha}dx$ tends to a finite limit or increases without limit according to whether $\alpha > 1$ or $\alpha > 1$. Consequently, the monotonic sequence of the numbers s_m is bounded or increases beyond all bounds according to whether $\alpha >$ or $\alpha > 1$, whence follows the theorem:

The series $\sum_{n=1}^{\infty}\frac{1}{n^\alpha}=\frac{1}{1^\alpha}+\frac{1}{2^\alpha}+\ldots$ converges - and, of course, absolutely - if and only if, $\alpha > 1$.

The divergence of the harmonic series, which we have proved previously in a different way, is an immediate consequence of this. In particular, we see that the series $\frac{1}{1^2}+\frac{1}{2^2}+\frac{1}{3^2}+\ldots$, $\frac{1}{1^3}+\frac{1}{2^3}+\frac{1}{3^3}+\ldots$,...................., converge.

The series $\sum_{\nu=1}^{\infty}\frac{1}{\nu^\alpha}$, the convergence of which we have just studied, frequently serve as comparison series in investigations of convergence. For example, we see at once that for $\alpha > 1$ the series $\sum_{\nu=1}^{\infty}\frac{c_\nu}{\nu^\alpha}$ converges absolutely if the absolute values $|c_n|$ of the coefficients remain less than a fixed bound independent of n.

Sequences and Series of Functions

The terms of the infinite series hitherto considered have been constants, whence these series (when convergent) always represented definite numbers. But both in theory and in applications, the series of outstanding importance are those in which the terms are functions of a variable, so that the sum of the series is also a function of the variable, as in the case of Taylor series.

We shall therefore consider a series $g_1(x) + g_2(x) + g_s(x) +...$, in which the functions $g_n(x)$ are defined in an interval $a > x > b$. We will denote the n-th partial sum $g_1(x) + g_2(x) +... + g_n(x)$, by $f_n(x)$. Then the sum $f(x)$ of our series, where it exists, is simply the limit $\lim_{n\to\infty} f_n(x)$.

We may therefore regard the sum of an infinite series of functions as the limit of a sequence of functions $f_1(x), f_2(x), f_n(x),...$ Conversely, for any such sequence of functions $f_1(x), f_2(x),...,$ we can form an equivalent series by setting $g_1(x) = f_1(x)...$ and $g_n(x) = f_n(x) - f_{n-1}(x)$ for $n > 1$. Hence, when it is convenient, we can pass from the consideration of series to that of sequences and conversely.

Limiting Processes with Functions and Curves

We shall now state exactly what we mean by saying that a function $f(x)$ is the limit of a sequence $f_1(x), f_2(x), \ldots, f_n(x)$, in an interval $a \} x \} b$. The definition is: The sequence $f_1(x), f_2(x),...$ converges in that interval to the limit function $f(x)$, if at each point x of the interval the values $f_n(x)$ converge in the usual sense to the value $f(x)$.. In this case, we shall write $\lim_{n\to\infty} f_n(x) = f(x)$

According to Cauchy's test, we can express the convergence of the sequence without necessarily knowing or stating the limit function $f(x)$. In fact, our sequence of functions will converge to a limit function if, and only if, at each point x in our interval and for every positive number e the quantity $|f_n(x) - f_m(x)|$ is less than e, provided that the numbers n and m are chosen large enough, i.e., larger than a certain number $N = N(\varepsilon)$. This number $N(\varepsilon)$ usually depends on e and x and increases beyond all bounds as ε tends to zero.

We have frequently encountered cases of limits of sequences of functions. We mention only the definition of the power x^a for irrational values of α by the equation

$$x^{\alpha} = \lim_{n\to\infty} x^{r_n}$$

where $r_1, r_2,..., r_n,...$ is a sequence of rational numbers tending to α or the equation

$$e^{\alpha} = \lim_{n\to\infty}\left(1+\frac{x}{n}\right)^n,$$

where the functions $f_n(x)$ on the right hand side are polynomials of degree n.

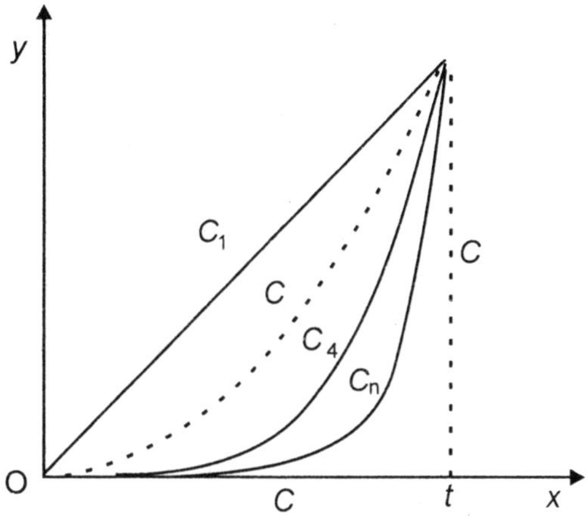

Fig. Limit curve and limit function

The graphical representation of functions by means of curves suggests that we can also speak of limits of sequences of curves, saying, for example, that the graphs of the above limit functions x^a and e^x are to be regarded as the limit curves of the graphs of the functions x^{r_n} *and* $\left(1+\frac{x}{n}\right)^n$ respectively. However, there is a fine distinction between the passages to the limit with functions and with curves. Until the middle of the Nineteenth Century, this distinction was not sufficiently recognized and only by having a clear idea of it can we avoid apparent paradoxes. We shall illustrate this point by an example.

Consider the functions

$$f_n(x) = x^n \ (n = 1, 2, ...)$$

in the interval $0 \leq x \leq 1$. All these functions are continuous and the limit function

$$\lim_{n\to\infty} f_n(x) = f(x)$$

exists. However, this limit function is not continuous. On the contrary, since for all values of n the value of the function $f_n(1) = 1$, the limit $f(1) = 1$,

while, on the other hand, for $0 \leq x < 1$, the limit $f(x) \lim_{n\to\infty} f_n(x) = 0$, as we have seen in 1.5.6. The function $f(x)$ is therefore a discontinuous function which at $x=1$ has the value 1 and for all other values of x in the interval has the value 0.

This discontinuity becomes clear, if we consider the graphs C_n of the functions $y = f_n(x)$. These are continuous curves, all of which pass through the origin and the point $x = 1$, $y = 1$ and which draw in closer and closer to the x-axis as n increases. The curves possess a limit curve C which is not at all discontinuous, but consists of the portion of the x-axis between $x=0$ and $x=1$ and the portion of the line $x=1$ between $y=0$ and $y=1$. The curves therefore converge to a continuous limit curve with a vertical portion, while the functions converge to a discontinuous limit function. We thus recognize that this discontinuity of the limit function expresses itself by the occurrence in the limit curve of a portion perpendicular to the x-axis. Such a portion must involve a discontinuity in the limit function and, in fact, such a portion is always present when the limit function is discontinuous.

This limit curve is not the graph of the limit function nor can any curve with a vertical portion be the graph of any single-valued function $y = f(x)$; in fact, corresponding to the value of x at which the vertical portion occurs, the curve gives an infinite number of values of y and the function only one. Hence the limit of the graphs of the functions $f_n(x)$ is not the same as the graph of the limit $f(x)$ of these functions. Naturally, corresponding statements also apply to infinite series.

Uniform and Non-uniform Convergence

The distinction between the concepts of convergence of functions and of convergence of curves introduces a phenomenon of which it is essential that the student should clearly recognize it. This is the so-called non-uniform convergence of sequences or infinite series of functions. Since it is well known that, as a rule, beginners encounter difficulties here, we should discuss the matter in some detail.

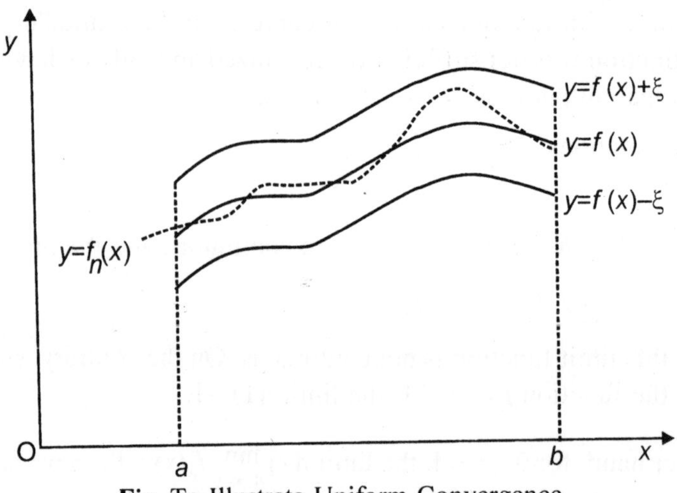

Fig. To Illustrate Uniform Convergence

By definition, a function $f(x)$ is the limit of a sequence $f_1(x), f_2(x),...$ in an interval $a \leq x \leq b$,only means that the usual limit relationship $f(x) = \lim_{n\to\infty} f(x)$ holds at each point x of the interval. From a naive point of view, one might expect that it would follow automatically from this concept of convergence that if we assign an arbitrary degree of accuracy, say ε=1/1000 or ε =1/100, then from a certain index N onwards all the functions $f_n(x)$ would lie between $f(x) + e$ and $f(x) - \varepsilon$ for all values of x, so that their graphs $y = f_n(x)$ will lie entirely in the strip indicated. Thus, for every positive ε, there is a corresponding number $N = N(\varepsilon)$, which, of course, will ordinarily increase beyond all bounds as $\varepsilon < 0$, such that, for $n > N$, the difference $|f(x) — f_n(x)| < \varepsilon$, no matter where x is chosen in the interval. If the accuracy of the approximation can be made at least equal to a pre-assigned number ε everywhere in the interval at the same time, i.e., by everywhere choosing the same number $N(\varepsilon)$ independent of x, we may say that the approximation is uniform. One is at first astonished to find that the naive assumption that convergence is necessarily uniform is altogether wrong; in other words, that convergence may very well be non-uniform.

A TEST OF UNIFORM CONVERGENCE

The preceding considerations show that the uniform convergence of a sequence or series is a special property which is not possessed by all sequences and series. We shall now reformulate the concept of uniform convergence. The convergent series $g_1(x) + g_2(x)$ +... is said to be uniformly convergent in an interval, if the sum $f(x)$ can be approximated to within ε (where ε is an arbitrarily small positive number) by taking a number of terms which is sufficiently large and which is the same throughout the interval.

First of all, we assume that the series $g_1(x) + g_2(x)$ +... converges at every point of a certain interval to a limit function $f(x)$; we denote by $f_n(x)$ the n-th partial sum of the series $f_n(x) = g_1(x) + g_n(x)$ and by $R_n(x)$ the remainder of the series after n terms:

$$R_n(x) = f(x) - f_n(x)$$

The series $g_1(x) + g_2(x)$ +... is said to be uniformly convergent in the interval if there corresponds to every positive number ε a number N, which only depends on e and not on x, such that for $n > N$ one has the inequality $|R(x)|=|f(x)-f_n(x)|<e$ for all values of x *in* the interval.

Expressed more figuratively, the partial sum $f_n(x)$ represents the sum $f(x)$ to within an error of less than e everywhere in the interval at the same time, provided only that n is chosen large enough. By Cauchy's test, we readily see that the series converges uniformly, if, and only if the difference $|f_n(x) - f_m(x)|$ can be made less than an arbitrary quantity e everywhere in the interval by choosing n and m larger than a number N independent of x. In fact, firstly, if the convergence is uniform, we can make both $|f_n(x) - f(x)|$ and $|f_m(x) - f(x)|$ less than $e/2$ by choosing n and m larger than a number N independent of x, whence it follows that $|f_n(x) - f_m(x)| < e$; and, secondly, if $|f_n(x) - f_m(x)| < e$, for all values of x whenever n and m are larger than N, then, on choosing any fixed value of $n > N$ and letting m increase beyond all bounds, we have

$$|f_n(x) - f(x)| = \lim_{m\to\infty} |f_n(x) - f_m(x)| \leqq \varepsilon,$$

for every value of x, whence the convergence is uniform.

If we wish to speak of the uniform convergence of a sequence of functions, we have only to make minute changes in the above definition; the sequence $f_1(x), f_2(x)$,... converges uniformly to $f(x)$ in an interval, if the difference $|f(x)-f_n(x)|$ can be made less than ε everywhere in the interval by choosing n larger than a number N independent of x. As above, a necessary and sufficient condition for the uniform convergence of the sequence is that $|f_n(x) - f_m(x)| < \varepsilon$ for all values of x when both n and m are larger than a certain number N, which depends on ε but not on x.

We shall soon see that it is just this condition of uniform convergence which makes infinite series and other limiting processes with functions into convenient and useful tools of analysis. Fortunately, in the limiting processes, usually encountered in the Calculus and its applications, non-uniform convergence is a sort of exceptional phenomenon which will scarcely trouble us in our present applications of analysis.

In most cases, the uniformity of the convergence of a series is established by means of the criterion:

If the terms of the series $\sum_{\nu=1}^{\infty} g_\nu(X)$ satisfy the condition $|g_\nu(X)| \leqq a_\nu$, where the numbers a_ϖ are constants which form a convergent series $\sum_{\nu=1}^{\infty} a_\nu$ then the series $\sum_{\nu=1}^{\infty} g_\nu(X)$ converges uniformly (and, we incidentally remark, absolutely).

In fact, we have then $\left|\sum_{\nu=n}^{m} g_\nu(x)\right| \leqq \left|\sum_{\nu=n}^{m}\right| g_\nu(x)| \leqq \sum_{\nu=n}^{m} a_\nu$, and since, by Cauchy's test, the sum $\sum_{\nu=n}^{m} a_\nu$, can be made arbitrarily small by choosing n and $m > n$ large enough, this expresses exactly the necessary and sufficient condition for uniform convergence.A first example is offered by the geometric series $1 + x + x + \ldots$, where x is restricted to the interval $|x| \} q$, a positive number less than 1. The terms of the series are then numerically less than or equal to the terms of the convergent geometric series $\sum q^\nu$ A further example is the trigonometric series $\frac{c_1 \sin(x-\delta_1)}{1^2} + \frac{c_2 \sin(x-\delta_2)}{2^2} + \frac{c_1 \sin(x-\delta_3)}{3^3} + \ldots$, provided that $|c_n| < c$, a positive constant independent of n. Because we have then

$$g_n(x) = \frac{c_n \sin(x-\delta_n)}{n^2}, \text{ so that } |g_n(x)| < \frac{c}{n^2}$$

Hence the uniform and absolute convergence of the trigonometric series follows from the convergence of the series

Continuity of the Sum of a Uniformly Convergent Series of Continuous Functions

As we have already hinted, the significance of the uniform convergence of an infinite series lies in the fact that a uniformly convergent series in many respects behaves exactly like the sum of a finite number of functions.

For example, the sum of a finite number of continuous functions is itself continuous and, correspondingly, we have the theorem: If a series of continuous terms converges uniformly in an interval, its sum is also a continuous function. The proof is quite simple. We subdivide the series $f(x) = g_1(x) + g_2(x) + \ldots$ into the m-th partial sum $f_n(x)$ plus the remainder $R_n(x)$. As usual, $f_n(x) = g_1(x) + \ldots g_n(x)$.

If now any positive number ε is assigned, we can,by virtue of the uniform convergence, choose the number n so large that the remainder is less than $\varepsilon/4$ throughout the entire interval, whence $|R_n(x+h) - R_n(x)| < \frac{\varepsilon}{2}$ for every pair of numbers x and $x + h$ in the interval. The partial sum $f_n(x)$ consists of the sum of a finite number of continuous functions, whence it is continuous; thus, for each point x in the interval, we can choose a positive δ so small that $|f_n(x+h) - f_n(x)| < \frac{\varepsilon}{2}$, provided $|h| < \delta$ and the points x and $m + h$ lie in the interval. It then follows that

$$\begin{aligned}|f(x+h) - f(x)| &= |f_n(x+h) - f_n(x) + R_n(x+h) - R_n(x)| \\ &\leqq |f_n(x+h) - f_n(x)| + |R_n(x+h) - R_n(x)| < \varepsilon,\end{aligned}$$

which expresses the continuity of our function.

The significance of this theorem becomes clear when we recall that the sums of non-uniformly convergent series of continuous functions are not necessarily continuous, as our previous examples show. We may conclude from the preceding theorem that, if the sum of a convergent series of continuous functions has a point of discontinuity, then in every neighbourhood of this point the convergence is non-uniform, whence every representation of discontinuous functions by series of continuous functions is based on the use of non-uniformly convergent limiting processes.

INTEGRATION OF UNIFORMLY CONVERGENT SERIES

A sum of a finite number of continuous functions can be integrated term by term, i.e., the integral of the sum can be found by integrating each term separately and adding the integrals. In the case of a convergent infinite series, the same procedure is possible, provided that the series converges uniformly in the interval of integration.

A series $\sum_{\nu=1}^{\infty} g_\nu(X) = f(X)$ which converges uniformly in an interval can be integrated term by term in that interval, or, more precisely, if a and x are two numbers in the interval of uniform convergence, the series $\sum_{\nu=1}^{\infty} \int_a^X g_\nu(t)\,dt$ converges and, in fact, converges uniformly with respect to x for each fixed value of a, its sum being equal to $\int_a^X f(t)\,dt$..

In order to prove this we write, as before,

$$f(x) = \sum_{\nu=1}^{\infty} g_\nu(x) = f_n(x) + R_n(x)$$

We have assumed that the separate terms of the series are continuous, whence, by 8.4.3, the sum is also continuous and therefore integrable. Now, if ε is any positive number, we can find a number N so large that, for every $n > N$, one has $|R_n(x)| < \varepsilon$ for every value of x in the interval. By the first mean value theorem of the integral calculus, we have $\left|\int_a^x \{f(t) - f_n(t)\} dt\right| \leq \in l$,

where l is the length of the interval of integration. Since the integration of the finite sum $f_n(x)$ can be performed term by term, this yields $\left|\int_a^x f(t)dt - \sum_1^n \int_a^x g_v(t)dt\right| \leq \in l$

However, since $\varepsilon\ l$ can be made as small as we please, this yields $\sum_{v=1}^{\infty} \int_a^x g_v(t)dt$ $= \lim_{n\to\infty} \sum_{v=1}^{n} \int_a^x g_v(t)dt = \int_a^x f(t)dt$, as was to be proved.

If we wish to conside instead of infinite series sequences of functions, our result can be expressed as follows:

If in an interval the sequence of functions $f_1(x), f_2(x), \ldots$ tends uniformly to the limit function $f(x)$, then $\int_a^b f(x)dx = \lim_{n\to\infty} \int_a^b f_n(x)dx$ for every pair of numbers a and b in the interval; in other words, we can then interchange the order of the operations of integration and going to the limit.

This fact is far from being a triviality. It is true that, from a naive point of view such as prevailed in the Eighteenth Century, the interchangeabi]ity of the two processes is hardly to be doubted; however, shows that in the case of non-uniform convergence the above equation might not hold. We need only consider Example, in which the integral of the limit function is 0, while the integral of the function $f_n(x)$ over the interval $0 \} x \} 1$, i.e., the area of the triangle has the value $\int_a^b f_n(x)dx = n^{a-2}$, and, when $\alpha \} 2$, this does not tend to zero. We see here immediately from figure that the reason for the difference between $\int_0^1 f_n(x)dx = n^{\alpha-2}$, and $\lim_{n\to\infty} \int_0^1 f_n(x)dx$ lies in the non-uniformity of the convergence.

On the other hand, by considering values of α in the interval $1 \} \alpha \} 2$, we see that the equation $\lim_{n\to\infty} \int_0^1 f_n(x)dx = \int_0^1 f(x)dx$ can be valid, although the convergence is non-uniform. As a further example, we can integrate term by term the series $\sum_0^{\infty} g_n(x)$,, where $g_n(x) = x^n - x^{n-1}$ for $n \} 1$ and $g_0(x) = 1$, between 0 and 1, even though it does not converge uniformly. Thus, while uniformity of convergence is a sufficient condition for term-by-term integrability, it is by no means a necessary condition. Neglect of this aspect may easily lead to a misunderstanding.

Differentiation of Infinite Series

The behaviour of uniformly converging series or sequences with respect to differentiation is quite different from that with respect to integration. For example, the sequence of functions $f_n(x) = \dfrac{\sin n^2\pi}{n}$ certainly converges uniformly to the limit function $f(x) = 0$, but the derivative $f_n'(x) = n\cos n^2 x$ does not converge everywhere to the derivative of the limit function $f'(0) = 0$, as we see by considering $x = 0$. In spite of the uniformity of the convergence, we cannot therefore change the order of the process of differentiation and pass on to the limit.

Naturally, corresponding statements hold for infinite series. For example, the series $\sin x + \dfrac{\sin 2^4 x}{2^2} + \dfrac{\sin 3^4 x}{3^2} + \ldots$ is absolutely and uniformly convergent, because its terms are numerically not greater than the terms of the convergent series $\dfrac{1}{1^2} + \dfrac{1}{2^2} + \dfrac{1}{3^2} + \ldots$.

However, if we differentiate the series term by term, we obtain the series $\cos x + 2^2 \cos 2^4 x + 3^2 \cos 3^4 x + \ldots$, which obviously does not converge everywhere, since, for example, it diverges at $x = 0$.

The only useful criterion which assures us in special cases that term-by-term differentiation is permissible is given by the theorem:

If, on differentiating a convergent infinite series $\sum_{\nu=0}^{\infty} G_\nu(X) = F(X)$ term by term, we obtain a uniformly convergent series of continuous terms $\sum_{\nu=0}^{\infty} g_\nu(X) = f(X)$, then the sum of this last series is equal to the derivative of the sum of the first series.

Hence, this theorem expressly requires that, after differentiating the series term by term, we must still investigate whether the result of the differentiation is a uniformly convergent series or not.

The proof of the theorem is almost trivial, because, we can integrate term by term the series obtained by differentiation. Recalling that $g_n(t) = G_n'(t)$, we obtain

$$\int_a^x f(t)\,dt = \int_a^x \left(\sum_{\nu=0}^{\infty} g_\nu(t) \right) dt = \sum_{\nu=0}^{\infty} \int_a^x g_\nu(t)\,dt = \sum_{\nu=0}^{\infty} \left(G_\nu(x) - G_\nu(a) \right)$$

$$= F(x) - F(a)$$

Since this is true for every value of x in the interval of uniform convergence, it follows that $f(x) = f(a)$ as was to be proved.

Power Series

Among infinite series, power series occupy the main position. We mean by a power series a series of the type

$$P(x) = c_0 + c_1 x + c_2 x^2 + ... = \sum_{\nu=0}^{\infty} c_\nu x^\nu$$

(a power series in x) or, more generally,

$$P(x) = c_0 + c_1 x(x - x_0) + c_2(x - x_0)^2 + ... = \sum_{\nu=0}^{\infty} c_\nu (x - x_0)^\nu$$

(a power series in $(x - x_0)$), where x_0 is a fixed number. If we introduce into the last series $\xi = x - x_0$ as a new variable, it becomes a power series $\sum_{\nu=0}^{\infty} c_\nu \xi^\nu$ in the new variable ξ, whence, without any loss of generality, we can confine our attention to power series of the more special form $\sum_{\nu=0}^{\infty} c_\nu x^\nu$

We have considered the approximate representation of functions by polynomials and were thus led to the expansion of functions in Taylor series, which, in fact, are power series. We shall now study power series in somewhat greater detail and obtain expansions of the most important functions in series in simpler and more convenient ways than before.

CONVERGENCE PROPERTIES OF POWER SERIES

There are power series which converge for no value of x, except, of course, for $x = 0$, e.g., the series $x + 2^2x^2 + 3^3x^3 + ... + n^n x^n + ...$

In fact, if $x < 0$, we can find an integer N such that $|x| > 1/N$. Then the absolute values of all the terms $n^n x^n$, for which $n > N$, will be larger than 1 and, as n increases, $nn^n x^n$ will increase beyond all bounds, so that the series fails to converge.

On the other hand, there are series which converge for every value of x; for example, the power series for the exponential function

$$e^x = 1 + x + \frac{x^2}{2!} + \frac{x^3}{3!} + ...,$$

the convergence of which for every value of x follows at once from the ratio test. The $(n + 1)$-th term, divided by the n-th term, yields x/n and, whatever number x is chosen, this ratio tends to zero as n increases. The behaviour of power series with regard to convergence is expressed in the fundamental theorem:

If a power series in x converges for a value $x = \xi$, it converges absolutely for every value x such that $|x| < |\xi|$ and the convergence is uniform in every interval $|x| \leq \eta$, where η is any positive number less than $|\xi|$. Hence h may lie as close to $|\xi|$ as we please.

The proof is simple. If the series $\sum_{\nu=0}^{\infty} c_\nu \xi^\nu$ converges, its terms tend to 0 as n increases. From this follows the weaker statement that all the terms lie below a bound M independent of n, i.e., $|c_n \xi^n| < M$. If now q is any number such that $0<q<1$ and we restrict x to the interval $|x| \leq q|\xi|$, then $|c_n x^n| < |c_n x^n| q^\nu < Mq^\nu$. Hence, in this interval, the terms of the series are smaller in absolute value than the terms of the convergent geometric series ΣMq^ν, whence follows, by the theorem, the absolute and uniform convergence of the series in the interval $-q|\xi| < x < q|x|$.

If a power series does not converge everywhere, i.e., if there is a value $x = x$ for which it diverges, it must diverge for every value of x such that $|x| > |\xi|$. In fact, if it were convergent for such a value of x, by the theorem above, it would have to converge for any numerically smaller value of ξ.

Hence, we recognize that a power series, which converges for at least one value of x other than 0 and which diverges for at least one value of x, has an interval of convergence, i.e., there exists a definite positive number r such that the series diverges for $|x| > \rho$ and converges for $|x| < \rho$. No general statement can be made for $|x| - \rho$. The limiting cases, in which the series converges only for x=0 and that which converges everywhere, are expressed symbolically by writing $\rho = 0$ and $\rho = \infty$, respectively.

It is possible to find this interval of convergence directly from the coefficients c_ν of the series. If the limit $\lim_{n\to\infty} \sqrt[n]{|c_n|}$ exists, then

$$\rho = \frac{1}{\lim_{n\to\infty} \sqrt[n]{|c_n|}}$$

In general, ρ is given by the formula

$$\rho = \frac{1}{\overline{\lim}_{n\to\infty} \sqrt[n]{|c_n|}}$$

where $\overline{\lim}$ is the symbol for the upper limit, as it has been defined in A1.5.

For example, $\rho = 1$ for the geometric series $1 + x + x + \ldots$; at the end-points of the interval of convergence, the series diverges. Similarly, for the series for the inverse tangent

$$\text{arc tan } x = x - x^3/3 + x^5/5 - + \ldots,$$

we have $\rho = 1$ and at both of the interval's end-points $x = 1$ the series converges, as we recognize at once from Leibnitz's test.

We deduce from the uniform convergence the important fact that within its interval of convergence (if such an interval exists) a power series represents a continuous function.

INTEGRATION AND DIFFERENTIATION OF POWER SERIES

On account of the uniformity of convergence, it is always permissible to integrate a power series

$$f(x) = \sum_{\nu=0}^{\infty} c_\nu x^\nu$$

term by term over any closed interval lying entirely within the interval of convergence. We thus obtain the function

$$F(x) = c + \sum_{\nu=0}^{\infty} \frac{c_\nu}{\nu+1} x^{\nu+1},$$

for which

$$F'(x) = f(x)$$

Moreover, since $|c_\nu/(n+1)| \leq |c_\nu|$ for all values of ν, the series obtained by integration converges more rapidly than the original series.

We can also differentiate a power series term by term within its interval of convergence, thus obtaining the equation

$$f'(x) = \sum_{\nu=1}^{\infty} \nu c_\nu x^{\nu-1}$$

In order to prove this statement, we need only show that the series on the right hand side converges uniformly, if x is restricted to an interval lying entirely within the interval of convergence. Now, let x be a number, lying as close to ρ as we please, for which $\sum_{\nu=1}^{\infty} c_\nu \xi^\nu$ converges; then, as we have seen before, all the numbers $|c_n \xi_n|$ lie below a bound M independently of ν, so that $\left|c_\nu \xi^{\nu-1}\right| < \frac{M}{|\xi|} = N$ Now let q be any number such that $0 < q < 1$; if we restrict x to the interval $|x| \leq q|\xi|$, the terms of the series under discussion are not larger than those of the series $\sum_{\nu=1}^{\infty} \left|\nu c_\nu q^{\nu-1} \xi^{\nu-1}\right|$, and therefore less than those of the series $\sum_{\nu=1} N\nu q^{\nu-1}$ .However, in this last series, the ratio of the $(n+1)$-th term to the n-th term is $\{(n+1)/n\}q$, which tends to q as n increases. Since $0<q<1$, it follows by Criterion III*a* that this series converges. Hence the series, obtained by differentiation, converges uniformly and by the theorem at the end of last section represents the derivative $f'_n(x)$ of the function $f(x)$, which proves our statement.

If we apply this result again to the power series

$$f'(x) = \sum_{\nu=1}^{\infty} \nu c_\nu x^{\nu-1},$$

we find on differentiating term by term that

$$f''(x) = \sum_{\nu=2}^{\infty} \nu(\nu-1)c_\nu x^{\nu-2}$$

and, by continuing the process, we arrive at the theorem:

Every function represented by a power series can be differentiated as often as we please within the interval of convergence and the differentiation can be performed term by term.

As an explicit expression for the *k*-th derivative, we obtain

$$f^{(k)}(x) = \sum_{\nu=k}^{\infty} \nu(\nu-1)\dots(\nu-k+1)c_\nu x^{\nu-k},$$

or in a slightly different form

$$\frac{f^{(k)}(x)}{k!} = \sum_{\nu=k}^{\infty} \binom{\nu}{k} c_\nu x^{\nu-k} = \sum_{\nu=0}^{\infty} \binom{k+\nu}{k} c_{k+\nu} x^{\nu}$$

These two formulae are frequently useful.

OPERATIONS WITH POWER SERIES

The preceding theorems on the behaviour of power series are our justification for operating in the same way with power series as with polynomials. Obviously, two power series can be added or subtracted by adding or subtracting the corresponding coefficients. It is also clear that a power series, like any other convergent series, can be multiplied by a constant factor by multiplying each term by that factor. On the other hand, the multiplication and division of two power series requires a somewhat more detailed study, for which we refer the reader to the Appendix. We merely mention here without a proof that two power series $f(x) = \sum_{\nu=0}^{\infty} a_\nu x^\nu$ *and* $g(x) = \sum_{\nu=0}^{\infty} b_\nu x^\nu$ can be multiplied by each other like polynomials. In order to be specific, we have the theorem:

Throughout the common part of the interval of convergence of two power series, their product is given by the convergent power series $\sum_{\nu=0}^{\infty} c_\nu x^\nu$ with the coefficients

$$\begin{aligned}
c_0 &= a_0b_0, \\
c_1 &= a_0b_1 + a_1b_0, \\
c_2 &= a_0b_2 + a_1b_1 + a_2b_0, \\
&\dots\dots\dots\dots \\
c_n &= a_0b_n + a_1b_{n-1} + \dots + a_nb_0, \\
&\dots\dots\dots\dots
\end{aligned}$$

Theorem of Uniqueness of Power Series

In the theory of power series, the following fact is important:

If both power series $\sum_{\nu=0}^{\infty} a_\nu x^\nu$ and $\sum_{\nu=0}^{\infty} b_\nu x^\nu$ converge in an interval which contains the point $x = 0$ inside and if in that interval the two series represent the same function $f(x)$, then they are identical, that is, the equation $a_n = b_n$ is true for every value of n.

In other words:

If at all, a function $f(x)$ can be represented by a power series in x in only one way,

Briefly speaking, the representation of a function by a power series is unique.

In order to prove this, we need only note that the difference of the two power series, i.e., the power series $\phi(x) = \sum_{\nu=0}^{\infty} c_\nu x^\nu$ with the coefficients $c_\nu = a_\nu - b_\nu$ represents the function $\phi(x) = f(x) - f(x) = 0$ in the interval, i.e., this last power series converges to the limit 0 everywhere in the interval. In particular, for $x = 0$, the sum of the series must be 0, i.e., $c_0 = 0$, so that $a_0 = b_0$. We now differentiate the series in the interior of the interval and obtain $\phi'(x) = \sum_{\nu=1}^{\infty} \nu c_\nu x^{\nu-1}$ However, $\phi'(x)$ is also 0 throughout the interval, whence, in particular, we have for $x = 0$: $c_1 = 0$ or $a_1 = b_1$. Continuing this process of differentiation and then setting $x = 0$, we find successively that all the coefficients c_n are equal to zero, which proves the theorem.

Moreover, we see that we can draw from the above discussion the conclusion:

If we take the ν-th derivative of a series $f(x) = \sum a_\nu x^\nu$ and then set $x = 0$, we at once obtain

$$a_\nu = \frac{1}{\nu!} f^{(\nu)}(0)$$

i.e., Every power series which converges for points other than $x = 0$ is the Taylor series of the function which it represents.

The uniqueness of the expansion is here expressed by the fact that the coefficients are uniquely determined by the function itself.

Expansion of Given Functions in Power Series

Within its interval of convergence, every power series represents a continuous function with continuous derivatives of all orders. We shall now discuss the converse problem of the expansion of a given function in a power series. In theory, we can always do this by means of Taylor's theorem, however, in practice, we often encounter difficulties with the actual calculation of the n-th derivative and with the estimation of the remainder. But we can often reach our goal more simply by means of the device:

We first write down the statement $f(x) = \sum_{\nu=0}^{\infty} c_\nu x^\nu$, where, to begin with, the coefficients c_ν are unknown. Then we determine by some known property of the function $f(x)$ the coefficients and prove subsequently the convergence of the series. The series represents a function and there remains only to prove that this function is identical to $f(x)$. Due to the uniqueness of the expansion in power series, we know that no other series than the one just found can be the required expansion.

We shall now consider some examples of the use of this method. In fact, we have already obtained the series for artan x and log $(1+x)$ by a method, which forms part of the range of ideas. We simply integrated term by term the series for the derivatives of these functions, which we knew to be geometric series.

The Exponential Function: Our problem is to find a function for which $f'(x) = f(x)$ and $f(0) = 1$. If we write down the series with undetermined coefficients $f(x) = c_0 + c_1x + c_2x^2 + ...$, and differentiate it, we obtain $f'(x) = c_1 + 2c_2x + 3c_3x^2 + ...$

Since, by assumption, these two power series must be identical, we have the equation $nc_n = c_{n-1}$, which is true for all values of $n \geq 1$. If we observe that, due to the relation $f(0)=1$, the coefficient c_0 must have the value 1, we can compute all the coefficients successively and obtain the power series $f(x) = 1 + \frac{x}{1!} + \frac{x^2}{2!} + \frac{x^3}{3!} + ...$

As is easily shown by the ratio test, this series converges for all values of x and therefore represents a function for which the relations $f'(x) = f(x)$, $f(0) = 1$ are actually fulfilled. Now, the function e^x certainly possesses these properties; we readily deduce that the function $\phi(x)$ is identical to e^x. In fact, if we form the quotient $f(x)=f(x)/e^x$ and differentiate it, we find

$$\varphi(\xi) = \frac{e^\alpha f'(x) - e^\alpha f'(x)}{e^{2x}} = 0$$

Hence, the function $\phi(x)$ is a constant and since it has the value 1 for $x = 0$, it must be identically equal to 1, thus proving that our power series and the exponential function are identical.

THE BINOMIAL SERIES

We can now return to the binomial series, this time using the method of undetermined coefficients. We wish to expand the function $f(x) = (1+x)^\alpha$ in a power series, whence we write $f(x) = (1 + x)^\alpha = c_0 + c_1x + c_2x^2 + ...$, where the coefficients c_n are undetermined. We now note that obviously our function satisfies the relation $(1 + x) = f'(x) = \alpha f(x) = \sum_{\nu=0}^{\infty} \alpha c_\nu x^\nu$

On the other hand, if we differentiate the series for $f(x)$ term by term and multiply by $(1+x)$, we obtain $(1+x)f'(x) = c_1 + (2c_2 + c_1)x + (3c_3 + 2c_2)x^2 + \ldots$; since these two power series for $f'(x)$ must be identical, we have $\alpha c_0 = c_1$, $\alpha c_1 = 2c_2 + c_1$, $\alpha c_2 = 3c_3 + 2c_2$,....

It is now certain that $c_0 = 1$, since our series must have the value 1 for $x = 0$; we thus find in succession the expressions

$$c_1 = \alpha,\ c_1 = \frac{(\alpha-1)\alpha}{2},\ c_1 = \frac{(\alpha-2)(\alpha-1)\alpha}{3\cdot 2},\ldots,$$

for the coefficients, and, in general, as is readily established,

$$c_\nu = \frac{(\alpha-\nu-1)(\alpha-\nu+2)\ldots(\alpha-1)\alpha}{\nu(\nu-1)\ldots 2.1} = \binom{\alpha}{\nu}$$

Substituting these values for the coefficients, we have the series $\sum_{\nu=0}^{\infty}\binom{\alpha}{\nu}x^\nu$; we must yet investigate the convergence of this series and show that it actually represents $(1+x)^n$.

By the ratio test, we find that, when α is not a positive integer, the series converges if $|x| < 1$ and diverges if $|x| > 1$; in fact, then the ratio of the $(n+1)$-th term to the n-th term is $\{(\alpha - n + 1)/n\}x$ and the absolute value of this expression tends to $|x|$ as n increases beyond all bounds.

We state here without proof the exact conditions under which this series converges. If the index a is an integer, the series terminates and is therefore valid for all values of x (as it becomes the ordinary binomial theorem). For all other values of α, the series is absolutely convergent for $|x|<1$ and diverges for $|x|>1$. For $x = +1$, the series converges absolutely if $\alpha > 0$, conditionally if $-1 < \alpha < 0$, and it diverges if $\alpha = -1$. Finally, at $x = -1$, the series is absolutely convergent if $a > 0$, divergent if $a < 0$.

Hence, if $|x| < 1$, our series represents a function $f(x)$ which satisfies the condition $(1+x)f'(x) = af(x)$, as follows from the method of forming the coefficients. Moreover, $f(0) = 1$. But these two conditions ensure that the function $f(x)$ is identical to $(1+x)^\alpha$. In fact, setting

$$\phi(x) = f(x)/(1+x)^\alpha,$$

we find that

$$\phi(x) = \frac{(1+x)^\alpha f'(x) - \alpha(1+x)^{\alpha-1} f(x)}{(1+x)^{2\alpha}}$$

hence $f(x)$ is a constant and is always equal to 1, since $\phi(0) = 1$. We have thus proved that for $|x| < 1$

$$(1+x)^\alpha = \sum_{\nu=0}^{\infty}\binom{\alpha}{\nu}x^\nu$$

which is the binomial series.

We present here the following special cases of the binomial series:

THE GEOMETRIC SERIES

$\frac{1}{1+x} = (1+x)^{-1} = 1 - x + x^2 - x^3 + x^4 - + ... = \sum_{\nu=0}^{\infty}(-1)^{\nu} x^{\nu}$; the series $\frac{1}{(1+x)^2} = (1+x)^2$

$= 1 - 2x + 3x^2 - 4x^3 + - ... = \sum_{\nu=0}^{\infty}(-1)^{\nu}(\nu+1)x^{\nu}$, which may otherwise be obtained from the geometric series by differentiation and the

$$\sqrt{(1+x)} = (1+x)^{\frac{1}{2}} = 1 + \frac{1}{2}x - \frac{1}{2.4}x^2 + \frac{1.3}{2.4.6}x^3 - \frac{1.3.5}{2.4.6.8}x^4 + - ...,$$

$$\frac{1}{\sqrt{(1+x)}} = (1+x)^{-\frac{1}{2}} = 1 + \frac{1}{2}x + \frac{1.3}{2.4}x^2 - \frac{1.3.5}{2.4.6}x^3 \frac{1.3.5.7}{2.4.6.8}x^4 - + ...$$

series the first two or three terms of which are useful expressions.

The Series for arsin x

This series can be established readily by expansion of the expression $1/\sqrt{(1-t^2)}$ as a binomial series

$$\left(1-t^2\right)^{-\frac{1}{2}} = 1 + \frac{1}{2}t^2 + \frac{1.3}{2.4}t^4 + ...$$

This series converges if $|t| \succ 1$, whence it converges uniformly for $|t| < q < 1$. On integrating term by term between 0 and x, we obtain

$$\text{arc sin}x = x + \frac{1}{2}\frac{x^3}{3} + \frac{1.3}{2.4}\frac{x^5}{5} + ...;$$

by the ratio test, we find that this series converges if $|x| < 1$ and diverges if $|x| > 1$.

The derivation of this series by Taylor's theorem would be decidedly less convenient owing to the difficulty of estimating the remainder.

The Series for arsinh x = *log {*x + *(1 +* x*)}*

We obtain this expansion by a similar method. By the binomial theorem, we write down the series for the derivative of arsinh x:

$$\frac{1}{\sqrt{1+x^2}} = 1 - \frac{1}{2}x^2 + \frac{1.3}{2.4}x^4 - \frac{1.3.5}{2.4.6}x^6 + - ...,$$

and then integrate term by term to obtain the expansion

$$\text{arc sinh}x = x - \frac{1}{2}\frac{x^3}{3} + \frac{1.3}{2.4}\frac{x^5}{5} - + ...;$$

with the interval of convergence $-1 < x < 1$.

Example of Multiplication of Series

The expansion of the function $\dfrac{\log(1+x)}{1+x}$ is a simple example of the application of the rule for the multiplication of power series.

We have only to multiply the logarithmic series $\log(1+x) = x - \dfrac{x^2}{2} + \dfrac{x^3}{3} + \dfrac{x^4}{4} + - \ldots$ by the geometric series $\dfrac{1}{1+x} = 1 - x + x^2 - x^3 + x^4 - + \ldots$; the reader should verify that this yields the remarkable expansion

$$\frac{\log(1+x)}{1+x} = x - \left(1+\frac{1}{2}\right)x^2 + \left(1+\frac{1}{2}+\frac{1}{3}\right)x^3 - \left(1+\frac{1}{2}+\frac{1}{3}+\frac{1}{4}\right)x^4 + - \ldots$$

for $|x| < 1$.

Example of Term-by-Term Integration (Elliptic Integral)

In previous applications, we have encountered the elliptic integral

$$K = \int_0^{\pi/2} \frac{d\varphi}{\sqrt{(1-k^2\sin^2\varphi)}} \, (k^2 < 1)$$

(the period of oscillation of a pendulum. In order to evaluate the integral, we can first expand the integrand by the binomial theorem:

$$\frac{1}{\sqrt{(1-k^2\sin^2\varphi)}} = 1 + \frac{1}{2}k^2\sin^2\varphi + \frac{1.3}{2.4}k^4\sin^4\varphi + \frac{1.3.5}{2.4.6}k^6\sin^6\varphi + \ldots,$$

Since $k^2 \sin^2 \varphi$ is never larger than k^2, this series converges uniformly for all values of φ and we may integrate term by term:

$$K = \int_0^{\pi/2} \frac{d\varphi}{\sqrt{1-k^2\sin^2\varphi}} = \int_0^{\pi/2} d\varphi + \frac{1}{2}k^2 \int_0^{\pi/2} \sin^2\varphi \, d\varphi + \frac{1.3}{2.4}k^4$$

$$\int_0^{\pi/2} \sin^4\varphi \, d\varphi + \ldots$$

The integrals here have already been calculated in 4.4.4. Substitution of their values yields

$$K = \int_0^{\pi/2} \frac{d\varphi}{\sqrt{1-k^2\sin^2\varphi}} = \frac{\pi}{2}\left\{1 + \left(\frac{1}{2}\right)^2 k^2 + \left(\frac{1.3}{2.4}\right)^2 k^4 + \left(\frac{1.3.5}{2.4.6}\right) k^6 + \ldots\right\}$$

General Theory of functions of a Complex Variable

Although the purely formal point of view, indicated above, is by itself free from objection, it is still desirable to recognize in the above formula something more than a mere formal connection. In order to pursue this aim, we are led to the general theory of functions, as (for the sake of brevity) we call the general theory of the so-called analytic functions of a complex variable. In this, we may use as our starting point a general discussion of the theory of power series with complex variables and complex coefficients. The construction of such a theory of power series presents no difficulties once we have defined the concept of limit in the domain of complex numbers; in fact, it follows almost exactly the theory of real power series. But as we shall not make any use of these matters in what follows, we shall content ourselves here by stating certain facts and omit the proofs.

If a power series converges for any complex value $x = \xi$ whatever, then it converges absolutely for every value x for which $|x| < |\xi|$; if it diverges for a value $x = x$, then it diverges for every value x for which $|x| > |\xi|$. A power series, which does not converge everywhere, but does converge for some other point in addition to $x = 0$, has a circle of convergence, that is, there exists a number $\rho > 0$ such that the series converges absolutely for $|x| < \rho$and diverges for $|x| > \rho$.

Having established the concept of functions of a complex variable, represented by power series, and having developed the rules for operating with such functions, we can think of the functions e^x, sin x, cos x, artan x, etc., of the complex variable x as simply defined by the power series which represent them for real values of x. Then, all the above formal relationships reduce to trivialities.

We shall merely indicate by two examples, how this introduction of complex variables helps us to understand the elementary functions. The geometric series for $1/(1 + x$ýÿ$)$ ceases to converge when x leaves the interval $-1 \succ x \succ 1$ and so does the series for artan x, although there are no peculiarities in the behaviour of these functions at the ends of the interval of convergence; in fact, they and all their derivatives are continuous for all real values of x. On the other hand, we can readily understand that the series for $1/(1 - x)$ and log $(1 - x)$ cease to converge as x passes through the value 1, since they become infinite there. But this divergence of the series for the inverse tangent and the series $\sum_{\nu=0}^{\infty}(-1)^{\nu} x^{2\nu}$ *for* $|x| > 1$ becomes clear immediately, if we also consider complex values of x. In fact, we find that when $x = i$, the sum functions become infinite and so cannot be represented by a convergent series. Hence, by our theorem about the circle of convergence, the series must diverge for all values of x such that $|x| > |i|$; in particular, for real values of x, the series diverge outside the interval $-1 \succ x \succ 1$.

Another example is given by the function $f(x) = e^{-1/x2}$ for $x \, \psi 0$, which, in spite of its apparently regular behaviour, cannot be expanded in a Taylor series. As a matter of fact, this function ceases to be continuous, if we take pure imaginary values of $x = i\xi$ into account. The function then takes the form e^{1/ξ^2} and increases beyond all bounds as $\xi \succ 0$.

It is therefore clear that no power series in x can represent this function for all complex values of x in a neighbourhood of the origin, no matter how small a neighbourhood we choose.

These remarks on the theory of functions and power series of a complex variable must be sufficient for us here.

POWER SERIES WITH COMPLEX TERMS

Introduction of Complex Terms into Power Series: The similarity between certain power series representing functions which are apparently unrelated led Euler to set up a purely formal link between them, found by giving complex values, in particular, pure imaginary values to the variable x. We shall first do this formally, unhindered by questions of rigour and investigate the results of the process.

The first striking relation of this sort is obtained, if we replace the quantity x in the series for e^x by a pure imaginary $i\phi$, where ϕ is a real number. If we recall the fundamental equation for the imaginary unit i, that is, $i = -1$, from which it follows that $i^3 = -i$, $i^4 = 1$, $i^5 = i$,..., then, on separating the real and imaginary terms of the series, we obtain

$$e^{1\phi} = \left(1 - \frac{\phi^2}{2!} + \frac{\phi^4}{4!} + \frac{\phi^6}{6!} + - ...\right) + i\left(\phi - \frac{\phi^3}{3!} + \frac{\phi^5}{5!} - \frac{\phi^7}{7!} + - ...\right),$$

or, in another form, $e^{1\phi} = \cos\phi + i \sin\phi$

This is the well-known and important Euler formula. which is as yet purely formal. It is consistent with De Moivre's theorem which is expressed by the equation $(\cos\phi + i \sin\phi)(\cos\psi + i \sin\psi) = (\cos\phi + \psi) + i \sin(\phi + \psi)$.

By virtue of Euler's formula, this equation merely states that the relation $e^{x.} e^{v} = e^{x+v}$ continues to hold for pure imaginary values $x = i\phi$, $y = i\psi$.

If we replace the variable x in the power series for $\cos x$ by the pure imaginary ix, we obtain at once the series for $\cosh x$; this relation can be expressed by the equation

$$\cosh x = \cos ix.$$

In the same way, we obtain

$$\sinh x = \frac{1}{i} \sinh ix.$$

Since Euler's formula also yields

$$e^{-if} = \cos\phi - i \sin\phi,$$

we arrive at the exponential expressions for the trigonometric functions

$$\sin x = \frac{e^{ix} - e^{-ix}}{2i}, \qquad \cos x = \frac{e^{ix} + e^{-ix}}{2}$$

These are exactly analogous to the exponential expressions for the hyperbolic functions and are, in fact, transformed into them by the relations

$$\cosh x = \cos ix, \qquad \sinh x = \frac{1}{i}\sin ix$$

Naturally, corresponding formal relations can also be obtained for the functions tan x, tanh x, cot x and coth x, which are interrelated by the equations

$$\tanh x = \frac{1}{i}\tan ix \qquad \coth x = i\cot ix$$

Finally, similar relations can also be found for the inverse trigonometric and hyperbolic functions. For example, we immediately find from

$$y = \tan x = \frac{e^{ix}-e^{-ix}}{i\left(e^{ix}+e^{-ix}\right)} = \frac{e^{2ix}-1}{i\left(e^{2ix}+1\right)}$$

that

$$e^{2ix} = \frac{1+iy}{1-iy}$$

It we take the logarithms of both sides of this equation and then write x instead of y and artan x instead of x, we obtain the equation

$$\operatorname{artan} x = \frac{1}{2i}\log\frac{1+ix}{1-ix},$$

which expresses a remarkable connection between the inverse tangent and the logarithm. If we replace in the known power series for $\log(1+x)/(1-x)$ x by ix, we actually obtain the power series for artan x:

$$\operatorname{artan} x = \frac{1}{i}\left(ix+\frac{(ix)^3}{3}+\frac{(ix)^5}{5}+\ldots\right) = x-\frac{x^3}{3}+\frac{x^5}{5}-+\ldots$$

The above relations have as yet a purely formal character and naturally demand a more exact statement as to the meaning which they are intended to convey. In the next sub-section, we shall indicate how this can be given with the help of function theory.

However, we shall later on only require Euler's formula $e^{i\phi} = \cos\phi + i\sin\phi$ This being so, we can avoid here a thorough analysis. We need only regard the symbol $e^{i\phi}$ as a formal abbreviation $\cos\phi + i\sin\phi$, for in which case De Moivre's formula $e^{i\phi}\,.\,e^{i\psi} = e^{i(\phi+\psi)}$ appears merely as a consequence of the elementary addition theorem of trigonometry. From this formal point of view, in order to make the relation $e^x\,.\,e^y = e^{x+y}$ e*. e* remain valid for any complex arguments, we set up the further definition

$$e^x = e^{\xi}(\cos\eta + i\ \mathrm{xin}\,\eta)$$

where $x = \xi + i\eta$ (ξ, η real).

Chapter 10

Series Solutions to Differential Equations

In this chapter we will finally be looking at nonconstant coefficient differential equations. While we won't cover all possibilities in this chapter we will be looking at two of the more common methods for dealing with this kind of differential equation.-

The first method that we'll be taking a look at, series solutions, will actually find a series representation for the solution instead of the solution itself. You first saw something like this when you looked at Taylor series in your Calculus class. As we will see however, these won't work for every differential equation.-

The second method that we'll look at will only work for a special class of differential equations. This special case will cover some of the cases in which series solutions can't be used.

POWER SERIES

Before looking at series solutions to a differential equation we will first need to do a cursory review of power series. A power series is a series in the form,

$$f(x) = \sum_{n=0}^{\infty} a_n (x - x_0)^n$$

where, x_0 and a_n are numbers. We can see from this that a power series is a function of x. The function notation is not always included, but sometimes it is so we put it into the definition above.

Before proceeding with our review we should probably first recall just what series really are. Recall that series are really just summations. One way to write our power series is then,

$$\begin{aligned} f(x) &= \sum_{n=0}^{\infty} a_n (x - x_0)^n \\ &= a_0 + a_1(x - x_0) + a_2(x - x_0)^2 + a_3(x - x_0)^3 + \cdots \end{aligned}$$

Notice as well that if we needed to for some reason we could always write the power series as,

$$\begin{aligned} f(x) &= \sum_{n=0}^{\infty} a_n (x - x_0)^n \\ &= a_0 + a_1(x - x_0) + a_2(x - x_0)^2 + a_3(x - x_0)^3 + \cdots \\ &= a_0 + \sum_{n=0}^{\infty} a_n (x - x_0)^n \end{aligned}$$

All that we're doing here is noticing that if we ignore the first term (corresponding to $n = 0$) the remainder is just a series that starts at $n = 1$. When we do this we say that we've stripped out the $n = 0$, or first, term. We don't need to stop at the first term either. If we strip out the first three terms we'll get,

$$\sum_{n=0}^{\infty} a_n (x - x_0)^n = a_0 + a_1(x - x_0)^2 + \sum_{n=3}^{\infty} a_n (x_n - x_0)^n.$$

There are times when we'll want to do this so make sure that you can do it.

Now, since power series are functions of x and we know that not every series will in fact exist, it then makes sense to ask if a power series will exist for all x. This question is answered by looking at the convergence of the power series. We say that a power series converges for $x = c$ if the series,

$$\sum_{n=0}^{\infty} a_n (c - x_0)^n$$

converges. Recall that this series will converge if the limit of partial sums,

$$\lim_{n \to \infty} \sum_{n=0}^{\infty} a_n (c - x_0)^n$$

exists and is finite. In other words, a power series will converge for $x = c$ if

$$\sum_{n=0}^{\infty} a_n (c - x_0)^n$$

is a finite number.

Note that a power series will always converge if $x = x_0$. In this case the power series will become

$$\sum_{n=0}^{\infty} a_n (x_0 - x_0)^n = a_0.$$

With this we now know that power series are guaranteed to exist for at least one value of x. We have the following fact about the convergence of a power series.

FACT

Given a power series, there will exist a number $0 \leq \rho \leq \infty$so that the power series will converge for $|x - x_0| < \rho$ and diverge for $|x - x_0| < \rho$. This number is called the radius of convergence.

Determining the radius of convergence for most power series is usually quite simple if we use the ratio test.

RATIO TEST

Given a power series compute,

$$L = |x - x_0| \lim_{n\to\infty} \left| \frac{a_{n+1}}{a_n} \right|$$

then,

$L < 1 \Rightarrow$ the series converges

$L < 1 \Rightarrow$ the series diverges

$L = 1 \Rightarrow$ the series may converge or diverge

Let's take a quick look at how this can be used to determine the radius of convergence.

Example: Determine the radius of convergence for the following power series.

$$\sum_{n=0}^{\infty} \frac{(-3)}{n7^{3+1}} (x-5)^n.$$

Solution: So, in this case we have,

$$a_n \frac{(-3)}{n7^{3+1}} \qquad a_{n+1} = \frac{(-3)^{n+1}}{(n+1)7^{n+2}}.$$

Remember that to compute a_{n+1} all we do is replace all the n's in a_n with $n+1$. Using the ratio test then gives,

$$L = |x-5| \lim_{n\to\infty} \left| \frac{a_{n+1}}{a_n} \right|$$

$$= |x-5| \lim_{n\to\infty} \left| \frac{(-3)^{n+1}}{(n+1)7^{n+2}} \frac{n7^{n+n}}{(-3)^n} \right|$$

$$= |x-5| \lim_{n\to\infty} \left| \frac{-3}{(n+1)7} \frac{n}{1} \right| = \frac{3}{7} |x-5|.$$

Now we know that the series will converge if,

$$\frac{3}{7}|x-5|<1 \qquad \Rightarrow \qquad |x-5|<\frac{7}{3}$$

and the series will diverge if,

$$\frac{3}{7}|x-5|>1 \qquad \Rightarrow \qquad |x-5|>\frac{7}{3}$$

In other words, the radius of the convergence for this series is,

$$\rho=\frac{7}{3}.$$

As this last example has shown, the radius of convergence is found almost immediately upon using the ratio test. So, why are we worried about the convergence of power series? Well in order for a series solution to a differential equation to exist at a particular x it will need to be convergent at that x. If it's not convergent at a given x then the series solution won't exist at that x. So, the convergence of power series is fairly important.

Next we need to do a quick review of some of the basics of manipulating series. We'll start with addition and subtraction.

There really isn't a whole lot to addition and subtraction. All that we need to worry about is that the two series start at the same place and both have the same exponent of the x–x_0. If they do then we can perform addition and/or subtraction as follows,

$$\sum_{n=n_0}^{\infty} a_n(x-x_0)^n \pm \sum_{n=n_0}^{\infty} b_n(x-x_0)^n = \sum_{n=n_0}^{\infty} (a_n \pm b_n)(x-x_0)^n.$$

In other words all we do is add or subtract the coefficients and we get the new series.-

One of the rules that we're going to have when we get around to finding series solutions to differential equations is that the only x that we want in a series is the x that sits in. This means that we will need to be able to deal with series of the form,

$$(x-x_0)^c \sum_{n=0}^{\infty} a_n(x-x_0)^n$$

where c is some constant. These are actually quite easy to deal with.

$$\begin{aligned}(x-x_0)^c \sum_{n=0}^{\infty} a_n(x-x_0)^n &= (x-x_0)^c(a_0+a_1(x-x_0)+a_1(x-x_0)^2+\cdots)\\ &= a_0(x-x_0)^c+a_0(x-x_0)^{1+c}+a_1(x-x_0)^{2+c}+\cdots\\ &\sum_{n=0}^{\infty} a_n(x-x_0)^{n+c}.\end{aligned}$$

So, all we need to do is to multiply the term in front into the series and add exponents. Also note that in order to do this both the coefficient in front of the series and the term inside the series must be in the form $x\text{-}x_0$. If they are not the same we can't do this, we will eventually see how to deal with terms that aren't in this form.

Next we need to talk about differentiation of a power series. By looking it should be fairly easy to see how we will differentiate a power series. Since a series is just a giant summation all we need to do is differentiate the individual terms. The derivative of a power series will be,

$$f'(x) = a_1 + 2a_2(x - x_0) + 3a_3(x - x_0)^2 + \cdots$$

$$= \sum_{n=1}^{\infty} na_n(x - x_0)^{n-1}$$

$$= \sum_{n=1}^{\infty} na_n(x - x_0)^{n-1}.$$

So, all we need to do is just differentiate the term inside the series and we're done. Notice as well that there are in fact two forms of the derivative. Since the n=0 term of the derivative is zero it won't change the value of the series and so we can include it or not as we need to. In our work we will usually want the derivative to start at n=*1*, however there will be the occasion problem were it would be more convenient to start it at n=0.

Following how we found the first derivative it should make sense that the second derivative is,

$$f'(x) = \sum_{n=1}^{\infty} n(n-1)a_n(x - x_0)^{n-2}$$

$$= \sum_{n=1}^{\infty} n(n-1)a_n(x - x_0)^{n-2}$$

$$= \sum_{n=1}^{\infty} n(n-1)a_n(x - x_0)^{n-2}.$$

In this case since the n=0 and n=*1* terms are both zero we can start at any of three possible starting points as determined by the problem that we're working.

Next we need to talk about index shifts. As we will see eventually we are going to want our power series written in terms of $(x - x_0)^n (x - x_0)^n$ $(x - x_0)^n$ and they often won't, initially at least, be in that form. To get them into the form we need we will need to perform an index shift. Index shifts themselves really aren't concerned with the exponent on the x term, they instead are concerned with where the series starts as the following example shows.

Example: Write the following as a series that starts at n=0 instead of n=3.

$$\sum_{n=1}^{\infty} n^2 a_{n-1}(x+4)^{n+2}.$$

Solution: An index shift is a fairly simple manipulation to perform. First we will notice that if we define i=n-3 then when n=3 we will have i=0. So what we'll do is rewrite the series in terms of i instead of n. We can do this by noting that n=i+3. So, everywhere we see an n in the actual series term we will replace it with an i+3. Doing this gives,

$$\sum_{n=1}^{\infty} n^2 a_{n-1}(x+4)^{n+2} = \sum_{n=1}^{\infty} (i+3)^2 a_{i+3-1}(x+4)^{i+3+2}$$

$$= \sum_{n=1}^{\infty} (i+3)^2 a_{i+2}(x+4)^{i+5}.$$

The upper limit won't change in this process since infinity minus three is still infinity.

The final step is to realize that the letter we use for the index doesn't matter and so we can just switch back to n's.

$$\sum_{n=1}^{\infty} n^2 a_{n+1}(x+4)^{n+2} = \sum_{n=0}^{\infty} (n+3)^2 a_{n+2}(x+4)^{n+5}.$$

Now, we usually don't go through this process to do an index shift. All we do is notice that we dropped the starting point in the series by 3 and everywhere else we saw an n in the series we increased it by 3. In other words, all the n's in the series move in the opposite direction that we moved the starting point.

Example: *3* Write the following as a series that starts at n=5 instead of n=3.

$$\sum_{n=1}^{\infty} n^2 a_{n+1}(x+4)^{n+2}.$$

Solution: To start the series to start at n=5 all we need to do is notice that this means we will increase the starting point by 2 and so all the other n's will need to decrease by 2. Doing this for the series in the previous example would give,

$$\sum_{n=1}^{\infty} n^2 a_{n+1}(x+4)^{n+2} = \sum_{n=5}^{\infty} (n-2)^2 a_{n-3}(x+4)^{n}.$$

Now, as we noted when we started this discussion about index shift the whole point is to get our series into terms of. We can see in the previous example that we did exactly that with an index shift. The original exponent on the (x+4) was n+2. To get this down to an n we needed to decrease the exponent by 2. This can be done with an index that increases the starting point by 2.

Let's take a look at a couple of more examples of this.

Example: Write each of the following as a single series in terms of $(x - x_0)^n$.

1. $(x+2)^2 \sum_{n=3}^{\infty} na_n(x+2)^{n-4} - \sum_{n=1}^{\infty} na_n(x+2)^{n+1}$.

2. $x\sum_{n=0}^{\infty} (n-5)^2 b_{n+1}(x-3)^{n+3}$.

Solution: 1. $(x+2)^2 \sum_{n=3}^{\infty} na_n(x+2)^{n-4} - \sum_{n=3}^{\infty} na_n(x+2)^{n-1}$.

First, notice that there are two series here and the instructions clearly ask for only a single series. So, we will need to subtract the two series at some point in time. The vast majority of our work will be to get the two series prepared for the subtraction. This means that the two series can't have any coefficients in front of them (other than one of course...), they will need to start at the same value of n and they will need the same exponent on the x-x_0. We'll almost always want to take care of any coefficients first. So, we have one in front of the first series so let's multiply that into the first series. Doing this gives,

$$\sum_{n=3}^{\infty} na_n(x+2)^{n-2} - \sum_{n=3}^{\infty} na_n(x+2)^{n+1}.$$

Now, the instructions specify that the new series must be in terms of $(x - x_0)^n$, so that's the next thing that we've got to take care of. We will do this by an index shift on each of the series. The exponent on the first series needs to go up by two so we'll shift the first series down by 2. On the second series will need to shift up by 1 to get the exponent to move down by 1. Performing the index shifts gives us the following,

$$\sum_{n=3}^{\infty} (n+2)a_{n+2}(x+2) - \sum_{n=3}^{\infty} (n-1)a_{n-1}(x+2)^n.$$

Finally, in order to subtract the two series we'll need to get them to start at the same value of n. Depending on the series in the problem we can do this in a variety of ways. In this case let's notice that since there is an *n-1* in the second series we can in fact start the second series at *n=1* without changing its value. Also note that in doing so we will get both of the series to start at *n=1* and so we can do the subtraction. Our final answer is then,

$$\sum_{n=3}^{\infty} (n+2)a_{n+2}(x+2)^n - \sum_{n=1}^{\infty} (n-1)a_{n-1}(x+2)^n = \sum_{n=1}^{\infty} [(n+2)a_{n+2} - (n-1)a_{n-1}](x+2)^n$$

2. $x\sum_{n=3}^{\infty} (n-5)^2 b_{n+1}(x-3)^{n+3}$.

In this part the main issue is the fact that we can't just multiply the coefficient into the series this time since the coefficient doesn't have the same form as the term inside the series. Therefore, the first thing that we'll need to do is correct the coefficient so that we can bring it into the series. We do this as follows,

$$x\sum_{n=3}^{\infty}(n-5)^2 b_{n+1}(x-3)^{n+3} = (x-3+3)\sum_{n=0}^{\infty}(n-5)^2 b_{n+1}(x-3)^{n+3}$$

$$= (x-3)\sum_{n=0}^{\infty}(n-5)^2 b_{n+1}(x-3)^{n+3} + 3\sum_{n=0}^{\infty}(n-5)^2 b_{n+1}(x-3)^{n+3}.$$

We can now move the coefficient into the series, but in the process of we managed to pick up a second series. This will happen so get used to it. Moving the coefficients of both series in gives,

$$\sum_{n=0}^{\infty}(n-5)^2 b_{n+1}(x-3)^{n+4} + \sum_{n=0}^{\infty}3(n-5)^2 b_{n+1}(x-3)^{n+3}.$$

We now need to get the exponent in both series to be an n. This will mean shifting the first series up by 4 and the second series up by 3. Doing this gives,

$$\sum_{n=0}^{\infty}(n-9)^2 b_{n-3}(x-3)^{n} + \sum_{n=3}^{\infty}3(n-8)^2 b_{n-2}(x-3)^{n}.$$

In this case we can't just start the first series at $n=3$ because there is not an n-3 sitting in that series to make the $n=3$ term zero. So, we won't be able to do this part as we did in the first part of this example. What we'll need to do in this part is strip out the n=3 from the second series so they will both start at $n=4$. We will then be able to add the two series together. Stripping out the $n=3$ term from the second series gives,

$$\sum_{v=4}^{\infty}(n-9)^2 b_{n-3}(x-3)^{n} + 3(-5)^2 b_1(x-3)^3 + \sum_{n=4}^{\infty}3(n-8)^2 b_{n-2}(x-3)^{n}.$$

We can now add the two series together.

$$75b_1(x-3)^3 + \sum_{n=4}^{\infty}[(n-9)^2 b_{n-3} + 3(n-8)^2 b_{n-2}](x-3)^{n}.$$

This is what we're looking for. We won't worry about the extra term sitting in front of the series. When we finally get around to finding series solutions to differential equations we will see how to deal with that term there. There is one final fact that we need take care of before moving on. Before giving this fact for power series let's notice that the only way for

$$a + bx + cx^2 = 0$$

to be zero for all x is to have a=b=c=0.

We've got a similar fact for power series.

FACT

If,

$$\sum_{n=4}^{\infty} a_n (x - x_0)^n = 0$$

for all x then,

$$a_n = 0, \quad n = 0, 1, 2, \ldots$$

This fact will be key to our work with differential equations so don't forget it.-

TAYLOR SERIES

We are not going to be doing a whole lot with Taylor series once we get out of the review, but they are a nice way to get us back into the swing of dealing with power series. By time most students reach this stage in their mathematical career they've not had to deal with power series for at least a semester or two. Remembering how Taylor series work will be a very convenient way to get comfortable with power series before we start looking at differential equations.-

If $f(x)$ is an infinitely differential function then the Taylor Series of $f(x)$ about $x=x_0$ is,-

$$f(x) = \sum_{n=0}^{\infty} \frac{f^{(n)}(x_0)}{n!}(x - x_0)^n.$$

Recall that

$$f^{(0)}(x) = f(x) \quad f^{(n)}(x) = n^{\text{th}} \text{ derivative of } f(x).$$

Let's take a look at an example.

Example: Determine the Taylor series for $f(x) = e^x$ about $x = 0$.

Solution: This is probably one of the easiest functions to find the Taylor series for. We just need to recall that,

$$f^{(n)}(x) = e^x \; n = 0, 1, 2, \ldots$$

and so we get,

$$f^{(n)}(x) = e^x \; n = 0, 1, 2, \ldots$$

The Taylor series for this example is then,

$$e^x = \sum_{n=0}^{\infty} \frac{x^n}{n!}.$$

Of course, it's often easier to find the Taylor series about $x=0$ but we don't always do that.

Example: Determine the Taylor series for $f(x) = e^x$about $x = -4$.

Solution: This problem is virtually identical to the previous problem. In this case we just need to notice that,

$$f(n)\ (-4) = e^{-4}\ n = 0, 1, 2, \ldots$$

The Taylor series for this example is then,

$$e^x = \sum_{n=0}^{\infty} \frac{e^{-4}}{n!}(x+4)^n.$$

Let's now do a Taylor series that requires a little more work.

Example: Determine the Taylor series for $f(x) = \cos(x)$ about $x = 0$.

Solution: This time there is no formula that will give us the derivative for each n so let's start taking derivatives and plugging in $x = 0$.

$$\begin{aligned}
f^{(0)}(x) &= \cos(x) & f^{(0)}(0) &= 1 \\
f^{(1)}(x) &= -\sin(x) & f^{(1)}(0) &= 0 \\
f^{(2)}(x) &= -\cos(x) & f^{(2)}(0) &= -1 \\
f^{(3)}(x) &= \sin(x) & f^{(3)}(0) &= 0 \\
f^{(4)}(x) &= \cos(x) & f^{(4)}(0) &= 1 \\
&\vdots & &\vdots
\end{aligned}$$

Once we reach this point it's fairly clear that there is a pattern emerging here. Just what this pattern is has yet to be determined, but it does seem fairly clear that a pattern does exist.

Let's plug what we've got into the formula for the Taylor series and see what we get.

$$\begin{aligned}
\cos(x) &= \sum_{n=0}^{\infty} \frac{f^{(n)}(0)}{n!} x^n \\
&= \frac{f^{(n)}(0)}{0!} + \frac{f^{(1)}(0)}{1!}x + \frac{f^{(2)}(0)}{2!}x^2 + \frac{f^{(3)}(0)}{3!}x^3 + \cdots \\
&= \frac{1}{0!} + 0\frac{x^2}{2!} + 0 + \frac{x^4(0)}{4!} + 0 - \frac{x^6}{6!} + 0 + \frac{x^8}{8!} + \cdots
\end{aligned}$$

So, every other term is zero.

We would like to write this in terms of a series, however finding a formula that is zero every other term and gives the correct answer for those that aren't zero would be unnecessarily complicated. So, let's rewrite what we've got above and while were at it renumber the terms as follows,

$$\cos(x) = \underbrace{\frac{1}{0!}}_{n=0} - \underbrace{\frac{x^2}{2!}}_{n=1} + \underbrace{\frac{x^4}{4!}}_{n=2} - \underbrace{\frac{x^6}{6!}}_{n=3} + \underbrace{\frac{x^8}{8!}}_{n=4} + \cdots$$

With this "renumbering" we can fairly easily get a formula for the Taylor series of the cosine function about $x = 0$.

$$\cos(x) = \sum_{n=0}^{\infty} \frac{(-1)^n x^{2n}}{(2n)!}.$$

For practice you might want to see if you can verify that the Taylor series for the sine function about $x = 0$ is,

$$\sin(x) = \sum_{n=0}^{\infty} \frac{(-1)^n x^{2n+1}}{(2n+1)!}.$$

We need to look at one more example of a Taylor series. This example is both tricky and very easy.

Example: Determine the Taylor series for $f(x) = 3x^2 - 8x + 2$about $x = 2$.

Solution: There's not much to do here except to take some derivatives and evaluate at the point.

$$f(x) = 3x^2 - 8x + 2 \qquad f(2) = -2$$

$$f'(x) = 6x - 8 \qquad f'(2) = 4$$

$$f''(x) = 6 \qquad f''(2) = 6$$

$$f^{(")}(x) = 0, n \geq 3 \qquad f^{(")}(2) = 0, n \geq 3.$$

So, in this case the derivatives will all be zero after a certain order. That happens occasionally and will make our work easier. Setting up the Taylor series then gives,

$$3x^2 - 8x + 2 = \sum_{n=0}^{\infty} \frac{f^{(n)}(2)}{n!}(x-2)^n$$

$$= \frac{f^{(n)}(2)}{0!} + \frac{f^{(1)}(2)}{1!}(x-2) + \frac{f^{(2)}(2)}{2!}(x-2) + \frac{f^{(3)}(2)}{3!}(x-2)^3 + \cdots$$

$$= -2 + 4(x-2) + \frac{6}{2}(x-2)^2 + 0$$

$$= -2 + 4(x-2) + 3(x-2)^2.$$

In this case the Taylor series terminates and only had three terms. Note that since we are after the Taylor series we do not multiply the 4 through on the second term or square out the third term. All the terms with the exception of the constant should contain an $x-2$.

Note in this last example that if we were to multiply the Taylor series we would get our original polynomial. This should not be too surprising as both are polynomials and they should be equal.

We now need a quick definition that will make more sense to give here rather than in the next section were we actually need it since it deals with Taylor series.

Definition: A function, $f(x)$, is called analytic at $x = a$ if the Taylor series for $f(x)$ about $x=a$ has a positive radius of convergence and converges to $f(x)$.

We need to give one final note before proceeding into the next section. We started this section out by saying that we weren't going to be doing much with Taylor series after this section. While that is correct it is only correct because we are going to be keeping the problems fairly simple. For more complicated problems we would also be using quite a few Taylor series.

SERIES SOLUTIONS TO DIFFERENTIAL EQUATIONS

Before we get into finding series solutions to differential equations we need to determine when we can find series solutions to differential equations. So, let's start with the differential equation,

$$p(x)y''+q(x)y'+r(x)y=0.$$

This time we really do mean nonconstant coefficients. To this point we've only dealt with constant coefficients. However, with series solutions we can now have nonconstant coefficient differential equations. Also, in order to make the problems a little nicer we will be dealing only with polynomial coefficients.-

Now, we say that $x=x_0$ is an ordinary point if provided both

$$\frac{q(x)}{p(x)} \quad \text{and} \quad \frac{r(x)}{p(x)}$$

are analytic at $x=x_0$. That is to say that these two quantities have Taylor series around $x=x_0$. We are going to be only dealing with coefficients that are polynomials so this will be equivalent to saying that

$$p(x_0) \neq 0$$

for most of the problems.

If a point is not an ordinary point we call it a singular point. The basic idea to finding a series solution to a differential equation is to assume that we can write the solution as a power series in the form,

$$\sum_{n=0}^{\infty} a_n (x - x_0)^n$$

and then try to determine what the a_n's need to be. We will only be able to do this if the point $x=x_0$, is an ordinary point. We will usually say that series solution is around $x=x_0$.

Let's start with a very basic example of this. In fact it will be so basic that we will have constant coefficients. This will allow us to check that we get the correct solution.

Example: Determine a series solution for the following differential equation about $x_0 = 0$.

$$y'' + y = 0.$$

Solution: Notice that in this case $p(x)=1$ and so every point is an ordinary point. We will be looking for a solution in the form,

$$y(x) = \sum_{n=0}^{\infty} a_n x^n.$$

We will need to plug this into our differential equation so we'll need to find a couple of derivatives.

$$y'(x) = \sum_{n=0}^{\infty} n a_n x^{n-1} \qquad y''(x) = \sum_{n=0}^{\infty} n(n-1) a_n x^{n-2}.$$

Recall from the power series review section on power series that we can start these at $n=0$ if we need to, however it's almost always best to start them where we have here. If it turns out that it would have been easier to start them at $n=0$ we can easily fix that up when the time comes around.

So, plug these into our differential equation. Doing this gives,

$$\sum_{n=2}^{\infty} n(n-1)\, a_n x^{n-2} + \sum_{n=0}^{\infty} a_n x^n = 0.$$

The next step is to combine everything into a single series. To do this requires that we get both series starting at the same point and that the exponent on the x be the same in both series.

We will always start this by getting the exponent on the x to be the same. It is usually best to get the exponent to be an n. The second series already has the proper exponent and the first series will need to be shifted down by 2 in order to get the exponent up to an n. If you don't recall how to do this take a quick look at the first review section where we did several of these types of problems.

Shifting the first power series gives us,

$$\sum_{n=2}^{\infty}(n+2)(n+1)\,a_n x^n + \sum_{n=0}^{\infty} a_n x^n = 0.$$

Notice that in the process of the shift we also got both series starting at the same place. This won't always happen, but when it does we'll take it. We can now add up the two series. This gives,

$$\sum_{n=2}^{\infty}[(n+2)(n+1)\,a_n x^n + a_n]\,x^n = 0.$$

Now recalling the fact from the power series review section we know that if we have a power series that is zero for all x (as this is) then all the coefficients must have been zero to start with. This gives us the following,

$$(n+2)(n+1)a_{n+2} + a_n = 0, \qquad n = 0,\ 1,\ 2,...$$

This is called the recurrence relation and notice that we included the values of n for which it must be true. We will always want to include the values of n for which the recurrence relation is true since they won't always start at $n = 0$ as it did in this case. Now let's recall what we were after in the first place. We wanted to find a series solution to the differential equation. In order to do this we needed to determine the values of the a_n's. We are almost to the point where we can do that. The recurrence relation has two different a_n's in it so we can't just solve this for a_n and get a formula that will work for all n. We can however, use this to determine what all but two of the a_n's are. To do this we first solve the recurrence relation for the a_n that has the largest subscript. Doing this gives,

$$a_{n+2} = -\frac{a_n}{(n+2)(n+1)} \qquad n = 0,1,2,....$$

Now, at this point we just need to start plugging in some value of n and see what happens,

$$n=0 \quad a_2 = \frac{-a_0}{(2)(1)} \qquad\qquad n=1 \quad a_3 = \frac{-a_1}{(3)(2)}$$

$$n=2 \quad a_4 = -\frac{a_2}{(4)(3)} \qquad\qquad n=3 \quad a_5 = -\frac{-a_3}{(5)(4)}$$

$$= \frac{a_0}{(4)(3)(2)(1)} \qquad\qquad = \frac{a_1}{(5)(4)(3)(2)}\left(\frac{\pi}{2}-\theta\right)$$

$$n=4 \quad a_6 = \frac{a_4}{(6)(5)} \qquad\qquad n=5 \quad a_1 = -\frac{a_5}{(7)(6)}$$

$$= \frac{-a_0}{(6)(5)(4)(3)(2)(1)} \qquad\qquad \frac{-a_1}{(7)(6)(5)(4)(3)(2)}$$

$$\vdots \qquad\qquad \vdots \qquad\qquad \vdots$$

$$n=2k \quad a_{2k} = \frac{(-1)^k a_0}{(2k)!},\ k=1,2... \qquad n=2k+1 \quad a_{2k+1} = \frac{(-1)^k a_1}{(2k+1)!},\ k=1,2,...$$

Notice that at each step we always plugged back in the previous answer so that when the subscript was even we could always write the a_n in terms of a_0 and when the coefficient was odd we could always write the a_n in terms of a_1. Also notice that, in this case, we were able to find a general formula for a_n's with even coefficients and a_n's with odd coefficients. This won't always be possible to do.-

There's one more thing to notice here. The formulas that we developed were only for k=1,2,... however, in this case again, the will also work for k=0. Again, this is something that won't always work, but does here.-

Do not get excited about the fact that we don't know what a_0 and a_1 are. As you will see, we actually need these to be in the problem to get the correct solution.

Now that we've got formulas for the a_n's let's get a solution. The first thing that we'll do is write out the solution with a couple of the a_n's plugged in.

$$\begin{aligned} y(x) &= \sum_{n=0}^{\infty} a_n x^n \\ &= a_0 + a_1 x + a_1 x^2 + a_3 x^3 + \cdots + a_{2k} x^{2k} + a_{2k+1} x^{2k+1} + \cdots \\ &= a_0 + a_1 x - \frac{a_0}{2!} x^2 - \frac{a_1}{3!} x^3 + \cdots + \frac{(-1)^k a_0}{(2k)!} x^{2k} + \frac{(-1)^{k+1} a_1}{(2k+1)!} x^{2k+1} + \cdots \end{aligned}$$

The next step is to collect all the terms with the same coefficient in them and then factor out that coefficient.

$$\begin{aligned} y(x) &= a_0 \left\{ 1 - \frac{a_0}{2!} \cdots + \frac{(-1)^k x^{2k}}{(2k)!} + \cdots \right\} + a_1 \left\{ x - \frac{x^3}{3!} + \cdots + \frac{(-1)^{k+1}}{(2k+1)!} x^{2k+1} + \cdots \right\} \\ &= a_0 \sum_{k=0}^{\infty} \frac{(-1)^k x^{2k}}{(2k)!} + a_1 \sum_{k=0}^{\infty} \frac{(-1)^k x^{2k+1}}{(2k+1)!}. \end{aligned}$$

In the last step we also used the fact that we knew what the general formula was to write both portions as a power series. This is also our solution. We are done.-

Before working another problem let's take a look at the solution to the previous example. First, we started out by saying that we wanted a series solution of the form,-

$$y(x) = \sum_{n=0}^{\infty} a_n x^n$$

and we didn't get that. We got a solution that contained two different power series. Also, each of the solutions had an unknown constant in them. This is not a problem. In fact, it's what we want to have happen. From our work with second order constant coefficient differential equations we know that the solution to the differential equation in the last example is,

$$y(x) = c_1 \cos(x) + c_2 \sin(x).$$

Solutions to second order differential equations consist of two separate functions each with an unknown constant in front of them that are found by applying any initial conditions. So, the form of our solution in the last example is exactly what we want to get. Also recall that the following Taylor series,

$$\cos(x) = \sum_{(2n)!}^{\infty} \frac{(-1)^n x^{2n}}{(2n)!} \qquad \sin(x) = \sum_{n=0}^{\infty} \frac{(-1)^n x^{2n+1}}{(2n+1)!}.$$

Recalling these we very quickly see that what we got from the series solution method was exactly the solution we got from first principles, with the exception that the functions were the Taylor series for the actual functions instead of the actual functions themselves.

Now let's work an example with nonconstant coefficients since that is where series solutions are most useful.

Example: Find a series solution around $x_0 = 0$for the following differential equation.-

$$y'' - xy = 0.$$

Solution: As with the first example $p(x)=1$ and so again for this differential equation every point is an ordinary point. Now we'll start this one out just as we did the first example. Let's write down the form of the solution and get its derivatives.

$$y(x) = \sum_{n=0}^{\infty} a_n x^n \quad y'(x) = \sum_{n=1}^{\infty} n a_n x^{n-1} \quad y''(x) = \sum_{n=2}^{\infty} (n-1)\, a_n x^{n-2}.$$

Plugging into the differential equation gives,

$$\sum_{n=2}^{\infty} n(n-1) a_n x^{n-2} - x \sum_{n=1}^{\infty} a_n x^n = 0.$$

Unlike the first example we first need to get all the coefficients moved into the series.

$$\sum_{n=2}^{\infty} n(n-1) a_n x^{n-2} - \sum_{n=1}^{\infty} a_n x^{n-1} = 0.$$

Now we will need to shift the first series down by 2 and the second series up by 1 to get both of the series in terms of x^n.

$$\sum_{n=0}^{\infty} (n+2)(n+1) a_{n+2} x^n - \sum_{n=1}^{\infty} a_{n-1} x^n = 0$$

Next we need to get the two series starting at the same value of n. The only way to do that for this problem is to strip out the n=0 term.

$$(2)(1)a_2x^0 + \sum_{n=1}^{\infty}[(n+2)(n+1)a_{n+2} - a_{n-1}]x^n = 0$$

$$2a_2 + \sum_{n=0}^{\infty}[(n+2)(n+1)a_{n+2} - a_{n-1}]x^n = 0.$$

We now need to set all the coefficients equal to zero. We will need to be careful with this however. The n=0 coefficient is in front of the series and the n=1,2,3... are all in the series. So, setting coefficient equal to zero gives,

$$n = 0: \qquad 2a_2 = 0$$

$$n = 1,2,3,... \quad (n+2)(n+1)a_{n+2} - a_{n+2} = 0$$

Solving the first as well as the recurrence relation gives,

$$n = 0: \qquad a_2 = 0$$

$$n = 1,2,3,... \qquad a_{n+2} = \frac{a_{n-1}}{(n+2)(n+1)}.$$

Now we need to start plugging in values of n.

$$a_3 = \frac{a_0}{(3)(2)} \qquad a_4 = \frac{a_1}{(4)(3)} \qquad a_5 = \frac{a_2}{(5)(4)} = 0$$

$$a_6 = \frac{a_3}{(6)(5)} \qquad a_1 = \frac{a_1}{(7)(6)} \qquad a_8 = \frac{a_5}{(8)(7)} = 0$$

$$= \frac{a_0}{(6)(5)(3)(2)} \qquad = \frac{a_4}{(7)(6)(5)(4)}$$

$$\vdots \qquad \vdots \qquad \vdots$$

$$a_{3k} = \frac{a_0}{(2)(3)(5)(6)...(3k-1)(3k)} \qquad a_{3k+1} = \frac{a_1}{(3)(4)(6)(7)...(3k)(3k+1)} \qquad a_{3k+2} = 0$$

$$k = 1,2,3,.... \qquad k = 1,2,3,.... \qquad k = 1,2,3,....$$

There are a couple of things to note about these coefficients. First, every third coefficient is zero. Next, the formulas here are somewhat unpleasant and not all that easy to see the first time around.

Finally, these formulas will not work for k=0 unlike the first example.

Now, get the solution,

$$y(x) = a_0 + a_1x + a_2x^2 + a_3x^3 + a_4x^4 + \cdots + a_{3k}x^{3k} + a_{3k+1}x^{3k+1}$$

$$= a_0 + a_1x + \frac{a_0}{6}x^3 + \frac{a_1}{12}x^4 \cdots + \frac{a_0x^{3k}}{(2)(3)(5)(6)\cdots(3k-1)(3k)} + \frac{a_0x^{3k}}{(3)(4)(6)(7)\cdots(3k)(3k+1)} + \cdots$$

Again, collect up the terms that contain the same coefficient, factor the coefficient out and write the results as a new series,

$$y(x) = a_0\left\{1 + \sum_{k=1}^{\infty}\frac{x^{3k}}{(2)(3)(5)(6)\cdots(3k-1)(3k)}\right\} + a_1\left\{x + \sum_{\kappa=1}^{\infty}\frac{a_0x^{3k}}{(3)(4)(6)(7)\cdots(3k)(3k+1)}\right\}$$

We couldn't start our series at k=0 this time since the general term doesn't hold for k=0.

Now, we need to work an example in which we use a point other that x=0. In fact, let's just take the previous example and rework it for a different value of x_0. We're also going to need to change up the instructions a little for this example.

Example: Find the first four terms in each portion of the series solution around $x_0 = -2$for the following differential equation.

$$y'' = -xy = 0.$$

Solution: Unfortunately for us there is nothing from the first example that can be reused here. Changing to $x_0 = -2$completely changes the problem. In this case our solution will be,

$$y(x) = \sum_{n=0}^{\infty} a_n(x+2)^n.$$

The derivatives of the solution are,

$$y'(x) = \sum_{n=1}^{\infty} na_n(x+2)^{n-1} \qquad y''(x) = \sum_{n=2}^{\infty} n(n-1)a_n(x+2)^{n-2}.$$

Plug these into the differential equation.

$$y'(x) = \sum_{n=0}^{\infty} na_n(x+2)^{n+2} - x\sum_{n=0}^{\infty} a_n(x+2)^n = 0.$$

We now run into our first real difference between this example and the previous example. In this case we can't just multiply the x into the second series since in order to combine with the series it must be x+2. Therefore we will first need to modify the coefficient of the second series before multiplying it into the series.

$$\sum_{n=2}^{\infty} n(n-1)a_n(x+2)^{n-2} - (x+2-2)\sum_{n=0}^{\infty} a_n(x+2)^n = 0$$

$$\sum_{n=2}^{\infty} n(n-1)a_n(x+2)^{n-2} - (x+2)\sum_{n=0}^{\infty} a_n(x+2)^n + 2\sum_{n=0}^{\infty} a_n(x+2)^n = 0$$

$$\sum_{n=2}^{\infty} n(n-1)a_n(x+2)^{n-2} - \sum_{n=0}^{\infty} a_n(x+2)^{n+1} + \sum_{n=0}^{\infty} 2a_n(x+2)^n = 0.$$

We now have three series to work with. This will often occur in these kinds of problems. Now we will need to shift the first series down by 2 and the second series up by 1 the get common exponents in all the series.

$$\sum_{n=2}^{\infty} (n+2)(n+1)a_{n+2}(x+2)^n - \sum_{n=0}^{\infty} a_{n-1}(x+2)^n + \sum_{n=0}^{\infty} 2a_n(x+2)^n = 0.$$

In order to combine the series we will need to strip out the n=0 terms from both the first and third series.

$$2a_2 + 2a_0 + \sum_{n=2}^{\infty} (n+2)(n+1)a_{n+2}(x+2)^n - \sum_{n=1}^{\infty} a_{n-1}(x+2)^n + \sum_{n=1}^{\infty} 2a_n(x+2)^n = 0$$

$$2a_2 + 2a_0 + \sum_{n=2}^{\infty} [(n+2)(n+1)a_{n+2} - a_{n-1} + 2a_n](x+2)^n = 0.$$

Setting coefficients equal to zero gives,

$$n = 0 \qquad 2a_2 + 2a_0 = 0$$

$$n = 1,2,3,\ldots \qquad (n+2)(n+1)a_{n+2} - a_{n-1} + 2a_n = 0.$$

We now need to solve both of these. In the first case there are two options, we can solve for a_2 or we can solve for a_0. Out of habit I'll solve for a_0. In the recurrence relation we'll solve for the term with the largest subscript as in previous examples.

$$n = 0 \qquad a_2 = -a_0$$

$$n = 1,2,3,\ldots \qquad a_{n+2} = \frac{a_{n-1} - 2a}{(n+2)(n+1)}.$$

Notice that in this example we won't be having every third term drop out as we did in the previous example.

At this point we'll also acknowledge that the instructions for this problem are different as well. We aren't going to get a general formula for the a_n's this time so we'll have to be satisfied with just getting the first couple of terms for each portion of the solution. This is often the case for series solutions. Getting general formulas for the a_n's is the exception rather than the rule in these kinds of problems. To get the first four terms we'll just start plugging in terms until we've got the required number of terms. Note that we will already be starting with an a_0 and an a_1 from the first two terms of the solution so all we will need are three more terms with an a_0 in them and three more terms with an a_1 in them.

$$n=0 \quad a_2 = -a_0.$$

We've got two a_0's and one a_1.

$$n=1 \qquad a_3 = \frac{a_0 - 2a_1}{(3)(2)} = \frac{a_0}{6} - \frac{a_1}{3}.$$

We've got three a_0's and two a_1's.

$$n=2 \qquad a_4 = \frac{a_1 - 2a_2}{(4)(3)} = \frac{a_1 - 2(-a_0)}{(4)(3)} = \frac{a_0}{6} + \frac{a_1}{12}.$$

We've got four a_0's and three a_1's. We've got all the a_0's that we need, but we still need one more a_1'. So, we'll need to do one more term it looks like.

$$n=3 \qquad a_5 = \frac{a_2 - 2a_3}{(5)(4)} = \frac{a_0}{20} - \frac{1}{10}\left(\frac{a_0}{6} - \frac{a_1}{3}\right) = -\frac{31a_0}{60} + \frac{a_1}{30}.$$

We've got five a_0's and four a_1's. We've got all the terms that we need.

Now, all that we need to do is plug into our solution.

$$\begin{aligned} y(x) &= \sum_{n=0}^{\infty} a_n (x+2)^n \\ &= a_0 + a_1(x+2)a_2(x+2)^2 + a_3(x+2)^3 + a_4(x+2)^4 + a_5(x+2)^5 + \cdots \\ &= a_0 + a_1(x+2) - a_0(x+2)^2 + \left(\frac{a_0}{6} - \frac{a_1}{3}\right)(x+2)^3 + \left(\frac{a_0}{6} + \frac{a_1}{12}\right)(x+2)^4 \\ &\quad + \left(\frac{31a_0}{60} + \frac{a_1}{30}\right)(x+2)^5 + \cdots \end{aligned}$$

Finally collect all the terms up with the same coefficient and factor out the coefficient to get,

$$y(x) = a_0\left\{1 - (x+2)^2 + \frac{1}{6}(x+2)^3 + \frac{1}{6}(x+2)^4 - \frac{31}{60}(x+2) + \cdots\right\} +$$

$$a_1\left\{(x+2) - \frac{1}{3}(x+2)^3 + \frac{1}{12}(x+2)^4 + \frac{1}{30}(x+2)^5 + \cdots\right\}.$$

That's the solution for this problem as far as we're concerned. Notice that this solution looks nothing like the solution to the previous example. It's the same differential equation, but changing x_0 completely changed the solution.

Let's work one final problem.

Example: Find the first four terms in each portion of the series solution around for the following differential equation.

$$(x^2+1)y''-4xy'+6y=0.$$

Solution: We finally have a differential equation that doesn't have a constant coefficient for the second derivative.

$$p(x)=x^2+1 \qquad p(0)=1\neq 0.$$

So is an ordinary point for this differential equation. We first need the solution and its derivatives,

$$y(x)=\sum_{n=0}^{\infty}a_nx^n \quad y'(x)=\sum_{n=0}^{\infty}na_nx^{n-1} \quad y''(x)=\sum_{n=0}^{\infty}n(n-1)a_nx^{n-2}.$$

Plug these into the differential equation.

$$(x^2+1)\sum_{n=0}^{\infty}n(n-1)a_nx^{n-2}-4x\sum_{n=0}^{\infty}na_nx^{n-1}+6\sum_{n=0}^{\infty}a_nx^n=0.$$

Now, break up the first term into two so we can multiply the coefficient into the series and multiply the coefficients of the second and third series in as well.

$$\sum_{n=2}^{\infty}n(n-1)a_nx^n+\sum_{n=2}^{\infty}(n-1)a_nx^{n-2}-\sum_{n=1}^{\infty}4na_nx^n+\sum_{n=1}^{\infty}6a_nx^n=0.$$

We will only need to shift the second series down by two to get all the exponents the same in all the series.

$$\sum_{n=2}^{\infty}n(n-1)a_nx^n+\sum_{n=2}^{\infty}(n+2)(n+1)a_{n+2}x^n-\sum_{n=1}^{\infty}4na_nx^n+\sum_{n=1}^{\infty}6a_nx^n=0.$$

At this point we could strip out some terms to get all the series starting at $n=2$, but that's actually more work than is needed. Let's instead note that we could start the third series at $n=0$ if we wanted to because that term is just zero. Likewise the terms in the first series are zero for both $n=1$ and $n=0$ and so we could start that series at $n=0$. If we do this all the series will now start at $n=0$ and we can add them up without stripping terms out of any series.

$$\sum_{n=0}^{\infty}[n(n-1)a_n+(n+2)(n+1)a_{n+2}4na_n+6a_n]x^n=0.$$

$$\sum_{n=0}^{\infty}[(n^2-5n+6)a_n+(n+2)(n+1)a_{n+2}]x^n=0$$

$$\sum_{n=0}^{\infty}[(n^2-2)(n-3)a_n+(n+2)(n+1)a_{n+2}]x^n=0.$$

Now set coefficients equal to zero.

$$(n-2)(n-3)a_n + (n+2)(n+1)a_{n+1}, n = 0,1,2,...$$

Solving this gives,

$$a_{n+2} = -\frac{(n-2)(n-3)a_n}{(n+2)(n+1)}, \qquad n = 0,1,2,...$$

Now, we plug in values of *n*.

$$n = 0: \qquad a_2 = -3a_0$$

$$n = 1: \qquad a_3 = -\frac{1}{3}a_1$$

$$n = 2: \qquad a_4 = -\frac{0}{12}a_2 = 0$$

$$n = 3: \qquad a_5 = -\frac{0}{12}a_3 = 0.$$

Now, from this point on all the coefficients are zero. In this case both of the series in the solution will terminate. This won't always happen, and often only one of them will terminate. The solution in this case is,

$$y(x) = a_0\{1-3x^2\} + a_1\left\{x - \frac{1}{3}x^3\right\}.$$

EULER EQUATIONS

In this section we want to look for solutions to

$$ax^2y + bxy' + cy = 0$$

around $x_0 = 0$. These type of differential equations are called Euler Equations.

Recall from the previous section that a point is an ordinary point if the quotients,

$$\frac{bx}{ax^2} = \frac{b}{ax} \quad \text{and} \quad \frac{c}{ax^2}$$

have Taylor series around. However, because of the *x* in the denominator neither of these will have a Taylor series around and so is a singular point. So, the method from the previous section won't work since it required an ordinary point.

However, it is possible to get solutions to this differential equation that aren't series solutions. Let's start off by assuming that $x>0$ (the reason for this will be apparent after we work the first example) and that all solutions are of the form,

$$y(x) = x'.$$

Now plug this into the differential equation to get,

$$ax^2(r)(r-1)x^{r-2} + bx(r)x^{r-1} + cx^r = 0$$

$$ar(r-1)x^r + b(r)x^r + cx^r = 0$$

$$(ar(r-1) + b(r) + c)x^r = 0$$

Now, we assumed that $x>0$ and so this will only be zero if,

$$ar(r-1) + b(r) + c = 0.$$

This equation is a quadratic in r and so we will have three cases to look at: Real, Distinct Roots, Double Roots, and Quadratic Roots.

REAL, DISTINCT ROOTS

There really isn't a whole lot to do in this case. We'll get two solutions that will form a fundamental set of solutions (we'll leave it to you to check this) and so our general solution will be,

$$y(x) = c_1x^{r_1} + c_2x^{r_2}.$$

Example: Solve the following IVP

$$2x^2y'' + 3xy' - 15y = 0, \qquad y(1) = 0 \quad y'(1) = 1.$$

Solution: We first need to find the roots

$$2r(r-1) + 3r - 15 = 0$$

$$2r^2 + r - 15 = (2r-5)(r+3) = 0 \quad \Rightarrow \quad r_1 = \frac{5}{2}, r_2 = -3.$$

The general solution is then,

$$y(x) = c_1x^{\frac{5}{2}} + c_2x^{-3}.$$

To find the constants we differentiate and plug in the initial conditions as we did back in the second order differential equations chapter.

$$y'(x) = \frac{5}{2}c_1x^{\frac{5}{2}} - 3c_2x^{-4}$$

$$\left.\begin{aligned} 0 &= y(0) = c_1 + c_2 \\ 1 &= y'(0) = \frac{5}{2}c_1 - 3c_2 \end{aligned}\right\} \quad \Rightarrow \quad c_1 = \frac{2}{11}, c_2 = -\frac{2}{11}.$$

The actual solution is then,

$$y(x) = \frac{2}{11}x^{\frac{5}{2}} - \frac{2}{11}x^{-3}.$$

With the solution to this example we can now see why we required $x>0$. The second term would have division by zero if we allowed $x=0$ and the first term would give us square roots of negative numbers if we allowed $x<0$.

DOUBLE ROOTS

This case will lead to the same problem that we've had every other time we've run into double roots (or double eigenvalues). We only get a single solution and will need a second solution. In this case it can be shown that the second solution will be,

$$y_2(x) = x^r \ln x$$

and so the general solution in this case is,

$$y(x) = c_1 x^r + c_2 x^r + \ln x = x^r (c_1 + c_2 \ln x).$$

We can again see a reason for requiring $x>0$. If we didn't we'd have all sorts of problems with that logarithm.

Example: Find the general solution to the following differential equation.

$$x^2 y'' - 7xy' + 16y = 0.$$

Solution: First the roots

$$r(r-1) - 7r + 16 = 0$$

$$r^2 - 8r + 16 = 0$$

$$(r-4)^2 = 0 \Rightarrow r = 4$$

So the general solution is then,

$$y(x) = c_1 x^4 + c_2 x^4 \ln x.$$

COMPLEX ROOTS

In this case we'll be assuming that our roots are of the form,

$$r_{1,2} = \lambda \pm \mu i \qquad r_{12} = \lambda \pm \mu i r_{1,2} = \lambda \pm \mu i.$$

If we take the first root we'll get the following solution.

$$x^{\lambda+\mu i} \qquad x^{\lambda+\mu i} x^{\lambda+\mu i}.$$

This is a problem since we don't want complex solutions, we only want real solutions. We can eliminate this by recalling that,

$$x^r = e^{\ln x^r} = e^{r \ln x}.$$

Plugging the root into this gives,

$$x^{\lambda+\mu i} = e^{(\lambda+\mu i)\ln x} = e^{\lambda \ln x} e^{\mu i \ln x}$$

$$= e^{\ln x^{\lambda}} (\cos(\mu \ln x) + i \sin(\mu \ln x))$$

$$= x^{\gamma} \cos(\mu \ln x) + i x^{\lambda} \sin(\mu \ln x)$$

Note that we had to use Euler formula as well to get to the final step. Now, as we've done every other time we've seen solution like this we can take the real part and the imaginary part and use those for our two solutions.

So, in the case of complex roots the general solution will be,

$$y(x) = c_1 x^{\lambda} \cos(\mu \ln x) + c_2 x^{\lambda} \sin(\mu \ln x) = x^{\lambda}(c_1 \cos(\mu \ln x) + c_2 x \sin(\mu \ln x)).$$

Once again we can see why we needed to require $x>0$.

Example: Find the solution to the following differential equation.

$$x^2 y'' + 3xy' + 4y = 0.$$

Solution: Get the roots first as always.

$$r(r-1) + 3r + 4 = 0$$

$$r^2 + 2r + 4 = 0 \quad \Rightarrow \quad r_{1,2} = -1 \pm \sqrt{3}i.$$

The general solution is then,

$$y(x) = c_1 x^{-1} \cos\left(\sqrt{3} \ln x\right) + c_2 x^{-1} \sin\left(\sqrt{3} \ln x\right).$$

We should now talk about how to deal with $x<0$ since that is a possibility on occasion. To deal with this we need to use the variable transformation,

$$\eta = -x.$$

In this case since $x<0$ we will get $\varsigma>0$. Now, define,

$$u(\eta) = y(x) = y(-\eta).$$

Then using the chain rule we can see that,

$$u'(\eta) = -y'(x) \quad \text{and} \quad u''(\eta) = y''(x).$$

With this transformation the differential equation becomes,

$$a(-\eta)^2 u'' + b(-\eta)(-u') + cu = 0$$

$$a\eta^2 u'' + b\eta u' + cu = 0$$

In other words, since $\varsigma>0$ we can use the work above to get solutions to this differential equation. We'll also go back to x's by using the variable transformation in reverse.

$$\eta = -x.$$

Let's just take the real, distinct case first to see what happens.

$$u(\eta) = c_1 \eta^{r_1} + c_2 \eta^{r_2}$$

$$y(x) = c_1(-x)^{r_1} + c_2(-x)^{r_2}.$$

Now, we could do this for the rest of the cases if we wanted to, but before doing that let's notice that if we recall the definition of absolute value,

$$|x| = \begin{cases} x & \text{if } x \geq 0 \\ -x & \text{if } x < 0 \end{cases}$$

we can combine both of our solutions to this case into one and write the solution as,

$$y(x) = c_1 |x|^{r_1} + c_2 |x|^{r_2}, \quad x \neq 0.$$

Note that we still need to avoid x=0 since we could still get division by zero. However this is now a solution for any interval that doesn't contain $x = 0$.

We can do likewise for the other two cases and the following solutions for any interval not containing $x = 0$,.

$$y(x) = c_1 |x|^{r_1} + c_2 |x|^r \ln |x|$$

$$y(x) = c_1 |x|^\lambda \cos(\mu \ln |x|) + c_2 |x|^\lambda \sin (\mu \ln |x|).$$

We can make one more generalization before working one more example. A more general form of an Euler Equation is,

$$a(x - x_0)^2 y'' + b(x \quad x_0)y' + cy - 0$$

and we can ask for solutions in any interval not containing $x = x_0$. The work for generating the solutions in this case is identical to all the above work and so isn't shown here.

The solutions in this general case for any interval not containing x=a are,

$$y(x) = c_1 |x - a|^{r_1} + c_2 |x - a|^{r_2}$$

$$y(x) = |x - a|^r (c_1 + c_2 \ln |x - a|)$$

$$y(x) = |x - a|^\lambda (c_1 \cos(\mu \ln |x - a|) + c_2 \sin (\mu \ln |x - a|)).$$

Where the roots are solutions to

$$ar(r - 1) + b(r) + c = 0.$$

Example: Find the solution to the following differential equation on any interval not containing $x = -6$.

$$3(x + 6)^2 y'' + 25(x + 6)y' - 16y = 0.$$

Solution: So we get the roots from the identical quadratic in this case.

$$3r(r - 1) + 25r - 16 = 0$$

$$3r^2 + 22r - 16 = 0$$

$$(3r - 2)(r + 8) = 0 \quad \Rightarrow \quad r_1 = \frac{2}{3}, r_2 = -8.$$

The general solution is then,

$$y(x) = c_1 |x - a|^{\frac{2}{3}} + c_2 |x - a|^{-8}.$$

Encyclopaedia

of

PARTIAL DIFFERENTIAL EQUATION

Encyclopaedia of PARTIAL DIFFERENTIAL EQUATION

Volume 2

Dr. Rakesh Kumar Pandey

ANMOL PUBLICATIONS PVT. LTD.
NEW DELHI - 110 002 (INDIA)

ANMOL PUBLICATIONS PVT. LTD.

Regd. Office: 4360/4, Ansari Road, Daryaganj,
New Delhi-110 002 (India)
Ph.: 23278000, 23261597

Branch Office: No. 1015, Ist Main Road, BSK IIIrd Stage
IIIrd Phase, IIIrd Block,
Bangalore-560 085 (India)
Tel.: 080-41723429
Visit us at: www.anmolpublications.com

Encyclopaedia of Partial Differential Equation

First Edition, 2009

ISBN 978-81-261-4107-4 (Set)

PRINTED IN INDIA

Printed at Mehra Offset Press, Delhi.

Contents

Preface

Encyclopaedia of Partial Differential Equation is a unique Encyclopaedia among itself, aiming to provide a fairly complete account of the basic concepts required to build a strong foundation for a student endeavoring to study this subject. The Analytical Approach to the major concepts makes the book highly self-contained and comprehensive guide that succeeds in making the concepts easily understandable. All the Elementary Principles and Fundamental Concepts have been explained rigorously, leaving no scope for Confusion. The focus throughout the text has been on presenting the subject matter in a well-knit manner and lucid style, so that even a student with average Mathematical Skill would find it accessible to himself. In addition, the book provides numerous well-graded solved examples, generally set in various university and competitive examinations, which will facilitate easy understanding besides acquainting the students with a variety of questions.

The Principal objective of the book is solving Boundary Value Problems involving partial digfferential equations. Seperation of variable receives the greatest attention because it is widely used in applications and because it provides a uniform method for solving importanvt cases of Heat, Wave and Potential Equatioins. one technique is not enough of course. The wave equation is developed in parallel with the series solution, and the distrubuted-source solution is constructed for the Heat Equation. In addition, there are chapters on Periodic and Orthogonal Functions, Laplace Transforms, Infinite Series, Series Solutions to Differential Equations, Exact Differential Equations, Bernoulli Differential Equations, Nonhomogeneous Differential Equations, Boundary Value, Intervals of Validity, Solutions to Systems.

Author

List of Symbols

$\pm$	$\not\subset$	$\subset$	$\supset$	$\subseteq$	$\in$	$\notin$	∂
$\neq$	$\equiv$	$\approx$	$\angle$	∇	$\sum$	Π	Ω
ϕ	α	β	χ	ε	γ	η	λ
μ	ν	$\perp$	π	θ	ρ	σ	τ
$\Re$	ω	ξ	ψ	ζ	Υ	$\exists$	$\int$
$\iint$	$\iiint$	$\oint$	$\times$	$\sqrt{}$	$\therefore$	$\because$	Δ
$\geq$	$\leq$	$\Rightarrow$	$\Downarrow$	$\Leftrightarrow$	∞	$\propto$	$\mathbb{R}$
$\mathbb{Z}$	$\mathbb{N}$	$\mathbb{C}$	$\mathbb{Q}$				

Chapter 1

Exact Differential Equations

Before we get into the full details behind solving exact differential equations it's probably best to work an example that will help to show us just what an exact differential equation is. It will also show some of the behind the scenes details that we usually don't bother with in the solution process. The vast majority of the following example will not be done in any of the remaining examples and the work that we will put into the remaining examples will not be shown in this example. The whole point behind this example is to show you just what an exact differential equation is, how we use this fact to arrive at a solution and why the process works as it does. The majority of the actual solution details will be shown in a later example.

Example: Solve the following differential equation.

$$2xy - 9x^2 + (2y + x^2 + 1)\frac{dy}{dx} = 0$$

Solution: Let's start off by supposing that somewhere out there in the world is a function $\Psi(x,y)$ that we can find. For this example the function that we need is

$$\Psi(x, y) = y^2 + \left(x^2 + 1\right)y - 3x^2$$

Do not worry at this point about where this function came from and how we found it. Finding the function, $\Psi(x,y)$, that is needed for any particular differential equation is where the vast majority of the work for these problems lies. As stated earlier however, the point of this example is to show you why the solution process works rather than showing you the actual solution process. We will see how to find this function in the next example, so at this point do not worry about how to find it, simply accept that it can be found and that we've done that for this particular differential equation.

Now, take some partial derivatives of the function.

$$\Psi_x = 2xy - 9x^2$$

$$\Psi_y = 2y + x^2 + 1$$

Now, compare these partial derivatives to the differential equation and you'll notice that with these we can now write the differential equation as.

$$\Psi_x + \Psi_y \frac{dy}{dx} = 0$$

Now, recall from your multi-variable calculus is nothing more that the following derivative (you'll need the multi-variable chain rule for this...).

$$\frac{d}{dx}\left(\Psi\left(x, y(x)\right)\right)$$

So, the differential equation can now be written as

$$\frac{d}{dx}\left(\Psi\left(x, y(x)\right)\right) = 0$$

Now, if the ordinary (not partial...) derivative of something is zero, that something must have been a constant to start with. In other words, we've got to have $\Psi(x,y) = c$. Or,

$$y^2 + \left(x^2 + 1\right)y - 3x^3 = c$$

This then is an implicit solution for our differential equation! If we had an initial condition we could solve for c. We could also find an explicit solution if we wanted to, but we'll hold off on that until the next example.

Okay, so what did we learn from the last example? Let's look at things a little more generally. Suppose that we have the following differential equation.

$$M(x, y) + N(x, y)\frac{dy}{dx} = 0$$

Note that it's important that it be in this form! There must be an "= 0" on one side and the sign separating the two terms must be a "+". Now, if there is a function somewhere out there in the world, $\Psi(x,y)$, so that,

$$\Psi_x = M(x, y) \text{ and } \Psi_y = N(x, y)$$

then we call the differential equation exact. In these cases we can write the differential equation as

$$\Psi_x + \Psi_y \frac{dy}{dx} = 0$$

Then using the chain rule from Calculus we can further reduce the differential equation to the following derivative,

$$\frac{d}{dx}\left(\Psi\left(x, y(x)\right)\right) = 0$$

The (implicit) solution to an exact differential equation is then

$$\Psi(x,y)=c$$

Well, it's the solution provided we can find *Ψ(x,y)* anyway. Therefore, once we have the function we can always just jump straight to get an implicit solution to our differential equation.

Finding the function *Ψ(x,y)* is clearly the central task in determining if a differential equation is exact and in finding it's solution. As we will see, finding *Ψ(x,y)* can be a somewhat lengthy process in which there is the chance of mistakes. Therefore, it would be nice if there was some simple test that we could use before even starting to see if a differential equation is exact or not. This will be especially useful if it turns out that the differential equation is not exact, since in this case *Ψ(x,y)* will not exist. It would be a waste of time to try and find a nonexistent function!

So, let's see if we can find a test for exact differential equations. Let's start with and assume that the differential equation is in fact exact. Since its exact we know that somewhere out there is a function *Ψ(x,y)* that satisfies

$$\Psi_x=M$$
$$\Psi_y=N$$

Now, provided *Ψ(x,y)* is continuous and it's first order derivatives are also continuous we know that

$$\Psi_{xy}=\Psi_{yx}$$

However, we also have the following.

$$\Psi_{xy}=(\Psi_x)_y=(M)_y=M_y$$
$$\Psi_{yx}=(\Psi_y)_x=(N)_x=N_x$$

Therefore, if a differential equation is exact and *Ψ(x,y)* meets all of its continuity conditions we must have.

$$M_y=N_x$$

Likewise if it is not true there is no way for the differential equation to be exact.

Therefore, we will use equation as a test for exact differential equations. If equation is true we will assume that the differential equation is exact and that *Ψ(x,y)* meets all of its continuity conditions and proceed with finding it. Note that for all the examples here the continuity conditions will be met and so this won't be an issue.

Okay, let's go back and rework the first example. This time we will use the example to show how to find *Ψ(x,y)*. We'll also add in an initial condition to the problem.

Example: Solve the following IVP and find the interval of validity for the solution.

$$2xy-9x^2+\left(2y+x^2+1\right)\frac{dy}{dx}=0, \qquad y(0)=-3$$

Solution: First identify M and N and check that the differential equation is exact.

$$M=2xy-9x^2 \qquad M_y=2x$$

$$N=2y+x^2+1 \qquad N_x=2x$$

So, the differential equation is exact according to the test. However, we already knew that as we have given you $\Psi(x,y)$. It's not a bad thing to verify it however and to run through the test at least once however.

Now, how do we actually find $\Psi(x,y)$? Well recall that

$$\Psi_x=M$$

$$\Psi_y=N$$

We can use either of these to get a start on finding $\Psi(x,y)$ by integrating as follows.

$$\Psi=\int M\,dx \qquad \text{OR} \qquad \Psi=\int N\,dy$$

However, we will need to be careful as this won't give us the exact function that we need. Often it doesn't matter which one you choose to work with while in other problems one will be significantly easier than the other. In this case it doesn't matter which one we use as either will be just as easy.

So, I'll use the first one.

$$\Psi(x,y)=\int 2xy-9x^2dx=x^2y-3x^2+h(y)$$

Note that in this case the "constant" of integration is not really a constant at all, but instead it will be a function of the remaining variable(s), y in this case.

Recall that in integration we are asking what function we differentiated to get the function we are integrating. Since we are working with two variables here and talking about partial differentiation with respect to x, this means that any term that contained only constants or y's would have differentiated away to zero, therefore we need to acknowledge that fact by adding on a function of y instead of the standard c.

Okay, we've got most of $\Psi(x,y)$ we just need to determine $h(y)$ and we'll be done. This is actually easy to do. We used to find most of $\Psi(x,y)$ so we'll use $\Psi_y=N$to find $h(y)$. Differentiate our $\Psi(x,y)$ with respect to y and set this equal to N (since they must be equal after all). Don't forget to "differentiate" $h(y)$! Doing this gives,

$$\Psi_y=x^2+h'(y)=2y+x^2+1=N$$

From this we can see that

$$h'(y) = 2y + 1$$

Note that at this stage *h(y)* must be only a function of *y* and so if there are any *x*'s in the equation at this stage we have made a mistake somewhere and it's time to go look for it.

We can now find *h(y)* by integrating.

$$h(y) = \int 2y + 1dy = y^2 + y + k$$

You'll note that we included the constant of integration, *k*, here. It will turn out however that this will end up getting absorbed into another constant so we can drop it in general.

So, we can now write down *Ψ(x,y)*.

$$\Psi(x,y) = x^2 y - 3x^2 + y^2 + y + k = y^2\left(x^2 + 1\right)y - 3x + k$$

With the exception of the *k* this is identical to the function that we used in the first example. We can now go straight to the implicit solution.

$$y^2 + (x^2 + 1)y - 3x^2 + k = c$$

We'll now take care of the *k*. Since both *k* and *c* are unknown constants all we need to do is subtract one from both sides and combine and we still have an unknown constant.

$$y^2 + \left(x^2 + 1\right)y - 3x^3 = c - k$$

$$y^2 + \left(x^2 + 1\right)y - 3x^3 = c$$

Therefore, we'll not include the *k* in anymore problems. This is where we left off in the first example. Let's now apply the initial condition to find *c*.

$$(-3)^2 + (0+1)(-3) - 3(0)^3 = c \qquad \Rightarrow \qquad c = 6$$

The implicit solution is then.

$$y^2 + \left(x^2 + 1\right)y - 3x^3 - 6 = 0$$

Now, as we saw in the separable differential equation section, this is quadratic in *y* and so we can solve for *y(x)* by using the quadratic formula.

$$y(x) = \frac{-\left(x^2+1\right) \pm \sqrt{\left(x^2+1\right)^2 - 4(1)\left(-3x^3-6\right)}}{2(1)}$$

$$= \frac{-\left(x^2+1\right) \pm \sqrt{x^4 + 12x^3 2x^2 + 25}}{2}$$

Now, reapply the initial condition to figure out which of the two signs in the ± ±±that we need.

$$-3 = y(0) = \frac{-1 \pm \sqrt{25}}{2} = \frac{-1 \pm 5}{2} = -3,2$$

So, it looks like the "-" is the one that we need. The explicit solution is then.

$$y(x) = \frac{-\left(x^2+1\right) - \sqrt{x^4+12x^3+2x^2+25}}{2}$$

Now, for the interval of validity. It looks like we might well have problems with square roots of negative numbers. So, we need to solve

$$x^4 + 12x^3 + 2x^2 + 25 = 0$$

Upon solving this equation is zero at $x = 11.81557624$ and $x = 1.396911133$. Note that you'll need to use some form of computational aid in solving this equation. Here is a graph of the polynomial under the radical.

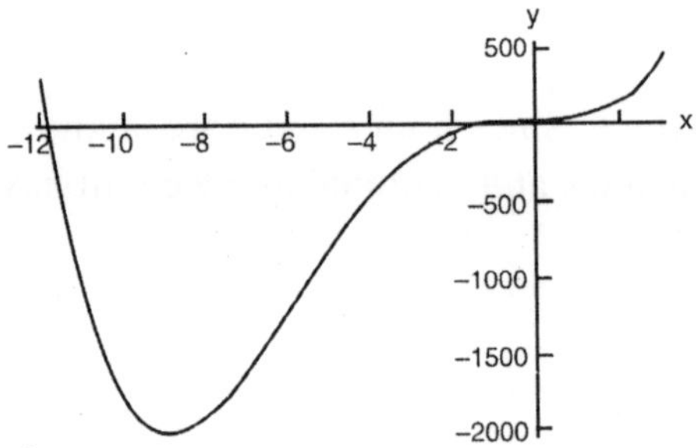

So, it looks like there are two intervals where the polynomial will be positive.

$$-\infty < x \leq -11.81557624$$

$$-1.396911133 \leq x < \infty$$

However, recall that intervals of validity need to be continuous intervals and contain the value of x that is used in the initial condition. Therefore the interval of validity must be.

$$-1.396911133 \geq x < 8$$

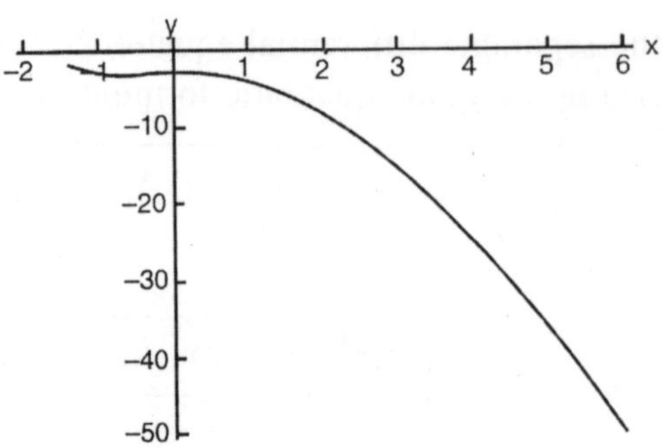

Here is a quick graph of the solution.

That was a long example, but mostly because of the initial explanation of how to find *Ψ(x,y)*. The remaining examples will not be as long.

Example: Find the solution and interval of validity for the following IVP.

$$2xy^2+4=2\left(3-x^2y\right)y' \qquad\qquad y(-1)=8$$

Solution: Here, we first need to put the differential equation into proper form before proceeding. Recall that it needs to be "= 0" and the sign separating the two terms must be a plus!

$$2xy^2+4-2\left(3-x^2y\right)y'=0$$

$$2xy^2+4+2\left(x^2y-3\right)y'=0$$

So we have the following

$$M=2xy^2+4 \qquad\qquad M_y=4xy$$

$$N=2x^2y-6 \qquad\qquad N_x=4xy$$

and so the differential equation is exact. We can either integrate M with respect to x or integrate N with respect to y. In this case either would be just as easy so we'll integrate N this time so we can say that we've got an example of both down here.

$$\Psi(x,y)=\int sx^2y-6dy=x^2y^2-6y+h(x)$$

This time, as opposed to the previous example, our "constant" of integration must be a function of x since we integrated with respect to y. Now differentiate with respect to x and compare this to M.

$$\Psi_x=2xy^2+h'(x)=2xy^2+4=M$$

So, it looks like

$$h'(x)=4 \qquad \Rightarrow \qquad h(x)=4x$$

Again, we'll drop the constant of integration that technically should be present in *h(x)* since it will just get absorbed into the constant we pick up in the next step. Also note that, *h(x)* should only involve x's at this point. If there are any y's left at this point a mistake has been made so go back and look for it.

Writing everything down gives us the following for *Ψ(x,y)*.

$$\Psi(x,y)=x^2y^2-6y+4x$$

So, the implicit solution to the differential equation is

$$x^2y^2 - 6y + 4x = c$$

Applying the initial condition gives,

$$64 - 48 - 4 = c \quad c = 12$$

The solution is then

$$x^2y^2 - 6y + 4x - 12 = 0$$

Using the quadratic formula gives us

$$y(x) = \frac{6 \pm \sqrt{36 - 4x^2(4x-12)}}{2x^2}$$

$$= \frac{6 \pm \sqrt{36 + 48x^2 - 16x^3}}{2x^2}$$

$$= \frac{6 \pm 2\sqrt{9 + 12x^2 - 4x^3}}{2x^2}$$

$$= \frac{3 \pm \sqrt{9 + 12x^2 - 4x^3}}{x^2}$$

Reapplying the initial condition shows that this time we need the "+" (we'll leave those details to you to check). Therefore, the explicit solution is

$$y(x) = \frac{3 + \sqrt{9 + 12x^2 - 4x^3}}{x^2}$$

Now let's find the interval of validity. We'll need to avoid $x = 0$ so we don't get division by zero. We'll also have to watch out for square roots of negative numbers so solve the following equation.

$$-4x^3 + 12x^2 + 9 = 0$$

The only real solution here is $x = 3.217361577$. Below is a graph of the polynomial.

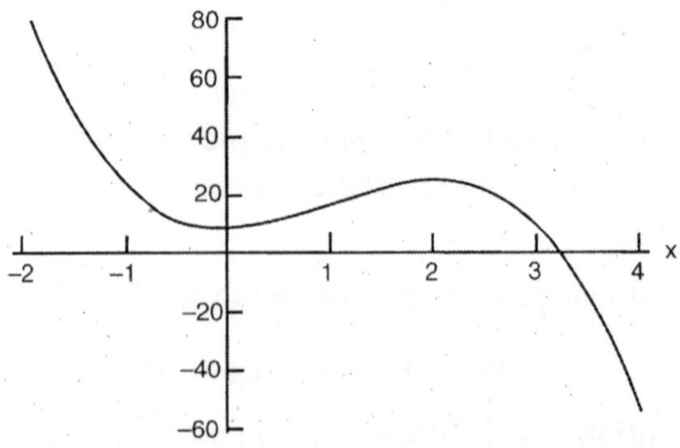

So, it looks like the polynomial will be positive, and hence okay under the square root on

$$-\infty < x < 3.217361577$$

Now, this interval can't be the interval of validity because it contains $x = 0$ and we need to avoid that point. Therefore, this interval actually breaks up into two different possible intervals of validity.

$$-\infty < x < 0$$
$$0 < x < 3.217361577$$

The first one contains $x = -1$, the x value from the initial condition. Therefore, the interval of validity for this problem is $-\infty < x < 0$. Here is a graph of the solution.

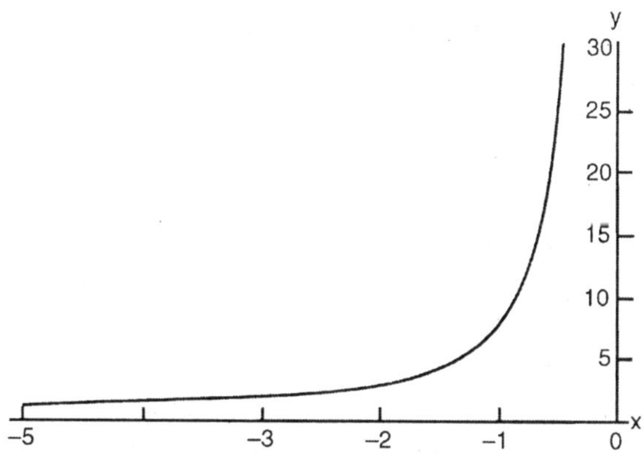

Example: Find the solution and interval of validity to the following IVP.

$$\frac{2ty}{t^2+1} - 2t - \left(2 - \text{In}\left(t^2+1\right)\right)y' = 0 \qquad y(5) = 0$$

Solution: So, first deal with that minus sign separating the two terms.

$$\frac{2ty}{t^2+1} - 2t + \left(In\left(t^2+1\right) - 2\right)y' = 0$$

Now, find M and N and check that it's exact.

$$M = \frac{2ty}{t^2+1} - 2t \qquad M_y = \frac{2t}{t^2+1}$$

$$N = In\left(t^2+1\right) - 2 \qquad N_t = \frac{2t}{t^2+1}$$

So, it's exact. We'll integrate the first one in this case.

$$\Psi(t, y) = \int \frac{2ty}{t^2+1} - 2t\,dt = yIn\left(t^2+1\right) - t^2 + h(y)$$

Differentiate with respect to y and compare to N.

$$\Psi_y = In\left(t^2+1\right)+h'(y)=In\left(t^2+1\right)-2=N$$

So, it looks like we've got.

$$h'(y)=-2 \qquad \Rightarrow \qquad h(y)=-2$$

This gives us

$$\Psi(t,y)=yIn\left(t^2+1\right)-t^2-2y$$

The implicit solution is then,

$$yIn\left(t^2+1\right)-t^2-2y=c$$

Applying the initial condition gives,

$$-25=c$$

The implicit solution is now,

$$y\left(In\left(t^2+1\right)-2\right)-t^2=-25$$

This solution is much easier to solve than the previous ones. No quadratic formula is needed this time, all we need to do is solve for y. Here's what we get for an explicit solution.

$$y(t)=\frac{t^2-25}{In\left(t^2+1\right)-2}$$

Alright, let's get the interval of validity. The term in the logarithm is always positive so we don't need to worry about negative numbers in that. We do need to worry about division by zero however. We will need to avoid the following point(s).

$$In\left(t^2+1\right)-2=0$$
$$In\left(t^2+1\right)=2$$
$$t^2+1=e^2$$
$$t^2=\pm\sqrt{e^2-1}$$

We now have three possible intervals of validity.

$$-\infty<t<-\sqrt{e^2-1}$$
$$-\sqrt{e^2-1}<t<\sqrt{e^2-1}$$
$$\sqrt{e^2-1}<t<\infty$$

The last one contains $t = 5$ and so is the interval of validity for this problem is $\sqrt{e^2 - 1} < t < \infty$. Here's a graph of the solution.

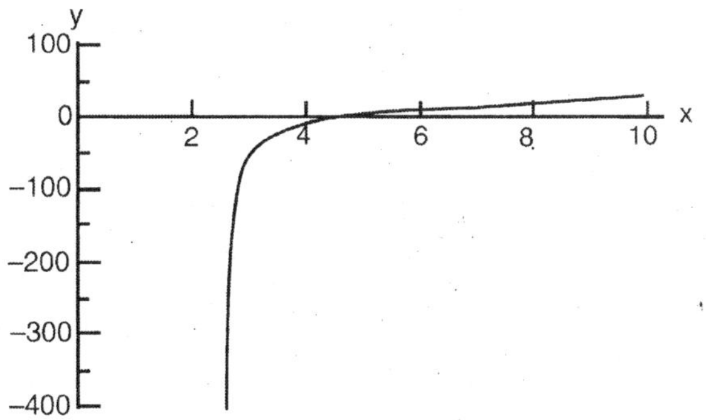

Example: Find the solution and interval of validity for the following IVP.

$$3y^3e3^{xy} - 1\left(2ye^{3xy} + 3xy^2e^{3xy}\right)y' = 0 \qquad y(0) = 1$$

Solution: Let's identify M and N and check that it's exact.

$$M = 3y^3e^{3xy} - 1 \qquad\qquad M_y = 9y^2e^{3xy} + 9xy^3e^{3xy}$$

$$N = 2ye^{3xy} + 3xy^2e^{3xy} \qquad\qquad N_x = 9y^2e^{3xy} + 9xy^3e^{3xy}$$

So, it's exact. With the proper simplification integrating the second one isn't too bad. However, the first is already set up for easy integration so let's do that one.

$$\Psi(x, y) = \int 3y^3e^{3xy} - 1dx = y^2e^{3xy} - x + h(y)$$

Differentiate with respect to y and compare to N.

$$\Psi_y = 2ye^{3xy} + 3xy^2e^{3xy} + h'(y) = 2ye^{3xy} + 3xy^2e^{3xy} = N$$

So, it looks like we've got

$$h'(y) = 0 \qquad \Rightarrow \qquad h(y) = 0$$

Recall that actually $h(y) = k$, but we drop the k because it will get absorbed in the next step. That gives us $h(y) = 0$. Therefore, we get.

$$\Psi(x, y) = y^2e^{3xy} - x$$

The implicit solution is then

$$y^2e^{3xy} - x = c$$

Apply the initial condition.

$$1 = c$$

The implicit solution is then

$$y^2 e^{3xy} - x = 1$$

This is as far as we can go. There is no way to solve this for *y* and get an explicit solution.

INTEGRAL OF RATIONAL EXPRESSIONS

In this section we are going to take a look at integrals of rational expressions of polynomials and once again let's start this section out with an integral that we can already do so we can contrast it with the integrals that we'll be doing in this section.

$$\int \frac{2x-1}{x^2-x+6}dx = \int \frac{1}{u}du \qquad \text{using} \quad u = x^2 - x + 6 \text{ and } du = (2x-1)\,dx$$

$$= \ln|x^2 - x + 6| + c$$

So, if the numerator is the derivative of the denominator (or a constant multiple of the derivative of the denominator) doing this kind of integral is fairly simple. However, often the numerator isn't the derivative of the denominator (or a constant multiple). For example, consider the following integral.

$$\int \frac{2x+11}{x^2-x+6}dx$$

In this case the numerator is definitely not the derivative of the denominator nor is it a constant multiple of the derivative of the denominator. Therefore, the simple substitution that we used above won't work. However, if we notice that the integrand can be broken up as follows,

$$\frac{3x+11}{x^2-x-6} = \frac{4}{x-3} - \frac{1}{x+2}$$

then the integral is actually quite simple.

$$\int \frac{3x+11}{x^2-x-6}dx = \int \frac{4}{x-3} - \frac{1}{x+2}dx$$

$$= 4\ln|x-3| - \ln|x+2| + c$$

This process of taking a rational expression and decomposing it into simpler rational expressions that we can add or subtract to get the original rational expression is called partial fraction decomposition. Many integrals involving rational expressions can be done if we first do partial fractions on the integrand.

So, let's do a quick review of partial fractions. We'll start with a rational expression in the form,

$$f(x) = \frac{P(x)}{Q(x)}$$

where both *P(x)* and *Q(x)* are polynomials and the degree of *P(x)* is smaller than the degree of $Q(x)$. Recall that the degree of a polynomial is the largest exponent in the polynomial. Partial fractions can only be done if the degree of the numerator is strictly less than the degree of the denominator. That is important to remember. So, once we've determined that partial fractions can be done we factor the denominator as completely as possible. Then for each factor in the denominator we can use the following table to determine the term(s) we pick up in the partial fraction decomposition.

Factor in Term in partial denominator fraction decomposition

$ax + b$	$\dfrac{A}{ax+b}$
$(ax + b)^k$	$\dfrac{A_1}{ax+b}+\dfrac{A_2}{(ax+b)^2}+\cdots+\dfrac{A_k}{(ax+b)^k}$, k=1,2,3,....
$ax^2 + bx + c$	$\dfrac{Ax+B}{ax^2+bx+c}$
$(ax^2 + bx + c)^k$	$\dfrac{A_1x+B_1}{ax^2+bx+c}+\dfrac{A_2x+B_1}{ax^2+bx+c}+\dfrac{A_2x+B_2}{(ax^2bx+c)^2}+\cdots+\dfrac{A_2x+B_k}{(ax^2bx+c)^k}$, k=1,2,3,....

Notice that the first and third cases are really special cases of the second and fourth cases respectively. There are several methods for determining the coefficients for each term and we will go over each of those in the following examples.

Let's start the examples by doing the integral above.

Example: Evaluate the following integral.

$$\int \frac{3x+11}{x^2-x-6}\,dx$$

Solution: The first step is to factor the denominator as much as possible and get the form of the partial fraction decomposition. Doing this gives,

$$\frac{3x+11}{(x-3)(x+2)}=\frac{A}{x-3}+\frac{B}{x+2}$$

The next step is to actually add the right side back up.

$$\frac{3x+11}{(x-3)(x+2)}=\frac{A(x+2)+B(x-3)}{(x-3)(x+2)}$$

Now, we need to choose A and B so that the numerators of these two are equal for every x. To do this we'll need to set the numerators equal.

$$3x + 11 = A(x + 2) + B\,(x - 3)$$

Note that in most problems we will go straight from the general form of the decomposition to this step and not bother with actually adding the terms back up. The only point to adding the terms is to get the numerator and we can get that without actually writing down the results of the addition. At this point we have one of two ways to proceed. One way will always work, but is often more work. The other, while it won't always work, is often quicker when it does work.

In this case both will work and so we'll use the quicker way for this example. We'll take a look at the other method in a later example. What we're going to do here is to notice that the numerators must be equal for *any* x that we would choose to use. In particular the numerators must be equal for $x = -2$ and $x = 3$. So, let's plug these in and see what we get.

$$x = -2 \qquad 5 = A(0) + B(-5) \quad \Rightarrow \quad B = -1$$
$$x = 3 \qquad 20 = A(5) + B(0) \quad \Rightarrow \quad A = 4$$

So, by carefully picking the x's we got the unknown constants to quickly drop out. Note that these are the values we claimed they would be above.

At this point there really isn't a whole lot to do other than the integral.

$$\begin{aligned}\int \frac{3x+11}{x^2-x-6}\,dx &= \int \frac{4}{x-3} - \frac{1}{x+2}\,dx \\ &= \int \frac{4}{x-3}\,dx - \int \frac{1}{x+2}\,dx \\ &= 4\ln|x-3| - \ln|x+2| + c\end{aligned}$$

Recall that to do this integral we first split it up into two integrals and then used the substitutions,

$$u = x - 3 \qquad v = x + 2$$

on the integrals to get the final answer.

Before moving onto the next example a couple of quick notes are in order here. First, many of the integrals in partial fractions problems come down to the type of integral seen above. Make sure that you can do those integrals. There is also another integral that often shows up in these kinds of problems so we may as well give the formula for it here since we are already on the subject.

$$\int \frac{1}{x^2+a^2}\,dx = \frac{1}{a}\tan^{-1}\left(\frac{x}{a}\right) + c$$

It will be an example or two before we use this so don't forget about it.

Now, let's work some more examples.

Example: Evaluate the following integral.

$$\int \frac{x^2+4}{3x^2+4x^2-4x}\,dx$$

Solution: We won't be putting as much detail into this solution as we did in the previous example. The first thing is to factor the denominator and get the form of the partial fraction decomposition.

$$\frac{x^2+4}{x(x+2)(3x-2)} = \frac{A}{x} + \frac{B}{x+2} + \frac{C}{3x-2}$$

The next step is to set numerators equal. If you need to actually add the right side together to get the numerator for that side then you should do so, however, it will definitely make the problem quicker if you can do the addition in your head to get,

$$x^2 + 4 = A\,(x+2)(3x-2) + Bx(3x-2) + Cx(x+2)$$

As with the previous example it looks like we can just pick a few values of x and find the constants so let's do that.

$$x = 0 \qquad 4 = A(2)(-2) \qquad \Rightarrow \quad A = -1$$

$$x = -2 \qquad 8 = B(-2)(-8) \qquad \Rightarrow \quad B = \frac{1}{2}$$

$$x = \frac{2}{3} \qquad \frac{40}{9} = C\left(\frac{2}{3}\right)\left(\frac{8}{3}\right) \qquad \Rightarrow \quad C = \frac{40}{16} = \frac{5}{2}$$

Note that unlike the first example most of the coefficients here are fractions. That is not unusual so don't get excited about it when it happens.

Now, let's do the integral.

$$\int \frac{x^2+4}{3x^3+4x^2-4x}\,dx = \int -\frac{1}{x} + \frac{\frac{1}{2}}{x+2} + \frac{\frac{1}{2}}{3x+2}\,dx$$

$$= -\ln|x| + \frac{1}{2}\ln|x+2| + \frac{5}{6}\ln|3x-2| + c$$

Again, as noted above, integrals that generate natural logarithms are very common in these problems so make sure you can do them.

Example: Evaluate the following integral.

$$\int \frac{x^2-29x+5}{(x-4)^2(x^2+3)}\,dx$$

Solution: This time the denominator is already factored so let's just jump right to the partial fraction decomposition.

$$\frac{x^2-29x+5}{(x-4)^2(x^2+3)} = \frac{A}{x-4} + \frac{B}{(x-4)^2} + \frac{Cx+D}{x^2+3}$$

Setting numerators gives,

$$x^2 - 29x + 5 = A(x-4)(x^2+3) + B(x^2+3) + (Cx+D)(x-4)^2$$

In this case we aren't going to be able to just pick values of x that will give us all the constants.

Therefore, we will need to work this the second (and often longer) way. The first step is to multiply out the right side and collect all the like terms together. Doing this gives,

$$x^2 - 29x + 5 = (A+C)x^3 + (-4A+B-8C+D)x^2(3A+16C-8D)x - 12A+3B+16D$$

Now we need to choose A, B, C, and D so that these two are equal. In other words we will need to set the coefficients of like powers of x equal. This will give a system of equations that can be solved.

$$\left.\begin{aligned} x^3 &: \quad A+C=0 \\ x^2 &: \quad -4A+B-8C+D=1 \\ x^1 &: \quad 3A+16C-8D=-29 \\ x^0 &: \quad -12A+3B+16D=5 \end{aligned}\right\} \Rightarrow A=1, B=-5, C=-1, D=2$$

Note that we used x^0 to represent the constants. Also note that these systems can often be quite large and have a fair amount of work involved in solving them. The best way to deal with these is to use some form of computer aided solving techniques.

Now, let's take a look at the integral.

$$\begin{aligned} \int \frac{x^2-29x+5}{(x-4)^2(x^2+3)}dx &= \int \frac{1}{x-4} - \frac{5}{(x-4)^2} + \frac{-x+2}{x^2+3}dx \\ &= \int \frac{1}{x-4} - \frac{5}{(x-4)^2} - \frac{x}{x^2+3} + \frac{2}{x^2+3}dx \\ &= \ln|x-4| + \frac{5}{x-4} - \frac{1}{4}\ln|x^2+3| + \frac{2}{\sqrt{3}}\tan^{-2}\left(\frac{x}{\sqrt{3}}\right) + c \end{aligned}$$

In order to take care of the third term we needed to split it up into two separate terms. Once we've done this we can do all the integrals in the problem. The first two use the substitution, $u = x-4$ the third uses the substitution $v = x^2+3$, and the fourth term uses the formula given above for inverse tangents.

Example: Evaluate the following integral.

$$\int \frac{x^2+10x^2+3x+36}{(x-1)(x^2+4)}dx$$

Solution: Let's first get the general form of the partial fraction decomposition.

$$\frac{x^3 10x+3x+36}{(x-1)(x^2+4)^2}=\frac{A}{x-1}+\frac{Bx+C}{x^2+4}+\frac{Dx+E}{(x^2+4)}$$

Now, set numerators equal, expand the right side and collect like terms.

$$x^3+10x+3x+36=A(x^2+4)^2+(Bx+C)(x-1)(x^2+4)+(Dx+E)(x-1).$$

$$=(A+B)x^4+(C-B)x^3+(8A+4B-C+D)x^2+$$

$$(-4B+4C-D+E)x+16A-4C-E$$

Setting coefficient equal gives the following system.

$$\left.\begin{array}{lr} x^4: & A+B=0 \\ x^3: & C+D=1 \\ x^2: & 8A+4B-C+D=10 \\ x^1: & -4B+4C-D+E=3 \\ x^0: & -16A-4C-E=36 \end{array}\right\} \Rightarrow A=2, B=-2, C=-1, D=1, E=0$$

Don't get excited if some of the coefficients end up being zero. It happens on occasion. Here's the integral.

$$\int\frac{x^3+10x^2+3x+36}{(x-1)(x^2+4)^2}dx=\int\frac{2}{x-1}+\frac{-2x-1}{x^2+4}+\frac{x}{(x^2+4)^2}dx$$

$$=\int\frac{2}{x-1}-\frac{2x}{x^2+4}-\frac{1}{x^2+4}+\frac{x}{(x^2+4)}dx$$

$$=2\ln|x-1|-\ln|x^2+4|-\frac{1}{2}\tan^{-1}\left(\frac{x}{2}\right)-\frac{1}{2}\frac{1}{x^2+4}+c$$

To this point we've only looked at rational expressions where the degree of the numerator was strictly less that the degree of the denominator. Of course not all rational expressions will fit into this form and so we need to take a look at a couple of examples where this isn't the case.

Example: Evaluate the following integral.

$$\int\frac{x^4+5x^2+6x^2-18}{x^3-3x^2}dx$$

Solution: So, in this case the degree of the numerator is 4 and the degree of the denominator is 3. Therefore, partial fractions can't be done on this rational expression. To fix this up we'll need to do long division on this to get it into a form that we can deal with. Here is the work for that.

$$\begin{array}{r} x-2 \\ x^3-3x^2 \overline{\big)\, x^4-5x^3+6x^2-18} \\ \underline{-(x^4-3x^3)} \\ -2x^3+6x^2-18 \\ \underline{-(-2x^3+6x^2)} \\ -18 \end{array}$$

So, from the long division we see that,

$$\frac{x^4-5x^3+6x^2-18}{x^3-3x^2} = x-2-\frac{18}{x^3-3x^2}$$

and the integral becomes,

$$\int \frac{x^4-5x^3+6x^2-18}{x^3-3x^2}\,dx = \int x-2-\frac{18}{x^3-3x^2}\,dx$$

$$= \int x-2dx - \int \frac{18}{x^3-3x^2}\,dx$$

The first integral we can do easily enough and the second integral is now in a form that allows us to do partial fractions. So, let's get the general form of the partial fractions for the second integrand.

$$\frac{18}{x^2(x-3)} = \frac{A}{x}+\frac{B}{x^2}+\frac{C}{x-3}$$

Setting numerators equal gives us,

$$18 = Ax(x-3)+B(x-3)+Cx^2$$

Now, there is a variation of the method we used in the first couple of examples that will work here. There are a couple of values of x that will allow us to quickly get two of the three constants, but there is no value of x that will just hand us the third. What we'll do in this example is pick x's to get the two constants that we can easily get and then we'll just pick another value of x that will be easy to work with (*i.e.* it won't give large/messy numbers anywhere) and then we'll use the fact that we also know the other two constants to find the third.

$$\begin{array}{lll} x=0 & 18 = B(-3)(-2) & \Rightarrow \quad B=-6 \\ x=3 & 18 = C(9) & \Rightarrow \quad C=2 \\ x=1 & 18 = A(-2)+B(-2)+C=-2A+14 & \Rightarrow \quad A=-2 \end{array}$$

The integral is then,

$$\int \frac{x^4-5x^3+6x^2-18}{x^3-3x^2}\,dx = \int x-2dx - \int -\frac{2}{x}-\frac{6}{x^2}+\frac{2}{x-3}\,dx$$

$$= \frac{1}{2}x^2-2x+2\ln|x|-\frac{6}{x}-2\ln|x-3|+c$$

In the previous example there were actually two different ways of dealing with the x^2 in the denominator. One is to treat is as a quadratic which would give the following term in the decomposition

$$\frac{Ax+B}{r^2}$$

and the other is to treat it as a linear term in the following way,

$$x^2 = (x-0)^2$$

which gives the following two terms in the decomposition,

$$\frac{A}{x}+\frac{B}{x^2}$$

We used the second way of thinking about it in our example. Notice however that the two will give identical partial fraction decompositions. So, why talk about this? Simple. This will work for x^2, but what about x^3 or x^4? In these cases we really will need to use the second way of thinking about these kinds of terms.

$$x^3 \Rightarrow \frac{A}{x}+\frac{B}{x^2}+\frac{C}{x^3} \qquad x^4 \Rightarrow \frac{A}{x}+\frac{B}{x^2}+\frac{C}{x^3}+\frac{D}{x^4}$$

Let's take a look at one more example.

Example: Evaluate the following integral.

$$\int \frac{x^2}{x^2-1}\,dx$$

Solution: In this case the numerator and denominator have the same degree. As with the last example we'll need to do long division to get this into the correct form. I'll leave the details of that to you to check.

$$\int \frac{x^2}{x^2-1}\,dx = \int 1+\frac{1}{x^2-1}\,dx = \int dx + \int \frac{1}{x^2-1}\,dx$$

So, we'll need to partial fraction the second integral. Here's the decomposition.

$$\frac{1}{(x-1)(x+1)} = \frac{A}{x-1}+\frac{B}{x-1}$$

Setting numerator equal gives, $1 = A(x+1) + B(x-1)$

Picking value of x gives us the following coefficients.

$$x=-1 \qquad 1 = B(-2) \quad \Rightarrow \quad B = -\frac{1}{2}$$

$$x=1 \qquad 1 = A(2) \quad \Rightarrow \quad A = \frac{1}{2}$$

The integral is then,

$$\int \frac{x^2}{x^2-1}dx = \int dx + \int dx \frac{\frac{1}{2}}{x-1} - \frac{\frac{1}{2}}{x-1}\,dx$$

$$= x + \frac{1}{2}\ln|x-1| - \frac{1}{2}\ln|x+1| + c$$

INTEGRALS INVOLVING QUADRATICS

To this point we've seen quite a few integrals that involve quadratics. A couple of examples are,

$$\int \frac{x}{x^2 \pm a}dx = \frac{1}{2}\ln|x^2 \pm a| + c \qquad \int \frac{1}{x^2+a^2}dx = \frac{1}{2}\tan^{-1}\left(\frac{x}{a}\right)$$

We also saw that integrals involving, $\sqrt{b^2x^2-a^2}, \sqrt{a^2-b^2x^2}$ and $\sqrt{a^2+b^2x^2}$ could be done with a trig substitution. Notice however that all of these integrals were missing an x term. They all consist of a quadratic term and a constant. Some integrals involving general quadratics are easy enough to do. For instance, the following integral can be done with a quick substitution.

$$\int \frac{2x+3}{4x^3+12x}dx = \frac{1}{4}\int \frac{1}{u}du \qquad (u = 4x^2+12x-1 \qquad du = 4(2x+3)dx)$$

$$= \frac{1}{4}\ln|4x^2+12x-1| + c$$

Some integrals with quadratics can be done with partial fractions. For instance,

$$\int \frac{10x-6}{3x^2+16x+5}dx = \int \frac{4}{x+5} - \frac{2}{3x+1}dx = 4\ln|x+5| - \frac{2}{3}\ln|3x+1| + c$$

Unfortunately, these methods won't work on a lot of integrals. A simple substitution will only work if the numerator is a constant multiple of the derivative of the denominator and partial fractions will only work if the denominator can be factored.

This section is how to deal with integrals involving quadratics when the techniques that we've looked at to this point simply won't work. Back in the Trig Substitution section we saw how to deal with square roots that had a general quadratic in them. Let's take a quick look at another one like that since the idea involved in doing that kind integral is exactly what we are going to need for the other integrals in this section.

Example: Evaluate the following integral.

$$\int \sqrt{x^2+4x+5}\,dx$$

Solution: Recall from the Trig Substitution section that in order to do a trig substitution here we first needed to complete the square on the quadratic. This gives,

$$x^2 + 4x + 5 = x^2 + 4x + 4 - 4 + 5 = (x+2)^2 + 1$$

After completing the square the integral becomes,

$$\int \sqrt{x^2 + 4x + 5}\,dx = \int \sqrt{(x+2)^2 + 1}\,dx$$

Upon doing this we can identify the trig substitution that we need. Here it is,

$$x + 2 = \tan\theta \qquad x = \tan\theta - 2 \qquad dx = \sec^2\theta\, d\theta$$

$$\sqrt{(x+2)^2 + 1} = \sqrt{\tan^2\theta + 1} = \sqrt{\sec^2\theta} = |\sec\theta| = \sec\theta$$

Recall that since we are doing an indefinite integral we can drop the absolute value bars. Using this substitution the integral becomes,

$$\int \sqrt{x^2 + 4x + 5}\,dx = \int \sec^3\theta\, d\theta$$

$$= \frac{1}{2}(\sec\theta\tan\theta + \ln|\sec\theta + \tan\theta|) + c$$

We can finish the integral out with the following right triangle.

$$\tan\theta = \frac{x+2}{1} \qquad \sec\theta = \frac{\sqrt{x^2 4x + 5}}{1} = \sqrt{x^2 + 4x + 5}$$

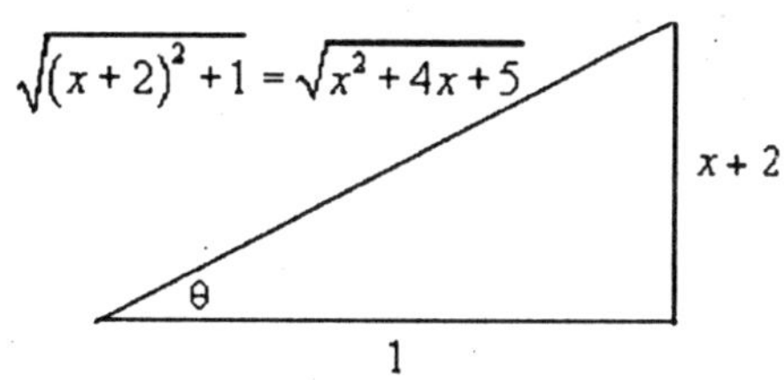

$$\int \sqrt{x^2 + 4x + 5}\,dx = \frac{1}{2}\left((x+2)\sqrt{x^2 + 4x + 5} + \ln\left|x + 2 + \sqrt{x^2 + 4x + 5}\right|\right) + c$$

So, by completing the square we were able to take an integral that had a general quadratic in it and convert it into a form that allowed use a known integration technique. Let's do a quick review of completing the square before proceeding. Here is the general completing the square formula that we'll use.

$$x^2 + bx + c = x^2 + bx + \left(\frac{b}{2}\right)^2 - \left(\frac{b}{2}\right)^2 + c = \left(x + \frac{b}{2}\right)^2 + c - \frac{b^2}{4}$$

This will always take a general quadratic and write it in terms of a squared term and a constant term. Recall as well that in order to do this we must have a coefficient of one in front of the x^2. If not we'll need to factor out the coefficient before completing the square. In other words,

$$ax^2 + bx + c = a\left(\underbrace{x^2 + \frac{b}{a}x + \frac{c}{a}}_{\text{complete the square on that}}\right)$$

Now, let's see how completing the square can be used to do integrals that we aren't able to do at this point.

Example: Evaluate the following integral.

$$\int \frac{1}{2x^2 - 3x + 2}\, dx$$

Solution: Okay, this doesn't factor so partial fractions just won't work on this. Likewise, since the numerator is just "1" we can't use the substitution. $u = 2x^2 - 3x + 8$. So, let's see what happens if we complete the square on the denominator.

$$2x^2 - 3x + 2 = 2\left(x^2 - \frac{3}{2}x + 1\right)$$

$$= 2\left(x^2 - \frac{3}{2}x + \frac{9}{16} - \frac{9}{16} + 1\right)$$

$$2\left(\left(x - \frac{3}{4}\right)^2 + \frac{7}{16}\right)$$

With this the integral is,

$$\int \frac{1}{2x^2 - 3x + 2}\, dx = \frac{1}{2}\int \frac{1}{\left(x - \frac{3}{4}\right)^2 + \frac{7}{16}}\, dx$$

Now this may not seem like all that great of a change. However, notice that we can now use the following substitution.

$$u = x - \frac{3}{4} \qquad du = dx$$

and the integral is now,

$$\int \frac{1}{2x^2 - 3x + 2}\, dx = \frac{1}{2}\int \frac{1}{u^2 + \frac{7}{16}}\, du$$

We can now see that this is an inverse tangent! So, using the formula from above we get,

$$\int \frac{1}{2x^2 - 3x + 2}\, dx = \frac{1}{2}\left(\frac{4}{\sqrt{7}}\right)\tan^{-1}\left(\frac{4u}{\sqrt{7}}\right) + c$$

$$= \frac{2}{\sqrt{7}}\tan^{-1}\left(\frac{4x - 3}{\sqrt{7}}\right) + c$$

Chapter 2

Bernoulli Differential Equations

In this section we are going to take a look at differential equations in the form,

$$y' + p(x)y = q(x)y^n$$

where $p(x)$ and $q(x)$ are continuous functions on the interval we're working on and n is a real number. Differential equations in this form are called Bernoulli Equations.

First notice that if $n = 0$or $n = 1$then the equation is linear and we already know how to solve it in these cases. Therefore, in this section we're going to be looking at solutions for values of n other than these two.

In order to solve these we'll first divide the differential equation by $y^n$to get,

$$y^{-n}y' + p(x)y^{1-n} = q(x)$$

We are now going to use the substitution $v = y^{1-n}$to convert this into a differential equation in terms of v. As we'll see this will lead to a differential equation that we can solve. We are going to have to be careful with this however when it comes to dealing with the derivative, . We need to determine just what is in terms of our substitution. This is easier to do than it might at first look to be. All that we need to do is differentiate both sides of our substitution with respect to x. Remember that both v and y are functions of x and so we'll need to use the chain rule on the right side. So, taking the derivative gives us,

$$v' = (1-n)y^{-n}y'$$

Now, plugging this as well as our substitution into the differential equation gives,

$$\frac{1}{1-n}v' + p(x)v = q(x)$$

This is a linear differential equation that we can solve for v and once we have this in hand we can also get the solution to the original differential equation by plugging v back into our substitution and solving for y.

Example: Solve the following IVP and find the interval of validity for the solution.

$$y' + \frac{4}{x}y = x^3 y^2 \qquad y(2) = -1, \qquad x > 0$$

Solution: So, the first thing that we need to do is get this into the "proper" form and that means dividing everything by . Doing this gives,

$$y^{-2}y' + \frac{4}{x}y^{-1} = x^3$$

The substitution and derivative that we'll need here is,

$$v = y^{-4} \qquad v' = -y^{-2}y'$$

With this substitution the differential equation becomes,

$$-v' + \frac{4}{x}v = x^3$$

So, as noted above this is a linear differential equation that we know how to solve. We'll do the details on this one and then for the rest of the examples in this section we'll leave the details for you to fill in. If you need a refresher on solving linear differential equations then go back to that section for a quick review.

Here's the solution to this differential equation.

$$-v' + \frac{4}{x}v = x^3 \qquad \Rightarrow \qquad \mu(x) = e^{\int \frac{4}{x}dx} = e^{-4\ln|x|} = x^4$$

$$\int (x^{-4}v)' dx = \int -x^{-1} dx$$

$$x^{-4} = -\ln|x| + c \qquad \Rightarrow \qquad v(x) = cx^4 - x^4 \ln x$$

Note that we dropped the absolute value bars on the x in the logarithm because of the assumption that.

Now we need to determine the constant of integration. This can be done in one of two ways. We can can convert the solution above into a solution in terms of y and then use the original initial condition or we can convert the initial condition to an initial condition in terms of v and use that. Because we'll need to convert the solution to y's eventually anyway and it won't add that much work in we'll do it that way.

So, to get the solution in terms of y all we need to do is plug the substitution back in. Doing this gives,

$$y^{-1} = x^4(c - \ln x)$$

At this point we can solve for y and then apply the initial condition or apply the initial condition and then solve for y. We'll generally do this with the later approach so let's apply the initial condition to get,

$$(-1)^{-1} = c2^4 - 2^4 \text{In}2 \quad \Rightarrow \quad c = \text{In}2\text{-}\frac{1}{16}$$

Plugging in for c and solving for y gives,

$$y(x) = \frac{1}{x^4\left(\text{In}2 - \frac{1}{16} - \text{In}x\right)} = \frac{-16}{x^4(1+16\text{In}x - 16\text{In}2)} = \frac{-16}{x^4\left(1+16\text{In}\frac{x}{2}\right)}$$

Note that we did a little simplification in the solution. This will help with finding the interval of validity.

Before finding the interval of validity however, we mentioned above that we could convert the original initial condition into an initial condition for v. Let's briefly talk about how to do that. To do that all we need to do is plug into the substitution and then use the original initial condition. Doing this gives,

$$v(2) = y^{-1}(2) = (-1)^{-1} = -1$$

So, in this case we got the same value for v that we had for y. Don't expect that to happen in general if you chose to do the problems in this manner.

First we already know that and that means we'll avoid the problems of having logarithms of negative numbers and division by zero at . So, all that we need to worry about then is division by zero in the second term and this will happen where,

$$1 + 16\text{In}\frac{x}{2} = 0$$

$$In\frac{x}{2} = -\frac{1}{16}$$

$$\frac{x}{2} = e^{-\frac{1}{16}} \quad \Rightarrow \quad x = 2e^{-\frac{1}{16}} \approx 1.8788$$

The two possible intervals of validity are then,

$$0 < x < 2e^{-\frac{1}{16}} \qquad 2e^{-\frac{1}{16}} < x < \infty$$

and since the second one contains the initial condition we know that the interval of validity is then.

Here is a graph of the solution.

Let's do a couple more examples and as noted above we're going to leave it to you to solve the linear differential equation when we get to that stage.

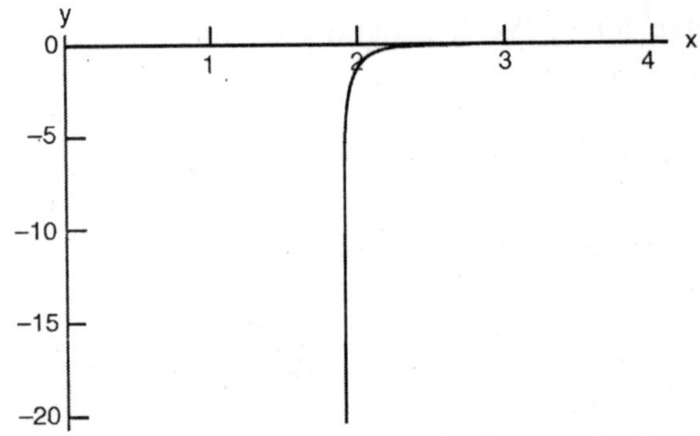

Example: Solve the following IVP and find the interval of validity for the solution.

$$y' = 5y + e^{-2x} y^{-2} \qquad y(0) = 2$$

Solution: The first thing we'll need to do here is multiply through by and we'll also do a little rearranging to get things into the form we'll need for the linear differential equation. This gives,

$$y^2 y' - 5y^3 = e^{-2x}$$

The substitution here and its derivative is,

$$v = y^3 \qquad v' = 3y^2 y'$$

Plugging the substitution into the differential equation gives,

$$\frac{1}{3}v' - 5v = e^{-2x} \quad \Rightarrow \quad v' - 15v = 3e^{-2x} \qquad \mu(x)e^{-15x}$$

We rearranged a little and gave the integrating factor for the linear differential equation solution. Upon solving we get,

$$v(x) = ce^{15x} - \frac{3}{17}e^{-2x}$$

Now go back to y's.

$$y^3 = ce^{15x} - \frac{3}{17}e^{-2x}$$

Applying the initial condition and solving for c gives,

$$8 = c - \frac{3}{17} \quad \Rightarrow \quad c = \frac{139}{17}$$

Plugging in c and solving for y gives,

$$y(x) = \left(\frac{139e^{15x} - 3e^{-2x}}{17} \right)^{\frac{1}{3}}$$

There are no problem values of x for this solution and so the interval of validity is all real numbers. Here's a graph of the solution.

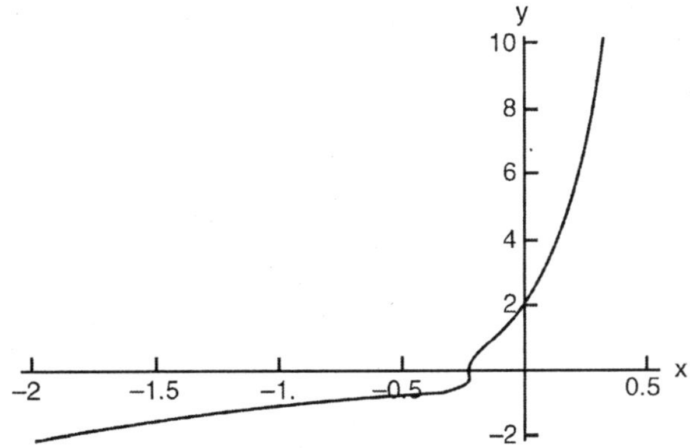

Example: Solve the following IVP and find the interval of validity for the solution.

$$6y' - 2y = xy^4 \qquad y(0) = -2$$

Solution: First get the differential equation in the proper form and then write down the substitution.

$$6y^{-4}y' - 2y^{-3} = x \qquad \Rightarrow \qquad v = y^{-3} \qquad v' = -3y^{-4}y'$$

Plugging the substitution into the differential equation gives,

$$-2v' - 2v = x \qquad \Rightarrow \qquad v' + v = -\frac{1}{2}x \qquad \mu(x) = e^x$$

Again, we've rearranged a little and given the integrating factor needed to solve the linear differential equation. Upon solving the linear differential equation we have,

$$v(x) = -\frac{1}{2}(x-1) + ce^{-x}$$

Now back substitute to get back into y's.

$$y^{-3} = -\frac{1}{2}(x-1) + ce^{-x}$$

Now we need to apply the initial condition and solve for c.

$$-\frac{1}{8} = \frac{1}{2} + c \qquad \Rightarrow \qquad c = -\frac{5}{8}$$

Plugging in c and solving for y gives,

$$y(x) = -\frac{2}{\left(4x - 4 + 5e^{-x}\right)^{\frac{1}{3}}}$$

Next, we need to think about the interval of validity. In this case all we need to worry about it is division by zero issues and using some form of computational aid (such as Maple or Mathematica) we will see that the denominator of our solution is never zero and so this solution will be valid for all real numbers.

Here is a graph of the solution.

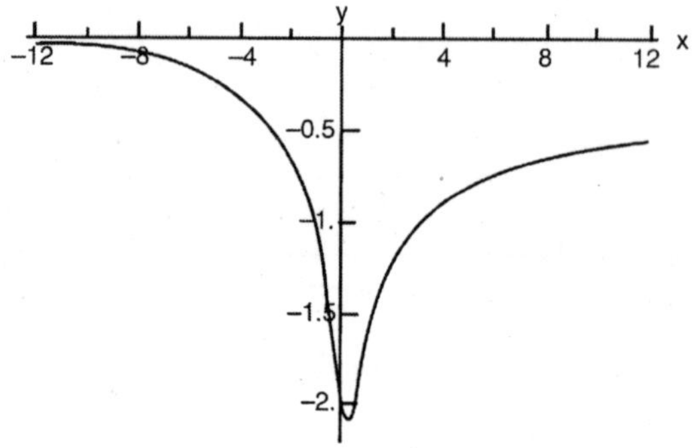

To this point we've only worked examples in which n was an integer (positive and negative) and so we should work a quick example where n is not an integer.

Example: Solve the following IVP and find the interval of validity for the solution.

$$y'+\frac{y}{x}-\sqrt{y}=0 \qquad y(1)=0$$

Solution: Let's first get the differential equation into proper form.

$$y'+\frac{1}{x}y=y^{\frac{1}{2}} \qquad \Rightarrow \qquad y^{-\frac{1}{2}}y'+\frac{1}{x}y^{\frac{1}{2}}=1$$

The substitution is then,

$$v=y^{\frac{1}{2}} \qquad v'=\frac{1}{2}y^{-\frac{1}{2}}y'$$

Now plug the substitution into the differential equation to get,

$$2v'+\frac{1}{x}v=1 \qquad \Rightarrow \qquad v'+\frac{1}{2x}v=\frac{1}{2} \qquad \mu(x)=x^{\frac{1}{2}}$$

As we've done with the previous examples we've done some rearranging and given the integrating factor needed for solving the linear differential equation. Solving this gives us,

$$v(x)=\frac{1}{3}x+cx^{-\frac{1}{2}}$$

In terms of *y* this is,

$$y^{\frac{1}{2}} = \frac{1}{3}x + cx^{-\frac{1}{2}}$$

Applying the initial condition and solving for c gives,

$$0 = \frac{1}{3} + c \qquad \Rightarrow \qquad c = -\frac{1}{3}$$

Plugging in for c and solving for y gives us the solution.

$$y(x) = \left(\frac{1}{3}x - \frac{1}{3}x^{-\frac{1}{2}}\right)^2 = \frac{x^3 - 2x^{\frac{3}{2}} + 1}{9x}$$

Note that we multiplied everything out and converted all the negative exponents to positive exponents to make the interval of validity clear here. Because of the root (in the second term in the numerator) and the x in the denominator we can see that we need to require in order for the solution to exist and it will exist for all positive x's and so this is also the interval of validity.

Here is the graph of the solution.

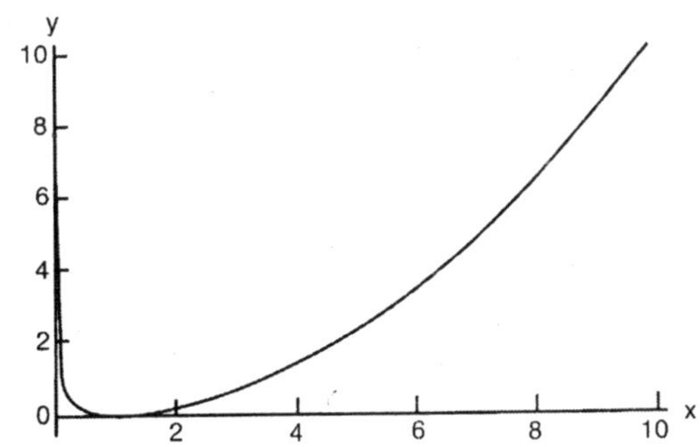

SUBSTITUTIONS

In the previous section we looked at Bernoulli Equations and saw that in order to solve them we needed to use the substitution $v = y^{1-n}$. Upon using this substitution we were able to convert the differential equation into a form that we could deal with (linear in this case). In this section we want to take a look at a couple of other substitutions that can be used to reduce some differential equations down to a solvable form.

The first substitution we'll take a look at will require the differential equation to be in the form,

$$y' = F\left(\frac{y}{x}\right)$$

First order differential equations that can be written in this form are called homogeneous differential equations. Note that we will usually have to do some rewriting in order to put the differential equation into the proper form. Once we have verified that the differential equation is a homogeneous differential equation and we've gotten it written in the proper form we will use the following substitution.

$$v(x) = \frac{y}{x}$$

We can then rewrite this as,

$$y = xv$$

and then remembering that both y and v are functions of x we can use the chain rule (recall that is implicit differentiation from Calculus I) to compute,

$$y' = v + xv'$$

Under this substitution the differential equation is then,

$$v + xv' = F(v)$$

$$xv' = F(v) - v \qquad \Rightarrow \qquad \frac{dv}{F(v) - v} = \frac{dx}{x}$$

As we can see with a small rewrite of the new differential equation we will have a separable differential equation after the substitution.

Example: Solve the following IVP and find the interval of validity for the solution.

$$xyy' + 4x^2 + y^2 + y^2 = 0 \qquad y(2) = -7, \qquad x > 0$$

Solution: Let's first divide both sides by to rewrite the differential equation as follows,

$$\frac{y}{x}y' = -4 - \frac{y^2}{x^2} = -4 - \left(\frac{y}{x}\right)^2$$

Now, this is not in the officially proper form as we have listed above, but we can see that everywhere the variables are listed they show up as the ratio, y/x and so this is really all the farther that we need to go. So, let's plug the substitution into this form of the differential equation to get,

$$v(v + xv') = -4 - v^2$$

Next, rewrite the differential equation to get everything separated out.

$$vxv' = -4 - 2v^2$$

$$xv' = -\frac{4 + 2v^2}{v}$$

$$\frac{v}{4 + 2v^2}dv = -\frac{1}{x}dx$$

Integrating both sides gives,

$$\frac{1}{4}\text{In}\left(4+2v^2\right)=-\text{In}(x)+c$$

We need to do a little rewriting using basic logarithm properties in order to be able to easily solve this for v.

$$\text{In}\left(4+2v^2\right)^{\frac{1}{4}}=-\text{In}(x)+c$$

Now exponentiate both sides and do a little rewriting

$$\left(4+2v^2\right)^{\frac{1}{4}}=e^{h(x)^{-1+c}}=e^c e^{\text{In}(x)^{-1}}=\frac{c}{x}$$

Note that because c is an unknown constant then so is and so we may as well just call this c as we did above.

Finally, let's solve for v and then plug the substitution back in and we'll play a little fast and loosc with constants again.

$$4+2v^2=\frac{c^4}{x^4}=\frac{c}{x^4}$$

$$v^2=\frac{1}{2}\left(\frac{c}{x^4}-4\right)$$

$$\frac{y^2}{x^2}=\frac{1}{2}\left(\frac{c-4x^4}{x^4}\right)$$

$$y^2=\frac{1}{2}x^2\left(\frac{c-4x^4}{x^2}\right)=\frac{c-4x^4}{2x^2}$$

At this point it would probably be best to go ahead and apply the initial condition. Doing that gives,

$$49=\frac{c-4(16)}{2(8)} \qquad \Rightarrow \qquad c=45$$

Note that we could have also converted the original initial condition into one for in terms of v and then applied it upon solving the separable differential equation. In this case however, it was probably a little easier to do it in terms of y given all the logarithms in the solution to the separable differential equation.

Finally, plug in c and solve for y to get,

$$y^2=\frac{228-2x^4}{x^2} \qquad \Rightarrow \qquad y(x)=\pm\sqrt{\frac{228-2x^4}{x^2}}$$

The initial condition tells us that the "—" must be the correct sign and so the actual solution is,

$$Y(x) = -\sqrt{\frac{228-2x^4}{x^2}}$$

For the interval of validity we can see that we need to avoid $x = 0$and because we can't allow negative numbers under the square root we also need to require that,

$$288-2x^4 \geq 0$$

$$x^4 \leq 114 \qquad \Rightarrow \qquad -3.2676 \leq x \leq 3.2676$$

So, we have two possible intervals of validity,

$$-3.2676 \leq x < 0 \qquad\qquad 0 < x \leq 3.2676$$

and the initial condition tells us that it must be

The graph of the solution is,

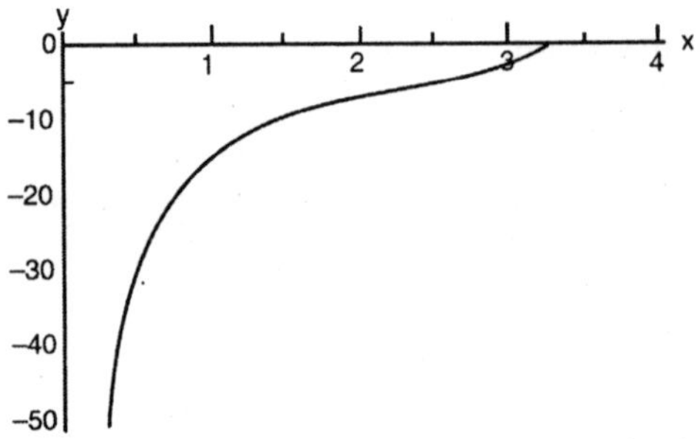

Example: Solve the following IVP and find the interval of validity for the solution.

$$xy' = y(\text{In}x - \text{In}y) \qquad y(1) = 4, \qquad x > 0$$

Solution: On the surface this differential equation looks like it won't be homogeneous. However, with a quick logarithm property we can rewrite this as,

$$y' = \frac{y}{x}\text{In}\left(\frac{x}{y}\right)$$

In this form the differential equation is clearly homogeneous. Applying the substitution and separating gives,

$$v + xv' = v\text{In}\left(\frac{1}{v}\right)$$

$$xv' = v\left(\text{In}\left(\frac{1}{v}\right) - 1\right)$$

$$\frac{dv}{v\left(\text{In}\left(\frac{1}{v}\right)-1\right)} = \frac{dx}{x}$$

Integrate both sides and do a little rewrite to get,

$$-\text{In}\left(\text{In}\left(\frac{1}{v}\right)-1\right)=\text{In}x+c$$

$$\text{In}\left(\text{In}\left(\frac{1}{v}\right)-1\right)=c-\text{In}x$$

You were able to do the integral on the left right? It used the substitution $u=\text{In}\left(\frac{1}{v}\right)-1$.

Now, solve for v and note that we'll need to exponentiate both sides a couple of times and play fast and loose with constants again.

$$\text{In}\left(\frac{1}{v}\right)-1=e^{c}e^{\text{In}(x)^{-4}}=\frac{c}{x}$$

$$\text{In}\left(\frac{1}{v}\right)=\frac{c}{x}+1$$

$$\frac{1}{v}=e^{\frac{c}{x}}+1 \qquad \Rightarrow \qquad v=e^{-\frac{c}{x}-1}$$

Plugging the substitution back in and solving for y gives,

$$\frac{y}{x}=e^{-\frac{c}{x}-1} \qquad \Rightarrow \qquad y(x)=xe^{-\frac{c}{x}-1}$$

Applying the initial condition and solving for c gives,

$$4=e^{-c-1} \qquad \Rightarrow \qquad c=-(1+\text{In}4)$$

The solution is then,

$$y(x)=xe^{\frac{1+h4}{x}-1}$$

We clearly need to avoid $x=0$to avoid division by zero and so with the initial condition we can see that the interval of validity is $x>0$.

The graph of the solution is,

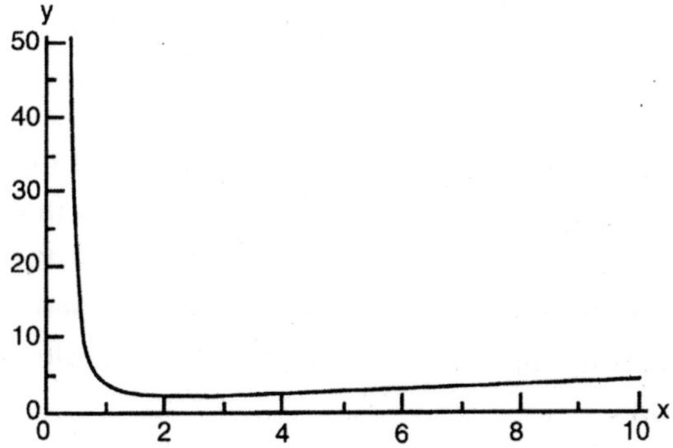

For the next substitution we'll take a look at we'll need the differential equation in the form,

$$y' = G(ax + by)$$

In these cases we'll use the substitution,

$$v = ax + by \qquad \Rightarrow \qquad v' = a + by'$$

Plugging this into the differential equation gives,

$$\frac{1}{b}(v' - a) = G(v)$$

$$v' = a + bG(v) \qquad \Rightarrow \qquad \frac{dv}{a + bG(v)} = dx$$

So, with this substitution we'll be able to rewrite the original differential equation as a new separable differential equation that we can solve.

Let's take a look at a couple of examples.

Example: Solve the following IVP and find the interval of validity for the solution.

$$y' - (4x - y + 1)^2 = 0 \qquad y(0) = 2$$

Solution: In this case we'll use the substitution.

$$v = 4x - y \qquad v' = 4 - y'$$

Note that we didn't include the "+1" in our substitution. Usually only the $ax + by$ part gets included in the substitution. There are times where including the extra constant may change the difficulty of the solution process, either easier or harder, however in this case it doesn't really make much difference so we won't include it in our substitution.

So, plugging this into the differential equation gives,

$$4 - v' - (v + 1)^2 = 0$$

$$v' = 4 - (v + 1)^2$$

$$\frac{dv}{(v + 1)^2 - 4} = -dx$$

As we've shown above we definitely have a separable differential equation. Also note that to help with the solution process we left a minus sign on the right side. We'll need to integrate both sides and in order to do the integral on the left we'll need to use partial fractions. We'll leave it to you to fill in the missing details and given that we'll be doing quite a bit of partial fraction work in a few chapters you should really make sure that you can do the missing details.

$$\int \frac{dv}{v^2+2v-3} = \int \frac{dv}{(v+3)(v-1)} = \int -dx$$

$$\frac{1}{4}\int \frac{1}{v-1} - \frac{1}{v+3} dv = \int -dx$$

$$\frac{1}{4}\left(In(v-1) - In(v+3)\right) = -x + c$$

$$In\left(\frac{v-1}{v+3}\right) = c - 4x$$

Note that we played a little fast and loose with constants above. The next step is fairly messy but needs to be done and that is to solve for v and note that we'll be playing fast and loose with constants again where we can get away with it and we'll be skipping a few steps that you shouldn't have any problem verifying.

$$\frac{v-1}{v+3} = e^{c-4x} = ce^{-4x}$$

$$v-1 = ce^{-4x}(v+3)$$

$$v\left(1-ce^{-4x}\right) = 1+3ce^{-4x}$$

At this stage we should back away a bit and note that we can't play fast and loose with constants anymore. We were able to do that in first step because the c appeared only once in the equation. At this point however the c appears twice and so we've got to keep them around. If we "absorbed" the 3 into the c on the right the "new" c would be different from the c on the left because the c on the left didn't have the 3 as well.

So, let's solve for v and then go ahead and go back into terms of y.

$$v = \frac{1+3ce^{-4x}}{1-ce^{-4x}}$$

$$4x - y = \frac{1+3ce^{-4x}}{1-ce^{-4x}}$$

$$y(x) = 4x - \frac{1+3ce^{-4x}}{1-ce^{-4x}}$$

The last step is to then apply the initial condition and solve for c.

$$2 = y(0) = \frac{1+3c}{1-c} \qquad \Rightarrow \qquad c = -3$$

The solution is then,

$$y(x) = 4x - \frac{1-9e^{-4x}}{1+3e^{-4x}}$$

Note that because exponentials exist everywhere and the denominator of the second term is always positive (because exponentials are always positive and adding a positive one onto that won't change the fact that it's positive) the interval of validity for this solution will be all real numbers.

Here is a graph of the solution.

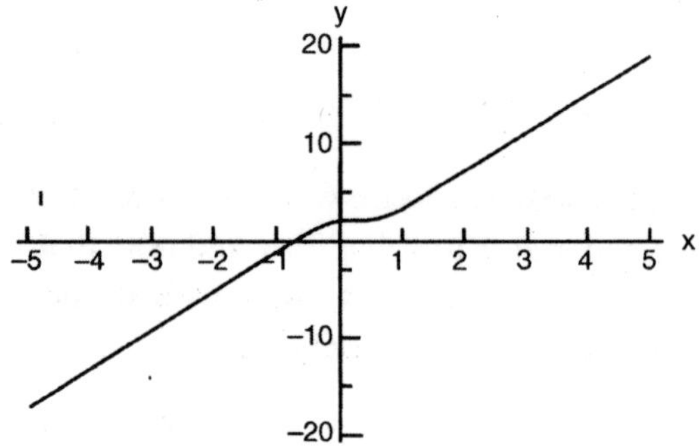

Example: Solve the following IVP and find the interval of validity for the solution.

$$y' = e^{9y-x} \qquad\qquad y(0=0)$$

Solution: Here is the substitution that we'll need for this example.

$$v = 9y - x \qquad\qquad v'9y' - 1$$

Plugging this into our differential equation gives,

$$\frac{1}{9}(v'+1) = e^{v}$$

$$v' = 9e^{v-1}$$

$$\frac{dv}{9e^{v}-1} = dx \qquad \Rightarrow \qquad \frac{e^{-v}dv}{9-e^{-v}} = dx$$

Note that we did a little rewrite on the separated portion to make the integrals go a little easier.

By multiplying the numerator and denominator by e^{-y} we can turn this into a fairly simply substitution integration problem. So, upon integrating both sides we get,

$$In\left(9-e^{-v}\right) = x + c$$

Solving for v gives,

$$9 - e^{-v} = e^{c}e^{x} = ce^{x}$$

$$e^{-v} = 9 - ce^{x}$$

$$v = -In\left(9 - ce^{x}\right)$$

Plugging the substitution back in and solving for y gives us,

$$y(x)=\frac{1}{9}\left(x-In\left(9-ce^{x}\right)\right)$$

Next, apply the initial condition and solve for c.

$$0=y(0)=-\frac{1}{9}In(9-c) \qquad \Rightarrow \qquad c=8$$

The solution is then,

$$y(x)=\frac{1}{9}\left(x-In\left(9-8e^{x}\right)\right)$$

Now, for the interval of validity we need to make sure that we only take logarithms of positive numbers as we'll need to require that,

$$9-8e^{x}>0 \qquad \Rightarrow \qquad e^{x}<\frac{9}{8} \qquad \Rightarrow x<In\frac{9}{8}=0.1178$$

Here is a graph of the solution.

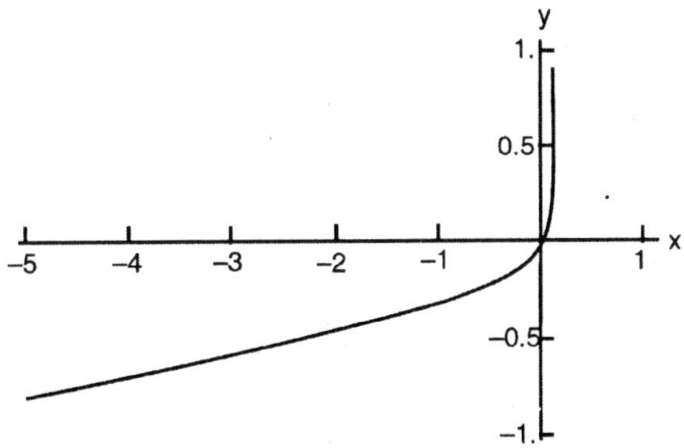

In both this section and the previous section we've seen that sometimes a substitution will take a differential equation that we can't solve and turn it into one that we can solve. This idea of substitutions is an important idea and should not be forgotten.

Not every differential equation can be made easier with a substitution and there is no way to show every possible substitution but remembering that a substitution may work is a good thing to do. If you get stuck on a differential equation you may try to see if a substitution if some kind will work for you.

Chapter 3

Nonhomogeneous Differential Equations

It's now time to start thinking about how to solve nonhomogeneous differential equations. A second order, linear nonhomogeneous differential equation is

$$y''+p(t)y'+q(t)y=g(t)$$

where $g(t)$ is a non-zero function. Note that we didn't go with constant coefficients here because everything that we're going to do in this section doesn't require it. Also, we're using a coefficient of 1 on the second derivative just to make some of the work a little easier to write down. It is not required to be a 1.

Before talking about how to solve one of these we need to get some basics out of the way, which is the point of this section. First, we will call

$$y''+p(t)y'+q(t)y=0$$

the associated homogeneous differential.

Now, let's take a look at the following theorem.

Theorem: Suppose that $Y_1(t)$ and $Y_2(t)$ are two solutions and that $y_1(t)$ and $y_2(t)$ are a fundamental set of solutions to the associated homogeneous differential equation then,

$$Y_1(t)-Y_2(t)$$

is a solution and it can be written as

$$Y_1(t)-Y_2(t)=c_1y_1(t)+c_2y_2(t).$$

Note the notation used here. Capital letters referred to solutions to (1) while lower case letters referred to solutions. This is a fairly common convention when dealing with nonhomogeneous differential equations. This theorem is easy enough to prove so let's do that. To prove that $Y_1(t)-Y_2(t)$ is a solution all we need to do is plug this into the differential equation and check it.

$$(Y_1-Y_2)''+p(t)(Y_1-Y_2)'+q(t)(Y_1-Y_2)=0$$

$$Y_1''+p(t)Y_1'+q(t)Y_1-(Y_2''+P(t)Y_2'+q(t)Y_2)=0$$

$$g(t) - g(t) = 0$$

$$0 = 0.$$

We used the fact that $Y_1(t)$ and $Y_2(t)$ are two solutions in the third step. Because they are solutions we know that

$$Y_1'' + p(t)Y_1' + q(t)Y_1 = g(t)$$

$$Y_1'' + p(t)Y_2' + q(t)Y_2 = g(t).$$

So, we were able to prove that the difference of the two solutions.

Proving that

$$Y_1(t) - Y_2(t) = c_1 y_1(t) + c_2 y_2(t)$$

is even easier. Since $y_1(t)$ and $y_2(t)$ are a fundamental set of solutions we know that they form a general solution and so any solution can be written in the form

$$y(t) = c_1 y_1(t) + c_2 y_2(t).$$

Well, $Y_1(t)$ - $Y_2(t)$ is a solution, as we've shown above, therefore it can be written as-

$$Y_1(t) - Y_2(t) = c_1 y_1(t) + c_2 y_2(t).$$

So, what does this theorem do for us? We can use this theorem to write down the form of the general solution. Let's suppose that $y(t)$ is the general solution and that $Y_P(t)$ is any solution that we can get our hands on. Then using the second part of our theorem we know that

$$y(t) - Y_p(t) = c_1 y_1(t) = c_1 y_1(t) + c_2 y_2(t)$$

where $y_1(t)$ and $y_2(t)$ are a fundamental set of solutions. Solving for $y(t)$ gives,

$$y(t) = c_1 y_1(t) + c_2 y_2(t) + Y_P(t).$$

We will call

$$y_c(t) = c_1 y_1(t) + c_2 y_2(t)$$

the complimentary solution and $Y_P(t)$ a particular solution. The general solution to a differential equation can then be written as.

$$y(t) = y_c(t) + Y_P(t).$$

So, to solve a nonhomogeneous differential equation, we will need to solve the homogeneous differential equation, which for constant coefficient differential equations is pretty easy to do, and we'll need a solution.

This seems to be a circular argument. In order to write down a solution we need a solution. However, this isn't the problem that it seems to be. There are ways to find a solution. They just won't, in general, be the general solution. In fact, the next two sections are devoted to exactly that, finding a particular solution to a nonhomogeneous differential equation.

There are two common methods for finding particular solutions: Undetermined Coefficients and Variation of Parameters. Both have their advantages and disadvantages as you will see in the next couple of sections.

UNDETERMINED COEFFICIENTS

In this section we will take a look at the first method that can be used to find a particular solution to a nonhomogeneous differential equation.

$$y''+p(t)y'+q(t)y=g(t).$$

One of the main advantages of this method is that it reduces the problem down to an algebra problem. The algebra can get messy on occasion, but for most of the problems it will not be terribly difficult. Another nice thing about this method is that the complimentary solution will not be explicitly required, although as we will see knowledge of the complimentary solution will be needed in some cases and so we'll generally find that as well.

There are two disadvantages to this method. First, it will only work for a fairly small class of *g(t)*'s. The class of *g(t)*'s for which the method works, does include some of the more common functions, however, there are many functions out there for which undetermined coefficients simply won't work. Second, it is generally only useful for constant coefficient differential equations.

The method is quite simple. All that we need to do is look at *g(t)* and make a guess as to the form of $Y_P(t)$ leaving the coefficient(s) undetermined (and hence the name of the method). Plug the guess into the differential equation and see if we can determine values of the coefficients. If we can determine values for the coefficients then we guessed correctly, if we can't find values for the coefficients then we guessed incorrectly.

It's usually easier to see this method in action rather than to try and describe it, so let's jump into some examples.

Example: Determine a particular solution to

$$y''-4y'-12y=3e^{5t}.$$

Solution: The point here is to find a particular solution, however the first thing that we're going to do is find the complimentary solution to this differential equation. Recall that the complimentary solution comes from solving,

$$y''-4y'-12y=0.$$

The characteristic equation for this differential equation and its roots are.

$$r^2-4r-12=(r-6)(r+2)=0 \quad \Rightarrow \quad r_1=-2,\ r_2=6.$$

The complimentary solution is then,

$$y_c(t) = c_1 e^{-2t} + c_2 e^{6t}.$$

At this point the reason for doing this first will not be apparent, however we want you in the habit of finding it before we start the work to find a particular solution. Eventually, as we'll see, having the complimentary solution in hand will be helpful and so it's best to be in the habit of finding it first prior to doing the work for undetermined coefficients.

Now, let's proceed with finding a particular solution. As mentioned prior to the start of this example we need to make a guess as to the form of a particular solution to this differential equation. Since *g(t)* is an exponential and we know that exponentials never just appear or disappear in the differentiation process it seems that a likely form of the particular solution would be

$$Y_P(t) = Ae^{5t}.$$

Now, all that we need to do is do a couple of derivatives, plug this into the differential equation and see if we can determine what *A* needs to be. Plugging into the differential equation gives

$$25Ae^{5t} - 4(5Ae^{5t}) - 12(Ae^{5t}) = 3e^{5t}$$

$$-7Ae^{5t} = 3e^{5t}.$$

So, in order for our guess to be a solution we will need to choose *A* so that the coefficients of the exponentials on either side of the equal sign are the same. In other words we need to choose *A* so that,

$$-7A = 3 \quad \Rightarrow \quad A = -\frac{3}{7}.$$

Okay, we found a value for the coefficient. This means that we guessed correctly. A particular solution to the differential equation is then,

$$Y_p(t) = -\frac{3}{7}e^{5t}.$$

Before proceeding any further let's again note that we started off the solution above by finding the complimentary solution. This is not technically part the method of Undetermined Coefficients however, as we'll eventually see, having this in had before we make our guess for the particular solution can save us a lot of work and/or headache. Finding the complimentary solution first is simply a good habit to have so we'll try to get you in the habit over the course of the next few examples. At this point do not worry about why it is a good habit. We'll eventually see why it is a good habit.

Now, back to the work at hand. Notice in the last example that we kept saying "a" particular solution, not "the" particular solution. This is because there are other possibilities out there for the particular solution we've just managed to find one of them. Any of them will work when it comes to writing down the general solution to the differential equation.

Speaking of which... This section is devoted to finding particular solutions and most of the examples will be finding only the particular solution. However, we should do at least one full blown IVP to make sure that we can say that we've done one.

Example: Solve the following IVP

$$y''-4y'-12y=3e^{5t} \quad y(0)=\frac{18}{7} \quad y'(0)=-\frac{1}{7}.$$

Solution: We know that the general solution will be of the form,

$$y(t)=y_c(y)+Y_P(t)$$

and we already have both the complimentary and particular solution from the first example so we don't really need to do any extra work for this problem.

One of the more common mistakes in these problems is.to find the complimentary solution and then, because we're probably in the habit of doing it, apply the initial conditions to the complimentary solution to find the constants. This however, is incorrect. The complimentary solution is only the solution to the homogeneous differential equation and we are after a solution to the nonhomogeneous differential equation and the initial conditions must satisfy that solution instead of the complimentary solution.

So, we need the general solution to the nonhomogeneous differential equation. Taking the complimentary solution and the particular solution that we found in the previous example we get the following for a general solution and its derivative.

$$y(t)=c_1e^{-2t}+c_2e^{6t}-\frac{3}{7}e^{5t}$$

$$y'(t)=-2c_1e^{-2t}+6c_2e^{6t}-\frac{15}{7}e^{5t}.$$

Now, apply the initial conditions to these.

$$\frac{18}{7}=y(0)=c_1+c_2-\frac{3}{7}$$

$$-\frac{1}{7}=y'(0)=-2c_1+6c_2-\frac{15}{7}.$$

Solving this system gives $c_1 = 2$ and $c_2 = 1$. The actual solution is then.

$$y(t)=2e^{-2t}+e^{6t}-\frac{3}{7}e^{5t}.$$

This will be the only IVP in this section so don't forget how these are done for nonhomogeneous differential equations!

Let's take a look at another example that will give the second type of $g(t)$ for which undetermined coefficients will work.

Example: Find a particular solution for the following differential equation.

$$y''-4y'-12y=\sin(2t).$$

Solution:

1. Again, let's note that we should probably find the complimentary solution before we proceed onto the guess for a particular solution. However, because the homogeneous differential equation for this example is the same as that for the first example we won't bother with that here.
2. Now, let's take our experience from the first example and apply that here. The first example had an exponential function in the $g(t)$ and our guess was an exponential. This differential equation has a sine so let's try the following guess for the particular solution.

$$Y_P(t)=A\sin(2t).$$

Differentiating and plugging into the differential equation gives,

$$-4A\sin(2t)-4(2A\cos(2t))-12(A\sin(2t))=\sin(2t).$$

Collecting like terms yields

$$-16A\sin(2t)-8A\cos(2t)=\sin(2t).$$

We need to pick A so that we get the same function on both sides of the equal sign. This means that the coefficients of the sines and cosines must be equal. Or,

$$\cos(2t):\qquad -8A=0\quad\Rightarrow\quad A=0$$

$$\sin(2t):\qquad -16A=0\quad\Rightarrow\quad A=-\frac{1}{16}.$$

Notice two things. First, since there is no cosine on the right hand side this means that the coefficient must be zero on that side.

More importantly we have a serious problem here. In order for the cosine to drop out, as it must in order for the guess to satisfy the differential equation, we need to set $A = 0$, but if $A = 0$, the sine will also drop out and that can't happen. Likewise, choosing A to keep the sine around will also keep the cosine around.

What this means is that our initial guess was wrong. If we get multiple values of the same constant or are unable to find the value of a constant then we have guessed wrong.

One of the nicer aspects of this method is that when we guess wrong our work will often suggest a fix. In this case the problem was the cosine that cropped up. So, to counter this let's add a cosine to our guess. Our new guess is

$$Y_P(t)=A\cos(2t)+B\sin(2t).$$

Plugging this into the differential equation and collecting like terms gives,

$$-4A\cos(2t) - 4B\sin(2t) - 4(-2A\sin(2t)) + 2B\cos(2t)) - 12(A\cos(2t) + B\sin(2t)) = \sin(2t)$$

$$(-4A - 8B - 12A)\cos(2t) + (-4B + 8A - 12B)\sin(2t) = \sin(2t)$$

$$(-16A - 8B)\cos(2t) + (8A - 16B)\sin(2t) = \sin(2t).$$

Now, set the coefficients equal

$$\cos(2t): \quad -16\text{A} - 8B = 0$$

$$\sin(2t): \quad 8\text{A} - 16B = 1.$$

Solving this system gives us

$$A = \frac{1}{40} \qquad B = -\frac{1}{40}.$$

We found constants and this time we guessed correctly. A particular solution to the differential equation is then,

$$Y_P(t) = \frac{1}{40}\cos(2t) - \frac{1}{20}\sin(2r).$$

Notice that if we had had a cosine instead of a sine in the last example then our guess would have been the same. In fact, if both a sine and a cosine had shown up we will see that the same guess will also work.

Let's take a look at the third and final type of basic *g(t)* that we can have. There are other types of *g(t)* that we can have, but as we will see they will all come back to two types that we've already done as well as the next one.

Example: Find a particular solution for the following differential equation.

$$y'' - 4y' - 12y = 2t^3 - t + 3.$$

Solution: Once, again we will generally want the complimentary solution in hand first, but again we're working with the same homogeneous differential equation so we'll again just refer to the first example. For this example *g(t)* is a cubic polynomial. For this we will need the following guess for the particular solution.

$$Y_P(t) = At^3 + Bt^3 + Ct + D.$$

Notice that even though *g(t)* doesn't have a t^2 in it our guess will still need one! So, differentiate and plug into the differential equation.

$$6At + 2B - 4(3At^2 + 2Bt + C) - 12(At^3 + Bt^2 + Ct + D) = 2t^3 - t + 3$$

$$-12At^3 + (-12A - 12B)t^2 + (6A - 8B - 12C)t + 2B - 4C - 12D = 2t^3 - t + 3.$$

Now, as we've done in the previous examples we will need the coefficients of the terms on both sides of the equal sign to be the same so set coefficients equal and solve.

$$t^3: \quad -12A = 2 \quad \Rightarrow \quad A = -\frac{1}{6}$$

$$t^2: \quad -12A - 12B = 0 \quad \Rightarrow \quad B = -\frac{1}{6}$$

$$t^1: \quad 6A - 8B - 12C = -1 \quad \Rightarrow \quad C = -\frac{1}{9}$$

$$t^0: \quad 2B - 4C - 12D = 3 \quad \Rightarrow \quad D = -\frac{5}{27}.$$

Notice that in this case it was very easy to solve for the constants. The first equation gave A. Then once we knew A the second equation gave B, *etc*. A particular solution for this differential equation is then

$$Y_P(t) = -\frac{1}{6}t^3 + \frac{1}{6}t^2 - \frac{1}{9}t - \frac{5}{27}.$$

Now that we've gone over the three basic kinds of functions that we can use undetermined coefficients on let's summarize.

$g(t)$	$Y_P(t)$ guess
$ae^{\beta t}$	$Ae^{\beta t}$
$a\cos(\beta t)$	$A\cos(\beta t) + B\sin(\beta t)$
$b\sin(\beta t)$	$A\cos(\beta t) + B\sin(\beta t)$
$a\cos(\beta t) + b\sin(\beta t)$	$A\cos(\beta t) + B\sin(\beta t)$
n^{th} degree polynomial	$A_n t^n + A_{n-1}t^{n-1} + \cdots A_1 t + A_0$.

Notice that there are really only three kinds of functions given above. If you think about it the single cosine and single sine functions are really special cases of the case where both the sine and cosine are present. Also, we have not yet justified the guess for the case where both a sine and a cosine show up. We will justify this later.

We now need move on to some more complicated functions. The more complicated functions arise by taking products and sums of the basic kinds of functions. Let's first look at products.

Example: Find a particular solution for the following differential equation.

$$y'' - 4y' - 12y = te^{4t}.$$

Solution: You're probably getting tired of the opening comment, but again find the complimentary solution first really a good idea but again we've already done the work in the first example so we won't do it again here. We promise that eventually you'll see why we keep using the same homogeneous problem and why we say it's a good idea to have the complimentary solution in hand first. At this point all we're trying to do is reinforce the habit of finding the complimentary solution first. Okay, let's start off by writing down the guesses for the individual pieces of the function. The guess for the t would be

$$At + B$$

while the guess for the exponential would be

$$Ce^{4t}.$$

Now, since we've got a product of two functions it seems like taking a product of the guesses for the individual pieces might work. Doing this would give

$$Ce^{4t}(At + B).$$

However, we will have problems with this. As we will see, when we plug our guess into the differential equation we will only get two equations out of this. The problem is that with this guess we've got three unknown constants. With only two equations we won't be able to solve for all the constants. This is easy to fix however. Let's notice that we could do the following

$$Ce^{4t}(At + B) = e^{4t}(ACt + BC).$$

If we multiply the C through, we can see that the guess can be written in such a way that there are really only two constants. So, we will use the following for our guess.

$$Y_P(t) = e^{4t}(At + B).$$

Notice that this is nothing more than the guess for the t with an exponential tacked on for good measure. Now that we've got our guess, let's differentiate, plug into the differential equation and collect like terms.

$$e^{4t}(16At + 16B + 8A) - 4(e^{4t}(4At + 4B + A)) - 12(e^{4t}(At + B)) = te^{4t}$$

$$(16A - 16A - 12A)te^{4t} + (16B + 8A - 16B - 4A - 12B)e^{4t} = te^{4t}$$

$$-12Ate^{4t} + (4A - 12B)e^{4t} = te^{4t}.$$

Note that when we're collecting like terms we want the coefficient of each term to have only constants in it. Following this rule we will get two terms when we collect like terms. Now, set coefficients equal.

$$te^{4t}: \quad -12A = 1 \quad \Rightarrow \quad A = -\frac{1}{12}$$

$$e^{4t}: \quad 4A - 12B = 0 \quad \Rightarrow \quad B = -\frac{1}{36}.$$

A particular solution for this differential equation is then

$$Y_P(t) = e^{4t}\left(-\frac{t}{12}-\frac{1}{36}\right) = -\frac{1}{36}(3t+1)\,e^{4t}.$$

This last example illustrated the general rule that we will follow when products involve an exponential. When a product involves an exponential we will first strip out the exponential and write down the guess for the portion of the function without the exponential, then we will go back and tack on the exponential without any leading coefficient.

Let's take a look at some more products. In the interest of brevity we will just write down the guess for a particular solution and not go through all the details of finding the constants. Also, because we aren't going to give an actual differential equation we can't deal with finding the complimentary solution first.

Example: Write down the form of the particular solution to

$$y'' + p(t)y' + q(t)y = g(t)$$

for the following *g(t)*'s.

1. $g(t) = 16e^{7t}\sin(10t)$
2. $g(t) = (9t^2 - 103t)\cos t$
3. $g(t) = -e^{-2t}(3-5t)\cos(9t).$

Solution:

1. $g(t) = 16e^{-7t}\sin(10t).$

So, we have an exponential in the function. Remember the rule. We will ignore the exponential and write down a guess for 16 sin(10*t*) then put the exponential back in.

The guess for the sine is

$$A\cos(10t) + B\sin(10t).$$

Now, for the actual guess for the particular solution we'll take the above guess and tack an exponential onto it. This gives,

$$Y_P(t) = e^{7t}(A\cos(10t) + B\sin(10t)).$$

One final note before we move onto the next part. The 16 in front of the function has absolutely no bearing on our guess. Any constants multiplying the whole function are ignored.

2. $g(t) = (9t^2 - 103t)\cos t.$

We will start this one the same way that we initially started the previous example. The guess for the polynomial is

$$At^2 + Bt + C$$

and the guess for the cosine is

$$D\cos t + E\sin t.$$

If we multiply the two guesses we get.

$$(At^2 + Bt + C)(D\cos t + E\sin t).$$

Let's simplify things up a little. First multiply the polynomial through as follows.

$$(At^2 + Bt + C)(D\cos t) + (At^2 + Bt + C)(E\sin t)$$

$$(ADt^2 + BDt + CD)\cos t + (AEt^2 + BEt + CE)\sin t.$$

Notice that everywhere one of the unknown constants occurs it is in a product of unknown constants. This means that if we went through and used this as our guess the system of equations that we would need to solve for the unknown constants would have products of the unknowns in them. These types of systems are generally very difficult to solve.

So, to avoid this we will do the same thing that we did in the previous example. Everywhere we see a product of constants we will rename it and call it a single constant. The guess that we'll use for this function will be.

$$Y_P(t) = (At^2 + Bt + C)\cos t + (Dt^2 + Et + F)\sin t.$$

This is a general rule that we will use when faced with a product of a polynomial and a trig function. We write down the guess for the polynomial and then multiply that by a cosine. We then write down the guess for the polynomial again, using different coefficients, and multiply this by a sine.

3. $g(t) = -e^{-2t}(3 - 5t)\cos(9t)$.

This final part has all three parts to it. First we will ignore the exponential and write down a guess for.

$$-(3 - 5t)\cos(9t).$$

The minus sign can also be ignored. The guess for this is

$$(At + B)\cos(9t) + (Ct + D)\sin(9t).$$

Now, tack an exponential back on and we're done.

$$Y_P(t) = e^{-2t}(At + B)\cos(9t) + e^{-2t}(Ct + D)\sin(9t).$$

Notice that we put the exponential on both terms.

There a couple of general rules that you need to remember for products.

If $g(t)$ contains an exponential, ignore it and write down the guess for the remainder. Then tack the exponential back on without any leading coefficient.

1. For products of polynomials and trig functions you first write down the guess for just the polynomial and multiply that by the appropriate cosine. Then add on a new guess for the polynomial with different coefficients and multiply that by the appropriate sine.

If you can remember these two rules you can't go wrong with products. Writing down the guesses for products is usually not that difficult. The difficulty arises when you need to actually find the constants. Now, let's take a look at sums of the basic components and/or products of the basic components. To do this we'll need the following fact.

FACT

If $Y_{P1}(t)$ is a particular solution for

$$y'' + p(t)y' + q(t)y = g_1(t)$$

and if $Y_{P2}(t)$ is a particular solution for

$$y'' + p(t)y' + q(t)y = g_2(t)$$

then $Y_{P1}(t) + Y_{P2}(t)$ is a particular solution for

$$y'' + p(t)y' + q(t)y = g_1(t) + g_2(t).$$

This fact can be used to both find particular solutions to differential equations that have sums in then and to write down guess for functions that have sums in them.

Example: Find a particular solution for the following differential equation.

$$y'' - 4y' - 12y = 3e^{5t} + \sin(2t) + te^{4t}.$$

Solution: This example is the reason that we've been using the same homogeneous differential equation for all the previous examples. There is nothing to do with this problem. All that we need to do it go back to the appropriate examples above and get the particular solution from that example and add them all together.

Doing this gives

$$Y_P(t) = -\frac{3}{7}e^{5t} + \frac{1}{40}\cos(2t) - \frac{1}{20}\sin(2t) - \frac{1}{36}(3t+1)\,e^{4t}.$$

Let's take a look at a couple of other examples. As with the products we'll just get guesses here and not worry about actually finding the coefficients.

Example: Write down the form of the particular solution to

$$y'' + p(t)y' + q(t)y = g(t)$$

for the following $g(t)$'s.

1. $g(t) = 4\cos(6t) - 9\sin(6t)$
2. $g(t) = -2\cos t + \sin(14t) - 5\sin(14t)$
3. $g(t) = e^{7t} + 6$
4. $g(t) = 6t^2 - 7\sin(3t) + 9$
5. $g(t) = 10e^{t} - 5te^{-8t} + 2e^{-8t}$
6. $g(t) = t^2\cos t - 5t\sin t$
7. $g(t) = 5e^{-3t} + e^{-3t}\cos(6t) - \sin(6t)$.

Solution:

1. $g(t) = 4\cos(6t) - 9\sin(6t)$.

This first one we've actually already told you how to do. This is in the table of the basic functions. However we wanted to justify the guess that we put down there. Using the fact on sums of function we would be tempted to write down a guess for the cosine and a guess for the sine. This would give.

$$\underbrace{A\cos(6t) + B\sin(6t)}_{\text{guess for the cosine}} + \underbrace{C\cos(6t) + D\sin(6t)}_{\text{guess for the sine}}.$$

So, we would get a cosine from each guess and a sine from each guess. The problem with this as a guess is that we are only going to get two equations to solve after plugging into the differential equation and yet we have 4 unknowns. We will never be able to solve for each of the constants.

To fix this notice that we can combine some terms as follows.

$$(A + C)\cos(6t) + (B + D)\sin(6t).$$

Upon doing this we can see that we've really got a single cosine with a coefficient and a single sine with a coefficient and so we may as well just use

$$Y_P(t) = A\cos(6t) + B\sin(6t).$$

The general rule of thumb for writing down guesses for functions that involve sums is to always combine like terms into single terms with single coefficients. This will greatly simplify the work required to find the coefficients.

2. $g(t) = -2\sin t + \sin(14t) - 5\cos(14t)$.

For this one we will get two sets of sines and cosines. This will arise because we have two different arguments in them. We will get on set for the sine with just a t as its argument and we'll get another set for the sine and cosine with the $14t$ as their arguments.

The guess for this function is

$$Y_P(t) = A\cos t + B\sin + C\cos(14t) + D\sin(14t)$$

3. $g(t) = e^{7t} + 6.$

The main point of this problem is dealing with the constant. But that isn't too bad. We just wanted to make sure that an example of that is somewhere in the notes. If you recall that a constant is nothing more than a zeroth degree polynomial the guess becomes clear.

The guess for this function is

$$Y_P(t) = Ae^{7t} + B.$$

4. $g(t) = 6t^2 - 7\sin(3t) + 9.$

This one can be a little tricky if you aren't paying attention. Let's first rewrite the function

$$g(t) = 6t^2 - 7\sin(3t) + 9 \text{ as}$$

$$g(t) = 6t^2 + 9 - 7\sin(3t).$$

All we did was move the 9. However upon doing that we see that the function is really a sum of a quadratic polynomial and a sine. The guess for this is then

$$Y_P = At^2 + Bt + C + D\cos(3t) + E\sin(3t).$$

If we don't do this and treat the function as the sum of three terms we would get

$$At^2 + Bt + C + D\cos(3t) + E\sin(3t) + G.$$

and as with the first part in this example we would end up with two terms that are essentially the same (the C and the G) and so would need to be combined. An added step that isn't really necessary if we first rewrite the function.

Look for problems where rearranging the function can simplify the initial guess.

5. $g(t) = 10e^t - 5te^{-8t} + 2e^{-8t}.$

So, this look like we've got a sum of three terms here. Let's write down a guess for that.

$$Ae^t + (Bt + C)e^{-8t} + De^{-8t}.$$

Notice however that if we were to multiply the exponential in the second term through we would end up with two terms that are essentially the same and would need to be combined. This is a case where the guess for one term is completely contained in the guess for a different term. When this happens we just drop the guess that's already included in the other term.

So, the guess here is actually.

$$Y_P(t) = Ae^t + (Bt + C)e^{-8t}.$$

Notice that this arose because we had two terms in our $g(t)$ whose only difference was the polynomial that sat in front of them. When this happens we look at the term that contains the largest degree polynomial, write down the guess for that and don't bother writing down the guess for the other term as that guess will be completely contained in the first guess.

6. $g(t) = t^2 \cos t - 5t \sin t.$

In this case we've got two terms whose guess without the polynomials in front of them would be the same. Therefore, we will take the one with the largest degree polynomial in front of it and write down the guess for that one and ignore the other term. So, the guess for the function is

$$Y_P(t) = (Ae^2 + Bt + C)\cos t + (Dt^2 + Et + F)\sin t.$$

7. $g(t) = 5e^{-3t} + e^{-3t} \cos(6t) - \sin(6t).$

This last part is designed to make sure you understand the general rule that we used in the last two parts. This time there really are three terms and we will need a guess for each term. The guess here is

$$Y_P(t) = Ae^{-3t} + e^{-3t}(B\cos(6t) + C\sin(6t)) + D\cos(6t) + E\sin(6t).$$

We can only combine guesses if they are identical up to the constant. So we can't combine the first exponential with the second because the second is really multiplied by a cosine and a sine and so the two exponentials are in fact different functions. Likewise, the last sine and cosine can't be combined with those in the middle term because the sine and cosine in the middle term are in fact multiplied by an exponential and so are different.

So, when dealing with sums of functions make sure that you look for identical guesses that may or may not be contained in other guesses and combine them. This will simplify your work later on.

We have one last topic in this section that needs to be dealt with. In the first few examples we were constantly harping on the usefulness of having the complimentary solution in hand before making the guess for a particular solution. We never gave any reason for this other that "trust us". It is now time to see why having the complimentary solution in hand first it useful. This is best shown with an example so let's jump into one.

Example: Find a particular solution for the following differential equation.

$$y'' - 4y' - 12y = e^{6t}.$$

Solution: This problem seems almost too simple to be given this late in the section.

This is especially true given the ease of finding a particular solution for $g(t)$'s that are just exponential functions. Also, because the point of this example is to illustrate why it is generally a good idea to have the complimentary solution in hand first we'll let's go ahead and recall the complimentary solution first. Here it is,

$$y_c(t) = c_1 e^{-2t} + c_2 e^{6t}.$$

Now, without worrying about the complimentary solution for a couple more seconds let's go ahead and get to work on the particular solution. There is not much to the guess here. From our previous work we know that the guess for the particular solution should be,

$$Y_P(y) = Ae^{6t}.$$

Plugging this into the differential equation gives,

$$\begin{aligned} 36Ae^{6t} - 24Ae^{6t} - 12Ae^{6t} &= e^{6t} \\ 0 &= e^{6t}. \end{aligned}$$

Clearly an exponential can't be zero. So, what went wrong? We finally need the complimentary solution. Notice that the second term in the complimentary solution (listed above) is exactly our guess for the form of the particular solution and now recall that both portions of the complimentary solution are solutions to the homogeneous differential equation,

$$y'' - 4y' - 12y = 0.$$

In other words, we had better have gotten zero by plugging our guess into the differential equation, it is a solution to the homogeneous differential equation!

So, how do we fix this? The way that we fix this is to add a t to our guess as follows.

$$Y_P(t) = Ate^{6t}.$$

Plugging this into our differential equation gives,

$$\begin{aligned} (12Ae^{6t} + 36Ate^{6t}) - 4(Ae^{6t} + 6Ate^{6t}) - 12Ate^{6t} &= e^{6t} \\ (36A - 24A - 12A)te^{6t} + (12A - 4A)e^{6t} &= e^{6t} \\ 8Ae^{6t} &= e^{6t}. \end{aligned}$$

Now, we can set coefficients equal.

$$8A = 1 \quad \Rightarrow \quad A = \frac{1}{8}.$$

So, the particular solution in this case is,

$$Y_P(t) = \frac{t}{8}e^{6t}.$$

So, what did we learn from this last example. While technically we don't need the complimentary solution to do undetermined coefficients, you can go through a lot of work only to figure out at the end that you needed to add in a t to the guess because it appeared in the complimentary solution. This work is avoidable if we first find the complimentary solution and comparing our guess to the complimentary solution and seeing if any portion of your guess shows up in the complimentary solution.

If a portion of your guess does show up in the complimentary solution then we'll need to modify that portion of the guess by adding in a t to the portion of the guess that is causing the problems. We do need to be a little careful and make sure that we add the t in the correct place however. The following set of examples will show you how to do this.

Example: Write down the guess for the particular solution to the given differential equation. Do not find the coefficients.

1. $y''+3y'-28y=7t+e^{-7t}-1$

4. $y''-100y=9t^2e^{10t}+\cos t-t\sin t$

3. $4y''-100y=e^{-2t}\sin\left(\frac{t}{2}\right)+6t\cos\left(\frac{t}{2}\right)$

4. $4y''+16y'+17y=e^{-2t}\sin\left(\frac{t}{2}\right)+6t\cos\left(\frac{t}{2}\right)$

5. $y''+8y'+16y=e^{-4t}+(t^2+5)e^{-4t}$.

Solution: In these solutions we'll leave the details of checking the complimentary solution to you.

1. $y''+3y'-28y=7t+e^{-7t}-1$.

The complimentary solution is

$$y_c(t)=c_1e^{4t}+c_2e^{-7t}.$$

Remembering to put the "-1" with the $7t$ gives a first guess for the particular solution.-

$$Y_P(t)=At+B+Ce^{-7t}.$$

Notice that the last term in the guess is the last term in the complimentary solution. The first two terms however aren't a problem and don't appear in the complimentary solution. Therefore, we will only add a t onto the last term. The correct guess for the form of the particular solution is.

$$Y_P(t)=At+B+Cte^{-7t}$$

(b) $y''-100y = 9t^2e^{10t} + \cos t - t\sin t.$

The complimentary solution is

$$y_c(t) = c_1e^{10t} + c_2e^{-10t}.$$

A first guess for the particular solution is

$$Y_P(t) = (At^2 + Bt + C)e^{10t} + (Et + F)\cos t + (Gt + H)\sin t.$$

Notice that if we multiplied the exponential term through the parenthesis that we would end up getting part of the complimentary solution showing up. Since the problem part arises from the first term the *whole* first term will get multiplied by t. The second and third terms are okay as they are.

The correct guess for the form of the particular solution in this case is.

$$Y_P(t) = t(At^2 + Bt + C)\,e^{10t} + (Et + F)\cos t + (Gt + H)\sin t.$$

So, in general, if you were to multiply out a guess and any part of a that shows up in the complimentary solution, then the whole term will get a t not just the problem portion of the term.

3. $4y''+ y = e^{-2t}\sin\left(\frac{t}{2}\right) + 6t\cos\left(\frac{t}{2}\right).$

The complimentary solution is

$$y_c(t) = c_1\cos\left(\frac{t}{2}\right) + c_2\sin\left(\frac{t}{2}\right).$$

A first guess for the particular solution is

$$Y_P(t) = e^{-2t}\left(A\cos\left(\frac{t}{2}\right) + B\sin\left(\frac{t}{2}\right)\right) + (Cr + D)\cos\left(\frac{t}{2}\right) + (Et + F)\sin\left(\frac{t}{2}\right).$$

In this case both the second and third terms contain portions of the complimentary solution. The first term doesn't however, since upon multiplying out, both the sine and the cosine would have an exponential with them and that isn't part of the complimentary solution. We only need to worry about terms showing up in the complimentary solution if the only difference between the complimentary solution term and the particular guess term is the constant in front of them.

So, in this case the second and third terms will get a t while the first won't

The correct guess for the form of the particular solution is.

$$Y_P(t) = e^{-2t}\left(A\cos\left(\frac{t}{2}\right) + B\sin\left(\frac{t}{2}\right)\right) + t(Ct + D)\cos\left(\frac{t}{2}\right) + t(Et + F)\sin\left(\frac{t}{2}\right).$$

4. $4y''+16y'+17y = e^{-2t}\sin\left(\frac{t}{2}\right) + 6t\cos\left(\frac{t}{2}\right).$

To get this problem we changed the differential equation from the last example and left the *g(t)* alone. The complimentary solution this time is

$$y_c(t) = c_1 e^{-2t}\cos\left(\frac{t}{2}\right) + c_2 e^{-2t}\sin\left(\frac{t}{2}\right).$$

As with the last part, a first guess for the particular solution is

$$Y_P(t) = e^{-2t}\left(A\cos\left(\frac{t}{2}\right) + B\sin\left(\frac{t}{2}\right)\right) + t(Ct+D)\cos\left(\frac{t}{2}\right) + (Et+F)\sin\left(\frac{t}{2}\right).$$

This time however it is the first term that causes problems and not the second or third. In fact, the first term is exactly the complimentary solution and so it will need a *t*. Recall that we will only have a problem with a term in our guess if it only differs from the complimentary solution by a constant. The second and third terms in our guess don't have the exponential in them and so they don't differ from the complimentary solution by only a constant.

The correct guess for the form of the particular solution is.

$$Y_P(t) = e^{-2t}\left(A\cos\left(\frac{t}{2}\right) + B\sin\left(\frac{t}{2}\right)\right) + (Ct+D)\cos\left(\frac{t}{2}\right) + (Et+F)\sin\left(\frac{t}{2}\right).$$

5. $y''+8y'+16y = e^{-4t} + (t^2+5)e^{-4t}.$

The complimentary solution is

$$y_c(t) = c_1 e^{-4t} + c_2 t e^{-4t}.$$

The two terms in *g(t)* are identical with the exception of a polynomial in front of them. So this means that we only need to look at the term with the highest degree polynomial in front of it. A first guess for the particular solution is

$$Y_P(t) = (Ar^2 + Bt + C)e^{-4t}.$$

Notice that if we multiplied the exponential term through the parenthesis the last two terms would be the complimentary solution. Therefore, we will need to multiply this whole thing by a *t*.

The next guess for the particular solution is then.

$$Y_P(t) = t(Ar^2 + Bt + C)e^{-4t}.$$

This still causes problems however. If we multiplied the *t* and the exponential through, the last term will still be in the complimentary solution. In this case, unlike the previous ones, a *t* wasn't sufficient to fix the problem. So, we will add in another *t* to our guess.

The correct guess for the form of the particular solution is.

$$Y_P(t) = t^2(Ar^2 + Bt + C)e^{-4t}.$$

Upon multiplying this out none of the terms are in the complimentary solution and so it will be okay.

As this last set of examples has shown, we really should have the complimentary solution in hand before even writing down the first guess for the particular solution. By doing this we can compare our guess to the complimentary solution and if any of the terms from your particular solution show up we will know that we'll have problems. Once the problem is identified we can add a t to the problem term(s) and compare our new guess to the complimentary solution. If there are no problems we can proceed with the problem, if there are problems add in another t and compare again.

Can you see a general rule as to when a t will be needed and when a t^2 will be needed for second order differential equations?

VARIATION OF PARAMETERS

In the last section we looked at the method of undetermined coefficients for finding a particular solution to

$$p(t)y'' + q(t)y' + r(t)y = g(t)$$

and we saw that while it reduced things down to just an algebra problem, the algebra could become quite messy. On top of that undetermined coefficients will only work for a fairly small class of functions.

The method of Variation of Parameters is a much more general method that can be used in many more cases. However, there are two disadvantages to the method. First, the complimentary solution is absolutely required to do the problem. This is in contrast to the method of undetermined coefficients where it was advisable to have the complimentary solution on hand, but was not required. Second, as we will see, in order to complete the method we will be doing a couple of integrals and there is no guarantee that we will be able to do the integrals. So, while it will always be possible to write down a formula to get the particular solution, we may not be able to actually find it if the integrals are too difficult or if we are unable to find the complimentary solution.

We're going to derive the formula for variation of parameters. We'll start off by acknowledging that the complimentary solution to the equation is

$$y_c(t) = c_1 y_1(t) + c_2 y_2(t).$$

Remember as well that this is the general solution to the homogeneous differential equation.

$$p(t)y'' + q(t)y' + r(t)y = 0.$$

Also recall that in order to write down the complimentary solution we know that $y_1(t)$ and $y_2(t)$ are a fundamental set of solutions. What we're going to do is see if we can find a pair of functions, $u_1(t)$ and $u_2(t)$ so that

$$Y_P(t) = u_1(t)y_1(t) + u_2(t)y_2(t)$$

will be a solution. We have two unknowns here and so we'll need two equations eventually. One equation is easy. Our proposed solution must satisfy the differential equation, so we'll get the first equation by plugging our proposed solution. The second equation can come from a variety of places. We are going to get our second equation simply by making an assumption that will make our work easier. We'll say more about this shortly.

So, let's start. If we're going to plug our proposed solution into the differential equation we're going to need some derivatives so let's get those. The first derivative is

$$Y_P'(t) = u_1'y_1 + u_1y_1' + u_2'y_2 + u_2y_2'.$$

Here's the assumption. Simply to make the first derivative easier to deal with we are going to assume that whatever $u_1(t)$ and $u_2(t)$ are they will satisfy the following.

$$u_1'y_1 + u_2'y_2 = 0.$$

Now, there is no reason ahead of time to believe that this can be done. However, we will see that this will work out. We simply make this assumption on the hope that it won't cause problems down the road and to make the first derivative easier so don't get excited about it. With this assumption the first derivative becomes.

$$Y_P'(t) = u_1y_1' + u_2y_2'.$$

The second derivative is then,

$$Y_p''(t) = u_1'y_1' + u_1y_1' + u_1y_1'' + u_2'y_2' + u_2y_2''.$$

Plug the solution and its derivatives into (1).

$$p(t)(u_1'y_1' + u_1y_1'' + u_2'y_2' + u_2y_1'') + q(t)(u_1y_1' + u_2y_2') + r(t)(u_1y_1 + u_2y_2) = g(t).$$

Rearranging a little gives the following.

$$p(t)(u_1'y_1' + u_2'y_2') + u_1(t)(p(t)y_1'' + q(t)y_1' + r(t)y_1) + u_2(t)(p(t)y_2''$$
$$+r(t)y_2) = g(t).$$

Now, both $y_1(t)$ and $y_2(t)$ are solutions and so the second and third terms are zero. Acknowledging this and rearranging a little gives us,

$$p(t)(u_1'y_1' + u_2'y_2') + u_1(t)(0) + u_2(t)(0) = g(t)$$

$$u_1'y_1' + u_2'y_2' = \frac{g(t)}{p(t)}.$$

We've almost got the two equations that we need. Before proceeding we're going to go back and make a further assumption. The last equation, is actually the one that we want, however, in order to make things simpler for us we are going to assume that the function *p(t)* = 1. In other words, we are going to go back and start working with the differential equation,

$$y''+q(t)y'+r(t)y=g(t).$$

If the coefficient of the second derivative isn't one divide it out so that it becomes a one. The formula that we're going to be getting will assume this! Upon doing this the two equations that we want so solve for the unknown functions are

$$u_1'y_1+u_2'y_2=0$$

$$u_1'y_1'+u_2'y_2'=g(t).$$

Note that in this system we know the two solutions and so the only two unknowns here are u_1' and u_1'. Solving this system is actually quite simple. First, solve for $u_1'\ u_1'\ u_1'$ and plug this and do some simplification.

$$u_1'=-\frac{u_2'y_2}{y_1}$$

$$\left(-\frac{u_2'y_2}{y_1}\right)y_1'+u_2'y_2'=g(t)$$

$$u_2'\left(y_2'-\frac{y_2y_1'}{y_1}\right)=g(t)$$

$$u_2'\left(\frac{y_1y_2'-y_2y_1'}{y_1}\right)=g(t)$$

$$u_2'=\frac{y_1g(t)}{y_1y_2'-y_2y_1'}.$$

So, we now have an expression for u_2'. Plugging this will give us an expression for u_1'.

$$u_1'=-\frac{y_2g(t)}{y_1y_2'-y_2y_1'}.$$

Next, let's notice that

$$W(y_1,y_2)=y_1y_2'-y_2y_1'\neq 0.$$

Recall that $y_1(t)$ and $y_2(t)$ are a fundamental set of solutions and so we know that the

Wronskian won't be zero! Finally, all that we need to do is integrate in order to determine what $u_1(t)$ and $u_2(t)$ are. Doing this gives,

$$u_1(t) = -\int \frac{y_2 g(t)}{W(y_1, y_2)} dt \qquad u_2(t) = \int \frac{y_1 g(t)}{W(y_1, y_2)} dt.$$

So, provided we can do these integrals, a particular solution to the differential equation is

$$\begin{aligned} Y_P(t) &= y_1 u_1 + y_2 u_2 \\ &= -y_1 \int \frac{y_2 g(t)}{W(y_1, y_2)} dt + y_2 \int \frac{y_1 g(t)}{W(y_1, y_2)} dt. \end{aligned}$$

So, let's summarize up what we've determined here.

VARIATION OF PARAMETERS

Consider the differential equation,

$$y'' + q(t)y' + r(t)y = g(t).$$

Assume that $y_1(t)$ and $y_2(t)$ are a fundamental set of solutions for

$$y' + q(t)y' + r(t)y = 0.$$

Then a particular solution to the nonhomogeneous differential equation is,

$$Y_P(t) = -y_1 \int \frac{y_2 g(t)}{W(y_1, y_2)} dt + y_2 \int \frac{y_1 g(t)}{W(y_1, y_2)} dt.$$

Depending on the person and the problem, some will find the formula easier to memorize and use, while others will find the process used to get the formula easier. The examples in this section will be done using the formula.

Before proceeding with a couple of examples let's first address the issues involving the constants of integration that will arise out of the integrals. Putting in the constants of integration will give the following.

$$\begin{aligned} Y_P(t) &= -y_1 \left(\int \frac{y_2 g(t)}{W(y_1, y_2)} dt + c \right) + y_2 \left(\int \frac{y_1 g(t)}{W(y_1, y_2)} dt + k \right). \\ &= -y_1 \int \frac{y_2 g(t)}{W(y_1, y_2)} dt + y_2 \int \frac{y_1 g(t)}{W(y_1, y_2)} dt + (-cy_1 + ky_2). \end{aligned}$$

The final quantity in the parenthesis is nothing more than the complimentary solution with $c_1 = -c$ and $c_2 = k$ and we know that if we plug this into the differential equation it will simplify out to zero since it is the solution to the homogeneous differential equation. In other words, these terms add nothing to the particular solution and so we will go ahead and assume that $c = 0$ and $k = 0$ in all the examples. One final note before we proceed with examples. Do not worry about which of your two solutions in the complimentary solution is $y_1(t)$ and which one is $y_2(t)$. It doesn't matter. You will get the same answer no matter which one you choose to be $y_1(t)$ and which one you choose to be $y_2(t)$.

Let's work a couple of examples now.

Example: Find a general solution to the following differential equation.

$$2y' + 18y = 6\tan(3t).$$

Solution: First, since the formula for variation of parameters requires a coefficient of a one in front of the second derivative let's take care of that before we forget. The differential equation that we'll actually be solving is

$$y'' + 9y = 3\tan(3t).$$

We'll leave it to you to verify that the complimentary solution for this differential equation is

$$y_c(t) = c_1\cos(3t) + c_2\sin(3t).$$

So, we have

$$y_1(t) = \cos(3t) \qquad y_2(t) = \sin(3t).$$

The Wronskian of these two functions is

$$W\begin{vmatrix} \cos(3t) & \sin(3t) \\ -3\sin(3t) & 3\cos(3t) \end{vmatrix} = 3\cos^2(3t) + 3\sin^2(3t) = 3.$$

The particular solution is then,

$$\begin{aligned}
Y_P(t) &= -\cos(3t)\int \frac{3\sin(3t)\tan(3t)}{3}\,dt + \sin(3t)\int \frac{3\cos(3t)\tan(3t)}{3}\,dt \\
&= -\cos(3t)\int \frac{3\sin^2(3t)}{\cos(3t)}\,dt + \sin(3t)\int \sin(3t)\,dt \\
&= -\cos(3t)\int \frac{1-\cos^2(3t)}{\cos(3t)}\,dt + \sin(3t)\int \sin(3t)\,dt \\
&= -\cos(3t)\int \sec(3t) - \cos(3t)\,dt + \sin(3t)\int \sin(3t)\,dt \\
&= -\frac{\cos(3t)}{3}(\ln|\sec(3t) + \tan(3t)| - \sin(3t)) + \frac{\sin(3t)}{3}(-\cos(3t)) \\
&= -\frac{\cos(3t)}{3}\ln|\sec(3t) + \tan(3t)|.
\end{aligned}$$

The general solution is,

$$y(t) = c_1 \cos(3t) + c_2 \sin(3t) - \frac{\cos(3t)}{3} \ln|\sec(3t) + \tan(3t)|.$$

Example: Find a general solution to the following differential equation.

$$y'' - 2y' + y = \frac{e^t}{t^2 + 1}.$$

Solution: We first need the complimentary solution for this differential equation. We'll leave it to you to verify that the complimentary solution is,

$$y_c(t) = c_1 e^t + c_2 t e^t.$$

So, we have

$$y_1(t) = e^t \qquad y_2(t) = te^t.$$

The Wronskian of these two functions is

$$W = \begin{vmatrix} e^t & te^t \\ e^t & e^t + te^t \end{vmatrix} = e^t(e^t + te^t) - e^t(te^t) = e^{2t}.$$

The particular solution is then,

$$Y_P = -e^t \int \frac{te^t e^t}{e^{2t}(t^2+1)} dt + te^t \int \frac{e^t e^t}{e^{2t}(t^2+1)} dt$$

$$= e^t \int \frac{t}{e^2 + 1} dt + te^t \int \frac{1}{t^2+1} dt$$

$$= \frac{1}{2} e^t \ln(1 + t^2) + te^t \tan^{-1}(t).$$

The general solution is,

$$y(t) = c_1 e^t + c_2 te^t - \frac{1}{2} e^t \ln(1 + t^2) + te^t \tan^{-1}(t).$$

This method can also be used on non-constant coefficient differential equations, provided we know a fundamental set of solutions for the associated homogeneous differential equation.

Example: Find the general solution to

$$ty'' - (t+1)y'' + y = t^2$$

given that

$$y_1(t) = e^t \qquad y_2(t) = t + 1$$

form a fundamental set of solutions for the homogeneous differential equation.

Solution: As with the first example, we first need to divide out by a t.

$$y''-\left(1+\frac{1}{t}\right)y'+\frac{1}{t}y=t.$$

The Wronskian for the fundamental set of solutions is

$$W=\begin{vmatrix} e^t & t+1 \\ e^t & 1 \end{vmatrix}=e^t-e^t(t+1)=-te^t.$$

The particular solution is.

$$\begin{aligned} Y_P &= -e^t\int\frac{(t+1)t}{-te^t}dt+(t+1)\int\frac{e^t(t)}{-te^t}dt \\ &= e^t\int(t+1)e^{-t}dt-(t+1)\int dt \\ &= e^t\left(-e^{-t}(t+2)\right)-(t+1)t \\ &= -t^2-2t-2. \end{aligned}$$

The general solution for this differential equation is.

$$y(t)=c_1e^t+c_2(t+1)-t^2-2t-2.$$

We need to address one more topic about the solution to the previous example. The solution can be simplified down somewhat if we do the following.

$$\begin{aligned} y(t) &= c_1e^t+c_2(t+1)-t^2-2t-2 \\ &= c_1e^t+c_2(t+1)-t^2-2(t+1) \\ &= c_1e^t+(c_2+2)(t+1)-t^2. \end{aligned}$$

Now, since is an unknown constant subtracting 2 from it won't change that fact. So we can just write the c_2-2as $c_2$and be done with it. Here is a simplified version of the solution for this example.

$$y(t)=c_1e^t+c_2(t+1)-t^2.$$

MECHANICAL VIBRATIONS

It's now time to take a look at an application of second order differential equations. We're going to take a look at mechanical vibrations. In particular we are going to look at a mass that is hanging from a spring.

Vibrations can occur in pretty much all branches of engineering and so what we're going to be doing here can be easily adapted to other situations, usually with just a change in notation.

Let's get the situation setup. We are going to start with a spring of length l, called the natural length, and we're going to hook an object with mass m up to it. When the object is attached to the spring the spring will stretch a length of L. We will call the equilibrium position the position of the center of gravity for the object as it hangs on the spring with no movement.

Below is sketch of the spring with and without the object attached to it.

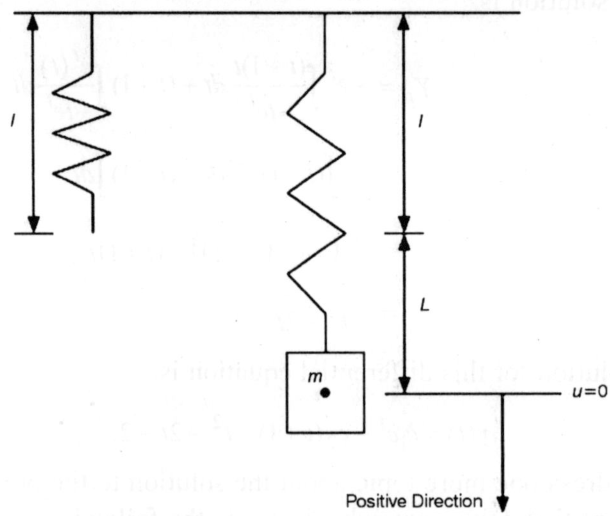

As denoted in the sketch we are going to assume that all forces, velocities, and displacements in the downward direction will be positive. All forces, velocities, and displacements in the upward direction will be negative.

Also, as shown in the sketch above, we will measure all displacement of the mass from its equilibrium position. Therefore, the $u = 0$ position will correspond to the center of gravity for the mass as it hangs on the spring and is at rest (*i.e.* no movement).

Now, we need to develop a differential equation that will give the displacement of the object at any time t. First, recall Newton's Second Law of Motion.

$$ma = F.$$

In this case we will use the second derivative of the displacement, u, for the acceleration and so Newton's Second Law becomes,

$$mu'' = F(t, u, u').$$

We now need to determine all the forces that will act upon the object. There are four forces that we will assume act upon the object. Two that will always act on the object and two that may or may not act upon the object.

Here is a list of the forces that will act upon the object.

Grasvity, F_g

The force due to gravity will always act upon the object of course. This force is

$$F_g = mg.$$

Spring, F_s

We are going to assume that Hooke's Law will govern the force that the spring exerts on the object. This force will always be present as well and is

$$F_s = -k(L+u).$$

Hooke's Law tells us that the force exerted by a spring will be the spring constant, $k > 0$, times the displacement of the spring from its natural length. For our set up the displacement from the springs natural length is $L + u$ and the minus sign is in there to make sure that the force always has the correct direction.

Let's make sure that this force does what we expect it to. If the object is at rest in its equilibrium position the displacement is L and the force is simply $F_s = kL$ which will act in the upward position as it should since the spring has been stretched from its natural length.

If the spring has been stretched further down from the equilibrium position then will be positive and F_s will be negative acting to pull the object back up as it should be.

Next, if the object has been moved up past it's equilibrium point, but not yet to it's natural length then u will be negative, but still less than L and so $L + u$ will be positive and once again F_s will be negative acting to pull the object up.

Finally, if the object has been moved upwards so that the spring is now compressed, then u will be negative and greater than L. Therefore, $L + u$ will be negative and now F_s will be positive acting to push the object down.

So, it looks like this force will act as we expect that it should.

Dampin'g, F_d

The next force that we need to consider is damping. This force may or may not be present for any given problem

Dampers work to counteract any movement. There are several ways to define a damping force. The one that we'll use is the following.

$$F_d = -\gamma u'$$

where, $\gamma > 0$ is the damping coefficient. Let's think for a minute about how this force will act. If the object is moving downward, then the velocity (u' $u'u'$) will be positive and so F_d will be negative and acting to pull the object back up. Likewise, if the object is moving upward, the velocity (u' $u'u'$) will be negative and so F_d will be positive and acting to push the object back down.

In other words, the damping force as we've defined it will always act to counter the current motion of the object and so will act to damp out any motion in the object.

External Forces, *F(t)*

This is the catch all force. If there are any other forces that we decide we want to act on our object we lump them in here and call it good. We typically call *F(t)* the forcing function. Putting all of these together gives us the following for Newton's Second Law.

$$mu'' = mg - k(L+u) - \gamma u' + F(t).$$

Or, upon rewriting, we get,

$$mu'' + \gamma u' + ku = mg - kL + F(t).$$

Now, when the object is at rest in its equilibrium position there are exactly two forces acting on the object, the force due to gravity and the force due to the spring. Also, since the object is at rest (*i.e.* not moving) these two forces must be canceling each other out. This means that we must have,

$$mg = kL.$$

Using this in Newton's Second Law gives us the final version of the differential equation that we'll work with.

$$mu'' + \gamma u' + ku = F(t).$$

Along with this differential equation we will have the following initial conditions.

$u'(0) = u_0$ Initial displacement from the equilibrium position

$u'(0) = u_0^{'}$ Initial velocity.

Note that we'll determine the spring constant, *k*.

FREE, UNDAMPED VIBRATIONS

This is the simplest case that we can consider. Free or unforced vibrations means that *F(t) = 0* and undamped vibrations means that $\gamma = 0$. In this case the differential equation becomes,

$$mu'' + ku = 0.$$

This is easy enough to solve in general. The characteristic equation has the roots,

$$r = \pm i\sqrt{\frac{k}{m}}.$$

This is usually reduced to,

$$r = \pm\, \omega_0\, i$$

where,

$$\omega_0 = \sqrt{\frac{k}{m}}.$$

and ω_0 is called the natural frequency. Recall as well that $m > 0$ and $k > 0$ and so we can guarantee that this quantity will be complex. The solution in this case is then

$$u(t) = c_1 \cos(\omega_0 t) + c_2 \sin(\omega_0 t).$$

We can write (4) in the following form,

$$u(t) = R\cos(\omega_0 t - \delta)$$

where R is the amplitude of the displacement and $\delta$is the phase shift or phase angle of the displacement.

So, assuming that we have c_1 and c_2 how do we determine R and δ? Let's start with equation and use a trig identity to write it as

$$u(t) = R\cos(\delta)\cos(\omega_0 t) + R\sin(\delta)\sin(\omega_0 t).$$

Now, R and $\delta$are constants and so if we compare we can see that

$$c_1 = R\cos\delta \qquad\qquad c_2 = R\cos\delta.$$

We can find R in the following way.

$$c_1^2 + c_2^2 = R^2\cos^2\delta + R^2\sin^2\delta = R^2.$$

Taking the square root of both sides and assuming that R is positive will give

$$R = \sqrt{c_1^2 + c_2^2}.$$

Finding $\delta$is just as easy. We'll start with

$$\frac{c_2}{c_1} = \frac{R\sin\delta}{R\cos\delta} = \tan\delta.$$

Taking the inverse tangent of both sides gives,

$$\delta = \tan^{-1}\left(\frac{c_2}{c_1}\right).$$

Before we work any examples let's talk a little bit about units of mass and the British vs. metric system differences.

Recall that the weight of the object is given by

$$W = mg$$

where m is the mass of the object and g is the gravitational acceleration. For the examples in this problem we'll be using the following values for g.

$$\text{British: } g = 32 \text{ ft/s}^2$$

$$\text{Metric: } g = 9.8 \text{ m/s}^2$$

This is not the standard 32.2 ft/s^2 or 9.81 m/s^2, but using these will make some of the numbers come out a little nicer.

In the metric system the mass of objects is given in kilograms (*kg*) and there is nothing for us to do. However, in the British system we tend to be given the weight of an object in pounds (yes, pounds are the units of weight not mass…) and so we'll need to compute the mass for these problems. At this point we should probably work an example of all this to see how this stuff works.

Example: A 16 lb object stretches a spring $\frac{8}{9}$ ft by itself. There is no damping and no external forces acting on the system. The spring is initially displaced 6 inches upwards from its equilibrium position and given an initial velocity of 1 ft/sec downward. Find the displacement at any time *t*, *u(t)*.

Solution: We first need to set up the IVP for the problem. This requires us to get our hands on m and k.

This is the British system so we'll need to compute the mass.

$$m = \frac{W}{g} = \frac{16}{32} = \frac{1}{2}.$$

Now, let's get k. We can use the fact that $mg = kL$ to find k. Don't forget that we'll need all of our length units the same. We'll use feet for the unit of measurement for this problem.

$$k = \frac{mg}{L} = \frac{16}{8/9} = 18.$$

We can now set up the IVP.

$$\frac{1}{2}u'' + 18u = 0 \qquad u(0) = -\frac{1}{2} \qquad u'(0) = 1.$$

For the initial conditions recall that upward displacement/motion is negative while downward displacement/motion is positive. Also, since we decided to do everything in feet we had to convert the initial displacement to feet.

Now, to solve this we can either go through the characteristic equation or we can just jump straight to the formula that we derived above. We'll do it that way. First, we need the natural frequency,

$$\omega_0 = \sqrt{\frac{18}{\frac{1}{2}}} = \sqrt{36} = 6.$$

The general solution, along with its derivative, is then,

$$u(t) = c_1 \cos(6t) + c_2 \sin(6t)$$

$$u'(t) = -c_1 6c_1 \sin(6t) + 6c_2 \cos(6t).$$

Applying the initial conditions gives

$$-\frac{1}{2} = u(0) = c_1 \qquad c_1 = -\frac{1}{2}$$

$$1 = u'(0) = 6c_2 \cos(6t) \quad c_1 = -\frac{1}{6}.$$

The displacement at any time t is then

$$u(t) = -\frac{1}{2}\cos(6t) + \frac{1}{6}\sin(6t).$$

Now, let's convert this to a single cosine. First let's get the amplitude, R.

$$R = \sqrt{\left(-\frac{1}{2}\right)^2 + \left(\frac{1}{6}\right)^2} = \frac{\sqrt{10}}{6} = 0.52705.$$

You can use either the exact value here or a decimal approximation. Often the decimal approximation will be easier. Now let's get the phase shift.

$$\delta = \tan^{-1}\left(\frac{\frac{1}{6}}{-\frac{1}{2}}\right) = -0.32175.$$

We need to be careful with this part. The phase angle found above is in Quadrant IV, but there is also an angle in Quadrant II that would work as well. We get this second angleπ $\pi\pi$by adding onto the first angle. So, we actually have two angles. They are

$$\delta_1 = -0.32175$$

$$\delta_2 = \delta_1 + \pi = 2.81984.$$

We need to decide which of these phase shifts is correct, because only one will be correct. To do this recall that

$$c_1 = R\cos\delta$$

$$c_2 = R\sin\delta.$$

Now, since we are assuming that R is positive this means that the sign of cosδ will be the same as the sign of c_1 and the sign of sinδ will be the same as the sign of c_2. So, for this particular case we must have $\cos\delta < 0$ and $\sin\delta > 0$. This means that the phase shift must be in Quadrant II and so the second angle is the one that we need.

So, after all of this the displacement at any time t is.

$$u(t) = 0.52705\cos(6t - 2.81984).$$

Here is a sketch of the displacement for the first 5 seconds.

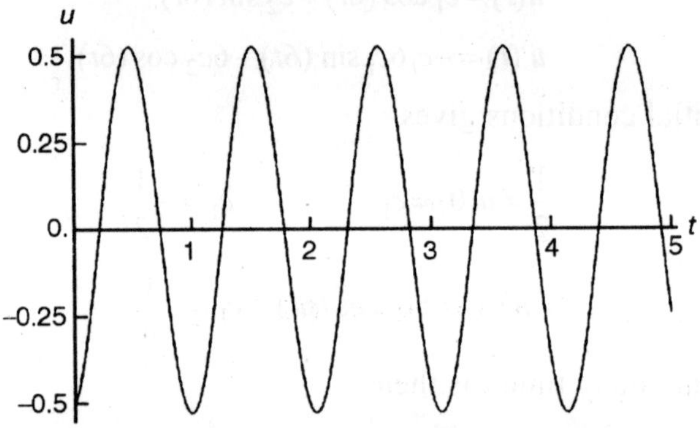

Now, let's take a look at a slightly more realistic situation. No vibration will go on forever. So let's add in a damper and see what happens now.

FREE, DAMPED VIBRATIONS

He are still going to assume that there will be no external forces acting on the system, with the exception of damping of course. In this case the differential equation will be.

$$mu'' + \gamma u' + ku = 0$$

where m, δ, and k are all positive constants. Upon solving for the roots of the characteristic equation we get the following.

$$r_{1,2} = \frac{-\gamma \pm \sqrt{\gamma^2 - 4mk}}{2m}.$$

We will have three cases here.

1. $\gamma^2 - 4mk = 0.$

In this case we will get a double root out of the characteristic equation and the displacement at any time t will be.

$$u(t) = c_1 e^{-\frac{\gamma t}{2m}} + c_1 t e^{-\frac{\gamma t}{2m}}.$$

Notice that as $t \to \infty$the displacement will approach zero and so the damping in this case will do what it's supposed to do.

This case is called critical damping and will happen when the damping coefficient is,

$$\gamma^2 - 4mk = 0$$
$$\gamma^2 = 4mk$$
$$\gamma = 2\sqrt{mk} = \gamma_{CR}.$$

The value of the damping coefficient that gives critical damping is called the critical damping coefficient and denoted by γ_{CR}.

2. $\gamma^2 - 4mk > 0.$

In this case let's rewrite the roots a little.

$$r_{1,2} = \frac{-\gamma \pm \sqrt{\gamma^2 - 4mk}}{2m}$$
$$= \frac{-\gamma \pm \gamma\sqrt{1 - \frac{4mk}{\gamma^2}}}{2m}$$
$$= -\frac{\gamma}{2m}\left(1 \pm \sqrt{1 - \frac{4mk}{\gamma^2}}\right).$$

Also notice that from our initial assumption that we have,

$$\gamma^2 > 4mk$$
$$1 > \frac{4mk}{\gamma^2}.$$

Using this we can see that the fraction under the square root above is less than one. Then if the quantity under the square root is less than one, this means that the square root of this quantity is also going to be less than one. In other words,

$$\sqrt{1 - \frac{4mk}{\gamma^2}} < 1.$$

Why is this important? Well, the quantity in the parenthesis is now one plus/minus a number that is less than one. This means that the quantity in the parenthesis is guaranteed to be positive and so the two roots in this case are guaranteed to be negative. Therefore the displacement at any time t is,

$$u(t) = c_1 e^{k_1 t} + c_2 e^{r_2 t}$$

and will approach zero as $t \to \infty$. So, once again the damper does what it is supposed to do.

This case will occur when

$$\gamma^2 > 4mk$$

$$\gamma > 2\sqrt{mk}$$

$$\gamma > \gamma_{CR}$$

and is called over damping.

3. $\gamma^2 - 4mk < 0.$

In this case we will get complex roots out of the characteristic equation.

$$r_{1,2} = \frac{-\gamma}{2m} \pm \frac{\sqrt{\gamma^2 - 4mk}}{2m} = \lambda \pm \mu i$$

where the real part is guaranteed to be negative and so the displacement is

$$\begin{aligned} u(t) &= c_1 e^{\lambda t} \cos(\mu t) + c_2 e^{\lambda t} \sin(\mu t) \\ &= e^{\lambda t}(c_1 \cos(\mu t) + c_2 \sin(\mu t)) \\ &= R e^{\lambda t} \cos(\mu t - \delta). \end{aligned}$$

Notice that we reduced the sine and cosine down to a single cosine in this case as we did in the undamped case. Also, since $\lambda < 0$ the displacement will approach zero as $t \to \infty$and the damper will also work as it's supposed to in this case.

We will get this case will occur when

$$\gamma^2 > 4mk$$

$$\gamma > 2\sqrt{mk}$$

$$\gamma > \gamma_{CR}$$

and is called under damping.

Let's take a look at a couple of examples here with damping.

Example: Take the spring and mass system from the first example and attach a damper to it that will exert a force of 12 lbs when the velocity is 2 ft/s. Find the displacement at any time *t*, *u(t)*.

Solution: The mass and spring constant were already found in the first example so we won't do the work here. We do need to find the damping coefficient however. To do this we will use the formula for the damping force given above with one modification. The original damping force formula is,

$$F_d = -\gamma u'.$$

However, remember that the force and the velocity are always acting in opposite directions. So, if the velocity is upward (*i.e.* negative) the force will be downward (*i.e.* positive) and so the minus in the formula will cancel against the minus in the velocity. Likewise, if the velocity is downward (*i.e.* positive) the force will be upwards (*i.e.* negative) and in this case the minus sign in the formula will cancel against the minus in the force. In other words, we can drop the minus sign in the formula and use

$$F_d = -\gamma u'$$

and then just ignore any signs for the force and velocity.

Doing this gives us the following for the damping coefficient

$$12 = \gamma(2) \quad \Rightarrow \quad \gamma = 6.$$

The IVP for this example is then,

$$\frac{1}{2}u'' + 6u' + 18u = 0 \qquad \text{u}(0) = -\frac{1}{2} \qquad u'(0) = 1.$$

Before solving let's check to see what kind of damping we've got. To do this all we need is the critical damping coefficient.

$$\gamma_{CR} = 2\sqrt{km} = 2\sqrt{(18)\left(\frac{1}{2}\right)} = 2\sqrt{9} = 6.$$

So, it looks like we've got critical damping. Note that this means that when we go to solve the differential equation we should get a double root.

Speaking of solving, let's do that. I'll leave the details to you to check that the displacement at any time t is.

$$u(t) = -\frac{1}{2}e^{-6t} - 2te^{-6t}.$$

Here is a sketch of the displacement during the first 3 seconds.

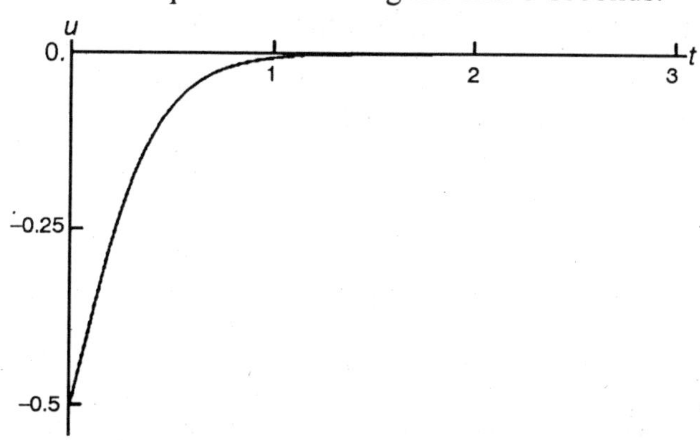

Notice that the "vibration" in the system is not really a true vibration as we tend to think of them. In the critical damping case there isn't going to be a real oscillation about the equilibrium point that we tend to associate with vibrations. The damping in this system is strong enough to force the "vibration" to die out before it ever really gets a chance to do much in the way of oscillation.

Example: Take the spring and mass system from the first example and this time let's attach a damper to it that will exert a force of 17 lbs when the velocity is 2 ft/s. Find the displacement at any time *t*, *u(t)*.

Solution: So, the only difference between this example and the previous example is damping force. So let's find the damping coefficient

$$17=\gamma(2) \quad \Rightarrow \quad \gamma=\frac{17}{2}=8.5>\gamma_{CR}.$$

So it looks like we've got over damping this time around so we should expect to get two real distinct roots from the characteristic equation and they should both be negative. The IVP for this example is,

$$\frac{1}{2}u''+\frac{17}{2}u'+18u=0 \qquad u(0)=-\frac{1}{2} \qquad u'(0)=1.$$

This one's a little messier than the previous example so we'll do a couple of the steps, leaving it to you to fill in the blanks. The roots of the characteristic equation are

$$r_{1,2}=\frac{-17\pm\sqrt{145}}{2}=-2.472,-14,4208.$$

In this case it will be easier to just convert to decimals and go that route. Note that, as predicted we got two real, distinct and negative roots. The general and actual solution for this example are then,

$$u(t)=c_1e^{-2.4792t}+c_2e^{-145208t}$$

$$u(t)=-0.5198e^{-2.4792t}+0.0199e^{-145208t}.$$

Here's a sketch of the displacement for this example.

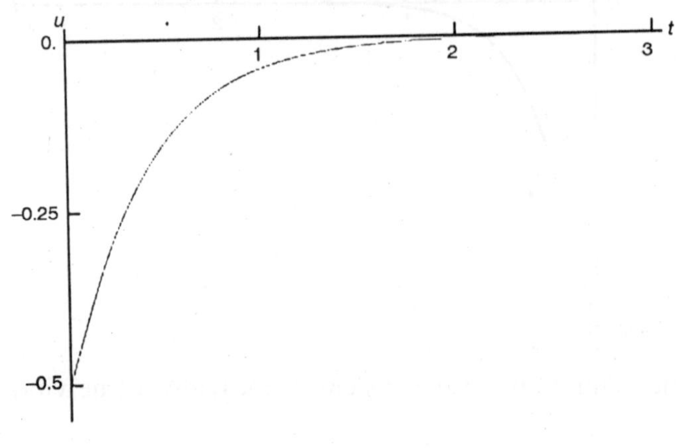

Notice an interesting thing here about the displacement here. Even though we are "over" damped in this case, it actually takes longer for the vibration to die out than in the critical damping case. Sometimes this happens, although it will not always be the case that over damping will allow the vibration to continue longer than the critical damping case.

Also notice that, as with the critical damping case, we don't get a vibration in the sense that we usually think of them. Again, the damping is strong enough to force the vibration do die out quick enough so that we don't see much, if any, of the oscillation that we typically associate with vibrations.

Let's take a look at one more example before moving on the next type of vibrations.

Example: Take the spring and mass system from the first example and for this example let's attach a damper to it that will exert a force of 5 lbs when the velocity is 2 ft/s. Find the displacement at any time *t*, *u(t)*.

Solution: So, let's get the damping coefficient.

$$s = \gamma(2) \qquad \Rightarrow \qquad \gamma = \frac{5}{2} = 2.5 < \gamma_{CR}.$$

So it's under damping this time. That shouldn't be too surprising given the first two examples. The IVP for this example is,

$$\frac{1}{2}u'' + \frac{5}{2}u' + 18u = 0 \qquad u(0) = -\frac{1}{2} \quad u'(0) = 1.$$

In this case the roots of the characteristic equation are

$$r_{1,2} = \frac{-5 \pm \sqrt{119}i}{2}.$$

They are complex as we expected to get since we are in the under damped case. The general solution and actual solution are

$$u(t) = e^{-\frac{5t}{2}}\left(c_1 \cos\left(\frac{\sqrt{119}}{2}t\right) + c_2 \sin\left(\frac{\sqrt{119}}{2}t\right)\right)$$

$$u(t) = e^{-\frac{5t}{2}}\left(-0.5\cos\left(\frac{\sqrt{119}}{2}t\right) - 0.04583\sin\left(\frac{\sqrt{119}}{2}t\right)\right).$$

Let's convert this to a single cosine as we did in the undamped case.

$$R = \sqrt{(-0.5)^2 + (-0.04583)^2} = 0.502096$$

$$\delta_1 = \tan^{-1}\left(\frac{-0.04583}{-0.5}\right) = 0.09051 \qquad \text{OR} \qquad \delta_2 = \delta_1 + \pi = 3.2321.$$

As with the undamped case we can use the coefficients of the cosine and the sine to determine which phase shift that we should use. The coefficient of the cosine (c_1) is negative and so $\cos\delta$ must also be negative. Likewise, the coefficient of the sine (c_2) is also negative and so $\sin\delta$ must also be negative. This means that δmust be in the Quadrant III and so the second angle is the one that we want.

The displacement is then

$$u(t) = 0.502096e^{-\frac{5t}{2}} \cos\left(\frac{\sqrt{119}}{2}t - 3.2321\right).$$

Here is a sketch of this displacement.

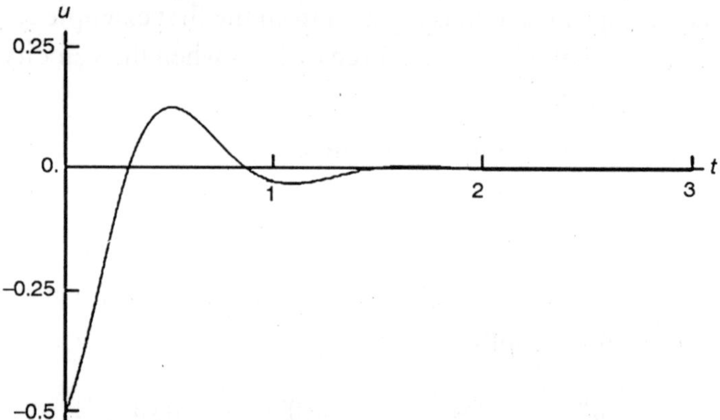

In this case we finally got want we usually consider to be a true vibration. In fact that is the point of critical damping. As we increase the damping coefficient, the critical damping coefficient will be the first one in which a true oscillation in the displacement will not occur. For all values of the damping coefficient larger than this (*i.e.* over damping) we will also not see a true oscillation in the displacement.

From a physical standpoint critical (and over) damping is usually preferred to under damping. Think of the shock absorbers in your car. When you hit a bump you don't want to spend the next few minutes bouncing up and down while the vibration set up by the bump die out. You would like there to be as little movement as possible. In other words, you will want to set up the shock absorbers in your car so get at the least critical damping so that you can avoid the oscillations that will arise from an under damped case.

It's now time to look at systems in which we allow other external forces to act on the object in the system.

UNDAMPED, FORCED VIBRATIONS

We will first take a look at the undamped case. The differential equation in this case is

$$mu'' + ku = F(t).$$

This is just a nonhomogeneous differential equation and we know how to solve these. The general solution will be

$$u(t) = u_c(t) + U_P(t)$$

where the complimentary solution is the solution to the free, undamped vibration case. To get the particular solution we can use either undetermined coefficients or variation of parameters depending on which we find easier for a given forcing function.

There is a particular type of forcing function that we should take a look at since it leads to some interesting results. Let's suppose that the forcing function is a simple periodic function of the form

$$F(t) = F_0 \cos(\omega t) \quad \text{OR} \quad F(t) = F_0 \sin(\omega t).$$

For the purposes of this discussion we'll use the first one. Using this, the IVP becomes,

$$mu'' + ku = F_0 \cos(\omega t).$$

The complimentary solution, as pointed out above, is just

$$u_c(t) = c_1 \cos(\omega_0 t) + c_2 \sin(\omega_0 t)$$

where ω_0 is the natural frequency.

We will need to be careful in finding a particular solution. The reason for this will be clear if we use undetermined coefficients. With undetermined coefficients our guess for the form of the particular solution would be,

$$U_P(t) = A\cos(\omega t) + B\sin(\omega t).$$

Now, this guess will be problems if $\omega_0 = \omega$. If this were to happen the guess for the particular solution is exactly the complimentary solution and so we'd need to add in a t. Of course if we don't have $\omega_0 = \omega$then there will be nothing wrong with the guess.

So, we will need to look at this in two cases.

2. $\omega_0 \neq \omega$

In this case our initial guess is okay since it won't be the complimentary solution. Upon differentiating the guess and plugging it into the differential equation and simplifying we get,

$$(-m\omega^2 A + kA)\cos(wt) + (-m\omega^2 B + kB)\sin(wt) = F_0 \cos(wt).$$

Setting coefficients equal gives us,

$$\cos(\omega t): \quad (-m\omega^2 + k)\,A = F_0 \quad \Rightarrow \quad A = \frac{F_0}{k - m\omega^2}$$

$$\sin(\omega t): \quad (-m\omega^2 + k)\,B = F_0 \quad \Rightarrow \quad B = 0$$

The particular solution is then

$$U_P(t) = \frac{F_0}{k - m\omega^2}\cos(\omega t)$$

$$= \frac{F_0}{m\left(\frac{k}{m} - \omega^2\right)}\cos(\omega t)$$

$$= \frac{F_0}{m\left(\omega_0^2 - \omega^2\right)}\cos(\omega t).$$

Note that we rearranged things a little. Depending on the form that you'd like the displacement to be in we can have either of the following.

$$u(t) = c_1 \cos(\omega_0 t) + c_2 \sin(\omega_0 t) + \frac{F_0}{m\left(\omega_0^2 - \omega^2\right)}\cos(\omega t)$$

$$u(t) = R\cos(\omega_0 t - \delta) + \frac{F_0}{m\left(\omega_0^2 - \omega^2\right)}\cos(\omega t).$$

If we used the sine form of the forcing function we could get a similar formula.

1. $\omega_0 = \omega$.

In this case we will need to add in a t to the guess for the particular solution.

$$U_P(t) = At\cos(\omega_0 t) + Bt\sin(\omega_0 t).$$

Note that we went ahead and acknowledge that $\omega_0 = \omega$in our guess. Acknowledging this will help with some simplification that we'll need to do later on. Differentiating our guess, plugging it into the differential equation and simplifying gives us the following.

$$(-m\omega_0^2 + k)At\cos(wt) + (-mw_0^2 + k)Bt\sin(wt) +$$
$$2m\omega_0 B\cos(\omega t) - 2m\omega_0 A\sin(wt) = F_0\cos(wt).$$

Before setting coefficients equal, let's remember the definition of the natural frequency and note that

$$-m\omega_0^2 + k = -m\left(\sqrt{\frac{k}{m}}\right)^2 + k = -m\left(\frac{k}{m}\right) + k = 0.$$

So, the first two terms actually drop out (which is a very good thing...) and this gives us,

$$2m\omega_0 B\cos(\omega t) - 2m\omega_0 A\sin(wt) = F_0\cos(wt).$$

Now let's set coefficient equal.

$$\cos(\omega t): \quad 2m\omega_0 B = F_0 \quad \Rightarrow \quad B = \frac{F}{2m\omega_0}$$

$$\sin(\omega t): \quad 2m\omega_0 A = F_0 \quad \Rightarrow \quad A = 0.$$

In this case the particular will be,

$$U_P(t) = \frac{F_0}{2m\omega_0} t \sin(\omega_0 t).$$

The displacement for this case is then

$$u(t) = c_1 \cos(\omega_0 t) + c_0 \sin(\omega_0 t) + \frac{F_0}{2m\omega_0} t \sin(\omega_0 t)$$

$$u(t) = R\cos(\omega_0 t - \delta) + \frac{F_0}{2m\omega_0} t \sin(\omega_0 t)$$

depending on the form that you prefer for the displacement.

So, what was the point of the two cases here? Well in the first case, $\omega_0 \neq \omega$our displacement function consists of two cosines and is nice and well behaved for all time.

In contrast, the second case, $\omega_0 = \omega$will have some serious issues at t increases. The addition of the t in the particular solution will mean that we are going to see an oscillation that grows in amplitude as t increases. This case is called resonance and we would generally like to avoid this at all costs.

In this case resonance arose by assuming that the forcing function was,

$$F(t) = F_0 \cos(\omega t).$$

We would also have the possibility of resonance if we assumed a forcing function of the form.

$$F(t) = F_0 \sin(\omega t).$$

We should also take care to not assume that a forcing function will be in one of these two forms. Forcing functions can come in a wide variety of forms. If we do run into a forcing function different from the one that used here you will have to go through undetermined coefficients or variation of parameters to determine the particular solution.

Example: A 3 kg object is attached to spring and will stretch the spring 392 mm by itself. There is no damping in the system and a forcing function of the form

$$F(t) = 10\cos(\omega t)$$

is attached to the object and the system will experience resonance. If the object is initially displaced 20 cm downward from its equilibrium position and given a velocity of 10 cm/sec upward find the displacement at any time t.

Solution: Since we are in the metric system we won't need to find mass as it's been given to us. Also, for all calculations we'll be converting all lengths over to meters.

The first thing we need to do is find *k*.

$$k = \frac{mg}{L} = \frac{(3)(9.8)}{0.392} = 75.$$

Now, we are told that the system experiences resonance so let's go ahead and get the natural frequency so we can completely set up the IVP.

$$\omega_0 = \sqrt{\frac{k}{m}} = \sqrt{\frac{75}{3}} = 5.$$

The IVP for this is then

$$3u'' + 75u = 10\cos(5t) \qquad u(0) = 0.2 \qquad u'(0) = -0.1.$$

Solution: wise there isn't a whole lot to do here. The complimentary solution is the free undamped solution which is easy to get and for the particular solution we can just use the formula that we derived above.

The general solution is then,

$$u(t) = c_1 \cos(5t) + c_2 \sin(5t) + \frac{10}{2(3)\,(5)} t \sin\,(5t)$$

$$u(t) = c_1 \cos(5t) + c_2 \sin(5t) + \frac{1}{3} t \sin\,(5t).$$

Applying the initial conditions gives the displacement at any time *t*. We'll leave the details to you to check.

$$u(t) = \frac{1}{5}\cos\,(5t) - \frac{1}{50}\sin\,(5t) + \frac{1}{3} t \sin\,(5t).$$

The last thing that we'll do is combine the first two terms into a single cosine.

$$R = \sqrt{\left(\frac{1}{5}\right)^2 + \left(-\frac{1}{50}\right)^2} = 0.200998$$

$$\delta_1 = \tan^{-1}\left(\frac{-1/50}{1/5}\right) = -0.099669 \qquad \delta_2 = \delta_1 + \pi = 3.041924.$$

In this case the coefficient of the cosine is positive and the coefficient of the sine is negative. This forces $\cos\delta$ to be positive and $\sin\delta$ to be negative.

This means that the phase shift needs to be in Quadrant IV and so the first one is the correct phase shift this time.

The displacement then becomes,

$$u(t) = 0.200998\cos(5t + 0.09969) + \frac{1}{3}t\sin(5t).$$

Here is a sketch of the displacement for this example.

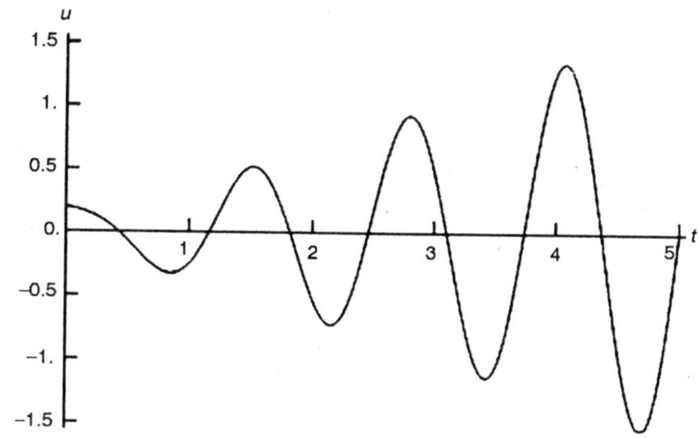

It's now time to look at the final vibration case.

FORCED, DAMPED VIBRATIONS

This is the full blown case where we consider every last possible force that can act upon the system. The differential equation for this case is,

$$mu'' + \gamma u' + ku = F(t).$$

The displacement function this time will be,

$$u(t) = u_c(t) + U_P(t)$$

where the complimentary solution will be the solution to the free, damped case and the particular solution will be found using undetermined coefficients or variation of parameter, whichever is most convenient to use.

There are a couple of things to note here about this case. First, from our work back in the free, damped case we know that the complimentary solution will approach zero as t increases. Because of this the complimentary solution is often called the transient solution in this case.

Also, because of this behavior the displacement will start too look more and more like the particular solution as t increases and so the particular solution is often called the steady state solution or forced response.

Let's work one final example before leaving this section. As with the previous examples, we're going to leave most of the details out for you to check.

Example: Take the system from the last example and add in a damper that will exert a force of 45 Newtons when then velocity is 50 cm/sec.

Solution: So, all we need to do is compute the damping coefficient for this problem then pull everything else down from the previous problem. The damping coefficient is

$$F_d = \gamma u'$$
$$45 = \gamma(0.5)$$
$$\gamma = 90.$$

The IVP for this problem is.

$$3u''+90u'+75u = 10\cos(5t) \qquad u(0) = 0.2 \qquad u'(0) = -0.1.$$

The complimentary solution for this example is

$$u_c(t) = c_1 e^{(-15+10\sqrt{2})t} + c_2 e^{(-15+10\sqrt{2})t}$$

$$u_c(t) = c_1 e^{-0.8579t} + c_2 e^{-29.1421t}.$$

For the particular solution we the form will be,

$$U_P(t) = A\cos(5t) + B\sin(5t).$$

Plugging this into the differential equation and simplifying gives us,

$$450B\cos(5t) - 450A\sin(5t) = 10\cos(5t).$$

Setting coefficient equal gives,

$$U_P(t) = \frac{1}{45}\sin(5t).$$

The general solution is then

$$u(t) = c_1 e^{-0.8579t} + c_2 e^{-29.1421t} + \frac{1}{45}\sin(5t).$$

Applying the initial condition gives

$$u(t) = 0.1986 e^{-0.8579t} + 0.00139 e^{-29.1421t} + \frac{1}{45}\sin(5t).$$

Here is a sketch of the displacement for this example.

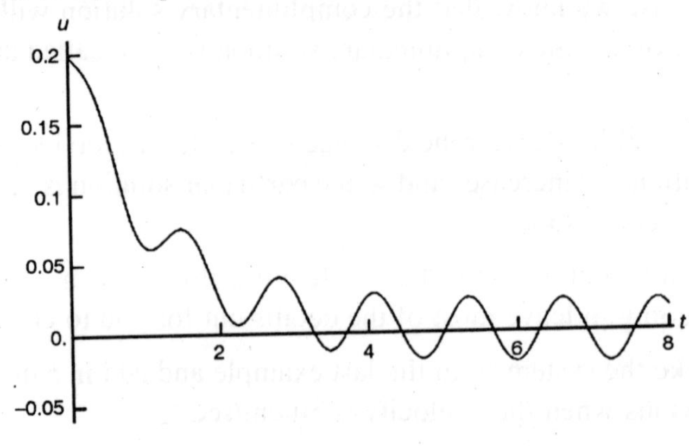

Chapter 4

Boundary Value

In this chapter we'll be taking a quick and very brief look at a couple of topics. The two main topics in this chapter are Boundary Value Problems and Fourier Series. We'll also take a look at a couple of other topics in this chapter. The main point of this chapter is to get some of the basics out of the way that we'll need in the next chapter where we'll take a look at one of the more common solution methods for partial differential equations.

It should be pointed out that both of these topics are far more in depth than what we'll be covering here. In fact you can do whole courses on each of these topics. What we'll be covering here are simply the basics of these topics that well need in order to do the work in the next chapter. There are whole areas of both of these topics that we'll not be even touching on.

BOUNDARY VALUE PROBLEMS

Before we start off this section we need to make it very clear that we are only going to scratch the surface of the topic of boundary value problems. There is enough material in the topic of boundary value problems that we could devote a whole class to it. The intent of this section is to give a brief (and we mean very brief) look at the idea of boundary value problems and to give enough information to allow us to do some basic partial differential equations in the next chapter.

Now, with that out of the way, the first thing that we need to do is to define just what we mean by a boundary value problem (BVP for short).

With initial value problems we had a differential equation and we specified the value of the solution and an appropriate number of derivatives at the same point (collectively called initial conditions). For instance for a second order differential equation the initial conditions are,

$$y(t_0) = y_0 \qquad y'(t_0) = y_0'.$$

With boundary value problems we will have a differential equation and we will specify the function and/or derivatives at *different* points, which we'll call boundary values. For second order differential equations, which will be looking at pretty much exclusively here, any of the following can, and will, be used for boundary conditions.

$$y(x_0) = y_0 \qquad y'(x_1) = y_1$$

$$y'(x_0) = y_0 \qquad y'(x_0) = y_1$$

$$y'(x_0) = y_0 \qquad y'(x_1) = y_1$$

$$y(x_0) = y_0 \qquad y'(x_1) = y_1.$$

As mentioned above we'll be looking pretty much exclusively at second order differential equations. We will also be restricting ourselves down to linear differential equations. So, for the purposes of our discussion here we'll be looking almost exclusively at differential equations in the form,

$$y'' + p(x)y' + q(x)y = g(x)$$

along with one of the sets of boundary conditions. We will, on occasion, look at some different boundary conditions but the differential equation will always be on that can be written in this form.

As we'll soon see much of what we know about initial value problems will not hold here. We can, of course, solve provided the coefficients are constant and for a few cases in which they aren't. None of that will change. The changes (and perhaps the problems) arise when we move from initial conditions to boundary conditions.

One of the first changes is a definition that we saw all the time in the earlier chapters. In the earlier chapters we said that a differential equation was homogeneous if $g(x) = 0$for all x. Here we will say that a boundary value problem is homogeneous if in addition to $g(x) = 0$we also have $y_0 = 0$and $y_1 = 0$(regardless of the boundary conditions we use). If any of these are not zero we will call the BVP nonhomogeneous.

It is important to now remember that when we say homogeneous (or nonhomogeneous) we are saying something not only about the differential equation itself but also about the boundary conditions as well.

The biggest change that we're going to see here comes when we go to solve the boundary value problem. When solving linear initial value problems a unique solution will be guaranteed under very mild conditions. We only looked at this idea for first order IVP's but the idea does extend to higher order IVP's. In that section we saw that all we needed to guarantee a unique solution was some basic continuity conditions. With boundary value problems we will often have no solution or infinitely many solutions even for very nice differential equations that would yield a unique solution if we had initial conditions instead of boundary conditions.

Before we get into solving some of these let's next address the question of why we're even talking about these in the first place. As we'll see in the next chapter in the process of solving some partial differential equations we will run into boundary value problems that will need to be solved as well. In fact, a large part of the solution process there will be in dealing with the solution to the BVP. In these cases the boundary conditions will represent things like the temperature at either end of a bar, or the heat flow into/out of either end of a bar. Or maybe they will represent the location of ends of a vibrating string. So, the boundary conditions there will really be conditions on the boundary of some process.

So, with some of basic stuff out of the way let's find some solutions to a few boundary value problems. Note as well that there really isn't anything new here yet. We know how to solve the differential equation and we know how to find the constants by applying the conditions. The only difference is that here we'll be applying boundary conditions instead of initial conditions.

Example: Solve the following BVP.

$$y''+4y=0 \qquad y(0)=-2 \qquad y\left(\frac{y}{4}\right)=10.$$

Solution: Okay, this is a simple differential equation so solve and so we'll leave it to you to verify that the general solution to this is,

$$y(x)=c_1\cos(2x)+c_2\sin(2x).$$

Now all that we need to do is apply the boundary conditions.

$$-2=y(0)=c_1$$

$$10=y\left(\frac{\pi}{4}\right)=c_2.$$

The solution is then,

$$y(x)=-2\cos(2x)+10\sin(2x)y(x)=-2\cos(2x)+10\sin(2x).$$

We mentioned above that some boundary value problems can have no solutions or infinite solutions we had better do a couple of examples of those as well here. This next set of examples will also show just how small of a change to the BVP it takes to move into these other possibilities.

Example: Solve the following BVP.

$$y''+4y=0 \qquad y(0)=-2 \qquad y(2\pi)=-2.$$

Solution: We're working with the same differential equation as the first example so we still have,

$$y(x)=c_1\cos(2x)+c_2\sin(2x).$$

Upon applying the boundary conditions we get,

$$-2=y(0)=c_1$$

$$-3=y(2\pi)=c_1.$$

So in this case, unlike previous example, both boundary conditions tell us that we have to have $c_1 = -2$and neither one of them tell us anything about c_2. Remember however that all we're asking for is a solution to the differential equation that satisfies the two given boundary conditions and the following function will do that,

$$y(x) = -2\cos(2x) + c_2 \sin(2x).$$

In other words, regardless of the value of we get a solution and so, in this case we get infinitely many solutions to the boundary value problem.

Example: Solve the following BVP.

$$y'' + 4y = 0 \qquad y(0) = -2 \qquad y(2\pi) = 3.$$

Solution: Again, we have the following general solution,

$$y(x) = c_1 \cos(2x) + c_2 \sin(2x).$$

This time the boundary conditions give us,

$$-2 = y(0) = c_1$$

$$3 = y(2\pi) = c_1.$$

In this case we have a set of boundary conditions each of which requires a different value of in order to be satisfied. This, however, is not possible and so in this case have no solution.

So, with Examples we can see that only a small change to the boundary conditions, in relation to each other and to Example, can completely change the nature of the solution. All three of these examples used the same differential equation and yet a different set of initial conditions yielded, no solutions, one solution, or infinitely many solutions.

Note that this kind of behavior is not always unpredictable however. If we use the conditions $y(0)$and $y(2\pi)$the only way we'll ever get a solution to the boundary value problem is if we have,

$$y(0) = a \qquad y(2\pi) = a$$

for any value of a. Also, note that if we do have these boundary conditions we'll in fact get infinitely many solutions.

All the examples we've worked to this point involved the same differential equation and the same type of boundary conditions so let's work a couple more just to make sure that we've got some more examples here. Also, note that with each of these we could tweak the boundary conditions a little to get any of the possible solution behaviors to show up (*i.e.* zero, one or infinitely many solutions).

Example: Solve the following BVP.

$$y'' + 3y = 0 \qquad y(0) = 7 \qquad y(2\pi) = 0.$$

Solution: The general solution for this differential equation is,

$$y(x) = c_1 \cos\left(\sqrt{3}x\right) + c_2 \sin\left(\sqrt{3}x\right).$$

Applying the boundary conditions gives,

$$7 = y(0) = c_1$$

$$0 = y(2\pi) = c_1 \cos(2\sqrt{2}\pi) + c_2 \sin(2\sqrt{3}\pi) \quad \Rightarrow \quad c_2 - = 7\cot(2\sqrt{3}\pi).$$

In this case we get a single solution,

$$y(x) = 7\cos\left(\sqrt{3}x\right) - 7\cot(2\sqrt{3}\pi)\sin\left(\sqrt{3}x\right).$$

Example: Solve the following BVP.

$$y'' + 25y = 0 \qquad y'(0) = 6 \qquad y'(\pi) = -9.$$

Solution: Here the general solution is,

$$y(x) = c_1 \cos(5x) + c_2 \sin(5x)$$

and we'll need the derivative to apply the boundary conditions,

$$y'(x) = -5c_1 \sin(5x) + 5c_2 \cos(5x).$$

Applying the boundary conditions gives,

$$6 = y'(0) = 5c_2 \quad \Rightarrow \quad c_1 = \tfrac{6}{5}$$

$$-9 = y'(\pi) = -5c_2 \quad \Rightarrow \quad c_1 = \tfrac{9}{5}.$$

This is not possible and so in this case have no solution.

All of the examples worked to this point have been nonhomogeneous because at least one of the boundary conditions have been non-zero. Let's work one nonhomogeneous example where the differential equation is also nonhomogeneous before we work a couple of homogeneous examples.-

Example: Solve the following BVP.

$$y'' + 9y = \cos x \qquad y'(0) = 5 \qquad y\left(\frac{\pi}{2}\right) = -\frac{5}{3}.$$

Solution: The complimentary solution for this differential equation is,

$$y_c(x) = c_1 \cos(3x) + c_2 \sin(3x).$$

Using Undetermined Coefficients or Variation of Parameters it is easy to show (we'll leave the details to you to verify) that a particular solution is,

$$Y_P(x) = \frac{1}{8}\cos x.$$

The general solution and its derivative (since we'll need that for the boundary conditions) are,

$$y(x) = c_1 \cos(3x) + c_2 \sin(3x) + \frac{1}{8}\cos x$$

$$y'(x) = -3c_1 \sin(3x) + 3c_2 \cos(3x) - \frac{1}{8}\sin x.$$

Applying the boundary conditions gives,

$$5 = y'(0) = 3c_2 \quad \Rightarrow \quad c_1 = \frac{5}{3}$$

$$-\frac{5}{3} = y\left(\frac{\pi}{2}\right) = -c_2 \quad \Rightarrow \quad c_1 = \frac{5}{3}.$$

The boundary conditions then tell us that we must have and they don't tell us anything about and so it is can be arbitrarily chosen. The solution is then,

$$y(x) = c_1 \cos(3x) + \frac{5}{3}\sin(3x) + \frac{1}{8}\cos x$$

and there will be infinitely many solutions to the BVP.

Let's now work a couple of homogeneous examples that will also be helpful to have worked once we get to the next section.

Example: Solve the following BVP.

$$y'' + 4y = 0 \qquad y(0) = 0 \qquad y(2\pi) = 0.$$

Solution: Here the general solution is,

$$y(x) = c_1 \cos(2x) + c_2 \sin(2x).$$

Applying the boundary conditions gives,

$$0 = y(0) = c_1$$

$$0 = y(2\pi) = c_1.$$

So is arbitrary and the solution is,

$$y(x) = c_2 \sin(2x)$$

and in this case we'll get infinitely many solutions.

Example: Solve the following BVP.

$$y'' + 3y = 0 \qquad y(0) = 0 \qquad y(2\pi) = 0.$$

Solution: The general solution in this case is,

$$y(x) = c_1 \cos\left(\sqrt{3}x\right) + c_2 \sin\left(\sqrt{3}x\right).$$

Applying the boundary conditions gives,

$$0 = y(0) = c_1$$

$$0 = y(2\pi) = c_2 \sin\left(2\sqrt{3}\pi\right) \quad \Rightarrow \quad c_2 = 0.$$

In this case we found both constants to be zero and so the solution is,

$$y(x) = 0.$$

In the previous example the solution was $y(x) = 0$. Notice however, that this will always be a solution to any homogenous system given and any of the (homogeneous) boundary conditions given. Because of this we usually call this solution the trivial solution. Sometimes, as in the case of the last example the trivial solution is the only solution however we generally prefer solutions to be non-trivial. This will be a major idea in the next section. -

Before we leave this section an important point needs to be made. In each of the examples, with one exception, the differential equation that we solved was in the form,

$$y'' + \lambda y = 0.$$

The one exception to this still solved this differential equation except it was not a homogeneous differential equation and so we were still solving this basic differential equation in some manner.

So, there are probably several natural questions that can arise at this point. Do all BVP's involve this differential equation and if not why did we spend so much time solving this one to the exclusion of all the other possible differential equations?

The answers to these questions are fairly simple. First, this differential equation is most definitely not the only one used in boundary value problems. It does however exhibit all of the behavior that we wanted to talk about here and has the added bonus of being very easy to solve. So, by using this differential equation almost exclusively we can see and discuss the important behavior that we need to discuss and frees us up from lots of potentially messy solution details and or messy solutions. We will, on occasion, look at other differential equations in the rest of this chapter, but we will still be working almost exclusively with this one.

There is another important reason for looking at this differential equation. When we get to the next chapter and take a brief look at solving partial differential equations we will see that almost every one of the examples that we'll work there come down to exactly this differential equation.

Also, in those problems we will be working some "real" problems that are actually solved in places and so are not just "made up" problems for the purposes of examples. Admittedly they will have some simplifications in them, but they do come close to realistic problem in some cases.

EIGENVALUES AND EIGENFUNCTIONS

As we did in the previous section we need to again note that we are only going to give a brief look at the topic of eigenvalues and eigenfunctions for boundary value problems. There are quite a few ideas that we'll not be looking at here. The intent of this section is simply to give you an idea of the subject and to do enough work to allow us to solve some basic partial differential equations in the next chapter.

Now, before we start talking about the actual subject of this section let's recall a topic from Linear Algebra that we briefly discussed previously in these notes. For a given square matrix, *A*, if we could find values of for which we could find nonzero solutions, *i.e.* $\vec{x} \neq \vec{0}$, to,

$$A\vec{x} = \lambda\vec{x}$$

then we called an eigenvalue of *A* and was its corresponding eigenvector.

It's important to recall here that in order for to be an eigenvalue then we had to be able to find nonzero solutions to the equation.

So, just what does this have to do with boundary value problems? In those two examples we solved homogeneous (and that's important!) BVP's in the form,

$$y'' + \lambda y = 0 \qquad y(0) = 0 \qquad y(2\pi) = 0.$$

In Example we had and we found nontrivial (*i.e.* nonzero) solutions to the BVP. In Example we used $\lambda = 3$and the only solution was the trivial solution (*i.e.* $y(t) = 0$). So, this homogeneous BVP (recall this also means the boundary conditions are zero) seems to exhibit similar behavior to the behavior in the matrix equation above. There are values of λthat will give nontrivial solutions to this BVP and values of λthat will only admit the trivial solution.

So, for those values of λthat give nontrivial solutions we'll call λan eigenvalue for the BVP and the nontrivial solutions will be called eigenfunctions for the BVP corresponding to the given eigenvalue.

We now know that for the homogeneous BVP given $\lambda = 4$is an eigenvalue (with eigenfunctions $y(x) = c_2 \sin(2x)$) and that $\lambda = 3$is not an eigenvalue.

Eventually we'll try to determine if there are any other eigenvalues, however before we do that let's comment briefly on why it is so important for the BVP to homogeneous in this discussion.

$$y'' + 4y = 0$$

with two different nonhomogeneous boundary conditions in the form,

$$y(0) = a \qquad y(2\pi) = b.$$

In these two examples we saw that by simply changing the value of *a* and/or *b* we were able to get either nontrivial solutions or to force no solution at all. In the discussion of eigenvalues/eigenfunctions we need solutions to exist and the only way to assure this behavior is to require that the boundary conditions also be homogeneous. In other words, we need for the BVP to be homogeneous.

There is one final topic that we need to discuss before we move into the topic of eigenvalues and eigenfunctions and this is more of a notational issue that will help us with some of the work that we'll need to do.

Let's suppose that we have a second order differential equations and its characteristic polynomial has two real, distinct roots and that they are in the form

$$r_1 = \alpha \qquad r_2 = -\alpha.$$

Then we know that the solution is,

$$y(x)c_1e^{r_1x} + c_1e^{r_2x} = c_1e^{ax} + c_2e^{-ax}.$$

While there is nothing wrong with this solution let's do a little rewriting of this. We'll start by splitting up the terms as follows,

$$\begin{aligned} y(x) &= c_1e^{ax} + c_2e^{-ax} \\ &= \frac{c_1}{2}e^{ax} + \frac{c_1}{2}e^{ax} + \frac{c_2}{2} + e^{-ax} + \frac{c_2}{2}e^{-ax}. \end{aligned}$$

Now we'll add/subtract the following terms (note we're "mixing" the and up in the new terms) to get,

$$y(x) = \frac{c_1}{2}e^{ax} + \frac{c_1}{2}e^{ax} + \frac{c_2}{2}e^{ax} + \frac{c_2}{2}e^{-ax} + \left(\frac{c_1}{2}e^{-ax} - \frac{c_1}{2}e^{-ax}\right) + \left(\frac{c_2}{2}e^{ax} - \frac{c_2}{2}e^{ax}\right).$$

Next, rearrange terms around a little,

$$y(x) = \frac{1}{2}\left(c_1e^{ax} + c_1e^{-ax} + c_2e^{ax} + c_2e^{-ax}\right) + \frac{1}{2}\left(c_1e^{-ax} - c_1e^{-ax} - c_2e^{ax} + c_2e^{-ax}\right).$$

Finally, the quantities in parenthesis factor and we'll move the location of the fraction as well. Doing this, as well as renaming the new constants we get,

$$\begin{aligned} y(x) &= (c_1 + c_2)\frac{e^{ax} + e^{-ax}}{2} + (c_1 - c_2)\frac{e^{ax} - e^{-ax}}{2}. \\ &= \bar{c}_1\frac{e^{ax} + e^{-ax}}{2} + \bar{c}_2\frac{e^{ax} - e^{-ax}}{2}. \end{aligned}$$

All this work probably seems very mysterious and unnecessary. However there really was a reason for it. In fact you may have already seen the reason, at least in part. The two "new" functions that we have in our solution are in fact two of the hyperbolic functions. In particular,

$$\cosh(x) = \frac{e^x + e^{-x}}{2} \qquad \sinh(x) = \frac{e^x + e^{-x}}{2}.$$

So, another way to write the solution to a second order differential equation whose characteristic polynomial has two real, distinct roots in the form $r_1 = a, r^2 = -\alpha$is,

$$y(x) = c_1 \cosh(ax) + c_2 \sinh(ax).$$

Having the solution in this form for some (actually most) of the problems we'll be looking will make our life a lot easier. The hyperbolic functions have some very nice properties that we can (and will) take advantage of.

First, since we'll be needing them later on, the derivatives are,

$$\frac{d}{dx}(\cosh(x)) = \sin h(x) \qquad \frac{d}{dx}(\cos h(x)) = \cosh(x).$$

Next let's take a quick look at the graphs of these functions.

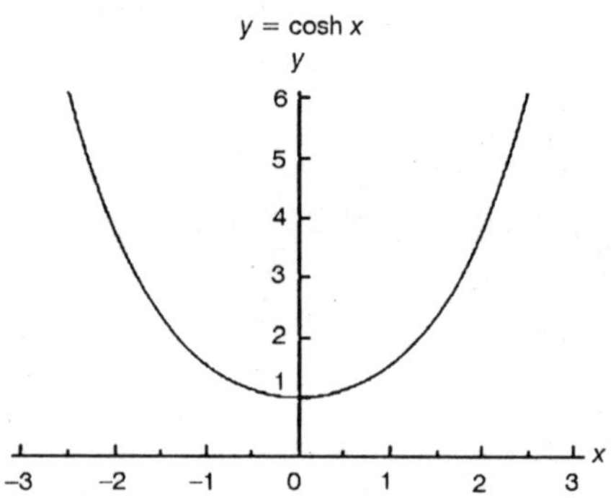

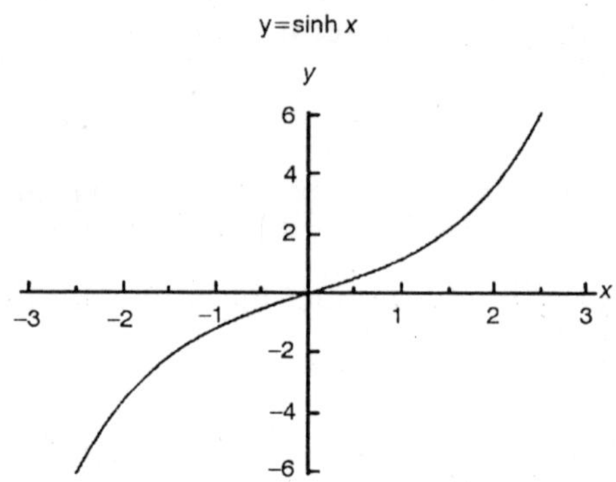

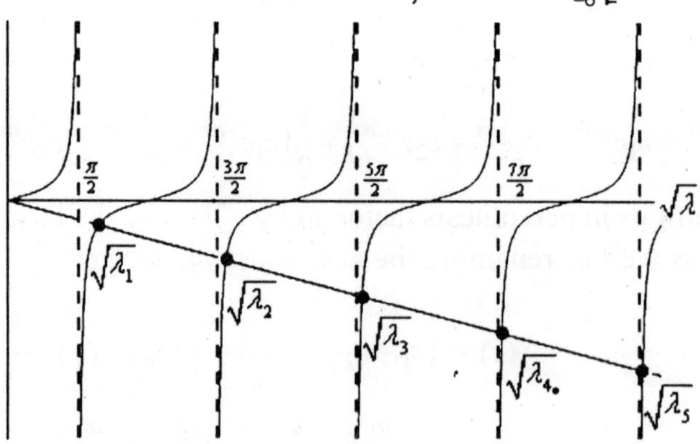

Note that cosh (0) = 1and sinh (0) = 0. Because we'll often be working with boundary conditions at $x = 0$these will be useful evaluations.

Next, and possibly more importantly, let's notice that cosh $(x) > 0$for all x and so the hyperbolic cosine will never be zero. Likewise, we can see that sinh $(x) = 0$only if $x = 0$. We will be using both of these facts in some of our work so we shouldn't forget them.

Okay, now that we've got all that out of the way let's work an example to see how we go about finding eigenvalues/eigenfunctions for a BVP.

Example: Find all the eigenvalues and eigenfunctions for the following BVP.

$$y''+\lambda y=0 \qquad y(0)=0 \qquad y(2\pi)=0.$$

Solution: We started off this section looking at this BVP and we already know one eigenvalue ($\lambda=4$) and we know one value of λ that is not an eigenvalue ($\lambda=3$). As we go through the work here we need to remember that we will get an eigenvalue for a particular value of λ if we get non-trivial solutions of the BVP for that particular value of λ.

In order to know that we've found all the eigenvalues we can't just start randomly trying values of λ to see if we get non-trivial solutions or not. Luckily there is a way to do this that's not too bad and will give us all the eigenvalues/eigenfunctions. We are going to have to do some cases however. The three cases that we will need to look at are : $\lambda>0$, $\lambda=0$, and $\lambda<0$. Each of these cases gives a specific form of the solution to the BVP which we can then apply the boundary conditions to see if we'll get non-trivial solutions or not. So, let's get started on the cases.

$\lambda > 0$

In this case the characteristic polynomial we get from the differential equation is,

$$r^2+\lambda=0 \qquad \Rightarrow \qquad r_{1,2}=\pm\sqrt{-\lambda}.$$

In this case since we know that $\lambda>0$ these roots are complex and we can write them instead as,

$$r_{1,2}=\pm\sqrt{-\lambda}i.$$

The general solution to the differential equation is then,

$$y(x)=c_1\cos\left(\sqrt{\lambda}x\right)+c_2\sin\left(\sqrt{\lambda}x\right).$$

Applying the first boundary condition gives us,

$$0=y(0)=c_1.$$

So, taking this into account and applying the second boundary condition we get,

$$0=y(2\pi)=c_2\sin\left(2\pi\sqrt{\lambda}\right).$$

This means that we have to have one of the following,

$$c_2=0 \qquad \text{or} \qquad \sin\left(2\pi\sqrt{\lambda}\right)=0.$$

However, recall that we want non-trivial solutions and if we have the first possibility we will get the trivial solution for all values of $\lambda > 0$. Therefore let's assume that $2\pi\sqrt{\lambda} > 0$. This means that we have,

$$\sin\left(2\pi\sqrt{\lambda}\right) = 0 \qquad \Rightarrow \qquad 2\pi\sqrt{\lambda} = n\pi \qquad n = 1, 2, 3, \ldots.$$

In other words, taking advantage of the fact that we know where sine is zero we can arrive at the second equation. Also note that because we are assuming that $\lambda > 0$we know that $2\pi\sqrt{\lambda} > 0$and so n can only be a positive integer for this case.

Now all we have to do is solve this for λ and we'll have all the positive eigenvalues for this BVP.

The positive eigenvalues are then,

$$\lambda_n = \left(\frac{n}{2}\right)^2 = \frac{n^2}{4} \qquad n = 1,2,3\ldots.$$

and the eigenfunctions that correspond to these eigenvalues are,

$$y_n(x) = \sin\left(\frac{nx}{2}\right) \qquad n = 1,2,3,\ldots$$

Note that we subscripted an n on the eigenvalues and eigenfunctions to denote the fact that there is one for each of the given values of n. Also note that we dropped the on the eigenfunctions. For eigenfunctions we are only interested in the function itself and not the constant in front of it and so we generally drop that.

Let's now move into the second case.

$\lambda = 0$

In this case the BVP becomes,

$$y'' = 0 \qquad y(0) = 0 \qquad y(2\pi) = 0$$

and integrating the differential equation a couple of times gives us the general solution,

$$y(x) = c_1 + c_2 x.$$

Applying the first boundary condition gives,

$$0 = y(0) = c_1.$$

Applying the second boundary condition as well as the results of the first boundary condition gives,

$$0 = y(2\pi) = 2c_2\pi$$

Here, unlike the first case, we don't have a choice on how to make this zero. This will only be zero if $c_2 = 0$.

Therefore, for this BVP (and that's important), if we have $\lambda = 0$the only solution is the trivial solution and so cannot be an eigenvalue for this BVP.

Now let's look at the final case.

$\lambda < 0$

In this case the characteristic equation and its roots are the same as in the first case. So, we know that,

$$r_{1,2} = \pm\sqrt{-\lambda}.$$

However, because we are assuming $\lambda < 0$here these are now two real distinct roots and so using our work above for these kinds of real, distinct roots we know that the general solution will be,

$$y(x) = c_1 \cosh\left(\sqrt{-\lambda}x\right) + c_2 \sinh\left(\sqrt{-\lambda}x\right).$$

Note that we could have used the exponential form of the solution here, but our work will be significantly easier if we use the hyperbolic form of the solution here.

Now, applying the first boundary condition gives,

$$0 = y(0) = c_1 \cosh(0) + c_1 \sinh(0) = c_1(1) + c_2(0) = c_1 \qquad \Rightarrow \qquad c_1 = 0.$$

Applying the second boundary condition gives,

$$0 = y(2\pi) = c_2 \sinh\left(2\pi\sqrt{-\lambda}\right).$$

Because we are assuming $\lambda < 0$ we know that $2\pi\sqrt{-\lambda}\neq 0$ and so we also know that $\sinh\left(2\pi\sqrt{-\lambda}\right) \neq 0$. Therefore, much like the second case, we must have $c_2 = 0$.

So, for this BVP (again that's important), if we have we only get the trivial solution and so there are no negative eigenvalues. In summary then we will have $\lambda < 0$ the following eigenvalues/eigenfunctions for this BVP.

$$\lambda_n = \frac{n^2}{4} \qquad y_n(x) = \sin\left(\frac{nx}{2}\right) \qquad n = 1,2,3,...$$

Let's take a look at another example with slightly different boundary conditions.

Example: Find all the eigenvalues and eigenfunctions for the following BVP.

$$y'' + \lambda y = 0 \qquad y'(0) = 0 \qquad y'(2\pi) = 0.$$

Solution: Here we are going to work with derivative boundary conditions. The work is pretty much identical to the previous example however so we won't put in quite as much detail here. We'll need to go through all three cases just as the previous example so let's get started on that.

$\lambda > 0$

The general solution to the differential equation is identical to the previous example and so we have,

$$y(x) = c_1 \cos(\sqrt{\lambda}x) + c_2 \sin(\sqrt{\lambda}x).$$

Applying the first boundary condition gives us,

$$0 = y'(0) = \sqrt{\lambda}c_2 \quad \Rightarrow \quad c_2 = 0.$$

Recall that we are assuming that $\lambda > 0$here and so this will only be zero if $c_2 = 0$. Now, the second boundary condition gives us,

$$0 = y'(2\pi) = -\sqrt{\lambda}c_1 \sin(s\pi\sqrt{\lambda}).$$

Recall that we don't want trivial solutions and that so we will only get non-trivial solution if we require that,

$$\sin(2\pi\sqrt{\lambda}) = \quad \Rightarrow \quad 2\pi\sqrt{\lambda} = n\pi \quad n = 1, 2, 3, ...$$

Solving for $\lambda$and we see that we get exactly the same positive eigenvalues for this BVP that we got in the previous example.

$$\lambda_n = \left(\frac{n}{2}\right)^2 = \frac{n^2}{4} \qquad n = 1,2,3,...$$

The eigenfunctions that correspond to these eigenvalues however are,

$$y_n(x) = \cos\left(\frac{nx}{2}\right) \qquad n = 1,2,3,...$$

So, for this BVP we get cosines for eigenfunctions corresponding to positive eigenvalues. Now the second case.

$\lambda = 0$

The general solution is,

$$y(x) = c_1 + c_2 x.$$

Applying the first boundary condition gives,

$$0 = y'(0) = c_2.$$

Using this the general solution is then,

$$y(x) = c_1$$

and note that this will trivially satisfy the second boundary condition,

$$0 = y'(2\pi) = 0.$$

Therefore, unlike the first example, $\lambda = 0$is an eigenvalue for this BVP and the eigenfunctions corresponding to this eigenvalue is,

$$y(x) = 1.$$

Again, note that we dropped the arbitrary constant for the eigenfunctions.

Finally let's take care of the third case.

$\lambda < 0$

The general solution here is,

$$y(x) = c_1 \cosh\left(\sqrt{-\lambda}x\right) + c_2 \sinh(\sqrt{-\lambda}x).$$

Applying the first boundary condition gives,

$$0 = y'(0) = \sqrt{-\lambda}c_1 \sinh(0) + \sqrt{-\lambda}c_2 \cosh(0) = \sqrt{-\lambda}c_2 \quad \Rightarrow \quad c_2 = 0.$$

Applying the second boundary condition gives,

$$0 = y'(2\pi) = \sqrt{-\lambda}c_1 \sinh\left(2\pi\sqrt{-\lambda}\right).$$

As with the previous example we again know that $2\pi\sqrt{-\lambda} \neq 0$and so sinh (2p$\sqrt{-\lambda}$) $\neq 0$. Therefore we must have $c_1 = 0$.

So, for this BVP we again have no negative eigenvalues.

In summary then we will have the following eigenvalues/eigenfunctions for this BVP.

$$\lambda_n = \frac{n^2}{4} \qquad y_n(x) = \cos\left(\frac{nx}{2}\right) \qquad n = 1,2,3,...$$

$$\lambda_0 = 0 \qquad y_0(x) = 1.$$

Notice as well that we can actually combine these if we allow the list of n's for the first one to start at zero instead of one. This will often not happen, but when it does we'll take advantage of it. So the "official" list of eigenvalues/eigenfunctions for this BVP is,

$$\lambda_n = \frac{n^2}{4} \quad y_n(x) = \cos\left(\frac{nx}{2}\right) \qquad n = 0,1,2,3,...$$

So, in the previous two examples we saw that we generally need to consider cases for λas different values will often lead to different general solutions. Do not get too locked into the cases we did here. We will mostly be solving this particular differential equation and so it will be tempting to assume that these are always the cases that we'll be looking at, but there are BVP's that will require other/different cases.

Also, as we saw in the two examples sometimes one or more of the cases will not yield any eigenvalues. This will often happen, but again we shouldn't read anything into the fact that we didn't have negative eigenvalues for either of these two BVP's. There are BVP's that will have negative eigenvalues.

Let's take a look at another example with a very different set of boundary conditions. These are not the traditional boundary conditions that we've been looking at to this point, but we'll see in the next chapter how these can arise from certain physical problems.

Example: Find all the eigenvalues and eigenfunctions for the following BVP.

$$y''+\lambda y=0 \qquad y(-\pi)=y(\pi) \qquad y'(-\pi)=y'(\pi).$$

Solution: So, in this example we aren't actually going to specify the solution or its derivative at the boundaries. Instead we'll simply specify that the solution must be the same at the two boundaries and the derivative of the solution must also be the same at the two boundaries. Also, this type of boundary condition will typically be on an interval of the form [-L,L] instead of [0,L] as we've been working on to this point.-

As mentioned above this kind of boundary conditions arise very naturally in certain physical problems and we'll see that in the next chapter.

As with the previous two examples we still have the standard three cases to look at.

$\lambda > 0$

The general solution for this case is,

$$y(x)=c_1\cos\left(\sqrt{\lambda}x\right)+c_2\sin\left(\sqrt{\lambda}x\right).$$

Applying the first boundary condition and using the fact that cosine is an even function (*i.e.*) and that sine is an odd function (*i.e.* sin(–*x*) = – (*x*)).gives us,

$$c_1\cos\left(-\pi\sqrt{\lambda}\right)+c_2\sin\left(-\pi\sqrt{\lambda}\right)=c_1\cos\left(\pi\sqrt{\lambda}\right)+c_2\sin\left(\pi\sqrt{\lambda}\right)$$
$$c_1\cos\left(\pi\sqrt{\lambda}\right)-c_2\sin\left(\pi\sqrt{\lambda}\right)=c_1\cos\left(\pi\sqrt{\lambda}\right)+c_2\sin\left(\pi\sqrt{\lambda}\right)$$
$$-c_1\sin\left(\pi\sqrt{\lambda}\right)=c_2\sin\left(\pi\sqrt{\lambda}\right)$$
$$0=2c_2\sin\left(\pi\sqrt{\lambda}\right).$$

This time, unlike the previous two examples this doesn't really tell us anything. We could have $\sin\left(\pi\sqrt{\lambda}\right)=0$but it is also completely possible, at this point in the problem anyway, for us to have $c_2=0$as well.

So, let's go ahead and apply the second boundary condition and see if we get anything out of that.

$$-\sqrt{\lambda}c_1\sin\left(-\pi\sqrt{\lambda}\right)+\sqrt{\lambda}\,c_2\cos\left(-\pi\sqrt{\lambda}\right)=-\sqrt{\lambda}\,c_1\sin\left(\pi\sqrt{\lambda}\right)+\sqrt{\lambda}\,c_2\cos\left(\pi\sqrt{\lambda}\right)$$
$$\sqrt{\lambda}\,c_1\sin\left(\pi\sqrt{\lambda}\right)+\sqrt{\lambda}\,c_2\cos\left(\pi\sqrt{\lambda}\right)=-\sqrt{\lambda}\,c_1\sin\left(\pi\sqrt{\lambda}\right)+\sqrt{\lambda}\,c_2\cos\left(\pi\sqrt{\lambda}\right)$$

$$\sqrt{\lambda}\, c_1 \sin\left(\pi\sqrt{\lambda}\right) = -\sqrt{\lambda}\, c_1 \sin\left(\pi\sqrt{\lambda}\right)$$

$$2\sqrt{\lambda}\, c_1 \sin\left(\pi\sqrt{\lambda}\right) = 0.$$

So, we get something very similar to what we got after applying the first boundary condition. Since we are assuming that $\lambda > 0$this tells us that either sin $(\pi\sqrt{\lambda})$or $c_1 = 0$.

Note however that if sin $(\pi\sqrt{\lambda}) \neq 0$then we will have to have $c_1 = c_2 = 0$and we'll get the trivial solution. We therefore need to require that sin $(\pi\sqrt{\lambda})= 0$ and so just as we've done for the previous two examples we can now get the eigenvalues,

$$\pi\sqrt{\lambda} = n\pi \qquad \Rightarrow \qquad \lambda = n^2 \qquad n = 1,2,3,...$$

Recalling that $\lambda > 0$and we can see that we do need to start the list of possible *n*'s at one instead of zero.

So, we now know the eigenvalues for this case, but what about the eigenfunctions. The solution for a given eigenvalue is,

$$y(x) = c_1 \cos(nx) + c_2 \sin(nx).$$

and we've got no reason to believe that either of the two constants are zero or non-zero for that matter. In cases like these we get two sets of eigenfunctions, one corresponding to each constant. The two sets of eigenfunctions for this case are,

$$y_n(x) = \cos(nx) \qquad y_n(x) = \sin(nx) \qquad n = 1,2,3,...$$

Now the second case.

$\lambda = 0$

The general solution is,

$$y(x) = c_1 + c_2 x.$$

Applying the first boundary condition gives,

$$c_1 + c_2(-\pi) = c_1 + c_2(\pi)$$

$$2\pi c_2 = 0 \qquad \Rightarrow \qquad c_2 = 0.$$

Using this the general solution is then,

$$y(x) = c_1$$

and note that this will trivially satisfy the second boundary condition just as we saw in the second example above. Therefore we again have as an eigenvalue for this BVP and the eigenfunctions corresponding to this eigenvalue is,

$$y(x) = 1.$$

Finally let's take care of the third case.

$\lambda < 0$

The general solution here is,

$$y(x) = c_1 \cosh\left(\sqrt{-\lambda}\,x\right) + c_2 \sinh\left(\sqrt{-\lambda}\,x\right).$$

Applying the first boundary condition and using the fact that hyperbolic cosine is even and hyperbolic sine is odd gives,

$$c_1 \cosh\left(-\pi\sqrt{-\lambda}\right) + c_2 \sinh\left(-\pi\sqrt{-\lambda}\right) = c_1 \cosh \pi\left(\pi\sqrt{-\lambda}\right) + c_2 \sinh\left(\pi\sqrt{-\lambda}\right)$$

$$-c_2 \sinh\left(-\pi\sqrt{-\lambda}\right) = c_2 \sinh\left(\pi\sqrt{-\lambda}\right)$$

$$2c_2 \sinh\left(\pi\sqrt{-\lambda}\right) = 0.$$

Now, in this case we are assuming that $\lambda < 0$and so we know that $\pi\sqrt{-\lambda} \neq 0$which in turn tells us that sinh $\left(\pi\sqrt{-\lambda}\right) \neq 0$. We therefore must have $c_2 = 0$.

Let's now apply the second boundary condition to get,

$$\sqrt{-\lambda}\, c_1 \sinh\left(-\pi\sqrt{-\lambda}\right) = \sqrt{-\lambda}\, c_1 \sinh\left(\pi\sqrt{-\lambda}\right)$$

$$2\sqrt{-\lambda}\, c_1 \sinh\left(-\pi\sqrt{-\lambda}\right) = 0 \qquad \Rightarrow \qquad c_1 = 0.$$

By our assumption on we again have no choice here but to have $c_1 = 0$.

Therefore, in this case the only solution is the trivial solution and so, for this BVP we again have no negative eigenvalues.

In summary then we will have the following eigenvalues/eigenfunctions for this BVP.

$$\lambda_n = n^2 \qquad y_n(x) = \sin(nx) \qquad n = 1, 2, 3, \ldots$$

$$\lambda_n = n^2 \qquad y_n(x) = \cos(nx) \qquad n = 1, 2, 3, \ldots \;\; \lambda_n = 0 \qquad y_0(x) = 1.$$

Note that we've acknowledged that for $\lambda > 0$we had two sets of eigenfunctions by listing them each separately. Also, we can again combine the last two into one set of eigenvalues and eigenfunctions. Doing so gives the following set of eigenvalues and eigenfunctions.

$$\lambda_n = n^2 \qquad y_n(x) = \sin(nx) \qquad n = 1, 2, 3, \ldots$$

$$\lambda_n = n^2 \qquad y_n(x) = \cos(nx) \qquad n = 0, 1, 2, 3, \ldots$$

Once again we've got an example with no negative eigenvalues. We can't stress enough that this is more a function of the differential equation we're working with than anything and there will be examples in which we may get negative eigenvalues.

Now, to this point we've only worked with one differential equation so let's work an example with a different differential equation just to make sure that we don't get too locked into this one differential equation.

Before working this example let's note that we will still be working the vast majority of our examples with the one differential equation we've been using to this point.

We're working with this other differential equation just to make sure that we don't get too locked into using one single differential equation.

Example: Find all the eigenvalues and eigenfunctions for the following BVP.

$$x^2 y'' + 3xy' + \lambda y = 0 \qquad \lambda(1) = 0 \qquad y(2) = 0.$$

Solution: This is an Euler differential equation and so we know that we'll need to find the roots of the following quadratic.

$$r(r-1) + 3r + \lambda = r^2 + 2r + \lambda = 0.$$

The roots to this quadratic are,

$$r_{1,2} = \frac{-2 \pm \sqrt{4 - 4\lambda}}{2} = -1 \pm \sqrt{1-\lambda}.$$

Now, we are going to again have some cases to work with here, however they won't be the same as the previous examples. The solution will depend on whether or not the roots are real distinct, double or complex and these cases will depend upon the sign/value of $1-\lambda$. So, let's go through the cases.

$$\underline{1-\lambda < 0, \lambda > 1.}$$

In this case the roots will be complex and we'll need to write them as follows in order to write down the solution.

$$r_{1,2} = -1 \pm \sqrt{1-\lambda} = -1 \pm \sqrt{-(\lambda-1)} = -1 \pm i\sqrt{\lambda-1}.$$

By writing the roots in this fashion we know that $\lambda - 1 > 0$and so $\sqrt{\lambda-1}$ is now a real number, which we need in order to write the following solution,

$$y(x) = c_1 x^{-1} \cos\left(\ln(x)\sqrt{\lambda-1}\right) + c_2 x^{-1} \sin\left(\ln(x)\sqrt{\lambda-1}\right).$$

Applying the first boundary condition gives us,

$$0 = y(1) = c_1 \cos(0) + c_2 \sin(0) = c_1, \quad \Rightarrow \quad c_1 = 0.$$

The second boundary condition gives us,

$$0 = y(2) = \frac{1}{2} c_2 \sin\left(\ln(2)\sqrt{\lambda-1}\right).$$

In order to avoid the trivial solution for this case we'll require,

$$\sin\left(\ln(2)\sqrt{\lambda-1}\right)=0 \qquad \Rightarrow \qquad \ln(2)\sqrt{\lambda-1}=n\pi \qquad n=1,2,3,\ldots.$$

This is much more complicated of a condition than we've seen to this point, but other than that we do the same thing. So, solving for λ gives us the following set of eigenvalues for this case.

$$\lambda_n = 1+\left(\frac{n\pi}{\ln 2}\right)^2 \qquad n=1,2,3,\ldots$$

Note that we need to start the list of n's off at one and not zero to make sure that we have $\lambda > 1$ as we're assuming for this case.

The eigenfunctions that correspond to these eigenvalues are,

$$y_n(x)=\sin\left(\frac{n\pi}{\ln 2}\ln(x)\right) \qquad n=1,2,3,\ldots$$

Now the second case.

$1-\lambda=0,\ \lambda=1$

In this case we get a double root of $r_{1,2}=-1$ and so the solution is,

$$y(x)=c_1x^{-1}+c_2x^{-1}\ln(x).$$

Applying the first boundary condition gives,

$$0=y(1)=c_1.$$

The second boundary condition gives,

$$0=y(2)=\frac{1}{2}c_2\ln(2) \qquad \Rightarrow \qquad c_2=0.$$

We therefore have only the trivial solution for this case and so $\lambda=1$ is not an eigenvalue. Let's now take care of the third (and final) case.

$1-\lambda>0,\ \lambda<1$

This case will have two real distinct roots and the solution is,

$$y(x)=c_1x^{-1+\sqrt{1-\lambda}}+c_2x^{-1-\sqrt{1-\lambda}}.$$

Applying the first boundary condition gives,

$$0=y(1)=c_1+c_2 \qquad \Rightarrow \qquad c_2=-c_1.$$

Using this our solution becomes,

$$y(x)=c_1x^{-1+\sqrt{1-\lambda}}-c_1x^{-1-\sqrt{1-\lambda}}.$$

Applying the second boundary condition gives,

$$0 = y(2) = c_1 2^{-1+\sqrt{1-\lambda}} - c_1 2^{-1-\sqrt{1-\lambda}} = c_1\left(2^{-1+\sqrt{1-\lambda}} - 2^{-1-\sqrt{1-\lambda}}\right).$$

Now, because we know that $\lambda \neq 1$ for this case the exponents on the two terms in the parenthesis are not the same and so the term in the parenthesis is not the zero. This means that we can only have,

$$c_1 = c_2 = 0$$

and so in this case we only have the trivial solution and there are no eigenvalues for which $\lambda < 1$. The only eigenvalues for this BVP then come from the first case.

So, we've now worked an example using a differential equation other than the "standard" one we've been using to this point. As we saw in the work however, the basic process was pretty much the same. We determined that there were a number of cases (three here, but it won't always be three) that gave different solutions. We examined each case to determine if non-trivial solutions were possible and if so found the eigenvalues and eigenfunctions corresponding to that case.-

We need to work one last example in this section before we leave this section for some new topics. The four examples that we've worked to this point we're all fairly simple (with simple being relative of course...), however we don't want to leave without acknowledging that many eigenvalue/eigenfunctions problems are so easy.

In many examples it is not even possible to get a complete list of all possible eigenvalues for a BVP. Often the equations that we need to solve to get the eigenvalues are difficult if not impossible to solve exactly. So, let's take a look at one example like this to see what kinds of things can be done to at least get an idea of what the eigenvalues look like in these kinds of cases.

Example: Find all the eigenvalues and eigenfunctions for the following BVP.

$$y'' + \lambda y = 0 \qquad y(0) = 0 \qquad y'(1) + y(1) = 0.$$

Solution: The boundary conditions for this BVP are fairly different from those that we've worked with to this point. However, the basic process is the same. So let's start off with the first case.

$\lambda > 0$

The general solution to the differential equation is identical to the first few examples and so we have,

$$y(x) = c_1 \cos\left(\sqrt{\lambda}x\right) + c_2 \sin\left(\sqrt{\lambda}x\right).$$

Applying the first boundary condition gives us,

$$0 = y(0) = c_1 \qquad \Rightarrow \qquad c_1 = 0.$$

The second boundary condition gives us,

$$0 = y(1) + y'(1) = c_2 \sin(\sqrt{\lambda}) + \sqrt{\lambda}\, c_2 \cos(\sqrt{\lambda}).$$

$$= c_2\left(\sin(\sqrt{\lambda}) + \sqrt{\lambda}\, c_2 \cos(\sqrt{\lambda})\right).$$

So, if we let $c_2 = 0$ we'll get the trivial solution and so in order to satisfy this boundary condition we'll need to require instead that,

$$0 = \sin(\sqrt{\lambda}) + \sqrt{\lambda}\cos(\sqrt{\lambda})$$

$$\sin(\sqrt{\lambda}) = -\sqrt{\lambda}\cos(\sqrt{\lambda})$$

$$\tan(\sqrt{\lambda}) = -\sqrt{\lambda}.$$

Now, this equation has solutions but we'll need to use some numerical techniques in order to get them. In order to see what's going on here let's graph tan ($\sqrt{\lambda}$)and $-\sqrt{\lambda}$ on the same graph. Here is that graph and note that the horizontal axis really is values of $\sqrt{\lambda}$ as that will make things a little easier to see and relate to values that we're familiar with.

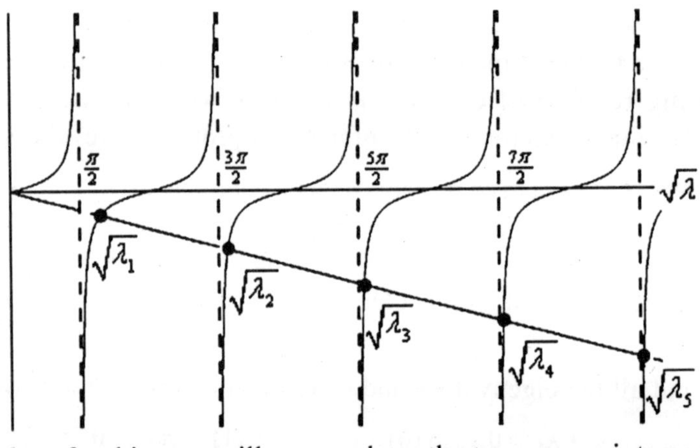

So, eigenvalues for this case will occur where the two curves intersect. We've shown the first five on the graph and again what is showing on the graph is really the square root of the actual eigenvalue as we've noted.

The interesting thing to note here is that the farther out on the graph the closer the eigenvalues come to the asymptotes of tangent and so we'll take advantage of that and say that for large enough n we can approximate the eigenvalues with the (very well known) locations of the asymptotes of tangent.

How large the value of n is before we start using the approximation will depend on how much accuracy we want, but since we know the location of the asymptotes and as n increases the accuracy of the approximation will increase so it will be easy enough to check for a given accuracy.

For the purposes of this example we found the first five numerically and then we'll use the approximation of the remaining eigenvalues. Here are those values/approximations.

$$\sqrt{\lambda_1} = 2.0288 \qquad \lambda_1 = 4.1160 \qquad (2.4674)$$

$$\sqrt{\lambda_2} = 4.9132 \qquad \lambda_2 = 24.1395 \qquad (22.2066)$$

$$\sqrt{\lambda_3} = 7.9787 \qquad \lambda_3 = 63.6597 \qquad (61.6850)$$

$$\sqrt{\lambda_4} = 11.0855 \qquad \lambda_4 = 122.8883 \qquad (120.9027)$$

$$\sqrt{\lambda_5} = 14.2074 \qquad \lambda_5 = 201.8502 \qquad (199.8595)$$

$$\sqrt{\lambda_n} \approx \frac{2n-1}{2}\pi \qquad \lambda_n \approx \frac{(2n-1)^2}{4}\pi^2 \qquad n = 6, 7, 8, \ldots$$

The number in parenthesis after the first five is the approximate value of the asymptote. As we can see they are a little off, but by the time we get to $n = 5$the error in the approximation is 0.9862%. So less than 1% error by the time we get to and it will only get better for larger value of n. The eigenfunctions for this case are,

$$y_n(x) = \sin\left(\sqrt{\lambda_n}\, x\right) \qquad n = 1,2,3,\ldots$$

where the values of $\lambda_n$are given above.

So, now that all that work is out of the way let's take a look at the second case.

$\lambda = 0$

The general solution is,

$$y(x) = c_1 + c_2 x.$$

Applying the first boundary condition gives,

$$0 = y(0) = c_1.$$

Using this the general solution is then,

$$y(x) = c_2 x.$$

Applying the second boundary condition to this gives,

$$0 = y'(1) + y(1) = c_2 + c_2 = 2c_2 \qquad \Rightarrow \qquad c_2 = 0.$$

Therefore for this case we get only the trivial solution and so $\lambda = 0$is not an eigenvalue. Note however that had the second boundary condition been $y'(1) - y(1) = 0$then $\lambda = 0$would have been an eigenvalue (with eigenfunctions $y(x) = x$) and so again we need to be careful about reading too much into our work here.

Finally let's take care of the third case.

$\lambda < 0$

The general solution here is,

$$y(x) = c_1 \cosh\left(\sqrt{-\lambda}\,x\right) + c_2 \sinh\left(\sqrt{-\lambda}\,x\right).$$

Applying the first boundary condition gives,

$$0 = y(0) = c_1 \cosh(0) + c_2 \sinh(0) = c_1 \qquad \Rightarrow \qquad c_1 = 0.$$

Using this the general solution becomes,

$$y(x) = c_2 \sinh\left(\sqrt{-\lambda}\,x\right).$$

Applying the second boundary condition to this gives,

$$\begin{aligned} 0 = y'(1) + y(1) &= \sqrt{-\lambda}\,c_2 \cosh\left(\sqrt{-\lambda}\right) + c_2 \sinh\left(\sqrt{-\lambda}\right) \\ &= c_2\left(\sqrt{-\lambda}\cosh\left(\sqrt{-\lambda}\right) + \sinh\left(\sqrt{-\lambda}\right)\right). \end{aligned}$$

Now, by assumption we know that $\lambda < 0$ and so $\sqrt{-\lambda} > 0$. This in turn tells us that $\sinh\left(\sqrt{-\lambda}\right) > 0$and we know that $\cosh(x) > 0$for all x. Therefore,

$$\sqrt{-\lambda}\cosh\left(\sqrt{-\lambda}\right) + \sinh\left(\sqrt{-\lambda}\right) \neq 0$$

and so we must have $c_2 = 0$and once again in this third case we get the trivial solution and so this BVP will have no negative eigenvalues.

In summary the only eigenvalues for this BVP come from assuming that $\lambda > 0$and they are given above.

So, we've worked several eigenvalue/eigenfunctions examples in this section. Before leaving this section we do need to note once again that there are a vast variety of different problems that we can work here and we've really only shown a bare handful of examples and so please do now walk away from this section believing that we've shown you everything.

Chapter 5

Intervals of Validity

This section is called Intervals of Validity because all of the examples will involve them. However, there is a lot more to this section. We will see a couple of theorems that will tell us when we can solve a differential equation. We will also see some of the differences between linear and nonlinear differential equations. First let's take a look at a theorem about linear first order differential equations. This is a very important theorem although we're not going to really use it for its most important aspect.

Theorem: Consider the following IVP.

$$y' + p(t)y = g(t) \qquad (t_0) = y_0$$

If *p(t)* and *g(t)* are continuous functions on an open interval $\alpha < t < \beta$ and the interval contains t_o, then there is a unique solution to the IVP on that interval.

So, just what does this theorem tell us? First, it tells us that for nice enough linear first order differential equations solutions are guaranteed to exist and more importantly the solution will be unique. We may not be able to find the solution, but do know that it exists and that there will only be one of them. This is the very important aspect of this theorem. Knowing that a differential equation has a unique solution is sometimes more important than actually having the solution itself!

Next, if the interval in the theorem is the largest possible interval on which *p(t)* and *g(t)* are continuous then the interval is the interval of validity for the solution. This means, that for linear first order differential equations, we won't need to actually solve the differential equation in order to find the interval of validity. Notice as well that the interval of validity will depend only partially on the initial condition. The interval must contain t_o, but the value of y_o, has no effect on the interval of validity. Let's take a look at an example.

Example: Without solving, determine the interval of validity for the following initial value problem.

$$\left(t^2 - 9\right)y' + 2y = \mathrm{In}\left|20 - 4t\right| \qquad y(4) = -3$$

Solution: First, in order to use the theorem to find the interval of validity we must write the differential equation in the proper form given in the theorem. So we will need to divide out by the coefficient of the derivative.

$$y' + \frac{2}{t^2 - 9}y = \frac{In|20 - 4t|}{t^2 - 9}$$

Next, we need to identify where the two functions are not continuous. This will allow us to find all possible intervals of validity for the differential equation. So, *p(t)* will be discontinuous at $t = \pm\, 3$since these points will give a division by zero. Likewise, *g(t)* will also be discontinuous at $t = \pm\, 3$as well as $t = 5$ since at this point we will have the natural logarithm of zero. Note that in this case we won't have to worry about natural log of negative numbers because of the absolute values.

Now, with these points in hand we can break up the real number line into four intervals where both *p(t)* and *g(t)* will be continuous. These four intervals are,

$$-\infty < t < -3 \qquad -3 < t < 3 \qquad 3 < t < 5 \qquad 5 < t < \infty$$

The endpoints of each of the intervals are points where at least one of the two functions is discontinuous. This will guarantee that both functions are continuous everywhere in each interval.

Finally, let's identify the actual interval of validity for the initial value problem. The actual interval of validity is the interval that will contain $t_o = 4$. So, the interval of validity for the initial value problem is.

$$3 < \mathrm{t} < 5$$

In this last example we need to be careful to not jump to the conclusion that the other three intervals cannot be intervals of validity. By changing the initial condition, in particular the value of t_o, we can make any of the four intervals the interval of validity.

The first theorem required a linear differential equation. There is a similar theorem for non-linear first order differential equations. This theorem is not as useful for finding intervals of validity as the first theorem was so we won't be doing all that much with it.

Here is the theorem.

Theorem: Consider the following IVP.

$$y' = f(t, y) \qquad\qquad y(t_0 = y_0)$$

If *f(t,y)* and are continuous functions in some rectangle $\alpha < \mathrm{t} < \beta\ \gamma < \mathrm{y} < \delta$, containing the point (t_o, y_o) then there is a unique solution to the IVP in some interval $t_o \_h < t < t_o + h$ that is contained in $\alpha < t < \beta$.

That's it. Unlike the first theorem, this one cannot really be used to find an interval of validity. So, we will know that a unique solution exists if the conditions of the theorem are met, but we will actually need the solution in order to determine its interval of validity. Note as well that for non-linear differential equations it appears that the value of y_0 may affect the interval of validity.

Here is an example of the problems that can arise when the conditions of this theorem aren't met.

Example: Determine all possible solutions to the following IVP.

$$y' = y^{\frac{1}{3}} \qquad y(0) = 0$$

Solution: First, notice that this differential equation does NOT satisfy the conditions of the theorem.

$$f(y) = y^{\frac{1}{3}} \qquad \frac{df}{dy} = \frac{1}{3y^{\frac{2}{3}}}$$

So, the function is continuous on any interval, but the derivative is not continuous at $y = 0$ and so will not be continuous at any interval containing $y = 0$. In order to use the theorem both must be continuous on an interval that contains $y_o = 0$ and this is problem for us since we do have $y_o = 0$.

Now, let's actually work the problem. This differential equation is separable and is fairly simple to solve.

$$\int y^{-\frac{1}{3}} dy = \int dt$$

$$\frac{3}{2} y^{\frac{2}{3}} = t + c$$

Applying the initial condition gives $c = 0$ and so the solution is.

$$\frac{3}{2} y^{\frac{2}{3}} = t$$

$$y^{\frac{2}{3}} = \frac{2}{3} t$$

$$y^2 = \left(\frac{2}{3} t\right)^3$$

$$y(t) = \pm\left(\frac{2}{3} t\right)^{\frac{3}{2}}$$

So, we've got two possible solutions here, both of which satisfy the differential equation and the initial condition. There is also a third solution to the IVP. $y(t) = 0$ is also a solution to the differential equation and satisfies the initial condition.

In this last example we had a very simple IVP and it only violated one of the conditions of the theorem, yet it had three different solutions. All the examples we've worked in the previous sections satisfied the conditions of this theorem and had a single unique solution to the IVP. This example is a useful reminder of the fact that, in the field of differential equations, things don't always behave nicely. It's easy to forget this as most of the problems that are worked in a differential equations class are nice and behave in a nice, predictable manner.

Let's work one final example that will illustrate one of the differences between linear and non-linear differential equations.

Example: Determine the interval of validity for the initial value problem below and give its dependence on the value of y_o

$$y' = y^2 \qquad\qquad y(0) = y_0$$

Solution: Before proceeding in this problem, we should note that the differential equation is non-linear and meets both conditions of the Theorem 2 and so there will be a unique solution to the IVP for each possible value of y_o.

Also, note that the problem asks for any dependence of the interval of validity on the value of y_o. This immediately illustrates a difference between linear and non-linear differential equations. Intervals of validity for linear differential equations do not depend on the value of y_o. Intervals of validity for non-linear differential can depend on the value of y_o as we pointed out after the second theorem.

So, let's solve the IVP and get some intervals of validity.

First note that if $y_o = 0$ then $y(t) = 0$ is the solution and this has an interval of validity of

$$-\infty < t < \infty$$

So for the rest of the problem let's assume that $y_0 \neq 0$. Now, the differential equation is separable so let's solve it and get a general solution.

$$\int y^{-2} dy = \int dt$$

$$-\frac{1}{y} = t + c$$

Applying the initial condition gives

$$c = -\frac{1}{y_0}$$

The solution is then.

$$-\frac{1}{y} = t - \frac{1}{y_0}$$

$$y(t) = \frac{1}{\frac{1}{y_0} - t}$$

$$y(t) = \frac{y_0}{1 - y_0 t}$$

Now that we have a solution to the initial value problem we can start finding intervals of validity. From the solution we can see that the only problem that we'll have is division by zero at

$$t = \frac{1}{y_0}$$

This leads to two possible intervals of validity.

$$-\infty < t < \frac{1}{y_0}$$

$$\frac{1}{y_0} < t < \infty$$

That actual interval of validity will be the interval that contains $t_o = 0$. This however, depends on the value of y_o. If $y_o < 0$ then $\frac{1}{y_0} < 0$ and so the second interval will contain $t_o = 0$. Likewise if $y_o > 0$ then $\frac{1}{y_0} < 0$ and in this case the first interval will contain $t_o = 0$.

This leads to the following possible intervals of validity, depending on the value of y_o.

If $y_0 > 0$ $\quad -\infty < t < \frac{1}{y_0}$ is the interval of validity

If $y_0 = 0$ $\quad -\infty < t < \infty$ is the interval of validity

If $y_0 < 0$ $\quad \frac{1}{y_0} < t < \infty$ is the interval of validity

On a side note, notice that the solution, in its final form, will also work if $y_o = 0$.

So what did this example show us about the difference between linear and non-linear differential equations?

First, as pointed out in the solution to the example, intervals of validity for non-linear differential equations can depend on the value of y_o, whereas intervals of validity for linear differential equations don't.

Second, intervals of validity for linear differential equations can be found from the differential equation with no knowledge of the solution. This is definitely not the case with non-linear differential equations. It would be very difficult to see how any of these intervals in the last example could be found from the differential equation. Knowledge of the solution was required in order for us to find the interval of validity.

MODELING WITH FIRST ORDER DIFFERENTIAL EQUATIONS

We now move into one of the main applications of differential equations both in this class and in general. Modeling is the process of writing a differential equation to describe a physical situation. Almost all of the differential equations that you will use in your job (for the engineers out there in the audience) are there because somebody, at some time, modeled a situation to come up with the differential equation that you are using.

This section is not intended to completely teach you how to go about modeling all physical situations. A whole course could be devoted to the subject of modeling and still not cover everything! This section is designed to introduce you to the process of modeling and show you what is involved in modeling. We will look at three different situations in this section : Mixing Problems, Population Problems, and Falling Bodies.

In all of these situations we will be forced to make assumptions that do not accurately depict reality in most cases, but without them the problems would be very difficult and beyond the scope of this discussion (and the course in most cases to be honest).

So let's get started.

MIXING PROBLEMS

In these problems we will start with a substance that is dissolved in a liquid. Liquid will be entering and leaving a holding tank. The liquid entering the tank may or may not contain more of the substance dissolved in it. Liquid leaving the tank will of course contain the substance dissolved in it. If $Q(t)$ gives the amount of the substance dissolved in the liquid in the tank at any time t we want to develop a differential equation that, when solved, will give us an expression for $Q(t)$. Note as well that in many situations we can think of air as a liquid for the purposes of these kinds of discussions and so we don't actually need to have an actual liquid, but could instead use air as the "liquid".

The main assumption that we'll be using here is that the concentration of the substance in the liquid is uniform throughout the tank. Clearly this will not be the case, but if we allow the concentration to vary depending on the location in the tank the problem becomes very difficult and will involve partial differential equations, which is not the focus of this course.

The main "equation" that we'll be using to model this situation is :

Rate of change of $Q(t)$	=	Rate at which $Q(t)$ enters the tank	—	Rate at which $Q(t)$ exits the tank

where,

Rate of change of $Q(t) = \dfrac{dQ}{dt} = Q'(t)$

Rate at which $Q(t)$ enters the tank = (flow rate of liquid entering) x (concentration of substance in liquid entering)

Rate at which $Q(t)$ exits the tank = (flow rate of liquid exiting) x (concentration of substance in liquid exiting)

Let's take a look at the first problem.

Example: A 1500 gallon tank initially contains 600 gallons of water with 5 lbs of salt dissolved in it. Water enters the tank at a rate of 9 gal/hr and the water entering the tank has a salt concentration of lbs/gal. If a well mixed solution leaves the tank at a rate of 6 gal/hr, how much salt is in the tank when it overflows?

Solution: First off, let's address the "well mixed solution" bit. This is the assumption that was mentioned earlier. We are going to assume that the instant the water enters the tank it somehow instantly disperses evenly throughout the tank to give a uniform concentration of salt in the tank at every point. Again, this will clearly not be the case in reality, but it will allow us to do the problem.

Now, to set up the IVP that we'll need to solve to get $Q(t)$ we'll need the flow rate of the water entering (we've got that), the concentration of the salt in the water entering (we've got that), the flow rate of the water leaving (we've got that) and the concentration of the salt in the water exiting (we don't have this yet).

So, we first need to determine the concentration of the salt in the water exiting the tank. Since we are assuming a uniform concentration of salt in the tank the concentration at any point in the tank and hence in the water exiting is given by,

$$\text{Concentration} = \frac{\text{Amount of salt in the tank at any time. } t}{\text{Volume of Water in the tank at any time } t}$$

The amount at any time t is easy it's just $Q(t)$. The volume is also pretty easy. We start with 600 gallons and every hour 9 gallons enters and 6 gallons leave. So, if we use t in hours, every hour 3 gallons enters the tank, or at any time t there is $600 + 3t$ gallons of water in the tank.

So, the IVP for this situation is,

$$Q'(t) = (9)\left(\frac{1}{5}(1+\cos(t))\right) - (6)\left(\frac{Q(t)}{600+3t}\right) \qquad Q(0)=5$$

$$Q'(t) = \frac{9}{5}(1+\cos(t)) - \frac{2Q(t)}{200+t} \qquad Q(0)=5$$

This is a linear differential equation and it isn't too difficult to solve (hopefully). We will show most of the details, but leave the description of the solution process out. If you need a refresher on solving linear first order differential equations go back and take a look at that section.

$$Q'(t) + \frac{2Q(t)}{200+t} = \frac{9}{5}(1+\cos(t))$$

$$\mu(t) = e^{\int \frac{2}{200} dt} = e^{2\ln(200+t)} = (200+t)^2$$

$$\int \left((200+t)^2 Q(t)\right)' dt = \int \frac{9}{5}(200+t)^2 (1+\cos(t)) dt$$

$$(200+t)^2 Q(t) = \frac{9}{5}\left(\frac{1}{3}(200+t)^3 + (200+t)^2 \sin(t) + 2(200+t)\cos(t) - 2\sin(t)\right) + c$$

$$Q(t) = \frac{9}{5}\left(\frac{1}{3}(200+t) + \sin(t) + \frac{2\cos(t)}{200+t} - \frac{2\sin(t)}{(200+t)^2}\right) + \frac{c}{(200+t)^2}$$

So, here's the general solution. Now, apply the initial condition to get the value of the constant, c.

$$5 = Q(0) = \frac{9}{5}\left(\frac{1}{3}(200) + \frac{2}{200}\right) + \frac{c}{(200)^2} \qquad c = -4600720$$

So, the amount of salt in the tank at any time t is.

$$Q(t) = \frac{9}{5}\left(\frac{1}{3}(200+t) + \sin(t) + \frac{2\cos(t)}{200+t} - \frac{2\sin(t)}{(200+t)^2}\right) - \frac{4600720}{(200+t)^2}$$

Now, the tank will overflow at $t = 300$ hrs. The amount of salt in the tank at that time is.

$$Q(300) = 279.797 \text{lbs}$$

Here's a graph of the salt in the tank before it overflows.

Note that the whole graph should have small oscillations in it as you can see in the range from 200 to 250. The scale of the oscillations however was small enough that the program used to generate the image had trouble showing all of them.

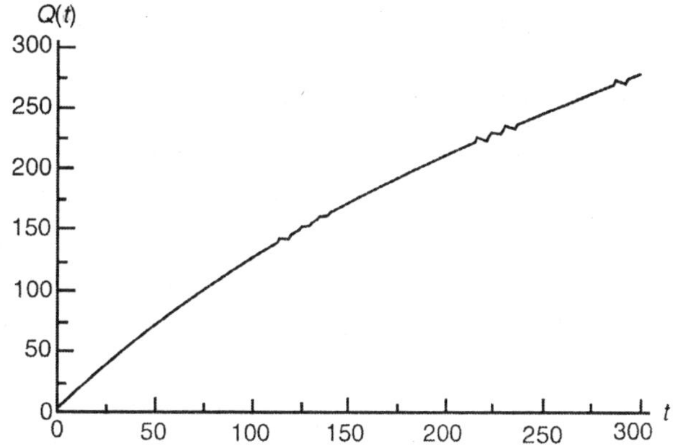

The work was a little messy with that one, but they will often be that way so don't get excited about it. This first example also assumed that nothing would change throughout the life of the process. That, of course will usually not be the case. Let's take a look at an example where something changes in the process.

Example: A 1000 gallon holding tank that catches runoff from some chemical process initially has 800 gallons of water with 2 ounces of pollution dissolved in it. Polluted water flows into the tank at a rate of 3 gal/hr and contains 5 ounces/gal of pollution in it. A well mixed solution leaves the tank at 3 gal/hr as well. When the amount of pollution in the holding tank reaches 500 ounces the inflow of polluted water is cut off and fresh water will enter the tank at a decreased rate of 2 gallons while the outflow is increased to 4 gal/hr. Determine the amount of pollution in the tank at any time t.

Solution: Okay, so clearly the pollution in the tank will increase as time passes. If the amount of pollution ever reaches the maximum allowed there will be a change in the situation. This will necessitate a change in the differential equation describing the process as well. In other words, we'll need two IVP's for this problem. One will describe the initial situation when polluted runoff is entering the tank and one for after the maximum allowed pollution is reached and fresh water is entering the tank.

Here are the two IVP's for this problem.

$$Q_1'(t) = (3)(5) - (3)\left(\frac{Q_1(t)}{800}\right) \qquad Q_1(0) = 2 \qquad 0 \le t \le t_m$$

$$Q_2'(t) = (2)(0) - (4)\left(\frac{Q_2(t)}{800 - 2(t - t_m)}\right) \qquad Q_2(t_m) = 500 \qquad t_m \le t \le t_e$$

The first one is fairly straight forward and will be valid until the maximum amount of pollution is reached. We'll call that time t_m. Also, the volume in the tank remains constant during this time so we don't need to do anything fancy with that this time in the second term as we did in the previous example.

We'll need a little explanation for the second one. First notice that we don't "start over" at $t = 0$. We start this one at t_m, the time at which the new process starts. Next, fresh water is flowing into the tank and so the concentration of pollution in the incoming water is zero. This will drop out the first term, and that's okay so don't worry about that.

Now, notice that the volume at any time looks a little funny. During this time frame we are losing two gallons of water every hour of the process so we need the "-2" in there to account for that. However, we can't just use t as we did in the previous example. When this new process starts up there needs to be 800 gallons of water in the tank and if we just use t there we won't have the required 800 gallons that we need in the equation. So, to make sure that we have the proper volume we need to put in the difference in times. In this way once we are one hour into the new process (*i.e* $t - t_m = 1$) we will have 798 gallons in the tank as required.

Finally, the second process can't continue forever as eventually the tank will empty. This is denoted in the time restrictions as t_e. We can also note that $t_e = t_m + 400$ since the tank will empty 400 hours after this new process starts up. Well, it will end provided something doesn't come along and start changing the situation again.

The explanations taken care of here's the simplified version of the IVP's that we'll be solving.

$$Q_1'(t) = 15 - \frac{3Q_1(t)}{800} \qquad Q_1(0) = 2 \qquad 0 \le t \le t_m$$

$$Q_2'(t) = -\frac{2Q_2(t)}{400 - (t - t_m)} \qquad Q_2(t_m) = 500 \quad t_m \le t \le t_e$$

The first IVP is a fairly simple linear differential equation so we'll leave the details of the solution to you to check. Upon solving you get.

$$Q_1(t) = 4000 - 3998e^{-\frac{3t}{800}}$$

Now, we need to find t_m. This isn't too bad all we need to do is determine when the amount of pollution reaches 500. So we need to solve.

$$Q_1(t) = 4000 - 3998e^{-\frac{3t}{800}} = 500 \qquad \Rightarrow \qquad t_m = 35.475$$

So, the second process will pick up at 35.475 hours. For completeness sake here is the IVP with this information inserted.

$$Q_2'(t) = -\frac{2Q_2(t)}{435.475 - t} \qquad Q_2(35.475) = 500 \qquad 35.475 \le t \le 435.475$$

This differential equation is both linear and separable and again isn't terribly difficult to solve so I'll leave the details to you again to check that we should get.

$$Q_2(t) = \frac{(435.476 - t)^2}{320}$$

So, a solution that encompasses the complete running time of the process is

$$Q(t) = \begin{cases} 4000 - 3998e^{-\frac{3t}{800}} & 0 \le t \le 35.475 \\ \dfrac{(435.476 - t)^2}{320} & 35.475 \le t \le 435.4758 \end{cases}$$

Here is a graph of the amount of pollution in the tank at any time t.

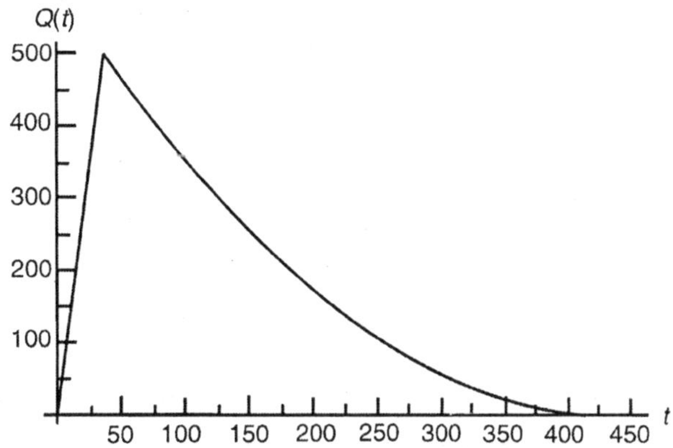

As you can surely see, these problems can get quite complicated if you want them to. Take the last example. A more realistic situation would be that once the pollution dropped below some predetermined point the polluted runoff would, in all likelihood, be allowed to flow back in and then the whole process would repeat itself. So, realistically, there should be at least one more IVP in the process.

Let's move on to another type of problem now.

POPULATION

These are somewhat easier than the mixing problems although, in some ways, they are very similar to mixing problems. So, if *P(t)* represents a population in a given region at any time *t* the basic equation that we'll use is identical to the one that we used for mixing. Namely,

Rate of change of P(t)	=	Rate at which P(t) enters the region	-	Rate at which P(t) exits the region

Here the rate of change of $P(t)$ is still the derivative. What's different this time is the rate at which the population enters and exits the region. For population problems all the ways for a population to enter the region are included in the entering rate. Birth rate and migration into the region are examples of terms that would go into the rate at which the population enters the region. Likewise, all the ways for a population to leave an area will be included in the exiting rate. Therefore things like death rate, migration out and predation are examples of terms that would go into the rate at which the population exits the area.

Example: A population of insects in a region will grow at a rate that is proportional to their current population. In the absence of any outside factors the population will triple in two weeks time. On any given day there is a net migration into the area of 15 insects and 16 are eaten by the local bird population and 7 die of natural causes. If there are initially 100 insects in the area will the population survive? If not, when do they die out?

Solution: Let's start out by looking at the birth rate. We are told that the insects will be born at a rate that is proportional to the current population. This means that the birth rate can be written as

$$rP$$

where r is a positive constant that will need to be determined. Now, let's take everything into account and get the IVP for this problem.

$$p' = (rp + 15) - (16 + 7) \qquad p(0) = 100$$
$$p' = rp - 8 \qquad p(0) = 100$$

Note that in the first line we used parenthesis to note which terms went into which part of the differential equation. Also note that we don't make use of the fact that the population will triple in two weeks time in the absence of outside factors here. In the absence of outside factors means that the ONLY thing that we can consider is birth rate. Nothing else can enter into the picture and clearly we have other influences in the differential equation.

So, just how does this tripling come into play? Well, we should also note that without knowing r we will have a difficult time solving the IVP completely. We will use the fact that the population triples in two week time to help us find r. In the absence of outside factors the differential equation would become.

$$p' = rp \qquad p(0) = 100 \qquad p(14) = 300$$

Note that since we used days as the time frame in the actual IVP I needed to convert the two weeks to 14 days. We could have just as easily converted the original IVP to weeks as the time frame, in which case there would have been a net change of 56 per week instead of the 8 per week that we are currently using in the original differential equation.

Okay back to the differential equation that ignores all the outside factors. This differential equation is separable and linear and is a simple differential equation to solve. I'll leave the detail to you to get the general solution.

$$p(t) = ce^{n}$$

Applying the initial condition gives $c = 100$. Now apply the second condition.

$$300 = p(14) = 100e^{14r} \qquad 300 = 100e^{14r}$$

We need to solve this for r. First divide both sides by 100, then take the natural log of both sides.

$$3 = \text{e } 14 \text{ r}$$

$$\text{In } 3 = \text{In } \text{e}^{14\text{r}}$$

$$\text{In } 3 = 14\,r$$

$$r = \frac{\text{In } 3}{14}$$

We made use of the fact that $\text{e}^{8(x)} = \text{g }(x)$here to simplify the problem. Now, that we have r we can go back and solve the original differential equation. We'll rewrite it a little for the solution process.

$$p' - \frac{\text{In}3}{14}p = -8 \qquad\qquad p(0) = 100$$

This is a fairly simple linear differential equation, but that coefficient of P always get people bent out of shape, so we'll go through at least some of the details here.

$$\mu(t) = e\int -\frac{\text{In}3}{14}dt = e^{-\frac{\text{In}3}{14}t}$$

Now, don't get excited about the integrating factor here. It's just like only this time the constant is a little more complicated than just a 2, but it is a constant! Now, solve the differential equation.

$$\int \left(pe^{-\frac{\text{In}3}{14}t} \right) dt = \int -8e^{-\frac{\text{In}3}{14}t}\, dt$$

$$pe^{-\frac{\text{In}}{14}t} = -8\left(-\frac{14}{\text{In}3} \right) e^{-\frac{\text{In}3}{14}t} + c$$

$$p(t) = \frac{112}{\text{In}3} + ce^{\frac{\text{In}3}{14}t}$$

Again, do not get excited about doing the right hand integral, it's just like integrating ! Applying the initial condition gives the following.

$$p(t) = \frac{112}{\text{In}3} + \left(100 - \frac{112}{\text{In}3}\right)e^{\frac{\text{In}3}{14}t} = \frac{112}{\text{In}3} - 1.94679e^{\frac{\text{In}3}{14}t}$$

Now, the exponential has a positive exponent and so will go to plus infinity as t increases. Its coefficient, however, is negative and so the whole population will go negative eventually. Clearly, population can't be negative, but in order for the population to go negative it must pass through zero. In other words, eventually all the insects must die. So, they don't survive and we can solve the following to determine when they die out.

$$0 = \frac{112}{\text{In}3} - 1.94679e^{\frac{\text{In}3}{14}t} \qquad \Rightarrow \qquad t = 50.4415$$

So, the insects will survive for around 7.2 weeks. Here is a graph of the population during the time in which they survive.

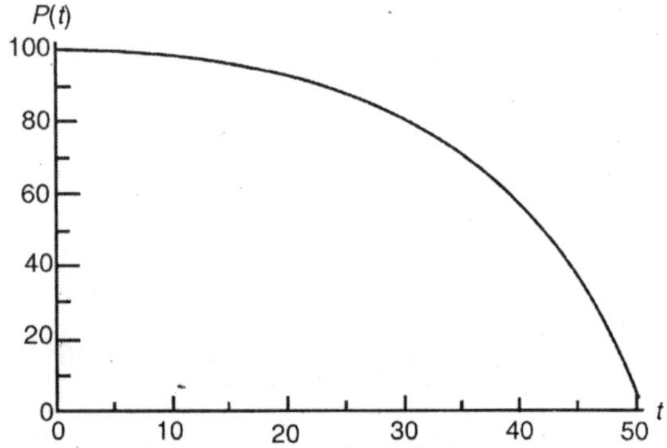

As with the mixing problems, we could make the population problems more complicated by changing the circumstances at some point in time. For instance, if at some point in time the local bird population saw a decrease due to disease they wouldn't eat as much after that point and a second differential equation to govern the time after this point.

Let's now take a look at the final type of problem that we'll be modeling in this section.

FALLING BODY

This will not be the first time that we've looked into falling bodies. If you recall, we looked at one of these when we were looking at Direction Fields. In that section we saw that the basic equation that we'll use is Newton's Second Law of Motion.

$$mv' = F(t, v)$$

The two forces that we'll be looking at here are gravity and air resistance. The main issue with these problems is to correctly define conventions and then remember to keep those conventions. By this we mean define which direction will be termed the positive direction and then make sure that all your forces match that convention. This is especially important for air resistance as this is usually dependent on the velocity and so the "sign" of the velocity can and does affect the "sign" of the air resistance force.

Example: A 50 kg mass is shot from a cannon straight up with an initial velocity of 10m/s off a bridge that is 100 meters above the ground. If air resistance is given by $5v$ determine the velocity of the mass when it hits the ground.

Solution: First, notice that when we say straight up, we really mean straight up, but in such a way that it will miss the bridge on the way back down. Here is a sketch of the situation.

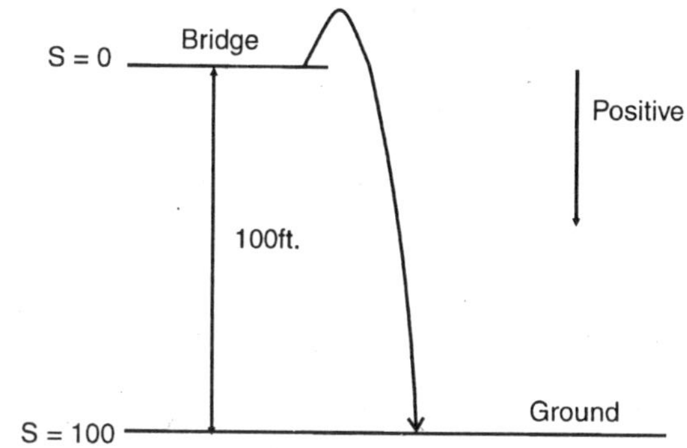

Notice the conventions that we set up for this problem. Since the vast majority of the motion will be in the downward direction we decided to assume that everything acting in the downward direction should be positive. Note that we also defined the "zero position" as the bridge, which makes the ground have a "position" of 100.

If you think about it we actually have two situations here. The initial phase in which the mass is rising in the air and the second phase when the mass is on its way down. We will need to examine both situations and set up an IVP for each. We will do this simultaneously. Here are the forces that are acting on the object on the way up and on the way down.

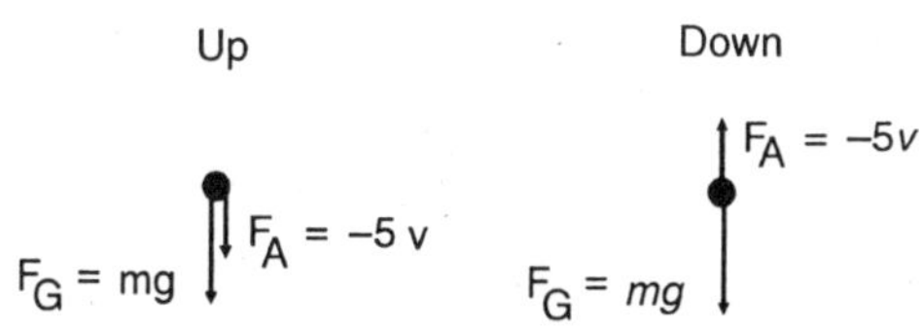

Notice that the air resistance force needs a negative in both cases in order to get the correct "sign" or direction on the force. When the mass is moving upwards the velocity (and hence v) is negative, yet the force must be acting in a downward direction. Therefore, the "-" must be part of the force to make sure that, overall, the force is positive and hence acting in the downward direction.

Likewise, when the mass is moving downward the velocity (and so v) is positive. Therefore, the air resistance must also have a "-" in order to make sure that it's negative and hence acting in the upward direction.

So, the IVP for each of these situations are.

Up	Down
$mv' = mg - 5v$	$mv' = mg - 5v$
$v(0) = -10$	$v(t_0) = 0$

In the second IVP, the t_0 is the time when the object is at the highest point and is ready to start on the way down. Note that at this time the velocity would be zero. Also note that the initial condition of the first differential equation will have to be negative since the initial velocity is upward. In this case, the differential equation for both of the situations is identical. This won't always happen, but in those cases where it does, we can ignore the second IVP and just let the first govern the whole process.

So, let's actually plug in for the mass and gravity (we'll be using $g = 9.8$ m/s^2 here). We'll go ahead and divide out the mass while we're at it since we'll need to do that eventually anyway.

$$v' = 9.8 - \frac{5v}{50} = 9.8 - \frac{v}{10} \qquad v(0) = -10$$

This is a simple linear differential equation to solve so we'll leave the details to you. Upon solving we arrive at the following equation for the velocity of the object at any time t.

$$v(t) = 98 - 10e^{-\frac{t}{10}}$$

Okay, we want the velocity of the ball when it hits the ground. Of course we need to know *when* it hits the ground before we can ask this. In order to find this we will need to find the position function. This is easy enough to do.

$$S(t) = \int v = (t)\,dt = \int 98 - 108e^{-\frac{t}{10}}\,dt = 98t + 1080e^{-\frac{t}{10}} + c$$

We can now use the fact that I took the convention that $s(0) = 0$ to find that $c = -1080$. The position at any time is then.

$$S(t) = 98t + 1080e^{-\frac{t}{10}} - 1080$$

To determine when the mass hits the ground we just need to solve.

$$S(t) = 98t + 1080e^{-\frac{t}{10}} - 1080 \qquad t = -3.32203, 5.98147$$

We've got two solutions here, but since we are starting things at $t = 0$, the negative is clearly the incorrect value. Therefore the mass hits the ground at $t = 5.98147$. The velocity of the object upon hitting the ground is then.

$$v(5.98147) = 38.61841$$

This last example gave us an example of a situation where the two differential equations needed for the problem ended up being identical and so we didn't need the second one after all. Be careful however to not always expect this. We could very easily change this problem so that it required two different differential equations.

For instance we could have had a parachute on the mass open at the top of its arc changing its air resistance. This would have completely changed the second differential equation and forced us to use it as well. Or, we could have put a river under the bridge so that before it actually hit the ground it would have first had to go through some water which would have a different "air" resistance for that phase necessitating a new differential equation for that portion. Or, we could be really crazy and have both the parachute and the river which would then require three IVP's to be solved before we determined the velocity of the mass before it actually hits the solid ground.

Before leaving this section let's work a couple examples illustrating the importance of remembering the conventions that you set up for the positive direction in these problems.

Awhile back I gave my students a problem in which a sky diver jumps out of a plane. Most of my students are engineering majors and following the standard convention from most of their engineering classes they defined the positive direction as upward, despite the fact that all the motion in the problem was downward. There is nothing wrong with this assumption, however, because they forgot the convention that up was positive they did not correctly deal with the air resistance which caused them to get the incorrect answer.

So, let's take a look at the problem and set up the IVP that will give the sky divers velocity at any time t.

Example: Set up the IVP that will give the velocity of a 60 kg sky diver that jumps out of a plane with no initial velocity and an air resistance of $0.8|v|$. For this example assume that the positive direction is upward.

Solution: Here are the forces that are acting on the sky diver

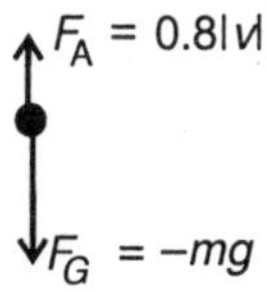

Because of the conventions the force due to gravity is negative and the force due to air resistance is positive. As set up, these forces have the correct sign and so the IVP is

$$mv' = -mg + 0.8|v| \qquad v(0) = 0 \text{ (Incorrect IVP!!)}$$

The problem arises when you go to remove the absolute value bars. In order to do the problem they do need to be removed. This is where most of the students made their mistake. Because they had forgotten about the convention and the direction of motion they just dropped the absolute value bars to get.

So, why is this incorrect? Well remember that the convention is that positive is upward. However in this case the object is moving downward and so v is negative! Upon dropping the absolute value bars the air resistance became a negative force and hence was acting in the downward direction!

To get the correct IVP recall that because v is negative then $|v| = -v$. Using this, the air resistance becomes $F_A = -0.8v$ and despite appearances this is a positive forces since the "-" cancels out against the velocity (which is negative) to get a positive force.

The correct IVP is then

$$mv' = -mg - 0.8v \qquad v(0) = 0$$

Plugging in the mass gives

$$v' = -9.8 - \frac{v}{75} \quad v(0) = 0$$

For the sake of completeness the velocity of the sky diver, at least until the parachute opens, which I didn't include in this problem is.

$$v(t) = -735 + 735e^{-\frac{t}{75}}$$

This mistake was made in part because the students were in a hurry and weren't paying attention, but also because they simply forgot about their convention and the direction of motion! Don't fall into this mistake. Always pay attention to your conventions and what is happening in the problems.

Just to show you the difference here is the problem worked by assuming that down is positive.

Example: Set up the IVP that will give the velocity of a 60 kg sky diver that jumps out of a plane with no initial velocity and an air resistance of $0.8|v|$. For this example assume that the positive direction is downward.

Solution: Here are the forces that are acting on the sky diver

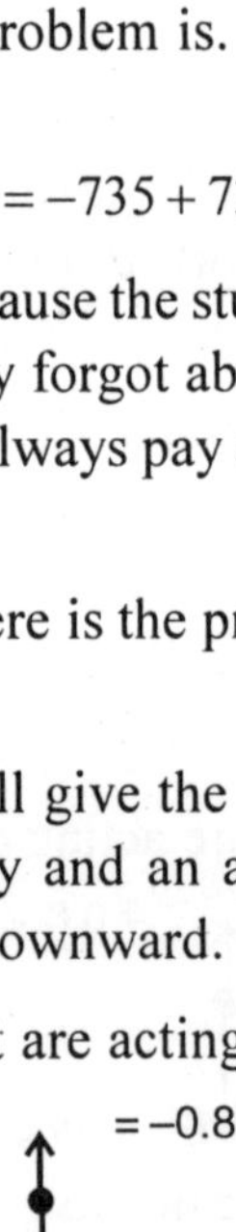

In this case the force due to gravity is positive since it's a downward force and air resistance is an upward force and so needs to be negative. In this case since the motion is downward the velocity is positive so $|v| = v$. The air resistance is then $F_A = -0.8v$. The IVP for this case is

$$mv' = mg - 0.8v \qquad v(0) = 0$$

Plugging in the mass gives

$$v' = 9.8 - \frac{v}{75} \qquad v(0) = 0$$

Solving this gives

$$v(t) = 735 - 735e^{-\frac{t}{75}}$$

This is the same solution as the previous example, except that it's got the opposite sign. This is to be expected since the conventions have been switched between the two examples.

EQUILIBRIUM SOLUTIONS

In the previous section we modeled a population based on the assumption that the growth rate would be a constant. However, in reality this doesn't make much sense. Clearly a population cannot be allowed to grow forever at the same rate. The growth rate of a population needs to depend on the population itself. Once a population reaches a certain point the growth rate will start reduce, often drastically.

A much more realistic model of a population growth is given by the logistic growth equation. Here is the logistic growth equation.

$$p' = r\left(1 - \frac{P}{K}\right)P$$

In the logistic growth equation r is the intrinsic growth rate and is the same r as in the last section. In other words, it is the growth rate that will occur in the absence of any limiting factors. K is called either the saturation level or the carrying capacity.

Now, we claimed that this was a more realistic model for a population. Let's see if that in fact is correct. To allow us to sketch a direction field let's pick a couple of numbers for r and K. We'll use $r = \frac{1}{2}$ and $K = 10$. For these values the logistics equation is.

$$p' = \frac{1}{2}\left(1 - \frac{P}{10}\right)P$$

If you need a refresher on sketching direction fields go back and take a look at that section. First notice that the derivative will be zero at $P = 0$ and $P = 10$. Also notice that these are in fact solutions to the differential equation.

These two values are called equilibrium solutions since they are constant solutions to the differential equation. We'll leave the rest of the details on sketching the direction field to you. Here is the direction field as well as a couple of solutions sketched in as well.

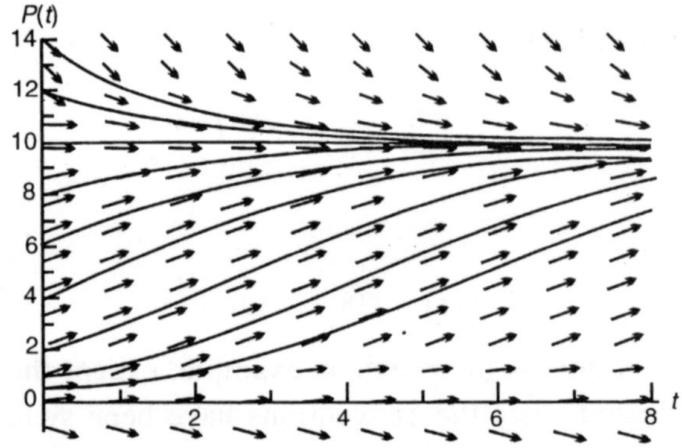

Note, that we included a small portion of negative P's in here even though they really don't make any sense for a population problem. The reason for this will be apparent down the road. Also, notice that a population of say 8 doesn't make all that much sense so let's assume that population is in thousands or millions so that 8 actually represents 8,000 or 8,000,000 individuals in a population.

Notice that if we start with a population of zero, there is no growth and the population stays at zero. So, the logistic equation will correctly figure out that. Next, notice that if we start with a population in the range $0 < P(0) < 10$ then the population will grow, but start to level off once we get close to a population of 10. If we start with a population of 10, the population will stay at 10. Finally if we start with a population that is greater than 10, then the population will actually die off until we start nearing a population of 10, at which point the population decline will start to slow down.

Now, from a realistic standpoint this should make some sense. Populations can't just grow forever without bound. Eventually the population will reach such a size that the resources of an area are no longer able to sustain the population and the population growth will start to slow as it comes closer to this threshold. Also, if you start off with a population greater than what an area can sustain there will actually be a die off until we get near to this threshold.

In this case that threshold appears to be 10, which is also the value of K for our problem. That should explain the name that we gave K initially. The carrying capacity or saturation level of an area is the maximum sustainable population for that area.

So, the logistics equation, while still quite simplistic, does a much better job of modeling what will happen to a population.

Now, let's move on to the point of this section. The logistics equation is an example of an autonomous differential equation. Autonomous differential equations are differential equations that are of the form.

$$\frac{dy}{dt} = f(y)$$

The only place that the independent variable, t in this case, appears is in the derivative.

Notice that if $f(y_0) = 0$ for some value then this will also be a solution to the differential equation. These values are called equilibrium solutions or equilibrium points. What we would like to do is classify these solutions. By classify we mean the following. If solutions start "near" an equilibrium solution will they move away from the equilibrium solution or towards the equilibrium solution? Upon classifying the equilibrium solutions we can then know what all the other solutions to the differential equation will do in the long term simply by looking at which equilibrium solutions they start near.

So, just what do I mean by "near"? Go back to our logistics equation.

$$p' = \frac{1}{2}\left(1 - \frac{p}{10}\right)p$$

As we pointed out there are two equilibrium solutions to this equation $P = 0$ and $P = 10$. If we ignore the fact that we're dealing with population these point break up the P number line into three distinct regions.

$$-\infty < p < 0 \qquad 0 < p < 10 \qquad 10 < p < \infty$$

We will say that a solution starts "near" an equilibrium solution if it starts in a region that is on either side of that equilibrium solution. So solutions that start "near" the equilibrium solution $P = 10$ will start in either

$$0 < p < 10 \qquad OR \qquad 10 < P < \infty$$

and solutions that start "near" $P = 0$ will start in either

$$-\infty < p < 0 \qquad OR \qquad 0 < P < 10$$

For regions that lie between two equilibrium solutions we can think of any solutions starting in that region as starting "near" either of the two equilibrium solutions as we need to.

Now, solutions that start "near" $P = 0$ all move away from the solution as t increases. Note that moving away does not necessarily mean that they grow without bound as they move away. It only means that they move away. Solutions that start out greater than $P = 0$ move away, but do stay bounded as t grows. In fact, they move in towards $P = 10$.

Equilibrium solutions in which solutions that start "near" them move away from the equilibrium solution are called unstable equilibriums points or unstable equilibrium solutions. So, for our logistics equation, $P = 0$ is an unstable equilibrium solution.

Next, solutions that start "near" $P = 10$ all move in toward $P = 10$ as t increases. Equilibrium solutions in which solutions that start "near" them move toward the equilibrium solution are called asymptotically stable equilibrium points or asymptotically stable equilibrium solutions. So, $P = 10$ is an asymptotically stable equilibrium solution.

There is one more classification, but I'll wait until we get an example in which this occurs to introduce it. So, let's take a look at a couple of examples.

Example: Find and classify all the equilibrium solutions to the following differential equation.

$$y' = y^2 - y - 6$$

Solution: First, find the equilibrium solutions. This is generally easy enough to do.

$$y^2 - y - 6 = (y-3)(y+2) = 0$$

So, it looks like we've got two equilibrium solutions. Both y = -2 and y = 3 are equilibrium solutions. Below is the sketch of some integral curves for this differential equation. A sketch of the integral curves or direction fields can simplify the process of classifying the equilibrium solutions.

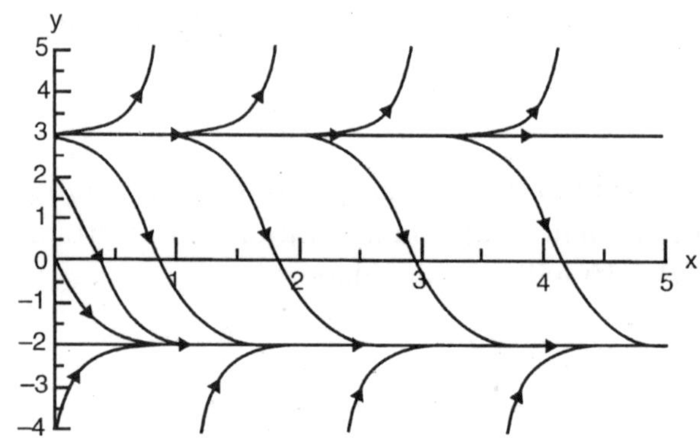

From this sketch it appears that solutions that start "near" y = -2 all move towards it as t increases and so y = -2 is an asymptotically stable equilibrium solution and solutions that start "near" y = 3 all move away from it as t increases and so y = 3 is an unstable equilibrium solution.

This next example will introduce the third classification that we can give to equilibrium solutions.

Example: Find and classify the equilibrium solutions of the following differential equation.

$$y' = (y^2 - 4)(y+1)^2$$

Solution: The equilibrium solutions are to this differential equation are y = -2, y = 2, and y = -1. Below is the sketch of the integral curves. We included the direction field this time as the integral curves where a little hard to interpret without it.

From this it is clear (hopefully) that y = 2 is an unstable equilibrium solution and y = -2 is an asymptotically stable equilibrium solution. However, y = -1 behaves differently from either of these two. Solutions that start above it move towards $y = -1$ while solutions that start below $y = -1$ move away as t increases.

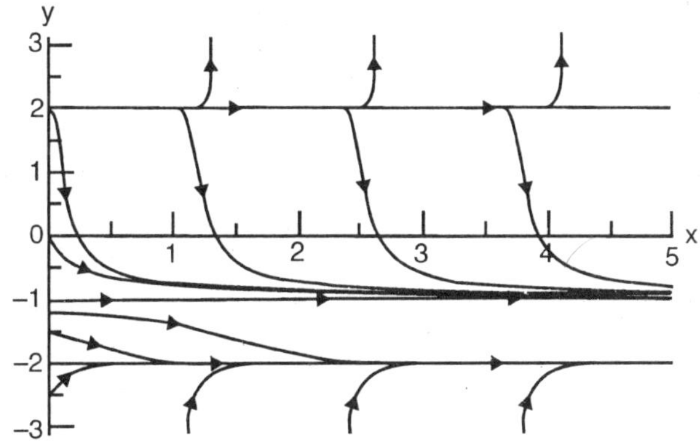

In cases where solutions on one side of an equilibrium solution move towards the equilibrium solution and on the other side of the equilibrium solution move away from it we call the equilibrium solution semi-stable.

So, $y = -1$ is a semi-stable equilibrium solution.

EULER'S METHOD

Up to this point practically every differential equation that we've been presented with could be solved. The problem with this is that these are the exceptions rather than the rule. The vast majority of first order differential equations can't be solved.

In order to teach you something about solving first order differential equations we've had to restrict ourselves down to the fairly restrictive cases of linear, separable, or exact differential equations or differential equations that could be solve with a set of very specific substitutions. Most first order differential equations however fall into none of these categories. In fact even those that are separable or exact cannot always be solved for an explicit solution. Without explicit solutions to these it would be hard to get any information about the solution.

So what do we do when faced with a differential equation that we can't solve? The answer depends on what you are looking for. If you are only looking for long term behavior of a solution you can always sketch a direction field. This can be done without too much difficulty for some fairly complex differential equations that we can't solve to get exact solutions.

The problem with this approach is that it's only really good for getting general trends in solutions and for long term behavior of solutions. There are times when we will need something more. For instance, maybe we need to determine how a specific solution behaves, including some values that the solution will take. There are also a fairly large set of differential equations that are not easy to sketch good direction fields for.

In these cases we resort to numerical methods that will allow us to approximate solutions to differential equations. There are many different methods that can be used to approximate solutions to a differential equation and in fact whole classes can be taught just dealing with the various methods. We are going to look at one of the oldest and easiest to use here. This method was originally devised by Euler and is called, oddly enough, Euler's Method. Let's start with a general first order IVP

$$\frac{dy}{dt} = f(t,y) \qquad\qquad y(t_0) = y_0$$

where $f(t,y)$ is a known function and the values in the initial condition are also known numbers. From the second theorem in the Intervals of Validity section we know that if f and f_y are continuous function then there is a unique solution to the IVP in some interval surrounding $t = t_0$. So, let's assume that everything is nice and continuous so that we know that a solution will in fact exist.

We'll start with the two pieces of information that we do know about the solution. First, we know the value of the solution at from the initial condition. Second, we also know the value of the derivative at. We can get this by plugging the initial condition into $f(t,y)$ into the differential equation itself. So, the derivative at this point is.

$$\left.\frac{dy}{dt}\right|_{t=t_0} = f(t_0, y_0)$$

Two pieces of information are enough for us to write down the equation of the tangent line to the solution at $t = t_0$. The tangent line is

$$y = y_0 + f(t_0, y_0)(t - t_0)$$

Take a look at the figure below

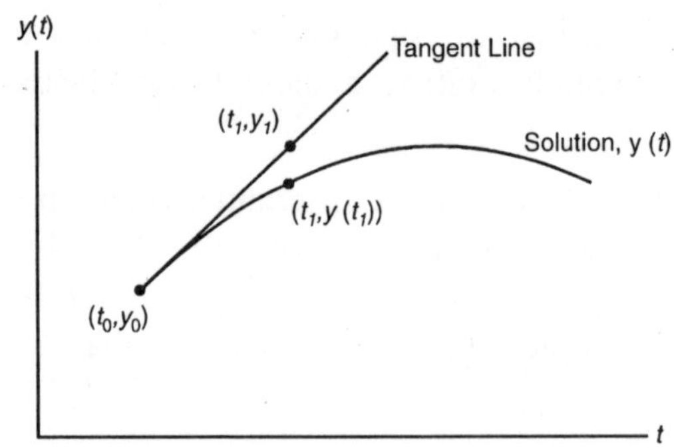

If t_1 is close enough to t_0 then the point y_1 on the tangent line should be fairly close to the actual value of the solution at t_1, or $y(t_1)$. Finding y_1 is easy enough. All we need to do is plug t_1 in the equation for the tangent line.

$$y = y_0 + f(t_0, y_0)(t_1 - t_0)$$

Now, we would like to proceed in a similar manner, but we don't have the value of the solution at t_1 and so we won't know the slope of the tangent line to the solution at this point. This is a problem. We can partially solve it however, by recalling that y_1 is an approximation to the solution at t_1. If y_1 is very good approximation to the actual value of the solution then we can use that to estimate the slope of the tangent line at t_1.

So, let's hope that y_1 is a good approximation to the solution and construct a line through the point (t_1, y_1) that has slope $f(t_1, y_1)$. This gives

$$y = y_1 + f(t_1, y_1)(t - t_1)$$

Now, to get an approximation to the solution at $t = t_2$ we will hope that this new line will be fairly close the actual solution at t_2 and use the value of the line at t_2 as an approximation to the actual solution. This gives.

$$y_2 = y_1 + f(t_1, y_1)(t_2 - t_1)$$

We can continue in this fashion. Use the previously computed approximation to get the next approximation. So,

$$\begin{aligned} y_3 &= y2 + f(t_2, y_2)(t_3 - t_2) \\ y_4 &= y_3 + f(t_3, y_3)(t_4 - t_3) \end{aligned} \quad \text{etc.}$$

In general, if we have t_n and the approximation to the solution at this point, y_n, and we want to find the approximation at t_{n+1} all we need to do is use the following.

$$y_{n+1} = y_n + f(t_n, y_n).(t_{n+1} - t_n)$$

If we define $f_n = f(t_n, y_n)$ we can simplify the formula to

$$y_{n+1} = y_n + f_n(t_{n+1} - t_n)$$

Often, we will assume that the step sizes between the points t_0, t_1, t_2, ... to a uniform size of h. In other words, we will often assume that

$$t_{n+1} - t_n = h$$

This doesn't have to be done and there are times when it's best that we not do this. However, if we do the formula for the next approximation becomes.

$$y_{n+1} = y_n = hf_n$$

So, how do we use Euler's Method? It's fairly simple. Then starting with (t_0, y_0) we repeatedly evaluate depending on whether we chose to use a uniform set size or not. We continue until we've gone the desired number of steps or reached the desired time. This will give us a sequence of numbers y_1, y_2, y_3, ... y_n that will approximate the value of the actual solution at t_1, t_2, t_3, ... t_n.

What do we do if we want value of the solution at some other point than those used here? One possibility is to go back and redefine our set of points to a new set that will include the points we are after and redo Euler's Method using this new set of points. However this is cumbersome and could take a lot of time especially if we had to make changes to the set of points more than once.

Another possibility is to remember how we arrived at the approximations in the first place. Recall that we used the tangent line

$$y = y_0 + f(t_0, y_0)(t - t_0)$$

to get the value of y_1. We could use this tangent line as an approximation for the solution on the interval $[t_0, t_1]$. Likewise, we used the tangent line

$$y = y_1 + f(t_1, y_1)(t - t_1)$$

to get the value of y_2. We could use this tangent line as an approximation for the solution on the interval $[t_1, t_2]$. Continuing in this manner we would get a set of lines that, when strung together, should be an approximation to the solution as a whole.

In practice you would need to write a computer program to do these computations for you. In most cases the function *f(t,y)* would be too large and/or complicated to use by hand and in most serious uses of Euler's Method you would want to use hundreds of steps which would make doing this by hand prohibitive. So, here is a bit of *pseudo-code* that you can use to write a program for Euler's Method that uses a uniform step size, *h*. define $f(t, y)$.

1. Input t_0 and y_0.
2. Input step size, *h* and the number of steps, *n*.
3. For *j* from 1 to *n* do
   (*a*) $m = f(t_0, y_0)$
   (*b*) $y_1 = y_0 + h*m$
   (*c*) $t_1 = t_0 + h$
   (*d*) Print t_1 and y_1
   (*e*) $t_0 = t_1$
   (*f*) $y_0 = y_1$
4. end

The *pseudo-code* for a non-uniform step size would be a little more complicated, but it would essentially be the same.

So, let's take a look at a couple of examples. We'll use Euler's Method to approximate solutions to a couple of first order differential equations. The differential equations that we'll be using are linear first order differential equations that can be easily solved for an exact solution.

Of course, in practice we wouldn't use Euler's Method on these kinds of differential equations, but by using easily solvable differential equations we will be able to check the accuracy of the method. Knowing the accuracy of any approximation method is a good thing. It is important to know if the method is liable to give a good approximation or not.

Example: For the IVP

$$y' + 2y = 2 - e^{-4t} \qquad y(0) = 1$$

Use Euler's Method with a step size of $h = 0.1$ to find approximate values of the solution at $t = 0.1, 0.2, 0.3, 0.4$, and 0.5. Compare them to the exact values of the solution as these points.

Solution: This is a fairly simple linear differential equation so we'll leave it to you to check that the solution is

$$y(t) = 1 + \frac{1}{2}e^{-4t} - \frac{1}{2}e^{-2t}$$

In order to use Euler's Method we first need to rewrite the differential equation into the form given in (1).

$$y' = 2 - e^{-4t} - 2y$$

From this we can see that $f(t, y) = 2 - e^{-4t} - 2y$. Also note that $t_0 = 0$ and $y_0 = 1$. We can now start doing some computations.

$$f_0 = f(0,1) = 2 - e^{-4(0)} - 2(1) = -1$$
$$y_1 = y_0 + hf_0 = 1 + (0.1)(-1) = 0.9$$

So, the approximation to the solution at $t_1 = 0.1$ is $y_1 = 0.9$.

At the next step we have

$$f_1 = f(0.1, 0.9) = 2 - e^{-4(0.1)} - 2(0.9) = -0.470320046$$
$$y_2 = y_1 + hf_1 = 0.9 + (0.1)(-0.470320046) = 0.852967995$$

Therefore, the approximation to the solution at $t_2 = 0.2$ is $y_2 = 0.852967995$.

I'll leave it to you to check the remainder of these computations.

$$f_2 = -0.155264954 \qquad y_3 = 0.837441500$$
$$f_3 = 0.023922788 \qquad y_4 = 0.839833779$$
$$f_4 = 0.1184359245 \qquad y_5 = 0.851677371$$

Here's a quick table that gives the approximations as well as the exact value of the solutions at the given points.

Time, t_n	*Approximation*	*Exact*	*Error*
$t_0 = 0$	$y_0 = 1$	$y(0) = 1$	0 %
$t_1 = 0.1$	$y_1 = 0.9$	$y(0.1) = 0.925794646$	2.79 %
$t_2 = 0.2$	$y_2 = 0.852967995$	$y(0.2) = 0.889504459$	4.11 %
$t_3 = 0.3$	$y_3 = 0.837441500$	$y(0.3) = 0.876191288$	4.42 %
$t_4 = 0.4$	$y_4 = 0.839833779$	$y(0.4) = 0.876283777$	4.16 %
$t_5 = 0.5$	$y_5 = 0.851677371$	$y(0.5) = 0.883727921$	3.63 %

We've also included the error as a percentage. It's often easier to see how well an approximation does if you look at percentages. The formula for this is,

$$Percent\,error = \frac{exact - approximate}{exact} \times 100$$

We used absolute value in the numerator because we really don't care at this point if the approximation is larger or smaller than the exact. We're only interested in how close the two are. The maximum error in the approximations from the last example was 4.42%, which isn't too bad, but also isn't all the great of an approximation. So, provided we aren't after very accurate approximations this didn't do too badly. This kind of error is generally unacceptable in almost all real applications however. So, how can we get better approximations?

Recall that we are getting the approximations by using a tangent line to approximate the value of the solution and that we are moving forward in time by steps of h. So, if we want a more accurate approximation, then it seems like one way to get a better approximation is to not move forward as much with each step. In other words, take smaller h's.

Example: Repeat the previous example only this time give the approximations at $t = 1$, $t = 2$, $t = 3$, $t = 4$, and $t = 5$. Use $h = 0.1$, $h = 0.05$, $h = 0.01$, $h = 0.005$, and $h = 0.001$ for the approximations.

Solution: Below are two tables, one gives approximations to the solution and the other gives the errors for each approximation. We'll leave the computational details to you to check.

Approximations

Time	*Exact*	$h = 0.1$	$h = 0.05$	$h = 0.01$	$h = 0.005$	$h = 0.001$
$t = 1$	0.9414902	0.9313244	0.9364698	0.9404994	0.9409957	0.9413914
$t = 2$	0.9910099	0.9913681	0.9911126	0.9910193	0.9910139	0.9910106
$t = 3$	0.9987637	0.9990501	0.9988982	0.9987890	0.9987763	0.9987662
$t = 4$	0.9998323	0.9998976	0.9998657	0.9998390	0.9998357	0.9998330
$t = 5$	0.9999773	0.9999890	0.9999837	0.9999786	0.9999780	0.9999774

Percentage Errors

Time	*h* = *0.1*	*h* = *0.05*	*h* = *0.01*	*h* = *0.005*	*h* = *0.001*
$t = 1$	1.08 %	0.53 %	0.105 %	0.053 %	0.0105 %
$t = 2$	0.036 %	0.010 %	0.00094 %	0.00041 %	0.0000703 %
$t = 3$	0.029 %	0.013 %	0.0025 %	0.0013 %	0.00025 %
$t = 4$	0.0065 %	0.0033 %	0.00067 %	0.00034 %	0.000067 %
$t = 5$	0.0012 %	0.00064 %	0.00013 %	0.000068 %	0.000014 %

We can see from these tables that decreasing h does in fact improve the accuracy of the approximation as we expected.

There are a couple of other interesting things to note from the data. First, notice that in general, decreasing the step size, h, by a factor of 10 also decreased the error by about a factor of 10 as well.

Also, notice that as t increases the approximation actually tends to get better. This isn't the case completely as we can see that in all but the first case the $t = 3$ error is worse than the error at $t = 2$, but after that point, it only get better. This should not be expected in general. In this case this is more a function of the shape of the solution. Below is a graph of the solution (the line) as well as the approximations (the dots) for $h = 0.1$.

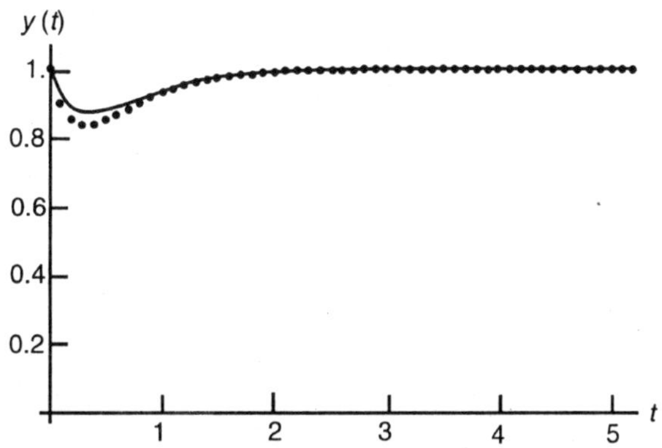

Notice that the approximation is worst where the function is changing rapidly. This should not be too surprising. Recall that we're using tangent lines to get the approximations and so the value of the tangent line at a given t will often be significantly different than the function due to the rapidly changing function at that point. Also, in this case, because the function ends up fairly flat as t increases, the tangents start looking like the function itself and so the approximations are very accurate. This won't always be the case of course.

Example: For the IVP

$$y' - y = -\frac{1}{2}e^{\frac{t}{2}}\sin(5t) + 5e^{\frac{t}{2}}\cos(5t) \qquad y(0) = 0$$

Use Euler's Method to find the approximation to the solution at $t = 1$, $t = 2$, $t = 3$, $t = 4$, and $t = 5$. Use $h = 0.1$, $h = 0.05$, $h = 0.01$, $h = 0.005$, and $h = 0.001$ for the approximations.

Solution:

I'll leave it to you to check the details of the solution process. The solution to this linear first order differential equation is.

$$y(t) = e^{\frac{t}{2}} \sin(5t)$$

Here are two tables giving the approximations and the percentage error for each approximation.

Approximations

Time	*Exact*	*h = 0.1*	*h = 0.05*	*h = 0.01*	*h = 0.005*	*h = 0.001*
$t = 1$	−1.58100	−0.97167	−1.26512	−1.51580	−1.54826	−1.57443
$t = 2$	−1.47880	0.65270	−0.34327	−2.18657	−1.35810	−1.45453
$t = 3$	2.91439	7.30209	5.34682	3.44488	3.18259	2.96851
$t = 4$	6.74580	15.56128	11.84839	7.89808	7.33093	6.86429
$t = 5$	−1.61237	21.95465	12.24018	1.56056	0.0018864	−1.28498

Percentage Errors

Time	$h = 0.1$	$h = 0.05$	$h = 0.01$	$h = 0.005$	$h = 0.001$
$t = 1$	38.54 %	19.98 %	4.12 %	2.07 %	0.42 %
$t = 2$	144.14 %	76.79 %	16.21 %	8.16 %	1.64 %
$t = 3$	150.55 %	83.46 %	18.20 %	9.20 %	1.86 %
$t = 4$	130.68 %	75.64 %	17.08 %	8.67 %	1.76 %
$t = 5$	1461.63 %	859.14 %	196.79 %	100.12 %	20.30 %

So, with this example Euler's Method does not do nearly as well as it did on the first IVP. Some of the observations we made in Example are still true however. Decreasing the size of h decreases the error as we saw with the last example and would expect to happen. Also, as we saw in the last example, decreasing h by a factor of 10 also decreases the error by about a factor of 10.

However, unlike the last example increasing t sees an increasing error. This behavior is fairly common in the approximations. We shouldn't expect the error to decrease as t increases as we saw in the last example. Each successive approximation is found using a previous approximation. Therefore, at each step we introduce error and so approximations should, in general, get worse as t increases.

Below is a graph of the solution (the line) as well as the approximations (the dots) for $h = 0.05$.

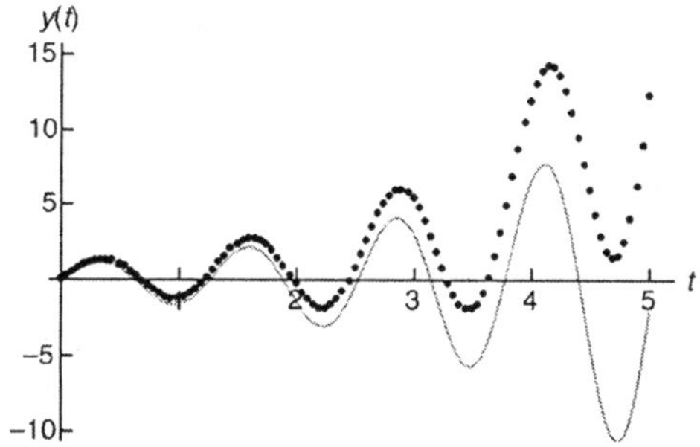

As we can see the approximations do follow the general shape of the solution, however, the error is clearly getting much worse as t increases.

So, Euler's method is a nice method for approximating fairly nice solutions that don't change rapidly. However, not all solutions will be this nicely behaved. There are other approximation methods that do a much better job of approximating solutions. These are not the focus of this course however, so I'll leave it to you to look further into this field if you are interested.

Also notice that we don't generally have the actual solution around to check the accuracy of the approximation. We generally try to find bounds on the error for each method that will tell us well an approximation should do. These error bounds are again not really the focus of this course, so I'll leave these to you as well if you're interested in looking into them.

Chapter 6

Solutions to Systems

Now that we've got some of the basic out of the way for systems of differential equations it's time to start thinking about how to solve a system of differential equations. We will start with the homogeneous system written in matrix form,

$$\vec{x}' = A\vec{x}$$

where, A is an n x n matrix and is a vector whose components are the unknown functions in the system.

Now, if we start with $n = 1$ then the system reduces to a fairly simple linear (or separable) first order differential equation.

$$x' = ax$$

and this has the following solution,

$$x(t) = ce^{at}.$$

So, let's use this as a guide and for a general n let's see if

$$\vec{x}(t) = \vec{\eta}e^{rt}$$

will be a solution. Note that the only real difference here is that we let the constant in front of the exponential be a vector. All we need to do then is plug this into the differential equation and see what we get. First notice that the derivative is,

$$\vec{x}(t) = r\vec{\eta}e^{rt}.$$

So upon plugging the guess into the differential equation we get,

$$\begin{aligned} r\vec{\eta}e^{rt} &= A\vec{\eta}e^{rt} \\ (A\eta - r\vec{\eta})e^{rt} &= \vec{0} \\ (A - rI)\vec{\eta}e^{rt} &= \vec{0} \end{aligned}$$

Now, since we know that exponentials are not zero we can drop that portion and we then see that in order to a solution then we must have

$$(A - rI)\vec{\eta} = \vec{0}.$$

Or, in order to be a solution, r and $\vec{\eta}$ must be an eigenvalue and eigenvector for the matrix A.

Therefore, in order to solve first find the eigenvalues and eigenvectors of the matrix A and then we can form solutions using. There are going to be three cases that we'll need to look at. The cases are real, distinct eigenvalues, complex eigenvalues and repeated eigenvalues.

None of this tells us how to completely solve a system of differential equations. We'll need the following couple of facts to do this.

FACT

1. If $\vec{x}_1(t)$and $\vec{x}_2(t)$are two solutions to a homogeneous system, then

$$c_1\vec{x}_1(t)+c_2\vec{x}_2(t)$$

   is also a solution to the system.

2. Suppose that A is an n x n matrix and suppose that $\vec{x}_1(t)$, $\vec{x}_2(t),\ldots,\vec{x}_n(t)$ are solutions to a homogeneous system. Define,

$$X=(\vec{x}_1\vec{x}_2\cdots\vec{x}_n).$$

In other words, X is a matrix whose *ith* column is the ith solution. Now define,

$$W=\det(X).$$

We call W the Wronskian. If then the solutions form a fundamental set of solutions and the general solution to the system is,

$$\vec{x}(t)=c_1\vec{x}_1(t)+c_2\vec{x}_2(t)+\cdots+c_n\vec{x}_n(t).$$

Note that if we have a fundamental set of solutions then the solutions are also going to be linearly independent. Likewise, if we have a set of linearly independent solutions then they will also be a fundamental set of solutions since the Wronskian will not be zero.

PHASE PLANE

Before proceeding with actually solving systems of differential equations there's one topic that we need to take a look at. This is a topic that's not always taught in a differential equations class but in case you're in a course where it is taught we should cover it so that you are prepared for it. Let's start with a general homogeneous system,

$$\vec{x}'=A\vec{x}.$$

Notice that

$$\vec{x}=\vec{0}$$

is a solution to the system of differential equations. What we'd like to ask is, do the other solutions to the system approach this solution as t increases or do they move away from this solution? We did something similar to this when we classified equilibrium solutions in a previous section. In fact, what we're doing here is simply an extension of this idea to systems of differential equations.

The solution $\vec{x} = \vec{0}$ is called an equilibrium solution for the system. As with the single differential equations case, equilibrium solutions are those solutions for which

$$A\vec{x} = \vec{0}.$$

We are going to assume that A is a nonsingular matrix and hence will have only one solution,

$$\vec{x} = \vec{0}$$

and so we will have only one equilibrium solution.

Back in the single differential equation case recall that we started by choosing values of y and plugging these into the function $f(y)$ to determine values of. We then used these values to sketch tangents to the solution at that particular value of y. From this we could sketch in some solutions and use this information to classify the equilibrium solutions.

We are going to do something similar here, but it will be slightly different as well. First we are going to restrict ourselves down to the 2 × 2 case. So, we'll be looking at systems of the form,

$$\begin{aligned} x_1' &= ax_1 + bx_2 \\ x_1' &= cx_1 + dx_2 \end{aligned} \quad \Rightarrow \quad \vec{x}' = \begin{pmatrix} a & b \\ c & d \end{pmatrix} \vec{x}.$$

Solutions to this system will be of the form, .

$$\begin{pmatrix} x_1 & (t) \\ x_d & (t) \end{pmatrix}.$$

and our single equilibrium solution will be,

$$\vec{x} = \begin{pmatrix} 0 \\ 0 \end{pmatrix}.$$

In the single differential equation case we were able to sketch the solution, $y(t)$ in the y–t plane and see actual solutions. However, this would somewhat difficult in this case since our solutions are actually vectors. What we're going to do here is think of the solutions to the system as points in the x_1–x_2 plane and plot these points. Our equilibrium solution will correspond to the origin of x_1–x_2 plane and the x_1–x_2 plane is called the phase plane.

To sketch a solution in the phase plane we can pick values of t and plug these into the solution. This gives us a point in the x_1–x_2 or phase plane that we can plot. Doing this for many values of t will then give us a sketch of what the solution will be doing in the phase plane. A sketch of a particular solution in the phase plane is called the trajectory of the solution. Once we have the trajectory of a solution sketched we can then ask whether or not the solution will approach the equilibrium solution as t increases.

We would like to be able to sketch trajectories without actually having solutions in hand. There are a couple of ways to do this. We'll look at one of those here and we'll look at the other in the next couple of sections.

One way to get a sketch of trajectories is to do something similar to what we did the first time we looked at equilibrium solutions. We can choose values of $\vec{x}$ (note that these will be points in the phase plane) and compute $A\vec{x}$. This will give a vector that represents $\vec{x}$ at that particular solution. As with the single differential equation case this vector will be tangent to the trajectory at that point. We can sketch a bunch of the tangent vectors and then sketch in the trajectories.

This is a fairly work intensive way of doing these and isn't the way to do them in general. However, it is a way to get trajectories without doing any solution work. All we need is the system of differential equations. Let's take a quick look at an example.

Example: Sketch some trajectories for the system,

$$\begin{aligned} x_1' &= x_1 + 2x_2 \\ x_1' &= 3x_1 + 2x_2 \end{aligned} \qquad \Rightarrow \qquad \vec{x}\,' = \begin{pmatrix} 1 & 2 \\ 3 & 2 \end{pmatrix} \vec{x}.$$

Solution: So, what we need to do is pick some points in the phase plane, plug them into the right side of the system. We'll do this for a couple of points.

$$\vec{x} = \begin{pmatrix} -1 \\ 1 \end{pmatrix} \qquad \Rightarrow \qquad \vec{x}\,' = \begin{pmatrix} 1 & 2 \\ 3 & 2 \end{pmatrix} \begin{pmatrix} -1 \\ 0 \end{pmatrix} = \begin{pmatrix} 1 \\ -1 \end{pmatrix}$$

$$\vec{x} = \begin{pmatrix} 2 \\ 0 \end{pmatrix} \qquad \Rightarrow \qquad \vec{x}\,' = \begin{pmatrix} 1 & 2 \\ 3 & 2 \end{pmatrix} \begin{pmatrix} 2 \\ 0 \end{pmatrix} = \begin{pmatrix} 2 \\ 6 \end{pmatrix}$$

$$\vec{x} = \begin{pmatrix} -3 \\ -2 \end{pmatrix} \qquad \Rightarrow \qquad \vec{x}\,' = \begin{pmatrix} 1 & 2 \\ 3 & 2 \end{pmatrix} \begin{pmatrix} -3 \\ -2 \end{pmatrix} = \begin{pmatrix} -7 \\ -13 \end{pmatrix}.$$

So, what does this tell us? Well at the point (-1, 1) in the phase plane there will be a vector pointing in the direction $\langle 1,-1 \rangle$. At the point (2,0) there will be a vector pointing in the direction $\langle 2,6 \rangle$. At the point (-3,-2) there will be a vector pointing in the direction $\langle -7,-13 \rangle$.

Doing this for a large number of points in the phase plane will give the following sketch of vectors.

Now all we need to do is sketch in some trajectories. To do this all we need to do is remember that the vectors in the sketch above are tangent to the trajectories.

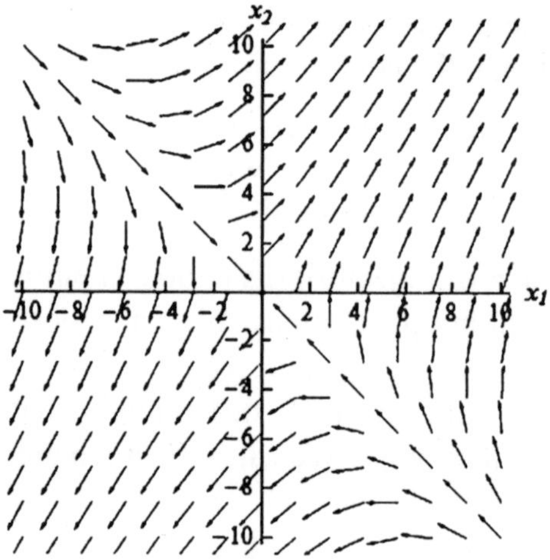

Also the direction of the vectors give the direction of the trajectory as t increases so we can show the time dependence of the solution by adding in arrows to the trajectories.

Doing this gives the following sketch.

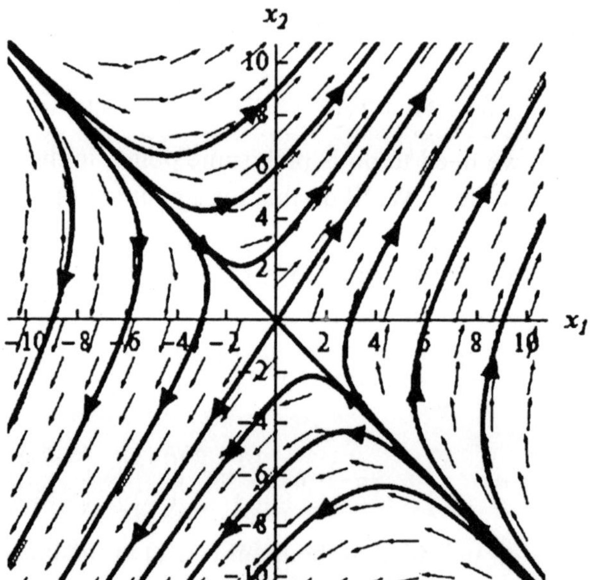

This sketch is called the phase portrait. Usually phase portraits only include the trajectories of the solutions and not any vectors. All of our phase portraits form this point on will only include the trajectories.

In this case it looks like most of the solutions will start away from the equilibrium solution then as t starts to increase they move in towards the equilibrium solution and then eventually start moving away from the equilibrium solution again.

There seem to be four solutions that have slightly different behaviors. It looks like two of the solutions will start at (or near at least) the equilibrium solution and them move straight away from it while two other solution start away from the equilibrium solution and then move straight in towards the equilibrium solution.

In these kinds of cases we call the equilibrium point a saddle point and we call the equilibrium point in this case unstable since all but two of the solutions are moving away from it as t increases.

As we noted earlier this is not generally the way that we will sketch trajectories. All we really need to get the trajectories are the eigenvalues and eigenvectors of the matrix A. We will see how to do this over the next couple of sections as we solve the systems.

Here are a few more phase portraits so you can see some more possible examples. We'll actually be generating several of these throughout the course of the next couple of sections.

Node-Unstable

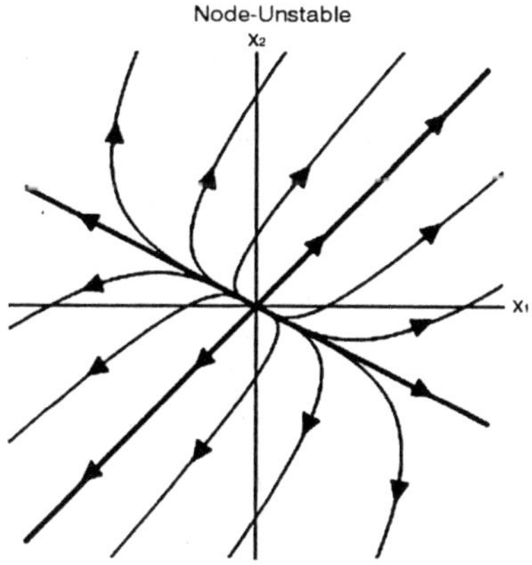

Node-Asymptotically Stable

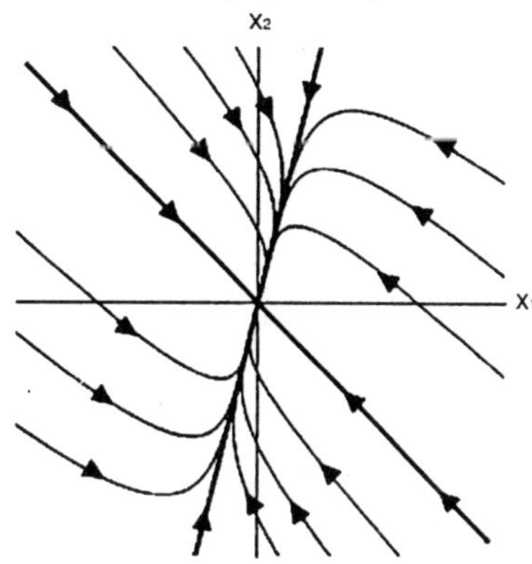

Improper Node-Unstable

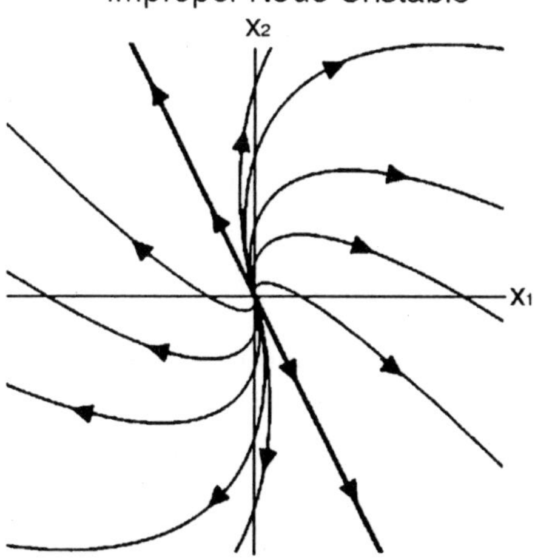

Improper Node - Asymptotically Stable

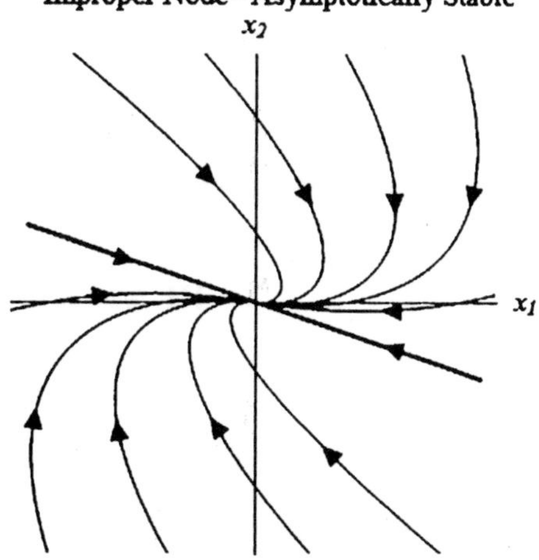

Not all possible phase portraits have been shown here. These are here to show you some of the possibilities. Make sure to notice that several kinds can be either asymptotically stable or unstable depending upon the direction of the arrows.

Saddle Point - Unstable

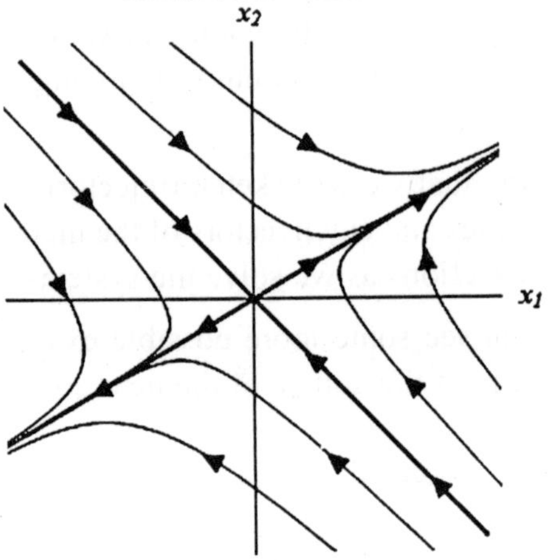

Center - Stable

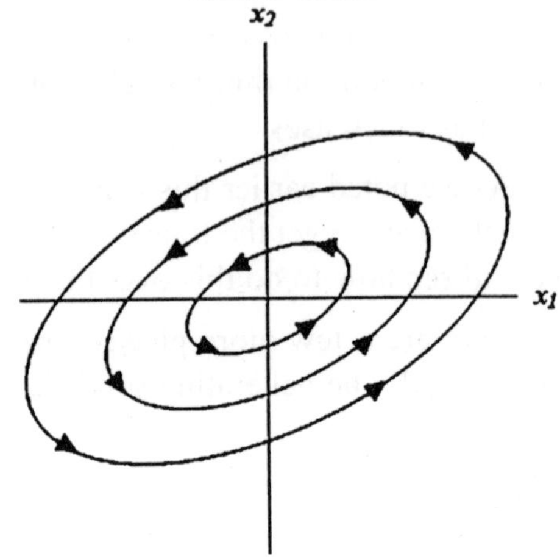

Spiral - Unstable

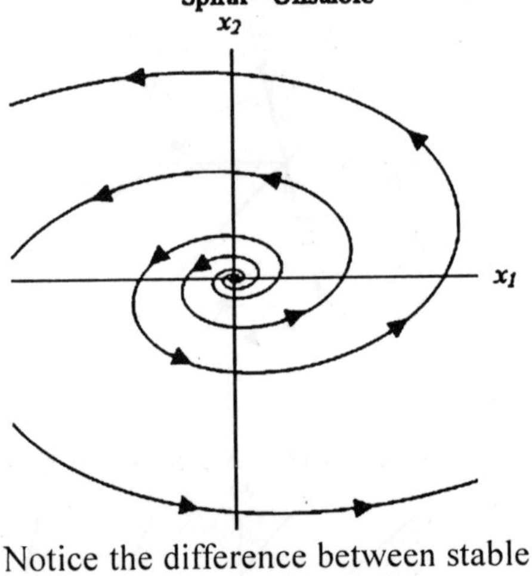

Spiral - Asymptotically Stable

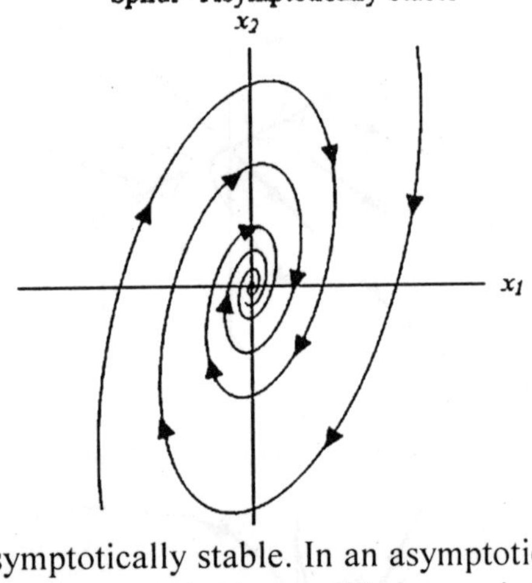

Notice the difference between stable and asymptotically stable. In an asymptotically stable node or spiral all the trajectories will move in towards the equilibrium point as t increases whereas, a center (which is always stable) trajectories will just move around the equilibrium point but never actually move in towards it.

REAL, DISTINCT EIGENVALUES

It's now time to start solving systems of differential equations. We've seen that solutions to the system,

$$\vec{x} = A\vec{x}$$

will be of the form

$$\vec{x} = \vec{\eta}e^{\lambda t}$$

where λ and $\vec{\eta}$ are eigenvalues and eigenvectors of the matrix A. We will be working with 2 x 2 systems so this means that we are going to be looking for two solutions, $\vec{x}_1(t)$ and $\vec{x}_2(t)$, where the determinant of the matrix,

$$X = (\vec{x}_2\ \vec{x}_2)$$

is nonzero.

We are going to start by looking at the case where our two eigenvalues, λ_1 and λ_2 are real and distinct. In other words they will be real, simple eigenvalues. Recall as well that the eigenvectors for simple eigenvalues are linearly independent. This means that the solutions we get from these will also be linearly independent. If the solutions are linearly independent the matrix X must be nonsingular and hence these two solutions will be a fundamental set of solutions. The general solution in this case will then be,

$$\vec{x}(t) = c_1 e^{\lambda_1 t}\vec{\eta}^{(1)} + c_2 e^{\lambda_2 t}\vec{\eta}^{(2)}$$

Note that each of our examples will actually be broken into two examples. The first example will be solving the system and the second example will be sketching the phase portrait for the system. Phase portraits are not always taught in a differential equations course and so we'll strip those out of the solution process so that if you haven't covered them in your class you can ignore the phase portrait example for the system.

Example: Solve the following IVP.

$$\vec{x}' = \begin{pmatrix} 1 & 2 \\ 3 & 2 \end{pmatrix}\vec{x}, \qquad \vec{x}(0) = \begin{pmatrix} 0 \\ -4 \end{pmatrix}.$$

Solution: So, the first thing that we need to do is find the eigenvalues for the matrix.

$$\det(A - \lambda I) = \begin{vmatrix} 1-\lambda & 2 \\ 3 & 2-\lambda \end{vmatrix}$$

$$= \lambda^2 - 3\lambda - 4 = (\lambda + 1)(\lambda - 4) \quad \Rightarrow \quad \lambda_1 = -1,\ \lambda_2 = 4.$$

Now let's find the eigenvectors for each of these.

$$\lambda_1 = -1.$$

We'll need to solve,

$$\begin{pmatrix} 2 & 2 \\ 3 & 3 \end{pmatrix}\begin{pmatrix} \eta_1 \\ \eta_2 \end{pmatrix} = \begin{pmatrix} 0 \\ 0 \end{pmatrix} \quad \Rightarrow \quad 2\eta_1 + 2\eta_2 = 0 \quad \Rightarrow \quad \eta_1 = -\eta_2.$$

The eigenvector in this case is,

$$\vec{\eta} = \begin{pmatrix} -\eta_1 \\ \eta_2 \end{pmatrix} \quad \Rightarrow \quad \vec{\eta}^{(1)} = \begin{pmatrix} -1 \\ 1 \end{pmatrix}, \quad \eta_2 = 1$$

$\lambda_2 = 4$: We'll need to solve,

$$\begin{pmatrix} -3 & 2 \\ 3 & -2 \end{pmatrix}\begin{pmatrix} \eta_1 \\ \eta_2 \end{pmatrix} = \begin{pmatrix} 0 \\ 0 \end{pmatrix} \Rightarrow 3\eta_1 + 2\eta_2 = 0 \Rightarrow \eta_1 = \frac{2}{3}\eta_2.$$

The eigenvector in this case is,

$$\vec{\eta} = \begin{pmatrix} \frac{2}{3}\eta_2 \\ \eta_2 \end{pmatrix} \Rightarrow \vec{\eta}^{(2)} = \begin{pmatrix} 2 \\ 3 \end{pmatrix}, \; \eta_2 = 3.$$

Then general solution is then,

$$\vec{x}(t) = c_1 e^{-t}\begin{pmatrix} -1 \\ 1 \end{pmatrix} + c_1 e^{4t}\begin{pmatrix} 2 \\ 3 \end{pmatrix}.$$

Now, we need to find the constants. To do this we simply need to apply the initial conditions.

$$\begin{pmatrix} 0 \\ -4 \end{pmatrix} = \vec{x}(0) = c_1\begin{pmatrix} -1 \\ 1 \end{pmatrix} + c_2\begin{pmatrix} 2 \\ 3 \end{pmatrix}.$$

All we need to do now is multiply the constants through and we then get two equations (one for each row) that we can solve for the constants. This gives,

$$\left.\begin{matrix} -c_1 + 2c_2 = 0 \\ c_1 + 3c_2 = -4 \end{matrix}\right\} \Rightarrow c_1 = -\frac{8}{5}, \; c_2 = -\frac{4}{5}.$$

The solution is then,

$$\vec{x}(t) = -\frac{8}{5}e^{-t}\begin{pmatrix} -1 \\ 1 \end{pmatrix} - \frac{4}{5}e^{4t}\begin{pmatrix} 2 \\ 3 \end{pmatrix}.$$

Now, let's take a look at the phase portrait for the system.

Example: Sketch the phase portrait for the following system.

$$\vec{x}\,' = \begin{pmatrix} 1 & 2 \\ 3 & 2 \end{pmatrix}\vec{x}.$$

Solution: From the last example we know that the eigenvalues and eigenvectors for this system are,

$$\lambda_1 = -1 \quad \vec{\eta}^{(1)} = \begin{pmatrix} -1 \\ 1 \end{pmatrix}$$

$$\lambda_2 = 4 \quad \vec{\eta}^{(2)} = \begin{pmatrix} 2 \\ 3 \end{pmatrix}.$$

It turns out that this is all the information that we will need to sketch the direction field. We will relate things back to our solution however so that we can see that things are going correctly.

Well start by sketching lines that follow the direction of the two eigenvectors. This gives,

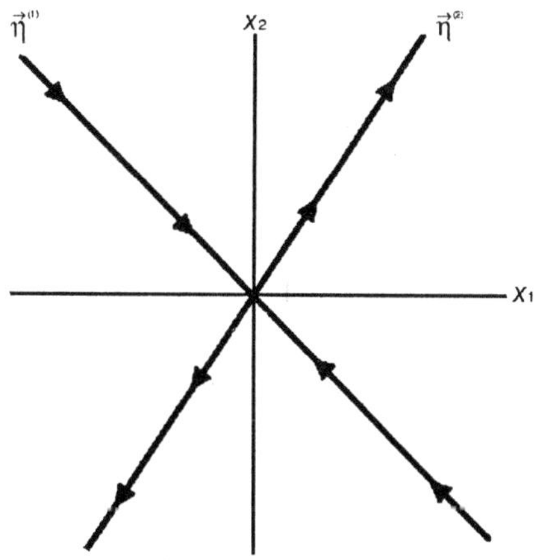

Now, from the first example our general solution is

$$x(t) = c_1 e^{-t}\begin{pmatrix}-1\\1\end{pmatrix} + c_2 e^{4t}\begin{pmatrix}2\\0\end{pmatrix}.$$

If we have $c_2 = 0$ then the solution is an exponential times a vector and that all the exponential does is affect the magnitude of the vector and the constant c_1 will affect both the sign and the magnitude of the vector. In other words, the trajectory in this case will be a straight line that is parallel to the vector, $\vec{\eta}^{(1)}$. Also notice that as t increases the exponential will get smaller and smaller and hence the trajectory will be moving in towards the origin. If $c_1 > 0$ the trajectory will be in Quadrant II and if $c_1 < 0$ the trajectory will be in Quadrant IV. So the line in the graph above marked with $\vec{\eta}^{(1)}$ will be a sketch of the trajectory corresponding to $c_2 = 0$ and this trajectory will approach the origin as t increases.

If we now turn things around and look at the solution corresponding to have $c_1 = 0$ we will have a trajectory that is parallel to $\vec{\eta}^{(2)}$. Also, since the exponential will increase as t increases and so in this case the trajectory will now move away from the origin as t increases. We will denote this with arrows on the lines in the graph above.

Notice that we could have gotten this information with actually going to the solution. All we really need to do is look at the eigenvalues. Eigenvalues that are negative will correspond to solutions that will move towards the origin as t increase in a direction that is parallel to its eigenvector. Likewise, eigenvalues that are positive move away from the origin as t increases in a direction that will be parallel to its eigenvector.

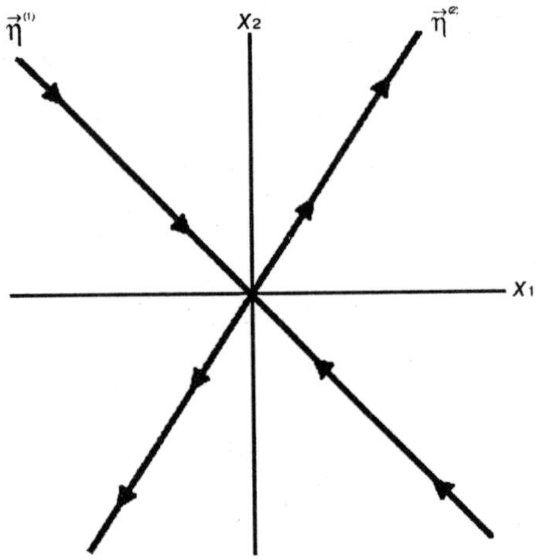

If both constants are in the solution we will have a combination of these behaviors. For large negative t's the solution will be dominated by the portion that has the negative eigenvalue since in these cases the exponent will be large and positive. Trajectories for large negative t's will be parallel to $\vec{\eta}^{(1)}$ and moving in the same direction.

Solutions for large positive t's will be dominated by the portion with the positive eigenvalue. Trajectories in this case will be parallel to $\vec{\eta}^{(2)}$ and moving in the same direction.

In general, it looks like trajectories will start "near" $\vec{\eta}^{(1)}$, move in towards the origin and then as they get closer to the origin they will start moving towards $\vec{\eta}^{(2)}$ and then continue up along this vector. Sketching some of these in will give the following phase portrait. Here is a sketch of this with the trajectories corresponding to the eigenvectors marked in blue.

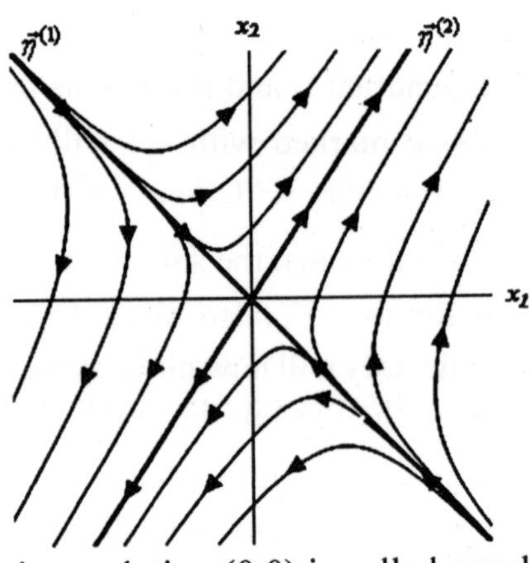

In this case the equilibrium solution (0,0) is called a saddle point and is unstable. In this case unstable means that solution move away from it as t increases.

So, we've solved a system in matrix form, but remember that we started out without the systems in matrix form. Now let's take a quick look at an example of a system that isn't in matrix form initially.

Example: Find the solution to the following system.

$$x_1' = x_1 + 2x_2 \qquad x_1(0) = 0$$

$$x_2' = 3x_1 + 2x_2 \qquad x_2(0) = -4.$$

Solution: We first need to convert this into matrix form. This is easy enough. Here is the matrix form of the system.

$$\vec{x} = \begin{pmatrix} 1 & 2 \\ 3 & 2 \end{pmatrix}\vec{x}, \qquad \vec{x}(0) = \begin{pmatrix} 0 \\ -4 \end{pmatrix}.$$

This is just the system from the first example and so we've already got the solution to this system. Here it is.

$$\vec{x}(t) = -\frac{8}{5}e^{-t}\begin{pmatrix} -1 \\ 1 \end{pmatrix} - \frac{4}{5}e^{4t}\begin{pmatrix} 2 \\ 3 \end{pmatrix}.$$

Now, since we want to solution to the system not in matrix form let's go one step farther here. Let's multiply the constants and exponentials into the vectors and then add up the two vectors.

$$\vec{x}(t) = \begin{pmatrix} \frac{8}{5}e^{-t} \\ -\frac{8}{5}e^{-t} \end{pmatrix} - \begin{pmatrix} \frac{8}{5}e^{-t} \\ \frac{12}{5}e^{4t} \end{pmatrix} = \begin{pmatrix} \frac{8}{5}e^{-t} & \frac{8}{5}e^{4t} \\ -\frac{8}{5}e^{-t} & -\frac{12}{5}e^{4t} \end{pmatrix}.$$

Now, recall,

$$\vec{x}(t) = \begin{pmatrix} x_1 & (t) \\ x_2 & (t) \end{pmatrix}.$$

So, the solution to the system is then,

$$x_1(t) = \frac{8}{5}e^{-t} - \frac{8}{5}e^{4t}$$

$$x_2(t) = -\frac{8}{5}e^{-t} - \frac{12}{5}e^{4t}.$$

Let's work another example.

Example: Solve the following IVP.

$$\vec{x} = \begin{pmatrix} -5 & 1 \\ 4 & -2 \end{pmatrix}\vec{x}, \qquad \vec{x}(0) = \begin{pmatrix} 1 \\ 2 \end{pmatrix}.$$

Solution: So, the first thing that we need to do is find the eigenvalues for the matrix.

$$\det(A-\lambda_1) = \begin{vmatrix} -5-1 & 1 \\ 4 & -2-\lambda \end{vmatrix} = \lambda^2+7\lambda+6$$

$$= (\lambda+1)(\lambda+6) \quad \Rightarrow \quad \lambda_1 = -1, \lambda_2 = -6.$$

Now let's find the eigenvectors for each of these.

$$\lambda_1 = -1:$$

We'll need to solve,

$$\begin{pmatrix} -4 & 1 \\ 4 & -1 \end{pmatrix}\begin{pmatrix} \eta_1 \\ \eta_2 \end{pmatrix} = \begin{pmatrix} 0 \\ 0 \end{pmatrix} \quad \Rightarrow \quad -4\eta_1+\eta_2=0 \quad \Rightarrow \quad \eta_2 = 4\eta_1.$$

The eigenvector in this case is,

$$\vec{\eta} = \begin{pmatrix} \eta_1 \\ 4\eta_1 \end{pmatrix} \quad \Rightarrow \quad \vec{\eta}^{(1)} = \begin{pmatrix} 1 \\ 4 \end{pmatrix}, \quad \eta_1 = 1.$$

$$\lambda_2 = -6:$$

We'll need to solve,

$$\begin{pmatrix} 1 & 1 \\ 4 & 4 \end{pmatrix}\begin{pmatrix} \eta_1 \\ \eta_1 \end{pmatrix} = \begin{pmatrix} 0 \\ 0 \end{pmatrix} \quad \Rightarrow \quad \eta_1+\eta_2=0 \quad \Rightarrow \quad \eta_1 = -\eta_2.$$

The eigenvector in this case is,

$$\vec{\eta} = \begin{pmatrix} -\eta_1 \\ \eta_1 \end{pmatrix} \quad \Rightarrow \quad \vec{\eta}^{(2)} \begin{pmatrix} -1 \\ 0 \end{pmatrix}, \quad \eta_2 = 1.$$

Then general solution is then,

$$\vec{x}(t) = c_1 e^{-t}\begin{pmatrix} 1 \\ 4 \end{pmatrix} \quad \Rightarrow \quad c_1 e^{-6t}\begin{pmatrix} -1 \\ 0 \end{pmatrix}.$$

Now, we need to find the constants. To do this we simply need to apply the initial conditions.

$$\begin{pmatrix} 1 \\ 2 \end{pmatrix} = \vec{x}(0) = c_1\begin{pmatrix} 1 \\ 4 \end{pmatrix} + c_2\begin{pmatrix} -1 \\ 1 \end{pmatrix}.$$

Now solve the system for the constants.

$$\left.\begin{aligned} c_1 - c_2 &= 1 \\ 4c_1 + c_2 &= 2 \end{aligned}\right\} \quad \Rightarrow \quad c_1 = \frac{3}{5}, c_2 = -\frac{2}{5}.$$

The solution is then,

$$\vec{x}(t) = \frac{3}{5}e^{-t}\begin{pmatrix}1\\4\end{pmatrix} - \frac{2}{5}e^{-6t}\begin{pmatrix}-1\\1\end{pmatrix}.$$

Now let's find the phase portrait for this system.

Example: Sketch the phase portrait for the following system.

$$\vec{x}' = \begin{pmatrix}-5 & 1\\4 & -2\end{pmatrix}\vec{x}.$$

Solution: From the last example we know that the eigenvalues and eigenvectors for this system are,

$$\lambda_1 = -1 \qquad \vec{\eta}^{(1)} = \begin{pmatrix}1\\4\end{pmatrix}$$

$$\lambda_2 = -6 \qquad \vec{\eta}^{(2)} = \begin{pmatrix}-1\\1\end{pmatrix}.$$

This one is a little different from the first one. However it starts in the same way. We'll first sketch the trajectories corresponding to the eigenvectors. Notice as well that both of the eigenvalues are negative and so trajectories for these will move in towards the origin as t increases. When we sketch the trajectories we'll add in arrows to denote the direction they take as t increases. Here is the sketch of these trajectories.

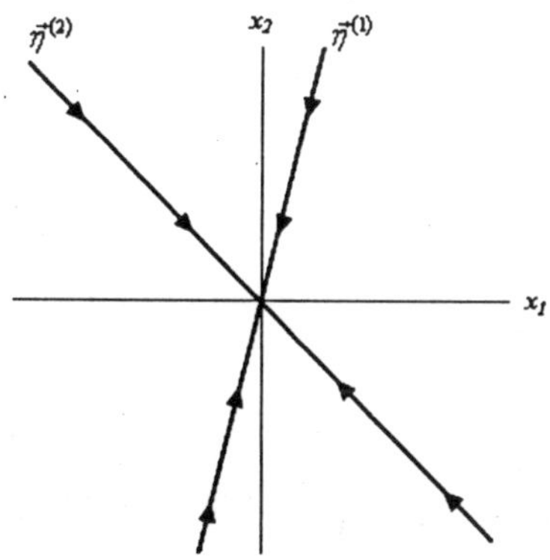

Now, here is where the slight difference from the first phase portrait comes up. All of the trajectories will move in towards the origin as t increases since both of the eigenvalues are negative. The issue that we need to decide upon is just how they do this. This is actually easier than it might appear to be at first.

The second eigenvalue is larger than the first. For large and positive t's this means that the solution for this eigenvalue will be smaller than the solution for the first eigenvalue. Therefore, as t increases the trajectory will move in towards the origin and do so parallel to $\vec{\eta}^{(1)}$. Likewise, since the second eigenvalue is larger than the first this solution will dominate for large and negative t's. Therefore, as we decrease t the trajectory will move away from the origin and do so parallel to $\vec{\eta}^{(2)}$.

Adding in some trajectories gives the following sketch.

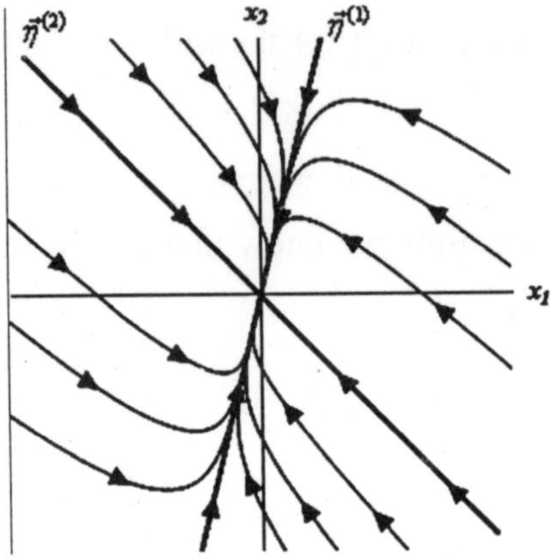

In these cases we call the equilibrium solution (0,0) a node and it is asymptotically stable. Equilibrium solutions are asymptotically stable if all the trajectories move in towards it as t increases.

Note that nodes can also be unstable. In the last example if both of the eigenvalues had been positive all the trajectories would have moved away from the origin and in this case the equilibrium solution would have been unstable.

Before moving on to the next section we need to do one more example. When we first started talking about systems it was mentioned that we can convert a higher order differential equation into a system. We need to do an example like this so we can see how to solve higher order differential equations using systems.

Example:Convert the following differential equation into a system, solve the system and use this solution to get the solution to the original differential equation.

$$2y''+5y'-3y=0, \qquad y(0)=-4 \;\; y'(0)=9.$$

Solution: So, we first need to convert this into a system. Here's the change of variables,

$$x_1 = y \qquad x_1' = y' = x_2$$

$$x_2 = y' \qquad x_2' = y'' = \frac{3}{2}y - \frac{5}{2}y' = \frac{3}{2}x_1 - \frac{5}{2}x_2.$$

The system is then,

$$\vec{x}' = \begin{pmatrix} 0 & 1 \\ \frac{5}{2} & -\frac{5}{2} \end{pmatrix}\vec{x} \qquad \vec{x}(0) = \begin{pmatrix} -4 \\ 9 \end{pmatrix}$$

where,

$$\vec{x}(t)=\begin{pmatrix} x_1(t) \\ x_2(t) \end{pmatrix}=\begin{pmatrix} y(t) \\ y'(t) \end{pmatrix}.$$

Now we need to find the eigenvalues for the matrix.

$$\det(A-\lambda I) = \begin{vmatrix} -\lambda & 1 \\ \frac{3}{2} & -\frac{5}{2}-\lambda \end{vmatrix}$$

$$= \lambda^2+\tfrac{5}{2}\lambda-\tfrac{3}{2}$$

$$= \tfrac{1}{2}(\lambda+3)(2\lambda-1) \qquad \lambda_1=-3,\ \lambda_2=\frac{1}{2}.$$

Now let's find the eigenvectors.

$$\lambda_1=-3:$$

We'll need to solve,

$$\begin{pmatrix} 3 & 1 \\ \frac{3}{2} & \frac{1}{2} \end{pmatrix}\begin{pmatrix} \eta_1 \\ \eta_2 \end{pmatrix}=\begin{pmatrix} 0 \\ 0 \end{pmatrix} \quad\Rightarrow\quad 3\eta_1+\eta_2=0 \quad\Rightarrow\quad \eta_2=-3\eta_1.$$

The eigenvector in this case is,

$$\vec{\eta}=\begin{pmatrix} \eta_1 \\ 3\eta_2 \end{pmatrix} \quad\Rightarrow\quad \vec{\eta}^{(1)}-\begin{pmatrix} 1 \\ -3 \end{pmatrix}, \quad \eta_1=1$$

$$\lambda_2=\tfrac{1}{2}:$$

We'll need to solve,

$$\begin{pmatrix} -\frac{1}{2} & 1 \\ \frac{3}{2} & -3 \end{pmatrix}\begin{pmatrix} \eta_1 \\ \eta_2 \end{pmatrix}=\begin{pmatrix} 0 \\ 0 \end{pmatrix} \quad\Rightarrow\quad -\frac{1}{2}\eta_1+\eta_2=0 \quad\Rightarrow\quad \eta_2=\frac{1}{2}\eta_1.$$

The eigenvector in this case is,

$$\vec{\eta}=\begin{pmatrix} \eta_1 \\ \frac{1}{2}\eta_1 \end{pmatrix} \quad\Rightarrow\quad \vec{\eta}^{(2)}=\begin{pmatrix} 2 \\ 1 \end{pmatrix}, \quad \eta_1=2.$$

The general solution is then,

$$\vec{x}(t)=c_1e^{-3t}\begin{pmatrix} 1 \\ -3 \end{pmatrix}+c_2e^{\frac{t}{2}}\begin{pmatrix} 2 \\ 1 \end{pmatrix}.$$

Apply the initial condition.

$$\begin{pmatrix}-4\\9\end{pmatrix}=\vec{x}(0)=c_1\begin{pmatrix}1\\-3\end{pmatrix}+c_2\begin{pmatrix}2\\1\end{pmatrix}.$$

This gives the system of equations that we can solve for the constants.

$$\left.\begin{array}{l}c_1+2c_2=-4\\-3c_1+c_2=9\end{array}\right\}\quad\Rightarrow\quad c_1=-\frac{22}{7},\ c_2=-\frac{3}{7}.$$

The actual solution to the system is then,

$$\vec{x}(t)=-\frac{22}{7}e^{-2t}\begin{pmatrix}1\\-3\end{pmatrix}-\frac{3}{7}e^{\frac{t}{2}}\begin{pmatrix}2\\1\end{pmatrix}.$$

Now recalling that,

$$\vec{x}(t)=\begin{pmatrix}y(t)\\y'(t)\end{pmatrix}.$$

we can see that the solution to the original differential equation is just the top row of the solution to the matrix system. The solution to the original differential equation is then,

$$y(t)=-\frac{22}{7}e^{-3t}-\frac{6}{7}e^{\frac{t}{2}}.$$

COMPLEX EIGENVALUES

In this section we will look at solutions to

$$\vec{x}'=A\vec{x}$$

where the eigenvalues of the matrix A are complex. With complex eigenvalues we are going to have the same problem that we had back when we were looking at second order differential equations. We want our solutions to only have real numbers in them, however since our solutions to systems are of the form,

$$\vec{x}=\vec{\eta}e^{\lambda t}$$

we are going to have complex numbers come into our solution from both the eigenvalue and the eigenvector. Getting rid of the complex numbers here will be similar to how we did it back in the second order differential equation case, but will involve a little more work this time around. It's easiest to see how to do this in an example.

Example: Solve the following IVP.

$$\vec{x}=\begin{pmatrix}3&-9\\4&-3\end{pmatrix}\vec{x}\qquad\vec{x}(0)=\begin{pmatrix}2\\-4\end{pmatrix}.$$

Solution: We first need the eigenvalues and eigenvectors for the matrix.

$$\det(A-\lambda I) = \begin{vmatrix} 3-\lambda & -9 \\ 4 & -3-\lambda \end{vmatrix}$$

$$= \lambda^2 + 27 \qquad \lambda_{1,2} = \pm\sqrt{3}i.$$

So, now that we have the eigenvalues recall that we only need to get the eigenvector for one of the eigenvalues since we can get the second eigenvector for free from the first eigenvector.

$$\lambda_1 = \sqrt{3}i:$$

We need to solve the following system.

$$\begin{pmatrix} 3-3\sqrt{3}i & -9 \\ 4 & -3-3\sqrt{3}i \end{pmatrix}\begin{pmatrix} \eta_1 \\ \eta_2 \end{pmatrix} = \begin{pmatrix} 0 \\ 0 \end{pmatrix}.$$

Using the first equation we get,

$$\left(3-3\sqrt{3}i\right)\eta_1 - 9\eta_2 = 0$$

$$h_2 = \frac{1}{3}\left(1-\sqrt{3}i\right)\eta_1.$$

So, the first eigenvector is,

$$\vec{\eta} = \begin{pmatrix} \eta_1 \\ \frac{1}{3}\left(1-\sqrt{3}i\right)\eta_1 \end{pmatrix}$$

$$\vec{\eta}^{(1)} = \begin{pmatrix} 3 \\ 1-\sqrt{3}i \end{pmatrix} \qquad \eta_1 = 3.$$

When finding the eigenvectors in these cases make sure that the complex number appears in the numerator of any fractions since we'll need it in the numerator later on.

Also try to clear out any fractions by appropriately picking the constant. This will make our life easier down the road.

Now, the second eigenvector is,

$$\vec{\eta}^{(2)} = \begin{pmatrix} 3 \\ 1+\sqrt{3}i \end{pmatrix}.$$

However, as we will see we won't need this eigenvector.

The solution that we get from the first eigenvalue and eigenvector is,

$$\vec{x}_1(t) = e^{3\sqrt{3}it}\begin{pmatrix} 3 \\ 1-\sqrt{3}i \end{pmatrix}.$$

So, as we can see there are complex numbers in both the exponential and vector that we will need to get rid of in order to use this as a solution. Recall from the complex roots section of the second order differential equation chapter that we can use Euler's formula to get the complex number out of the exponential. Doing this gives us,

$$\vec{x}_1(t) = \left(\cos\left(3\sqrt{3}t\right) + i\sin\left(3\sqrt{3}t\right)\right)\begin{pmatrix} 3 \\ 1-\sqrt{3}i \end{pmatrix}.$$

The next step is to multiply the cosines and signs into the vector.

$$\vec{x}_1(t) = \begin{pmatrix} 3\cos\left(3\sqrt{3}t\right) + 3i\sin\left(3\sqrt{3}t\right) \\ \left(\cos\left(3\sqrt{3}t\right) + i\sin\left(3\sqrt{3}t\right) - \sqrt{3}i\cos\left(3\sqrt{3}t\right) + \sqrt{3}\sin\left(3\sqrt{3}t\right)\right) \end{pmatrix}.$$

Now combine the terms with an "*i*" in them and split these terms off from those terms that don't contain an "*i*". Also factor the "*i*" out of this vector.

$$\begin{aligned}\vec{x}_1(t) &= \begin{pmatrix} 3\cos\left(3\sqrt{3}t\right) \\ \cos\left(3\sqrt{3}t\right) + \sqrt{3}\sin\left(3\sqrt{3}t\right) \end{pmatrix} + i\begin{pmatrix} 3\sin\left(3\sqrt{3}t\right) \\ \cos\left(3\sqrt{3}t\right) - \sqrt{3}\cos\left(3\sqrt{3}t\right) \end{pmatrix}. \\ &= \vec{u}\,(t) + i\vec{v}(t).\end{aligned}$$

Now, it can be shown (we'll leave the details to you) that $\vec{u}\,(t)$and $\vec{v}\,(t)$are two linearly independent solutions to the system of differential equations. This means that we can use them to form a general solution and they are both real solutions.

So, the general solution to a system with complex roots is

$$\vec{x}\,(t) = c_1\vec{u}(t) + c_2\vec{v}(t)$$

where $\vec{u}(t)$ and $\vec{v}(t)$ are found by writing the first solution as

$$\vec{x}\,(t) = \vec{u}(t) + i\vec{v}(t).$$

For our system then, the general solution is,

$$\vec{x}(t) = c_1\begin{pmatrix} 3\cos\left(3\sqrt{3}t\right) \\ \cos\left(3\sqrt{3}t\right) + \sqrt{3}\sin\left(3\sqrt{3}t\right) \end{pmatrix} + c_2\begin{pmatrix} 3\sin\left(3\sqrt{3}t\right) \\ \cos\left(3\sqrt{3}t\right) - \sqrt{3}\cos\left(3\sqrt{3}t\right) \end{pmatrix}.$$

We now need to apply the initial condition to this to find the constants.

$$\begin{pmatrix} 2 \\ -4 \end{pmatrix} = \vec{x}(0) = c_1 \begin{pmatrix} 3 \\ 1 \end{pmatrix} + c_2 \begin{pmatrix} 0 \\ -\sqrt{3} \end{pmatrix}.$$

This leads to the following system of equations to be solved,

$$\left.\begin{aligned} 3c_1 &= 2 \\ c_1 - \sqrt{2}c_2 &= -4 \end{aligned}\right\} \quad \Rightarrow \quad c_1 = \frac{2}{3},\ c_3 = \frac{14}{3\sqrt{3}}.$$

The actual solution is then,

$$\vec{x}(t) = \frac{2}{3}\begin{pmatrix} 3\cos\left(3\sqrt{3}t\right) \\ \cos\left(3\sqrt{3}t\right) + \sqrt{3}\sin\left(3\sqrt{3}t\right) \end{pmatrix} + \frac{14}{3\sqrt{3}}\begin{pmatrix} 3\sin\left(3\sqrt{3}t\right) \\ \sin\left(3\sqrt{3}t\right) - \sqrt{3}\cos\left(3\sqrt{3}t\right) \end{pmatrix}.$$

As we did in the last section we'll do the phase portraits separately from the solution of the system in case phase portraits haven't been taught in your class.

Example: Sketch the phase portrait for the system.

$$\vec{x} = \begin{pmatrix} 3 & -9 \\ 4 & -3 \end{pmatrix}\vec{x}.$$

Solution: When the eigenvalues of a matrix A are purely complex, as they are in this case, the trajectories of the solutions will be circles or ellipses that are centered at the origin. The only thing that we really need to concern ourselves with here are whether they are rotating in a clockwise or counterclockwise direction.

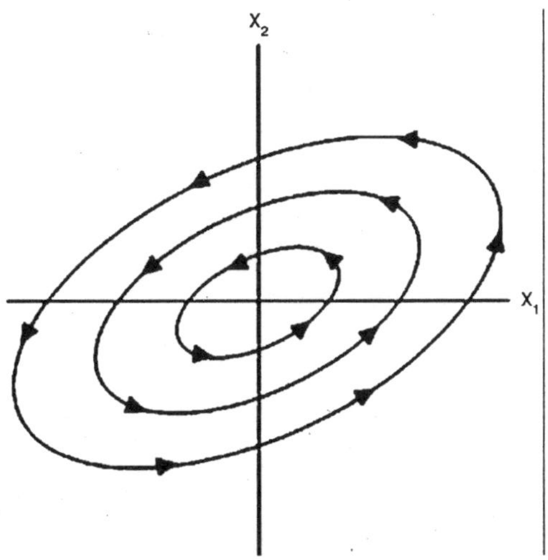

This is easy enough to do. Recall when we first looked at these phase portraits a couple of sections ago that if we pick a value of $\vec{x}(t)$ and plug it into our system we will get a vector that will be tangent to the trajectory at that point and pointing in the direction that the trajectory is traveling.. So, let's pick the following point and see what we get.

$$\vec{x}\begin{pmatrix}1\\0\end{pmatrix} \quad \Rightarrow \quad \vec{x} = \begin{pmatrix}3 & -9\\4 & -3\end{pmatrix}\begin{pmatrix}1\\0\end{pmatrix} = \begin{pmatrix}3\\4\end{pmatrix}.$$

Therefore at the point (1,0) in the phase plane the trajectory will be point in a upwards direction. The only way that this can be is if the trajectories are traveling in a counterclockwise direction. Here is the sketch of some of the trajectories for this problem.

The equilibrium solution in the case is called a center and is stable.

Note in this last example that the equilibrium solution is stable and not asymptotically stable. Asymptotically stable refers to the fact that the trajectories are moving in toward the equilibrium solution as t increases. In this example the trajectories are simply revolving around the equilibrium solution and not moving in towards it. The trajectories are also not moving away from the equilibrium solution and so they aren't unstable. Therefore we call the equilibrium solution stable.

Not all complex eigenvalues will result in centers so let's take a look at an example where we get something different.

Example: Solve the following IVP.

$$\vec{x} = \begin{pmatrix}3 & -9\\4 & -3\end{pmatrix}\vec{x} \qquad \vec{x}(0) = \begin{pmatrix}1\\-10\end{pmatrix}.$$

Solution: Let's get the eigenvalues and eigenvectors for the matrix.

$$\det(A - \lambda I) = \begin{vmatrix}3-\lambda & -13\\5 & 1-\lambda\end{vmatrix}$$

$$= \lambda^2 - 4\lambda + 68 \qquad \lambda_{1,2} = 2 \pm 8i.$$

Now get the eigenvector for the first eigenvalue.

$$\lambda^2 = 2 + 8i:$$

We need to solve the following system.

$$\begin{pmatrix}1-8i & -13\\5 & -1-8i\end{pmatrix}\begin{pmatrix}\eta_1\\\eta_2\end{pmatrix} = \begin{pmatrix}0\\0\end{pmatrix}.$$

Using the second equation we get,

$$5\eta_1 + (-1-8i)\eta_2 = 0$$

$$\eta_1 = \frac{1}{5}(1+8i)\eta_2.$$

So, the first eigenvector is,

$$\vec{\eta} = \begin{pmatrix} \frac{1}{5}(1+8i)\eta_2 \\ \eta_2 \end{pmatrix}$$

$$\vec{\eta}^{(1)} = \begin{pmatrix} 1+8i \\ 5 \end{pmatrix} \qquad \eta_2 = 5.$$

The solution corresponding the this eigenvalue and eigenvector is

$$\begin{aligned} \vec{x}_1(t) &= e^{(2+8i)t}\begin{pmatrix} 1+8i \\ 5 \end{pmatrix} \\ &= e^{2t}e^{8it}\begin{pmatrix} 1+8i \\ 5 \end{pmatrix} \\ &= e^{2t}\left(\cos(8t)+i\sin(8t)\right)\begin{pmatrix} 1+8i \\ 5 \end{pmatrix}. \end{aligned}$$

As with the first example multiply cosines and signs into the vector and split it up. Don't forget about the exponential that is in the solution this time.

$$\begin{aligned} \vec{x}_1(t) &= e^{2t}\begin{pmatrix} \cos(8t)-8\sin(8t) \\ 5\cos(8t) \end{pmatrix} + ie^{2t}\begin{pmatrix} \cos(8t)-8\sin(8t) \\ 5\cos(8t) \end{pmatrix} \\ &= \vec{u}(t)+i\vec{v}(t). \end{aligned}$$

The general solution to this system then,

$$c_1e^{2t}\begin{pmatrix} \cos(8t)-8\sin(8t) \\ 5\cos(8t) \end{pmatrix} + c_2e^{2t}\begin{pmatrix} \cos(8t)-8\sin(8t) \\ 5\cos(8t) \end{pmatrix}.$$

Now apply the initial condition and find the constants.

$$\begin{pmatrix} 3 \\ -10 \end{pmatrix} = \vec{x}(0) = c_1\begin{pmatrix} 1 \\ 5 \end{pmatrix} + c_2e^{2t}\begin{pmatrix} 8 \\ 0 \end{pmatrix}$$

$$\left.\begin{aligned} c_1+8c_2 &= 3 \\ 5c_1 &= -10 \end{aligned}\right\} \quad \Rightarrow \quad c_1=-2,\ c_2=\frac{5}{8}.$$

The actual solution is then,

$$\vec{x}(t) = -2e^{2t}\begin{pmatrix} \cos(8t)-8\sin(8t) \\ 5\cos(8t) \end{pmatrix} + \frac{5}{8}e^{2t}\begin{pmatrix} \cos(8t)-8\sin(8t) \\ 5\cos(8t) \end{pmatrix}.$$

Let's take a look at the phase portrait for this problem.

Example: Sketch the phase portrait for the system.

$$\vec{x}' = \begin{pmatrix} 3 & -13 \\ 5 & 1 \end{pmatrix} \vec{x}.$$

Solution: When the eigenvalues of a system are complex with a real part the trajectories will spiral into or out of the origin. We can determine which one it will be by looking at the real portion. Since the real portion will end up being the exponent of an exponential function (as we saw in the solution to this system) if the real part is positive the solution will grow very large as t increases. Likewise, if the real part is negative the solution will die out as t increases.

So, if the real part is positive the trajectories will spiral out from the origin and if the real part is negative they will spiral into the origin. We determine the direction of rotation in the same way that we did for the center. In our case the trajectories will spiral out from the origin since the real part is positive and

$$\vec{x}' = \begin{pmatrix} 3 & -13 \\ 5 & 1 \end{pmatrix} \begin{pmatrix} 1 \\ 0 \end{pmatrix} = \begin{pmatrix} 3 \\ 5 \end{pmatrix}.$$

will rotate in the counterclockwise direction as the last example did.

Here is a sketch of some of the trajectories for this system.

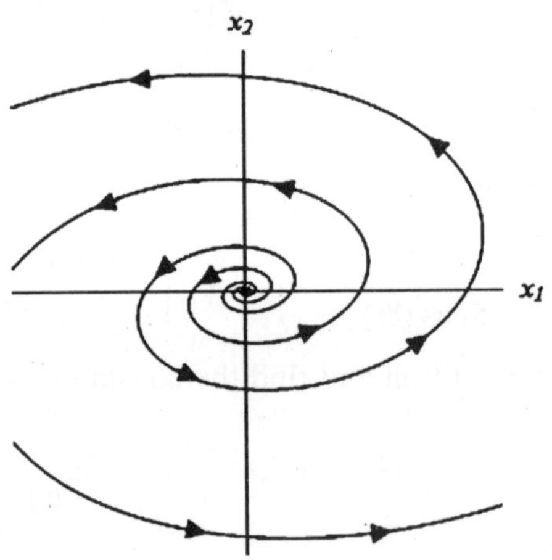

Here we call the equilibrium solution a spiral and in this case it's unstable since the trajectories move away from the origin. If the real part of the eigenvalue is negative the trajectories will spiral into the origin and in this case the equilibrium solution will be asymptotically stable.

REPEATED EIGENVALUES

This is the final case that we need to take a look at. In this section we are going to look at solutions to the system,

$$\vec{x}' = A\vec{x}.$$

where the eigenvalues are repeated eigenvalues. Since we are going to be working with systems in which A is a 2×2 matrix we will make that assumption from the start. So the system will have a double eigenvalue, λ.

This presents us with a problem. We want two linearly independent solutions so that we can form a general solution. However, with a double eigenvalue we will have only one,

$$\vec{x}_1 = \vec{\eta}e^{\lambda t}.$$

So, we need to come up with a second solution. Recall that when we looked at the double root case with the second order differential equations we ran into a similar problem. In that section we simply added a t to the solution and were able to get a second solution. Let's see if the same thing will work in this case as well. We'll see if

$$\vec{x} = te^{\lambda t}\vec{\eta}$$

will also be a solution.

To check all we need to do is plug into the system. Don't forget to product rule the proposed solution when you differentiate!

$$\vec{\eta}e^{\lambda t} + \lambda\vec{\eta}te^{\lambda t} = A\vec{\eta}te^{\lambda t}.$$

Now, we got two functions here on the left side, an exponential by itself and an exponential times a t. So, in order for our guess to be a solution we will need to require,

$$A\vec{\eta} = \lambda\vec{\eta} \quad \Rightarrow \quad (A - \lambda I)\vec{\eta} = \vec{0}$$

$$\vec{\eta} = \vec{0}.$$

The first requirement isn't a problem since we this just says that λ is an eigenvalue and it's eigenvector is $\vec{\eta}$. We already knew this however so there's nothing new there. The second however is a problem. Since $\vec{\eta}$ is an eigenvector we know that it can't be zero, yet in order to satisfy the second condition it would have to be.

So, our guess was incorrect. The problem seems to be that there is a lone term with just an exponential in it so let's see if we can't fix up our guess to correct that. Let's try the following guess.

$$\vec{x} = te^{\lambda t}\vec{\eta} + e^{\lambda t}\vec{\rho}$$

where $\vec{\rho}$ is an unknown vector that we'll need to determine.

As with the first guess let's plug this into the system and see what we get.

$$\vec{\eta}e^{\lambda t} + \lambda\vec{\eta}te^{\lambda t} + \lambda\vec{\rho}e^{\lambda t} = A\left(\vec{\eta}te^{\lambda t} + \vec{\rho}e^{\lambda t}\right)$$

$$\left(\vec{\eta} + \lambda\vec{\rho}\right)e^{\lambda t} + \lambda\vec{\eta}te^{\lambda t} = A\vec{\eta}te^{\lambda t} + A\vec{\rho}e^{\lambda t}.$$

Now set coefficients equal again,

$$\lambda\vec{\eta} = A\vec{\eta} \quad \Rightarrow \quad (A-\lambda I)\vec{\eta} = \vec{0}$$
$$\vec{\eta} + \lambda\vec{\rho} = A\vec{\rho} \quad \Rightarrow \quad (A-\lambda I)\vec{\rho} = \vec{\eta}.$$

As with our first guess the first equation tells us nothing that we didn't already know. This time the second equation is not a problem. All the second equation tells us is that $\vec{\rho}$ must be a solution to this equation.

It looks like our second guess worked. Therefore,

$$\vec{x}_2 = te^{\lambda t}\vec{\eta} + e^{\lambda t}\vec{\rho}$$

will be a solution to the system provided $\vec{\rho}$ is a solution to

$$(A-\lambda I)\,\vec{\rho} = \vec{\eta}.$$

Also this solution and the first solution are linearly independent and so they form a fundamental set of solutions and so the general solution in the double eigenvalue case is,

$$\vec{x} = c_1 e^{\lambda t}\vec{\eta} + c_2\left(te^{\lambda t}\vec{\eta} + e^{\lambda t}\vec{\rho}\right).$$

Let's work an example.

Example: Solve the following IVP.

$$\vec{x} = \begin{pmatrix} 7 & 1 \\ -4 & 3 \end{pmatrix}\vec{x} \qquad \vec{x}(0) = \begin{pmatrix} 2 \\ -5 \end{pmatrix}.$$

Solution: First find the eigenvalues for the system.

$$\det(A-\lambda I) = \begin{vmatrix} 7-\lambda & 1 \\ -4 & 3-\lambda \end{vmatrix}$$
$$= \lambda^2 - 10\lambda + 25$$
$$= (\lambda - 5) \Rightarrow \lambda_{1,2} = 5.$$

So, we got a double eigenvalue. Of course that shouldn't be too surprising given the section that we're in. Let's find the eigenvector for this eigenvalue.

$$\begin{pmatrix} 2 & 1 \\ -4 & -2 \end{pmatrix}\begin{pmatrix} \eta_1 \\ \eta_2 \end{pmatrix} = \begin{pmatrix} 0 \\ 0 \end{pmatrix} \quad \Rightarrow \quad \eta_1 + \eta_2 = 0 \qquad \eta_2 = -2\eta_1$$

$$\begin{pmatrix} 2 & 1 \\ -4 & -2 \end{pmatrix}\begin{pmatrix} \eta_1 \\ \eta_2 \end{pmatrix} = \begin{pmatrix} 0 \\ 0 \end{pmatrix} \quad \Rightarrow \quad 2\eta_1 + \eta_2 = 0 \qquad \eta_2 = -2\eta_1.$$

The eigenvector is then,

$$\vec{\eta} = \begin{pmatrix} \eta_1 \\ -2\eta_1 \end{pmatrix} \qquad \eta_1 \neq 0$$

$$\vec{\eta}^{(1)} = \begin{pmatrix} 1 \\ -2 \end{pmatrix} \qquad \eta_1 = 1$$

The next step is find. To do this we'll need to solve,

$$\begin{pmatrix} 2 & 1 \\ -4 & -2 \end{pmatrix}\begin{pmatrix} \rho_1 \\ \rho_2 \end{pmatrix} = \begin{pmatrix} 1 \\ -2 \end{pmatrix} \quad \Rightarrow \quad 2\rho_1 + \rho_2 = 1 \qquad \rho_2 = 1 - 2\rho_1.$$

Note that this is almost identical to the system that we solve to find the eigenvalue. The only difference is the right hand side. The most general possible $\vec{\rho}$ is

$$\vec{\rho} = \begin{pmatrix} \rho_1 \\ 1 - 2\rho_1 \end{pmatrix} \quad \Rightarrow \quad \vec{\rho} = \begin{pmatrix} 0 \\ 1 \end{pmatrix} \quad \text{if } \rho_1 = 0.$$

In this case, unlike the eigenvector system we can choose the constant to be anything we want, so we might as well pick it to make our life easier. This usually means picking it to be zero.

We can now write down the general solution to the system.

$$\vec{x}(t) = c_1 e^{5t}\begin{pmatrix} 1 \\ -2 \end{pmatrix} + c_2\left(e^{5t} t\begin{pmatrix} 1 \\ -2 \end{pmatrix} + e^{5t}\begin{pmatrix} 0 \\ 1 \end{pmatrix}\right).$$

Applying the initial condition to find the constants gives us,

$$\begin{pmatrix} 1 \\ -5 \end{pmatrix} = \vec{x}(0) = c_1\begin{pmatrix} 1 \\ -2 \end{pmatrix} + c_2\begin{pmatrix} 0 \\ 1 \end{pmatrix}.$$

$$\left.\begin{aligned} c &= 2 \\ -2c_1 + c_2 &= -5 \end{aligned}\right\} \quad \Rightarrow \quad c_1 = 2,\ c_2 = -1.$$

The actual solution is then,

$$\begin{aligned} \vec{x}(t) &= \begin{pmatrix} 1 \\ -2 \end{pmatrix} - \left(te^{5t}\begin{pmatrix} 1 \\ -2 \end{pmatrix} + e^{5t}\begin{pmatrix} 0 \\ 1 \end{pmatrix}\right) \\ &= \begin{pmatrix} 2 \\ -4 \end{pmatrix} - e^{5t} t\begin{pmatrix} 1 \\ -2 \end{pmatrix} - e^{5t}\begin{pmatrix} 0 \\ 1 \end{pmatrix} \\ &= e^{5t}\begin{pmatrix} 2 \\ -5 \end{pmatrix} + e^{5t} t\begin{pmatrix} 1 \\ -2 \end{pmatrix}. \end{aligned}$$

Note that we did a little combining here to simplify the solution up a little.

So, the next example will be to sketch the phase portrait for this system.

Example: Sketch the phase portrait for the system.

$$\vec{x}' = \begin{pmatrix} 7 & 1 \\ -4 & 3 \end{pmatrix} \vec{x}.$$

Solution: These will start in the same way that real, distinct eigenvalue phase portraits start. We'll first sketch in a trajectory that is parallel to the eigenvector and note that since the eigenvalue is positive the trajectory will be moving away from the origin.

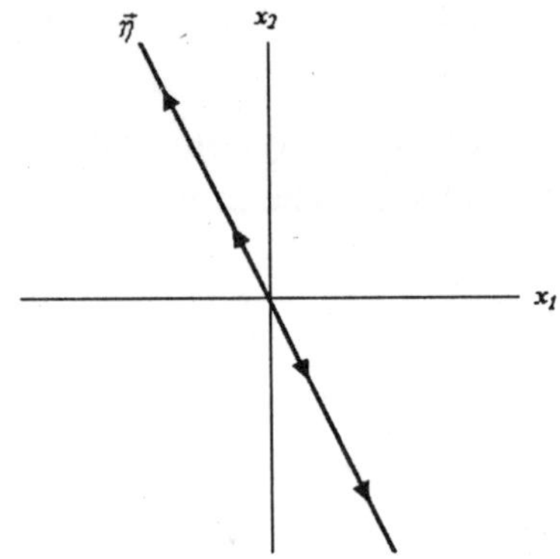

Now, it will be easier to explain the remainder of the phase portrait if we actually have one in front of us. So here is the full phase portrait with some more trajectories sketched in.

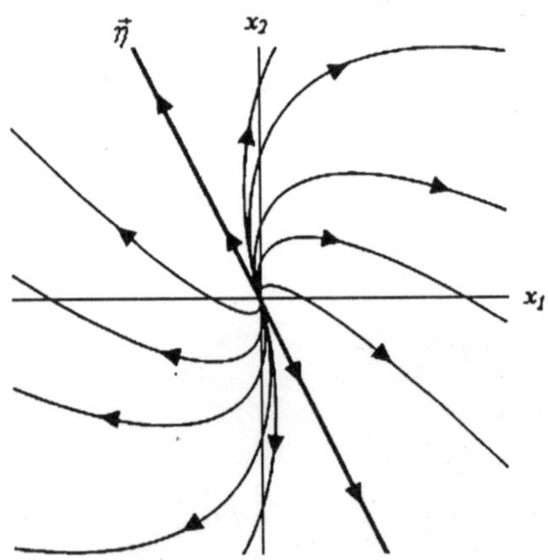

Trajectories in these cases always emerge from (or move into) the origin in a direction that is parallel to the eigenvector. Likewise they will start in one direction before turning around and moving off into the other direction. The directions in which they move are opposite depending on which side of the trajectory corresponding to the eigenvector we are on. Also, as the trajectories moves away from the origin it should start becoming parallel to the trajectory corresponding to the eigenvector.

So, how do we determine the direction? We can do the same thing that we did in the complex case. We'll plug in (1,0) into the system and see which direction the trajectories are moving at that point. Since this point is directly to the right of the origin the trajectory at that point must have already turned around and so this will give the direction that it will traveling after turning around. Doing that for this problem to check our phase portrait gives,

$$\begin{pmatrix} 7 & 1 \\ -4 & 3 \end{pmatrix}\begin{pmatrix} 1 \\ 0 \end{pmatrix} = \begin{pmatrix} 7 \\ -4 \end{pmatrix}.$$

This vector will point down into the fourth quadrant and so the trajectory must be moving into the fourth quadrant as well. This does match up with our phase portrait.

In these cases the equilibrium is called a node and is unstable in this case. Note that sometimes you will hear nodes for the repeated eigenvalue case called degenerate nodes or improper nodes. Let's work one more example.

Example:Solve the following IVP.

$$\vec{x}\,' = \begin{pmatrix} -1 & \frac{3}{2} \\ -\frac{1}{6} & -2 \end{pmatrix}\vec{x} \qquad \vec{x}(2) = \begin{pmatrix} 1 \\ 0 \end{pmatrix}.$$

Solution:First the eigenvalue for the system.

$$\det(A - \lambda I) = \begin{vmatrix} -1-\lambda & \frac{3}{2} \\ -\frac{1}{6} & -2-\lambda \end{vmatrix} = \lambda^2 + 3\lambda + \frac{9}{4}$$

$$= \left(\lambda + \frac{3}{2}\right)^2 \quad \Rightarrow \quad \lambda_{1,2} = -\frac{3}{2}.$$

Now let's get the eigenvector.

$$\begin{pmatrix} \frac{1}{2} & \frac{3}{2} \\ -\frac{1}{6} & -\frac{1}{2} \end{pmatrix}\begin{pmatrix} \eta_1 \\ \eta_2 \end{pmatrix} = \begin{pmatrix} 0 \\ 0 \end{pmatrix} \quad \Rightarrow \quad \frac{1}{2}\eta_1 + \frac{3}{2}\eta_2 = 0 \quad \eta_1 = -3\eta_2$$

$$\vec{\eta} = \begin{pmatrix} -3\eta_2 \\ \eta_2 \end{pmatrix} \qquad \eta_2 \neq 0$$

$$\vec{\eta}^{(1)} = \begin{pmatrix} -3\eta_2 \\ \eta_2 \end{pmatrix} \qquad \eta_2 = 1$$

Now find $\vec{\rho}$,

$$\begin{pmatrix} \frac{1}{2} & \frac{3}{2} \\ -\frac{1}{6} & -\frac{1}{2} \end{pmatrix}\begin{pmatrix} \rho_1 \\ \rho_2 \end{pmatrix} = \begin{pmatrix} -3 \\ 1 \end{pmatrix} \quad \Rightarrow \quad \frac{1}{2}\rho_1 + \frac{3}{2}\rho_2 = -3 \quad \rho_1 = -6 - 3\rho_2$$

$$\vec{\rho}\begin{pmatrix} -6-3\rho_2 \\ \rho_2 \end{pmatrix} \quad \Rightarrow \quad \vec{\rho} = \begin{pmatrix} -6 \\ 0 \end{pmatrix} \quad \text{if} \rho_2 = 0.$$

The general solution for the system is then,

$$\vec{x}(t) = c_1 e^{\frac{3t}{2}} \begin{pmatrix} -3 \\ 1 \end{pmatrix} + c_2 \left(te^{-\frac{3t}{2}} \begin{pmatrix} -3 \\ 1 \end{pmatrix} + e^{-\frac{3t}{2}} \begin{pmatrix} -6 \\ 0 \end{pmatrix} \right).$$

Applying the initial condition gives,

$$\begin{pmatrix} 1 \\ 0 \end{pmatrix} = \vec{x}(2) = c_1 e^{-3} \begin{pmatrix} -3 \\ 1 \end{pmatrix} + c_2 \left(2e^{-3} \begin{pmatrix} -3 \\ 1 \end{pmatrix} + e^{-3} \begin{pmatrix} -6 \\ 0 \end{pmatrix} \right).$$

Note that we didn't use $t=0$ this time! We now need to solve the following system,

$$\left. \begin{aligned} -3e^{-3}c_1 - 12e^{-3}c_2 &= 1 \\ e^{-3}c_1 + 2e^{-3}c_2 &= 0 \end{aligned} \right\} \Rightarrow \quad c_1 = \frac{e^3}{3}, \quad c_2 = -\frac{e^3}{3}.$$

The actual solution is then,

$$\vec{x}(t) = \frac{e^3}{3} e^{-\frac{3t}{2}} \begin{pmatrix} -3 \\ 1 \end{pmatrix} - \frac{e^3}{6} \left(te^{\frac{3t}{2}} \begin{pmatrix} -3 \\ 1 \end{pmatrix} + e^{-\frac{3t}{2}} \begin{pmatrix} -6 \\ 0 \end{pmatrix} \right) = e^{-\frac{3t}{2}} \begin{pmatrix} 0 \\ \frac{1}{3} \end{pmatrix} + te^{-\frac{3t}{2}+3} \begin{pmatrix} \frac{1}{2} \\ -\frac{1}{6} \end{pmatrix}.$$

And just to be consistent with all the other problems that we've done let's sketch the phase portrait.

Example: Sketch the phase portrait for the system.

$$\vec{x}' = \begin{pmatrix} -1 & \frac{3}{2} \\ -\frac{1}{6} & -2 \end{pmatrix} \vec{x}.$$

Solution: Let's first notice that since the eigenvalue is negative in this case the trajectories should all move in towards the origin. Let's check the direction of the trajectories at (1,0)

$$\begin{pmatrix} -1 & \frac{3}{2} \\ -\frac{1}{6} & -2 \end{pmatrix} \begin{pmatrix} 1 \\ 0 \end{pmatrix} = \begin{pmatrix} -1 \\ -\frac{1}{6} \end{pmatrix}.$$

So it looks like the trajectories should be pointing into the third quadrant at (1,0). This gives the following phase portrait.

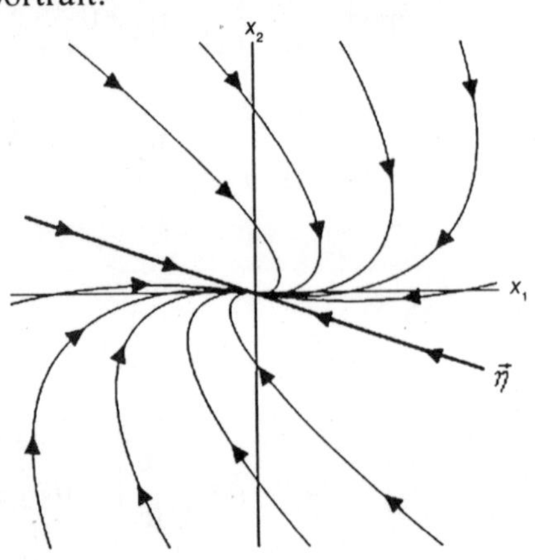

Chapter 7

Solving Partial Differential Equations

In this chapter we are going to take a very brief look at one of the more common methods for solving simple partial differential equations. The method we'll be taking a look at is that of Separation of Variables.

We need to make it very clear before we even start this chapter that we are going to be doing nothing more than barely scratching the surface of not only partial differential equations but also of the method of separation of variables. It would take several classes to cover most of the basic techniques for solving partial differential equations. The intent of this chapter is to do nothing more than to give you a feel for the subject and if you'd like to know more taking a class on partial differential equations should probably be your next step.

Also note that in several sections we are going to be making heavy use of some of the results from the previous chapter. That in fact was the point of doing some of the examples that we did there. Having done them will, in some cases, significantly reduce the amount of work required in some of the examples we'll be working in this chapter. When we do make use of a previous result we will make it very clear where the result is coming from.

THE HEAT EQUATION

Before we get into actually solving partial differential equations and before we even start discussing the method of separation of variables we want to spend a little bit of time talking about the two main partial differential equations that we'll be solving later on in the chapter. We'll look at the first one in this section and the second one in the next section.

The first partial differential equation that we'll be looking at once we get started with solving will be the heat equation, which governs the temperature distribution in an object. We are going to give several forms of the heat equation for reference purposes, but we will only be really solving one of them.

We will start out by considering the temperature in a 1-D bar of length L. What this means is that we are going to assume that the bar starts off at $x = 0$and ends when we reach

$x = L$. We are also going to so assume that at any location, x the temperature will be constant an every point in the cross section at that x. In other words, temperature will only vary in x and we can hence consider the bar to be a 1-D bar. Note that with this assumption the actual shape of the cross section (*i.e.* circular, rectangular, *etc.*) doesn't matter.

Note that the 1-D assumption is actually not all that bad of an assumption as it might seem at first glance. If we assume that the lateral surface of the bar is perfectly insulated (*i.e.* no heat can flow through the lateral surface) then the only way heat can enter or leave the bar as at either end. This means that heat can only flow from left to right or right to left and thus creating a 1-D temperature distribution.

The assumption of the lateral surfaces being perfectly insulated is of course impossible, but it is possible to put enough insulation on the lateral surfaces that there will be very little heat flow through them and so, at least for a time, we can consider the lateral surfaces to be perfectly insulated.

Okay, let's now get some definitions out of the way before we write down the first form of the heat equation.

$u(x,t) =$ Temperature at any point x and any time t

$c(x) =$ Specific Heat

$\rho(x) =$ Mass Density

$\phi(x,t) =$ Heat Flux

$Q(x,t) =$ Heat enery generated per unit volume per unit time

We should probably make a couple of comments about some of these quantities before proceeding.

The specific heat, $c(x) > 0$, of a material is the amount of heat energy that it takes to raise one unit of mass of the material by one unit of temperature. As indicated we are going to assume, at least initially, that the specific heat may not be uniform throughout the bar. Note as well that in practice the specific heat depends upon the temperature. However, this will generally only be an issue for large temperature differences (which in turn depends on the material the bar is made out of) and so we're going to assume for the purposes of this discussion that the temperature differences are not large enough to affect our solution.

The mass density, $\rho(x)$,is the mass per unit volume of the material. As with the specific heat we're going to initially assume that the mass density may not be uniform throughout the bar.

The heat flux, $\phi(x,t)$, is the amount of thermal energy that flows to the right per unit surface area per unit time. The "flows to the right" bit simply tells us that if $\phi(x,t) > 0$ for some x and t then the heat is flowing to the right at that point and time. Likewise if $\phi(x,t) < 0$ then the heat will be flowing to the left at that point and time.

The final quantity we defined above is $Q(x, y)$ and this is used to represent any external sources or sinks (*i.e.* heat energy taken out of the system) of heat energy. If $Q(x,t) > 0$ then heat energy is being added to the system at that location and time and if $Q(x,t) < 0$ then heat energy is being removed from the system at that location and time.

With these quantities the heat equation is,

$$c(x)\rho(x)\frac{\partial u}{\partial t} = -\frac{\partial \varphi}{\partial x} + Q(x,t)$$

While this is a nice form of the heat equation it is not actually something we can solve. In this form there are two unknown functions, u and φ, and so we need to get rid of one of them. With Fourier's law we can easily remove the heat flux from this equation.

Fourier's law states that,

$$\varphi(x,t) = -K_0(x)\frac{\partial u}{\partial x}$$

where $K_0(x) > 0$ is the thermal conductivity of the material and measures the ability of a given material to conduct heat. The better a material can conduct heat the larger $K_0(x)$ will be. As noted the thermal conductivity can vary with the location in the bar. Also, much like the specific heat the thermal conductivity can vary with temperature, but we will assume that the total temperature change is not so great that this will be an issue and so we will assume for the purposes here that the thermal conductivity will not vary with temperature.

Fourier's law does a very good job of modeling what we know to be true about heat flow. First, we know that if the temperature in a region is constant, *i.e.* $\frac{\partial u}{\partial x} = 0$, then there is no heat flow.

Next, we know that if there is a temperature difference in a region we know the heat will flow from the hot portion to the cold portion of the region. For example, if it is hotter to the right then we know that the heat should flow to the left. When it is hotter to the right then we also know that $\frac{\partial u}{\partial x} > 0$ (*i.e.* the temperature increases as we move to the right) and so we'll have $\varphi < 0$and so the heat will flow to the left as it should. Likewise, if (*i.e.* it is hotter to the left) then we'll have $\varphi < 0$and heat will flow to the right as it should.

Finally, the greater the temperature difference in a region (*i.e.* the larger $\frac{\partial u}{\partial x}$ is) then the greater the heat flow.

So, if we plug Fourier's law, we get the following form of the heat equation,

$$c(x)\rho(x)\frac{\partial u}{\partial t}=\frac{\partial}{\partial x}\left(K_0(x)\frac{\partial u}{\partial x}\right)+Q(x,t)$$

Note that we factored the minus sign out of the derivative to cancel against the minus sign that was already there. We cannot however, factor the thermal conductivity out of the derivative since it is a function of x and the derivative is with respect to x.

It is quite difficult due to the non uniform nature of the thermal properties and the mass density. So, let's now assume that these properties are all constant, *i.e.*,

$$c(x)=c \qquad \rho(x)=\rho \qquad K_0(x)=K_0$$

where c, ρ and K_0 are now all fixed quantities. In this case we generally say that the martial in the bar is uniform. Under these assumptions the heat equation becomes,

$$c\rho\frac{\partial u}{\partial t}=K_0\frac{\partial^2 u}{\partial x^2}+Q(x,t)$$

For a final simplification to the heat equation let's divide both sides by c ρ and define the thermal diffusivity to be,

$$K\frac{K_0}{c\rho}$$

The heat equation is then,

$$\frac{\partial u}{\partial t}=K\frac{\partial^2 u}{\partial x^2}+\frac{Q(x,t)}{c\rho}$$

To most people this is what they mean when they talk about the heat equation and in fact it will be the equation that we'll be solving. Well, actually we'll be solving with no external sources, *i.e.* $Q(x,t)=0$, but we'll be considering this form when we start discussing separation of variables in a couple of sections. We'll only drop the sources term when we actually start solving the heat equation.

Now that we've got the 1-D heat equation taken care of we need to move into the initial and boundary conditions we'll also need in order to solve the problem. If you go back to any of our solutions of ordinary differential equations that we've done in previous sections you can see that the number of conditions required always matched the highest order of the derivative in the equation.

In partial differential equations the same idea holds except now we have to pay attention to the variable we're differentiating with respect to as well. So, for the heat equation we've got a first order time derivative and so we'll need one initial condition and a second order spatial derivative and so we'll need two boundary conditions.

The initial condition that we'll use here is,

$$u(x,0)=f(x)$$

and we don't really need to say much about it here other than to note that this just tells us what the initial temperature distribution in the bar is.

The boundary conditions will tell us something about what the temperature and/or heat flow is doing at the boundaries of the bar. There are four of them that are fairly common boundary conditions.

The first type of boundary conditions that we can have would be the prescribed temperature boundary conditions, also called Dirichlet conditions. The prescribed temperature boundary conditions are,

$$u(0,t)=g_1(t) \qquad u(L,t)=g_2(t)$$

The next type of boundary conditions are prescribed heat flux, also called Neumann conditions. Using Fourier's law these can be written as,

$$-K_0(0)\frac{\partial u}{\partial x}(0,t)=\varphi_1(t) \qquad\qquad -K_0(L)\frac{\partial u}{\partial x}(L,t)=\varphi_2(t)$$

If either of the boundaries are perfectly insulated, *i.e.* there is no heat flow out of them then these boundary conditions reduce to,

$$\frac{\partial u}{\partial x}(0,t)=0 \qquad \frac{\partial u}{\partial x}(L,t)=0$$

and note that we will often just call these particular boundary conditions insulated boundaries and drop the "perfectly" part.

The third type of boundary conditions use Newton's law of cooling and are sometimes called Robins conditions. These are usually used when the bar is in a moving fluid and note we can consider air to be a fluid for this purpose.

Here are the equations for this kind of boundary condition.

$$-K_0(0)\frac{\partial u}{\partial x}(0,t)=-H\left[u(0,t)-g_1(t)\right] \qquad -K_0(L)\frac{\partial u}{\partial x}(L,t)=H\left[u(L,t)-g_2(t)\right]$$

where H is a positive quantity that is experimentally determined and $g_1(t)$ and $g_2(t)$ give the temperature of the surrounding fluid at the respective boundaries.

Note that the two conditions do vary slightly depending on which boundary we are at. At $x = 0$ we have a minus sign on the right side while we don't at $x = L$. To see why this is let's first assume that at $x = 0$ we have $u(0.t)>g_1(t)$. In other words the bar is hotter than the surrounding fluid and so at $x = 0$ the heat flow (as given by the left side of the equation) must be to the left, or negative since the heat will flow from the hotter bar into the cooler surrounding liquid. If the heat flow is negative then we need to have a minus sign on the right side of the equation to make sure that it has the proper sign.

If the bar is cooler than the surrounding fluid at $x = 0$, *i.e.* $u(0.t) < g_1(t)$ we can make a similar argument to justify the minus sign. We'll leave it to you to verify this.

If we now look at the other end, $x = L$, and again assume that the bar is hotter than the surrounding fluid or, $u(L,t) > g_2(t)$ .In this case the heat flow must be to the right, or be positive, and so in this case we can't have a minus sign. Finally, we'll again leave it to you to verify that we can't have the minus sign at $x = L$is the bar is cooler than the surrounding fluid as well.

Note that we are not actually going to be looking at any of these kinds of boundary conditions here. These types of boundary conditions tend to lead to boundary value problems in the Eigenvalues and Eigenfunctions section of the previous chapter. As we saw in that example it is often very difficult to get our hands on the eigenvalues and as we'll eventually see we will need them.

It is important to note at this point that we can also mix and match these boundary conditions so to speak. There is nothing wrong with having a prescribed temperature at one boundary a prescribed flux at the other boundary for example so don't always expect the same boundary condition to show up at both ends. This warning is more important that it might seem at this point because once we get into solving the heat equation we are going to have the same kind of condition on each end to simplify the problem somewhat.

The final type of boundary conditions that we'll need here are periodic boundary conditions. Periodic boundary conditions are,

$$u(-L,t) = u(L,t) \qquad \frac{\partial u}{\partial x}(-L,t) = \frac{\partial u}{\partial x}(L,t)$$

Note that for these kinds of boundary conditions the left boundary tends to be $x = -L$ instead of $x = 0$as we were using in the previous types of boundary conditions. The periodic boundary conditions will arise very naturally from a couple of particular geometries that we'll be looking at down the road.

We will now close out this section with a quick look at the 2-D and 3-D version of the heat equation. However, before we jump into that we need to introduce a little bit of notation first.

The del operator is defined to be,

$$\nabla = \frac{\partial}{\partial x}\vec{i} + \frac{\partial}{\partial y}\vec{j} \qquad \nabla = \frac{\partial}{\partial x}\vec{i} + \frac{\partial}{\partial y}\vec{j} + \frac{\partial}{\partial k}\vec{k}$$

depending on whether we are in 2 or 3 dimensions. Think of the del operator as a function that takes functions as arguments (instead of numbers as we're used to). Whatever function we "plug" into the operator gets put into the partial derivatives.

So, for example in 3-D we would have,

$$\nabla f = \frac{\partial f}{\partial x}\vec{i} + \frac{\partial f}{\partial y}\vec{j} + \frac{\partial f}{\partial k}\vec{k}$$

This of course is also the gradient of the function $f(x, y, z)$.

The del operator also allows us to quickly write down the divergence of a function. So, again using 3-D as an example the divergence of $f(x, y, z)$ can be written as the dot product of the del operator and the function. Or,

$$\nabla \bullet f = \frac{\partial f}{\partial x} + \frac{\partial f}{\partial y} + \frac{\partial f}{\partial k}$$

Finally, we will also see the following show up in the our work,

$$\nabla \bullet (\nabla f) = \frac{\partial}{\partial x}\left(\frac{\partial f}{\partial x}\right) + \frac{\partial}{\partial y}\left(\frac{\partial f}{\partial y}\right) + \frac{\partial}{\partial z}\left(\frac{\partial f}{\partial k}\right) = \frac{\partial^2 f}{\partial x^2} + \frac{\partial^2 f}{\partial y^2} + \frac{\partial^2 f}{\partial z^2}$$

This is usually denoted as,

$$\nabla^2 f = \frac{\partial^2 f}{\partial x^2} + \frac{\partial^2 f}{\partial y^2} + \frac{\partial^2 f}{\partial z^2}$$

and is called the Laplacian. The 2-D version of course simply doesn't have the third term.

Okay, we can now into the 2-D and 3-D version of the heat equation and where ever the del operator and or Laplacian appears assume that it is the appropriate dimensional version.

The higher dimensional version is,

$$c\rho \frac{\partial u}{\partial t} = -\nabla \bullet \varphi + Q$$

and note that the specific heat, c, and mass density, ρ, are may not be uniform and so may be functions of the spatial variables. Likewise, the external sources term, Q, may also be a function of both the spatial variables and time.

Next, the higher dimensional version of Fourier's law is,

$$\varphi = -K_0 \nabla u$$

where the thermal conductivity, K_0, is again assumed to be a function of the spatial variables.

If we plug this we get the heat equation for a non uniform bar (*i.e.* the thermal properties may be functions of the spatial variables) with external sources/sinks,

$$c\rho\frac{\partial u}{\partial t} = \nabla \bullet (K_0 \nabla u) + Q$$

If we now assume that the specific heat, mass density and thermal conductivity are constant (*i.e.* the bar is uniform) the heat equation becomes,

$$\frac{\partial u}{\partial t} = k\nabla^2 u + \frac{Q}{c\rho}$$

where we divided both sides by $c\rho$ to get the thermal diffusivity, k in front of the Laplacian.

The initial condition for the 2-D or 3-D heat equation is,

$$u(x, y, t) = f(x, y) \qquad or \qquad u(x, y, t) = f(x, y, z)$$

depending upon the dimension we're in.

The prescribed temperature boundary condition becomes,

$$u(x, y, t) = T(x, y, t) \qquad or \qquad u(x, y, z, t) = T(x, y, z, t)$$

where (x, y) or (x, y, z),depending upon the dimension we're in, will range over the portion of the boundary in which we are prescribing the temperature.

The prescribed heat flux condition becomes,

$$-K_0 \nabla u \bullet \vec{n} = \varphi(t)$$

where the left side is only being evaluated at points along the boundary and $\vec{n}$ is the outward unit normal on the surface.

Newton's law of cooling will become,

$$-K_0 \nabla u \bullet \vec{n} = H(u - u_B)$$

where H is a positive quantity that is experimentally determine, u_B is the temperature of the fluid at the boundary and again it is assumed that this is only being evaluated at points along the boundary.

We don't have periodic boundary conditions here as they will only arise from specific 1-D geometries.

THE WAVE EQUATION

In this section we want to consider a vertical string of length L that has been tightly stretched between two points at $x = 0$and $x = L$.

Because the string has been tightly stretched we can assume that the slope of the displaced string at any point is small. So just what does this do for us? Let's consider a point x on the string in its equilibrium position, *i.e.* the location of the point at $t = 0$. As the string vibrates this point will be displaced both vertically and horizontally, however, if we assume that at any point the slope of the string is small then the horizontal displacement will be very small in relation to the vertical displacement. This means that we can now assume that at any point x on the string the displacement will be purely vertical. So, let's call this displacement $u(x,t)$.

We are going to assume, at least initially, that the string is not uniform and so the mass density of the string, $\rho(x)$ may be a function of x.

Next we are going to assume that the string is perfectly flexible. This means that the string will have no resistance to bending. This in turn tells us that the force exerted by the string at any point x on the endpoints will be tangential to the string itself. This force is called the tension in the string and its magnitude will be given by $T(x.t)$.

Finally, we will let $Q(x.t)$ represent the vertical component per unit mass of any force acting on the string.

Provided we again assume that the slope of the string is small the vertical displacement of the string at any point is then given by,

$$\rho(x)\frac{\partial^2 u}{\partial t^2} = \frac{\partial}{\partial x}\left(T(x,t)\frac{\partial u}{\partial x}\right) + \rho(x)Q(x,t)$$

This is a very difficult partial differential equation to solve so we need to make some further simplifications.

First, we're now going to assume that the string is perfectly elastic. This means that the magnitude of the tension, $T(x,t)$, will only depend upon how much the string stretches near x. Again, recalling that we're assuming that the slope of the string at any point is small this means that the tension in the string will then very nearly be the same as the tension in the string in its equilibrium position. We can then assume that the tension is a constant value, $T(x,t) = T_0$.

Further, in most cases the only external force that will act upon the string is gravity and if the string light enough the effects of gravity on the vertical displacement will be small and so will also assume that $Q(x,t) = 0$. This leads to

$$\rho\frac{\partial^2 u}{\partial t^2} = T_0\frac{\partial^2 u}{\partial x^2}$$

If we know divide by the mass density and define,

$$c^2 = \frac{T_0}{\rho}$$

we arrive at the 1-D wave equation,

$$\frac{\partial^2 u}{\partial t^2} = c^2 \frac{\partial^2 u}{\partial x^2}$$

In the previous section when we looked at the heat equation he had a number of boundary conditions however in this case we are only going to consider one type of boundary conditions. For the wave equation we are only going to consider the boundary condition will that of prescribed location of the boundaries or,

$$u(0,t) = h_1(t) \quad u(L,t) = h_2(t)$$

The initial conditions (and yes we meant more than one...) will also be a little different here from what we saw with the heat equation. Here we have a 2nd order time derivative and so we'll also need two initial conditions. At any point we will specify both the initial displacement of the string as well as the initial slope of the string. The initial conditions are then,

$$u(x,0) = f(x) \quad \frac{\partial u}{\partial t}(x,0) = g(x)$$

For the sake of completeness we'll close out this section with the 2-D and 3-D version of the wave equation. We'll not actually be solving this at any point, but since we gave the higher dimensional version of the heat equation (in which we will solve a special case) we'll give this as well.

The 2-D and 3-D version of the wave equation is,

$$\frac{\partial^2 u}{\partial t^2} c^2 \nabla^2 u$$

where ∇^2 is the Laplacian.

TERMINOLOGY

We've got one more section that we need to take care of before we actually start solving partial differential equations. This will be a fairly short section that will cover some of the basic terminology that we'll need in the next section as we introduce the method of separation of variables.

Let's start off with the idea of an operator. An operator is really just a function that takes a function as an argument instead of numbers as we're used to dealing with in functions. You already know of a couple of operators even if you didn't know that they were operators. Here are some examples of operators.

$$L = \frac{d}{dx} \qquad L = \int dx \qquad L = \int_a^b dx \qquad L = \frac{\partial}{\partial t}$$

Or, if we plug in a function, say $u(x)$, into each of these we get,

$$L(u) = \frac{du}{dx} \qquad L(u) = \int u(x)\,dx \qquad L(u) = \int_a^b u(x)\,dx \qquad L(u) = \frac{\partial u}{\partial t}$$

These are all fairly simple examples of operators but the derivative and integral are operators. A more complicated operator would be the heat operator. We get the heat operator from a slight rewrite of the heat equation without sources. The heat operator is,

$$L = \frac{\partial}{\partial t} - k\frac{\partial^2}{\partial x^2}$$

Now, what we really want to define here is not an operator but instead a linear operator. A linear operator is any operator that satisfies,

$$L(c_1u_1 \_ c_2u_2) = c_1L(u_1) + c_2L(u_2)$$

The heat operator is an example of a linear operator and this is easy enough to show using the basic properties of the partial derivative so let's do that.

$$\begin{aligned}
L(c_1u_1 + c_2u_2) &= \frac{\partial}{\partial t}(c_1u_1 + c_2u_2) - k\frac{\partial^2}{\partial x^2}(c_1u_1 + c_2u_2) \\
&= \frac{\partial}{\partial t}(c_1u_1) + \frac{\partial}{\partial t}(c_2u_2) - k\left[\frac{\partial^2}{\partial x^2}(c_1u_1) + \frac{\partial^2}{\partial x^2}(c_2u_2)\right] \\
&= c_1\frac{\partial u_1}{\partial t} - c_2\frac{\partial u_2}{\partial t} - k\left[c_1\frac{\partial^2 u_1}{\partial x^2} + c_2\frac{\partial^2 u_2}{\partial x^2}\right] \\
& c_1\frac{\partial u_1}{\partial t} - kc_1\frac{\partial^2 u_1}{\partial x^2} + c_2\frac{\partial u_2}{\partial t} - kc_2\frac{\partial^2 u_2}{\partial x^2} \\
&= c_1\left[\frac{\partial u_1}{\partial t} - k\frac{\partial^2 u_1}{\partial x^2}\right] + c_2\left[\frac{\partial u_2}{\partial t} - k\frac{\partial^2 u_2}{\partial x^2}\right] \\
&= c_1L(u_1) + c_2L(u_2)
\end{aligned}$$

You might want to verify for yourself that the derivative and integral operators we gave above are also linear operators. In fact, in the process of showing that the heat operator is a linear operator we actually showed as well that the first order and second order partial derivative operators are also linear. The next term we need to define is a linear equation. A linear equation is an equation in the form,

$$L(u) = f$$

where L is a linear operator and f is a known function.

Here are some examples of linear partial differential equations.

$$\frac{\partial u}{\partial t} = k\frac{\partial^2 u}{\partial x^2} + \frac{Q(x,t)}{c\rho}$$

$$\frac{\partial^2 u}{\partial t^2} = c^2 \frac{\partial^2 u}{\partial x^2}$$

$$\frac{\partial^2 u}{\partial x^2} + \frac{\partial^2 u}{\partial y^2} + \frac{\partial^2 u}{\partial z^2} = \nabla^2 u = 0$$

$$\frac{\partial u}{\partial t} - 4\frac{\partial^2 u}{\partial t^2} = \frac{\partial^3 u}{\partial x^3} + 8u - g(x.t)$$

The first two from this list are of course the heat equation and the wave equation. The third uses the Laplacian and is usually called Laplace's Equation. We'll actually be solving the 2-D version of Laplace's Equation in a few sections. The fourth equation was just made up to give a more complicated example.

Notice as well with the heat equation and the fourth example above that the presence of the $Q(x,t)$ and $g(x,t)$ do not prevent these from being linear equations. The main issue that allows these to be linear equations is the fact that the operator in each is linear.

Now just to be complete here are a couple of examples of nonlinear partial differential equations.

$$\frac{\partial u}{\partial t} = k\frac{\partial^2 u}{\partial x^2} + u^2$$

$$\frac{\partial^2 u}{\partial t^2} - \frac{\partial u}{\partial x}\frac{\partial u}{\partial t} = u + f(x,t)$$

We'll leave it to you to verify that the operators in each of these are not linear however the problem term in the first is the u^2 while in the second the product of the two derivatives is the problem term.

Now, if we go back and suppose that $f = 0$ then we arrive at,

$$L(u) = 0$$

We call this a linear homogeneous equation (recall that L is a linear operator).

Notice that $u = 0$ will always be a solution to a linear homogeneous equation (go back to what it means to be linear and use $c_1 = c_2 = 0$ with any two solutions and this is easy to verify). We call $u = 0$ the trivial solution. In fact this is also a really nice way of determining if an equation is homogeneous. If L is a linear operator and we plug in $u = 0$ into the equation and we get $L(u) = 0$ then we will know that the operator is homogeneous.

We can also extend the ideas of linearity and homogeneous to boundary conditions. If we go back to the various boundary conditions we discussed for the heat equation for example we can also classify them as linear and/or homogeneous.

The prescribed temperature boundary conditions,

$$u(0,t)=g_1(t) \qquad\qquad u(L,t)=g_2(t)$$

are linear and will only be homogenous if $g_1(t)=0$ and $g_2(t)=0$.

The prescribed heat flux boundary conditions,

$$-K_0(0)\frac{\partial u}{\partial x}(0,t)=\varphi_1(t) \qquad\qquad -K_0(L)\frac{\partial u}{\partial x}(L,t)=\varphi_2(t)$$

are linear and will again only be homogeneous if $\varphi_1(t)=0$ and $\varphi_2(t)=0$.

Next, the boundary conditions from Newton's law of cooling,

$$-K_0(0)\frac{\partial u}{\partial x}(0,t)=-H\left[u(0.t)-g_1(t)\right]$$

$$-K_0(L)\frac{\partial u}{\partial x}(L,t)=H\left[u(L.t)-g_2(t)\right]$$

are again linear and will only be homogenous if $g_1(t)=0$ and $g_2(t)=0$.

The final set of boundary conditions that we looked at were the periodic boundary conditions,

$$u(-L,t)=u(L,t) \qquad\qquad \frac{\partial u}{\partial x}(-L,t)=\frac{\partial u}{\partial x}(L,t)$$

and these are both linear and homogeneous.

The final topic in this section is not really terminology but is a restatement of a fact that we've seen several times in these notes already.

PRINCIPLE OF SUPERPOSITION

If u_1 and u_2 are solutions to a linear homogeneous equation then so is $c_1u_2+c_1u_2$ for any values of c_1 and c_2.

Now, as stated earlier we've seen this several times this semester but we didn't really do much with it. However this is going to be a key idea when we actually get around to solving partial differential equations.

Without this fact we would not be able to solve all but the most basic of partial differential equations.

SEPARATION OF VARIABLES

Okay, it is finally time to at least start discussing one of the more common methods for solving basic partial differential equations. The method of Separation of Variables cannot always be used and even when it can be used it will not always be possible to get much past the first step in the method. However, it can be used to easily solve the 1-D heat equation with no sources, the 1-D wave equation, and the 2-D version of Laplace's Equation,

In order to use the method of separation of variables we must be working with a linear homogenous partial differential equations with linear homogeneous boundary conditions. At this point we're not going to worry about the initial condition(s) because the solution that we initially get will rarely satisfy the initial condition(s). As we'll see however there are ways to generate a solution that will satisfy initial condition(s) provided they meets some fairly simple requirements.

The method of separation of variables relies upon the assumption that a function of the form,

$$u(x,t)=\varphi(x)G(t)$$

will be a solution to a linear homogeneous partial differential equation in x and t. This is called a product solution and provided the boundary conditions are also linear and homogeneous this will also satisfy the boundary conditions. However, as noted above this will only rarely satisfy the initial condition, but that is something for us to worry about in the next section.

Now, before we get started on some examples there is probably a question that we should ask at this point and that is: Why? Why did we choose this solution and how do we know that it will work? This seems like a very strange assumption to make. After all there really isn't any reason to believe that a solution to a partial differential equation will in fact be a product of a function of only x's and a function of only t's. This seems more like a hope than a good assumption/guess. Unfortunately the best answer is that we chose it because it will work. As we'll see it works because it will reduce our partial differential equation down to two ordinary differential equations and provided we can solve those then we're in business and the method will allow us to get a solution to the partial differential equations. So, let's do a couple of examples to see how this method will reduce a partial differential equation down to two ordinary differential equations.

Example: Use Separation of Variables on the following partial differential equation.

$$\frac{\partial u}{\partial t}=k\frac{\partial^2 u}{\partial x^2}$$

$$u(x,0)=f(x) \quad u(0,t)=0 \qquad u(L,t)=0$$

Solution: So, we have the heat equation with no sources, fixed temperature boundary conditions (that are also homogeneous) and an initial condition. The initial condition is only here because it belongs here, but we will be ignoring it until we get to the next section.

The method of separation of variables tells us to assume that the solution will take the form of the product,

$$u(x,t) = \varphi(x)G(t)$$

so all we really need to do here is plug this into the differential equation and see what we get.

$$\frac{\partial}{\partial t}\left(\varphi(x)G(t)\right) = k\frac{\partial^2}{\partial x^2}\left(\varphi(x)G(t)\right)$$

$$\varphi(x)\frac{dG}{dt} = kG(t)\frac{d^2\varphi}{dx^2}$$

As shown above we can factor the $\varphi(x)$ out of the time derivative and we can factor the $G(t)$ out of the spatial derivative. Also notice that after we've factored these out we no longer have a partial derivative left in the problem.

In the time derivative we are now differentiating only $G(t)$ with respect to t and this is now an ordinary derivative. Likewise, in the spatial derivative we are now only differentiating $\varphi(x)$ with respect to x and so we again have an ordinary derivative.

At this point it probably doesn't seem like we've done much to simplify the problem. However, just the fact that we've gotten the partial derivatives down to ordinary derivatives is liable to be good thing even if it still looks like we've got a mess to deal with.

Speaking of that apparent mess, is it really the mess that it looks like? The idea is to eventually get all the t's on one side of the equation and all the x's on the other side. In other words we want to "separate the variables" and hence the name of the method. In this case let's notice that if we divide both sides by $\varphi(x)G(t)$ we get want we want and we should point out that it won't always be as easy as just dividing by the product solution. So, dividing out gives us,

$$\frac{1}{G}\frac{dG}{dt} = k\frac{1}{\varphi}\frac{d^2\varphi}{dx^2} \qquad \Rightarrow \qquad \frac{1}{kG}\frac{dG}{dt} = \frac{1}{\varphi}\frac{d^2\varphi}{dx^2}$$

Notice that we also divided both sides by k. This was done only for convenience down the road. It doesn't have to be done and nicely enough if it turns out to be a bad idea we can always come back to this step and put it back on the right side. Likewise, if we don't do it and it turns out to maybe not be such a bad thing we can always come back and divide it out. For the time being however, please accept our word that this was a good thing to do for this problem. We will discuss the reasoning for this after we're done with this example.

Now, while we said that this is what we wanted it still seems like we've got a mess. Notice however that the left side is a function of only t and the right side is a function only of x as we wanted. Also notice these two functions must be equal.

Let's think about this for a minute. How is it possible that a function of only t's can be equal to a function of only x's regardless of the choice of t and/or x that we have? This may seem like an impossibility until you realize that there is one way that this can be true. If both functions (*i.e.* both sides of the equation) were in fact constant and not only a constant, but the same constant then they can in fact be equal.

So, we must have,

$$\frac{1}{kG}\frac{dG}{dt} = \frac{1}{\varphi}\frac{d^2\varphi}{\varphi dx^2} = -\lambda$$

where the $-\lambda$ is called the separation constant and is arbitrary.

The next question that we should now address is why the minus sign? Again, much like the dividing out the k above, the answer is because it will be convenient down the road to have chosen this. The minus sign doesn't have to be there and in fact there are times when we don't want it there.

So how do we know it should be there or not? The answer to that is to proceed to the next step in the process and at that point we'll know if would be convenient to have it or not and we can come back to this step and add it in or take it our depending what we chose to do here.

Okay, let's proceed with the process. The next step is to acknowledge that we can take the equation above and split it into the following two ordinary differential equations.

$$\frac{dG}{dt} = -k\lambda G \qquad \frac{d^2\varphi}{dt^2} = -\lambda\varphi$$

Both of these are very simple differential equations, however because we don't know what $\lambda$is we actually can't solve the spatial one yet. The time equation however could be solved at this point if we wanted to, although that won't always be the case. At this point we don't want to actually think about solving either of these yet however.

The last step in the process that we'll be doing in this section is to also make sure that our product solution,, satisfies the boundary conditions so let's plug it into both of those.

$$u(0,t) = \varphi(0)G(t) = 0 \qquad u(L,t) = \varphi(L)G(t) = 0$$

Let's consider the first one for a second. We have two options here. Either $\varphi(0) = 0$ or

$G(t)=0$ for every t. However, if we have $G(t)=0$ for every t then we'll also have, $u(x,t)=0$, *i.e.* the trivial solution, and as we discussed in the previous section this is definitely a solution to any linear homogeneous equation we would really like a non-trivial solution.

Therefore we will assume that in fact we must have $\varphi(0)=0$. Likewise, from the second boundary condition we will get $\varphi(L)=0$ to avoid the trivial solution. Note as well that we were only able to reduce the boundary conditions down like this because they were homogeneous. Had they not been homogeneous we could not have done this.

So, after applying separation of variables to the given partial differential equation we arrive at a 1st order differential equation that we'll need to solve for $G(t)$ and a 2nd order boundary value problem that we'll need to solve for $\varphi(x)$. The point of this section however is just to get to this point and we'll hold off solving these until the next section.

Let's summarize everything up that we've determined here.

$$\frac{dG}{dt}=-k\lambda G \qquad \frac{d^2\varphi}{dx^2}+\lambda\varphi=0$$
$$\varphi(0)=0 \qquad \varphi(L)=0$$

and note that we don't have a condition for the time differential equation and is not a problem. Also note that we rewrote the second one a little.

Okay, so just what have we learned here? By using separation of variables we were able to reduce our linear homogeneous partial differential equation with linear homogeneous boundary conditions down to an ordinary differential equation for one of the functions in our product solution, $G(t)$ in this case, and a boundary value problem that we can solve for the other function, $\varphi(x)$ in this case.

Note as well that the boundary value problem is in fact an eigenvalue/eigenfunction problem. When we solve the boundary value problem we will be identifying the eigenvalues, λ, that will generate non-trivial solutions the their corresponding eigenfunctions. Again, we'll look into this more in the next section. At this point all we want to do is identify the two ordinary differential equations that we need to solve to get a solution.

Before we do a couple of other examples we should take a second to address the fact that we made two very arbitrary seeming decisions in the above work. We divided both sides of the equation by k at one point and chose to use $-\lambda$instead of as λ the separation constant.

Both of these decisions were made to simplify the solution to the boundary value problem we got from our work. The addition of the k in the boundary value problem would just have complicated the solution process with another letter we'd have to keep track of so we moved it into the time problem were it won't cause as many problems in the solution process. Likewise, we chose $-\lambda$because we've already solved that particular boundary value problem (albeit with a specific L, but the work will be nearly identical) when we first looked at finding eigenvalues and eigenfunctions. This by the way was the reason we rewrote the boundary value problem to make it a little clearer that we have in fact solved this one already.

We can now at least partially answer the question of how do we know to make these decisions. We wait until we get the ordinary differential equations and then look at them and decide of moving things like the k or which separation constant to use based on how it will affect the solution of the ordinary differential equations. There is also, of course, a fair amount of experience that comes into play at this stage. The more experience you have in solving these the easier it often is to make these decisions.

Again, we need to make clear here that we're not going to go any farther in this section than getting things down to the two ordinary differential equations. Of course we will need to solve them in order to get a solution to the partial differential equation but that is the topic of the remaining sections in this chapter. All we'll say about it here is that we will need to first solve the boundary value problem, which will tell us what λmust be and then we can solve the other differential equation. Once that is done we can then turn our attention to the initial condition.

Okay, we need to work a couple of other examples and these will go a lot quicker because we won't need to put in all the explanations. After the first example this process always seems like a very long process but it really isn't. It just looked that way because of all the explanation that we had to put into it.

So, let's start off with a couple of more examples with the heat equation using different boundary conditions.

Example: Use Separation of Variables on the following partial differential equation.

$$\frac{\partial u}{\partial t} = k\frac{\partial^2 u}{\partial x^2}$$

$$u(x,0) = f(x) \qquad \frac{\partial u}{\partial x}(0,t) = 0 \qquad \frac{\partial u}{\partial x}(L,t) = 0$$

Solution: In this case we're looking at the heat equation with no sources and perfectly insulated boundaries.

So, we'll start off by again assuming that our product solution will have the form,

$$u(x,t) = \varphi(x)G(t)$$

and because the differential equation itself hasn't changed here we will get the same result from plugging this in as we did in the previous example so the two ordinary differential equations that we'll need to solve are,

$$\frac{dG}{dt} = -k\lambda G \qquad \frac{d^2\varphi}{dx^2} = -\lambda\varphi$$

Now, the point of this example was really to deal with the boundary conditions so let's plug the product solution into them to get,

$$\frac{\partial\left(G(t)\varphi(x)\right)}{\partial x}(0,t) = 0 \qquad \frac{\partial\left(G(t)\varphi(x)\right)}{\partial x}(0,t) = 0$$

$$G(t)\frac{d\varphi}{dx}(0) = 0 \qquad G(t)\frac{d\varphi}{dx}(L) = 0$$

Now, just as with the first example if we want to avoid the trivial solution and so we can't have $G(t)= 0$ for every t and so we must have,

$$\frac{d\varphi}{dx}(0) = 0 \qquad \frac{d\varphi}{dx}(L) = 0$$

Here is a summary of what we get by applying separation of variables to this problem.

$$\frac{dG}{dt} = -k\lambda G \qquad \frac{d^2\varphi}{dx^2} + \lambda\varphi = 0$$

$$\frac{d\varphi}{dx}(0) = 0 \qquad \frac{d\varphi}{dx}(0) = 0$$

Next, let's see what we get if use periodic boundary conditions with the heat equation.

Example: Use Separation of Variables on the following partial differential equation.

$$\frac{\partial u}{\partial t} = k\frac{\partial^2 u}{\partial x^2} \qquad u(x,0) = f(x)$$

$$u(-L,t) = u(L,t) \qquad \frac{\partial u}{\partial x}(-L,t) = \frac{\partial u}{\partial x}(L,t)$$

Solution: First note that these boundary conditions really are homogeneous boundary conditions. If we rewrite them as,

$$u(-L,t) - u(L,t) = 0 \qquad \frac{\partial u}{\partial x}(-L,t) - \frac{\partial u}{\partial x}(L,t) = 0$$

it's a little easier to see.

Now, again we've done this partial differential equation so we'll start off with,

$$u(x,t) = \varphi(x)G(t)$$

and the two ordinary differential equations that we'll need to solve are,

$$\frac{dG}{dt} = -k\lambda G \qquad \frac{d^2\varphi}{dx^2} = -\lambda\varphi$$

Plugging the product solution into the rewritten boundary conditions gives,

$$G(t)\varphi(-L) - G(t)\varphi(L) = G(t)\left[\varphi(-L) - \varphi(L)\right] = 0$$

$$G(t)\frac{d\varphi}{dx}(-L) - G(t)\frac{d\varphi}{dx}(L) = G(t)\left[\frac{d\varphi}{dx}(-L) - \frac{d\varphi}{dx}(L)\right] = 0$$

and we can see that we'll only get non-trivial solution if,

$$\varphi(-L) - \varphi(L) = 0 \qquad \frac{d\varphi}{dx}(-L) - \frac{d\varphi}{dx}(L) = 0$$

$$\varphi(-L) = \varphi(L) \qquad \frac{d\varphi}{dx}(-L) = \frac{d\varphi}{dx}(L)$$

So, here is what we get by applying separation of variables to this problem.

$$\frac{dG}{dt} = -k\lambda G \qquad \frac{d^2\varphi}{dx^2} + \lambda\varphi = 0$$

$$\varphi(-L) = \varphi(L) \qquad \frac{d\varphi}{dx}(-L) = \frac{d\varphi}{dx}(L)$$

Let's now take a look at what we get by applying separation of variables to the wave equation with fixed boundaries.

Example: Use Separation of Variables on the following partial differential equation.

$$\frac{\partial^2 u}{\partial t^2} = c^2 \frac{\partial^2 u}{\partial x^2}$$

$$u(x,0) = f(x) \qquad \frac{\partial u}{\partial t}(x,0) = g(x)$$

$$u(0,t) = 0 \qquad u(L,t) = 0$$

Solution: Now, as with the heat equation the two initial conditions are here only because they need to be here for the problem. We will not actually be doing anything with them here and as mentioned previously the product solution will rarely satisfy them. We will be dealing with those in a later section when we actually go past this first step. Again, the point of this example is only to get down to the two ordinary differential equations that separation of variables gives.

So, let's get going on that and plug the product solution, $u(x,t) = \varphi(x)h(t)$ (we switched the G to an h here to avoid confusion with the g in the second initial condition) into the wave equation to get,

$$\frac{\partial^2}{\partial t^2}(\varphi(x)h(t)) = c^2\frac{\partial^2}{\partial x^2}(\varphi(x)h(t))$$

$$\varphi(x)\frac{d^2h}{dt^2} = c^2h(t)\frac{d^2\varphi}{dx^2}$$

$$\frac{1}{c^2h}\frac{d^2h}{dt^2} = \frac{1}{\varphi}\frac{d^2\varphi}{\varphi dx^2}$$

Note that we moved the $c^2$to the right side for the same reason we moved the k in the heat equation. It will make solving the boundary value problem a little easier.

Now that we've gotten the equation separated into a function of only t on the left and a function of only x on the right we can introduce a separation constant and again we'll use so we can arrive at a boundary value problem that we are familiar with. So, after introducing the separation constant we get,

$$\frac{1}{c^2h}\frac{d^2h}{dt^2} = \frac{1}{\varphi}\frac{d^2\varphi}{\varphi dx^2} = -\lambda$$

The two ordinary differential equations we get are then,

$$\frac{d^2h}{dt^2} = -\lambda c^2h \qquad \frac{d^2\varphi}{dx^2} = -\lambda\varphi$$

The boundary conditions in this example are identical to those from the first example and so plugging the product solution into the boundary conditions gives,

$$\varphi(0) = 0 \qquad \varphi(L) = 0$$

Applying separation of variables to this problem gives,

$$\frac{d^2h}{dt^2} = -\lambda c^2h \quad \frac{d^2\varphi}{dx^2} = -\lambda\varphi$$

$$\varphi(0) = 0 \qquad \varphi(L) = 0$$

Next, let's take a look at the 2-D Laplace's Equation.

Example: Use Separation of Variables on the following partial differential equation.

$$\frac{\partial^2 u}{\partial x^2} + \frac{\partial^2 u}{\partial y^2} = 0 \qquad 0 \le x \le L \qquad 0 \le y \le H$$

$$u(0,y) = g(y) \qquad u(L,y) = 0$$

$$u(x,u) = 0 \qquad u(x,H) = 0$$

Solution: This problem is a little (well actually quite a bit in some ways) different from the heat and wave equations. First, we no longer really have a time variable in the equation but instead we usually consider both variables to be spatial variables and we'll be assuming that the two variables are in the ranges shown above in the problems statement. Note that this also means that we no longer have initial conditions, but instead we now have two sets of boundary conditions, one for x and one for y.

Also, we should point out that we have three of the boundary conditions homogeneous and one nonhomogeneous for a reason. When we get around to actually solving this Laplace's Equation we'll see that this is in fact required in order for us to find a solution.

For this problem we'll use the product solution,

$$u(x,y)=h(x)\varphi(y)$$

It will often be convenient to have the boundary conditions in hand that this product solution gives before we take care of the differential equation. In this case we have three homogeneous boundary conditions and so we'll need to convert all of them. Because we've already converted these kind of boundary conditions we'll leave it to you to verify that these will become,

$$h(L)=0 \qquad \varphi(0)=0 \qquad \varphi(H)=0$$

Plugging this into the differential equation and separating gives,

$$\frac{\partial^2}{\partial x^2}\left(h(x)\varphi(y)\right)+\frac{\partial^2}{\partial y^2}\left(h(x)\varphi(y)\right)=0$$

$$\varphi(y)\frac{d^2h}{dx^2}+h(x)\frac{d^2\varphi}{dy^2}=0$$

$$\frac{1}{h}\frac{d^2h}{dx^2}=-\frac{1}{\varphi}\frac{d^2\varphi}{dy^2}$$

Okay, now we need to decide upon a separation constant. Note that every time we've chosen the separation constant we did so to make sure that the differential equation

$$\frac{d^2\varphi}{dx^2}+\lambda\varphi=0$$

would show up. Of course, the letters might need to be different depending on how we defined our product solution (as they'll need to be here). We know how to solve this eigenvalue/eigenfunction problem as we pointed out in the discussion after the first example. However, in order to solve it we need two boundary conditions.

So, for our problem here we can see that we've got two boundary conditions for $\varphi(y)$ but only one for $h(x)$ and so we can see that the boundary value problem that we'll have to solve will involve $\varphi(y)$ and so we need to pick a separation constant that will give use the boundary value problem we've already solved. In this case that means that we need to choose $\lambda$for the separation constant. If you're not sure you believe that yet hold on for a second and you'll soon see that it was in fact the correct choice here.

Putting the separation constant gives,

$$\frac{1}{h}\frac{d^2h}{dx^2} = -\frac{1}{\varphi}\frac{d^2\varphi}{dy^2} = \lambda$$

The two ordinary differential equations we get from Laplace's Equation are then,

$$\frac{d^2h}{dx^2} = \lambda h \qquad -\frac{d^2\varphi}{dy^2} = \lambda\varphi$$

and notice that if we rewrite these a little we get,

$$\frac{d^2h}{dx^2} - \lambda h = 0 \qquad \frac{d^2\varphi}{dy^2}\lambda\varphi = 0$$

We can now see that the second one does now look like one we've already solved (with a small change in letters of course, but that really doesn't change things).

So, let's summarize up here.

$$\frac{d^2h}{dx^2} - \lambda h = 0 \qquad \frac{d^2\varphi}{dy^2} + \lambda\varphi = 0$$

$$h(L) = 0 \qquad \varphi(0) = 0$$

$$\varphi(H) = 0$$

So, we've finally seen an example where the constant of separation didn't have a minus sign and again note that we chose it so that the boundary value problem we need to solve will match one we've already seen how to solve so there won't be much work to there.

All the examples worked in this section to this point are all problems that we'll continue in later sections to get full solutions for. Let's work one more however to illustrate a couple of other ideas. We will not however be doing any work with this in later sections however, it is only hear to illustrate a couple of points.

Example: Use Separation of Variables on the following partial differential equation.

$$\frac{\partial u}{\partial t} = k\frac{\partial^2 u}{\partial x^2} - u \qquad u(x,0) = f(x)$$

$$u(0,t) = 0 \qquad -\frac{\partial u}{\partial x}(L,t) = u(L,t)$$

Solution: Note that this is a heat equation with the source term of $Q(x,t) = -c\rho u$ and is both linear and homogenous. Also note that for the first time we've mixed boundary condition types. At $x = 0$ we've got a prescribed temperature and at $x = L$we've got a Newton's law of cooling type boundary condition. We should not come away from the first few examples with the idea that the boundary conditions at both boundaries always the same type. Having them the same type just makes the boundary value problem a little easier to solve in many cases.

So we'll start off with,

$$u(x,t) = \varphi(x)G(t)$$

and plugging this into the partial differential equation gives,

$$\frac{\partial}{\partial t}(\varphi(x)G(t)) = k\frac{\partial^2}{\partial x^2}(\varphi(x)G(t)) - \varphi(x)G(t)$$

$$\varphi(x)\frac{dG}{dt} = kG(t)\frac{d^2\varphi}{dx^2} - \varphi(x)G(t)$$

Now, the next step is to divide by $\varphi(x)G(t)$ and notice that upon doing that the second term on the right will become a one and so can go on either side. Theoretically there is no reason that the one can't be on either side, however from a practical standpoint we again want to keep things a simple as possible so we'll move it to the t side as this will guarantee that we'll get a differential equation for the boundary value problem that we've seen before.

So, separating and introducing a separation constant gives,

$$\frac{1}{k}\left(\frac{1}{G}\frac{dG}{dt} + 1\right) = \frac{1}{\varphi}\frac{d^2\varphi}{dx^2} = -\lambda$$

The two ordinary differential equations that we get are then (with some rewriting),

$$\frac{dG}{dt} = -(\lambda k + 1)G \qquad \frac{d^2\varphi}{dx^2} = -\lambda\varphi$$

Now let's deal with the boundary conditions.

$$G(t)\varphi(0) = 0$$

$$G(t)\frac{d\varphi}{dx}(L) + G(t)\varphi(L) = G(t)\left[\frac{d\varphi}{dx}(L) + \varphi(L)\right] = 0$$

and we can see that we'll only get non-trivial solution if,

$$\varphi(0) = 0 \qquad \frac{d\varphi}{dx}(L) + \varphi(L) = 0$$

So, here is what we get by applying separation of variables to this problem.

$$\frac{dG}{dt} = -(\lambda k + 1)G \qquad \frac{d^2\varphi}{dx^2} + \lambda\varphi = 0$$

$$\varphi(0) = 0 \qquad \frac{d\varphi}{dx}(L) + \varphi(L) = 0$$

On a quick side note we solved the boundary value problem in the example of the Eigenvalues and Eigenfunctions section and that example illustrates why separation of variables is not always so easy to use. As we'll see in the next section to get a solution that will satisfy any sufficiently nice initial condition we really need to get our hands on all the eigenvalues for the boundary value problem. However, as the solution to this boundary value problem shows this is not always possible to do. There are ways (which we won't be going into here) to use the information here to at least get approximations to the solution but we won't ever be able to get a complete solution to this problem.

Okay, that's it for this section. It is important to remember at this point that what we've done here is really only the first step in the separation of variables method for solving partial differential equations. Also, in the Laplace's Equation section the last two examples show pretty much the whole separation of variable process from defining the product solution to getting an actual solution. The only step that's missing from those two examples is the solving of a boundary value problem that will have been already solved at that point and so was not put into the solution given that they tend to be fairly lengthy to solve.

SOLVING THE HEAT EQUATION

Okay, it is finally time to completely solve a partial differential equation. In the previous section we applied separation of variables to several partial differential equations and reduced the problem down to needing to solve two ordinary differential equations. In this section we will now solve those ordinary differential equations and use the results to get a solution to the partial differential equation. We will be concentrating on the heat equation in this section and will do the wave equation and Laplace's equation in later sections.

The first problem that we're going to look at will be the temperature distribution in a bar with zero temperature boundaries. We are going to do the work in a couple of steps so we can take our time and see how everything works.

The first thing that we need to do is find a solution that will satisfy the partial differential equation and the boundary conditions. At this point we will not worry about the initial condition. The solution we'll get first will not satisfy the vast majority of initial conditions but as we'll see it can be used to find a solution that will satisfy a sufficiently nice initial condition.

Example: Find a solution to the following partial differential equation that will also satisfy the boundary conditions.

$$\frac{\partial u}{\partial t} = k\frac{\partial^2 u}{\partial x^2}$$
$$u(x,0) = f(x) \quad u(0,t) = 0 \qquad u(L,t) = 0$$

Solution: Okay the first thing we technically need to do here is apply separation of variables. Even though we did that in the previous section let's recap here what we did.

First, we assume that the solution will take the form,

$$u(x,t) = \varphi(x)G(t)$$

and we plug this into the partial differential equation and boundary conditions. We separate the equation to get a function of only t on one side and a function of only x on the other side and then introduce a separation constant. This leaves us with two ordinary differential equations.

We did all of this in Example of the previous section and the two ordinary differential equations are,

$$\frac{dG}{dt} = -k\lambda G \qquad \frac{d^2\varphi}{dx^2} + \lambda\varphi = 0$$
$$\varphi(0) = 0 \qquad \varphi(L) = 0$$

The time dependent equation can really be solved at any time, but since we don't know what $\lambda$is yet let's hold off on that one. Also note that in many problems only the boundary value problem can be solved at this point so don't always expect to be able to solve either one at this point.

The spatial equation is a boundary value problem and we know from our work in the previous chapter that it will only have non-trivial solutions (which we want) for certain values of λ, which we'll recall are called eigenvalues. Once we have those we can determine the non-trivial solutions for each, λ *i.e.* eigenfunctions.

Now, we actually solved the spatial problem,

$$\frac{d^2\varphi}{dx^2} + \lambda\varphi = 0$$
$$\varphi(0) = 0 \qquad \varphi(L) = 0$$

in Example of the Eigenvalues and Eigenfunctions section of the previous chapter for $L = 2\pi$. So, because we've solved this once for a specific L and the work is not all that much different for a general L we're not going to be putting in a lot of explanation here and if you need a reminder on how something works or why we did something go back to Example from the Eigenvalues and Eigenfunctions section for a reminder.

We've got three cases to deal with so let's get going.

$\lambda > 0$

In this case we know the solution to the differential equation is,

$$\varphi(x) = c_1 \cos\left(\sqrt{\lambda}\, x\right) + c_2 \sin\left(\sqrt{\lambda}\, x\right)$$

Applying the first boundary condition gives,

$$0 = \varphi(0) = c_1$$

Now applying the second boundary condition, and using the above result of course, gives,

$$0 = \varphi(L) = c_2 \sin\left(L\sqrt{\lambda}\right)$$

Now, we are after non-trivial solutions and so this means we must have,

$$\sin\left(L\sqrt{\lambda}\right) = 0 \quad \Rightarrow \quad L\sqrt{\lambda} = n\pi \qquad n = 1, 2, 3, \ldots$$

The positive eigenvalues and their corresponding eigenfunctions of this boundary value problem are then,

$$\lambda_n = \left(\frac{n\pi}{L}\right)^2 \qquad \varphi_n(x) = \sin\left(\frac{n\pi x}{L}\right) \qquad n = 1, 2, 3, \ldots$$

Note that we don't need the c_2 in the eigenfunction as it will just get absorbed into another constant that we'll be picking up later on.

$\lambda = 0$

The solution to the differential equation in this case is,

$$\varphi(x) = c_1 + c_2 x$$

Applying the boundary conditions gives,

$$0 = \varphi(0) = c_1 \qquad 0 = \varphi(L) = c_1 L \quad \Rightarrow \quad c_2 = 0$$

So, in this case the only solution is the trivial solution and so is not an eigenvalue for this boundary value problem.

$\lambda < 0$

Here the solution to the differential equation is,

$$\varphi(x) c_1 \cosh\left(\sqrt{-\lambda}\, x\right) + c_2 \sinh\left(\sqrt{-\lambda}\, x\right)$$

Applying the first boundary condition gives,

$$0 = \varphi(0) - c_1$$

and applying the second gives,

$$0 = \varphi(L) - c_2 \sinh\left(L\sqrt{-\lambda}\right)$$

So, we are assuming $\lambda < 0$ and so $L\sqrt{-\lambda} \neq 0$ and this means $\sinh\left(L\sqrt{-\lambda}\right) \neq 0$. We therefore we must have $c_2 = 0$ and so we can only get the trivial solution in this case.

Therefore, there will be no negative eigenvalues for this boundary value problem.

The complete list of eigenvalues and eigenfunctions for this problem are then,

$$\lambda_n = \left(\frac{n\pi}{L}\right)^2 \qquad \varphi_n(x) = \sin\left(\frac{n\pi x}{L}\right) \qquad n = 1, 2, 3, \ldots$$

Now let's solve the time differential equation,

$$\frac{dG}{dt} = -k\lambda_2 G$$

and note that even though we now know λ we're not going to plug it in quite yet to keep the mess to a minimum. We will however now use λ_n to remind us that we actually have an infinite number of possible values here.

This is a simple linear (and separable for that matter) 1st order differential equation and so we'll let you verify that the solution is,

$$G(t) = ce^{-k\lambda_n t} = ce^{-k\left(\frac{n\pi}{L}\right)^2 t}$$

Okay, now that we've gotten both of the ordinary differential equations solved we can finally write down a solution. Note however that we have in fact found infinitely many solutions since there are infinitely many solutions (*i.e.* eigenfunctions) to the spatial problem.

Our product solution are then,

$$u_n(x,t) = B_n \sin\left(\frac{n\pi x}{L}\right) e^{-k\left(\frac{n\pi}{L}\right)^2 t} \qquad n = 1, 2, 3, \ldots$$

We've denoted the product solution u_n to acknowledge that each value of *n* will yield a different solution. Also note that we've changed the *c* in the solution to the time problem to B_n to denote the fact that it will probably be different for each value of *n* as well and because had we kept the c_2 with the eigenfunction we'd have absorbed it into the *c* to get a single constant in our solution.

So, there we have it. The function above will satisfy the heat equation and the boundary condition of zero temperature on the ends of the bar.

The problem with this solution is that it simply will not satisfy almost every possible initial condition we could possibly want to use. That does not mean however, that there aren't at least a few that it will satisfy as the next example illustrates.

Example: Solve the following heat problem for the given initial conditions.

$$\frac{\partial u}{\partial t} = k\frac{\partial^2 u}{\partial x^2}$$

$$u(x,0) = f(x) \qquad u(0,t) = 0 \qquad u(L,t) = 0$$

1. $f(x) = 6\sin\left(\frac{\pi x}{L}\right)$

2. $f(x) = 12\sin\left(\frac{9\pi x}{L}\right) - 7\sin\left(\frac{4\pi x}{L}\right)$

Solution:

1. This is actually easier than it looks like. All we need to do is choose $n = 1$ and $B_1 = 6$ in the product solution above to get,

$$u(x,t) = 6\sin\left(\frac{\pi x}{L}\right)e^{-k\left(\frac{\pi}{L}\right)^2 t}$$

and we've got the solution we need. This is a product solution for the first example and so satisfies the partial differential equation and boundary conditions and will satisfy the initial condition since plugging in $t = 0$ will drop out the exponential.

2. This is almost as simple as the first part. Recall from the Principle of Superposition that if we have two solutions to a linear homogeneous differential equation (which we've got here) then their sum is also a solution. So, all we need to do is choose n and B_n as we did in the first part to get a solution that satisfies each part of the initial condition and then add them up. Doing this gives,

$$u(x,t) = f(x) = 12\sin\left(\frac{9\pi x}{L}\right)e^{-k\left(\frac{9\pi}{L}\right)^2 t} - 7\sin\left(\frac{4\pi x}{L}\right)e^{-k\left(\frac{4\pi}{L}\right)^2 t}$$

We'll leave it to you to verify that this does in fact satisfy the initial condition and the boundary conditions.

So, we've seen that our solution from the first example will satisfy at least a small number of highly specific initial conditions. Now, let's extend the idea out that we used in the second part of the previous example a little to see how we can get a solution that will satisfy any sufficiently nice initial condition. The Principle of Superposition is, of course, not restricted to only two solutions. For instance the following is also a solution to the partial differential equation.

$$u(x,t)=\sum_{n=1}^{M} B_n \sin\left(\frac{n\pi x}{L}\right) e^{-k\left(\frac{n\pi}{L}\right)^2 t}$$

and notice that this solution will not only satisfy the boundary conditions but it will also satisfy the initial condition,

$$u(x,0)=\sum_{n=1}^{M} B_n \sin\left(\frac{n\pi x}{L}\right)$$

Let's extend this out even further and take the limit as $M \to \infty$.Doing this our solution now becomes,

$$u(x,t)=\sum_{n=1}^{\infty} B_n \sin\left(\frac{n\pi x}{L}\right) e^{-k\left(\frac{n\pi}{L}\right)^2 t}$$

This solution will satisfy any initial condition that can be written in the form,

$$u(x,0)=f(x)=\sum_{n=1}^{\infty} B_n \sin\left(\frac{n\pi x}{L}\right)$$

This may still seem to be very restrictive, but the series on the right should look awful familiar to you after the previous chapter. The series on the left is exactly the Fourier sine series we looked at in that chapter. Also recall that when we can write down the Fourier sine series for any piecewise smooth function on $0 \le x \le L$.

So, provided our initial condition is piecewise smooth after applying the initial condition to our solution we can determine the $B_n$as if we were finding the Fourier sine series of initial condition. So we can either proceed as we did in that section and use the orthogonality of the sines to derive them or we can acknowledge that we've already done that work and know that coefficients are given by,

$$B_n=\frac{2}{L}\int_0^L f(x)\sin\left(\frac{n\pi x}{L}\right)dx \qquad n=1,2,3,\ldots$$

So, we finally can completely solve a partial differential equation.

Example: Solve the following BVP.

$$\frac{\partial u}{\partial t}=k\frac{\partial^2 u}{\partial x^2}$$

$$u(0,t)=0 \qquad u(0,t)=0 \qquad u(L,t)=0$$

Solution: There isn't really all that much to do here as we've done most of it in the examples and discussion above.

First, the solution is,

$$u(x,t)=\sum_{n=1}^{\infty} B_n \sin\left(\frac{n\pi x}{L}\right) e^{-k\left(\frac{n\pi}{L}\right)^2 t}$$

The coefficients are given by,

$$B_n=\frac{2}{L}20\sin\left(\frac{n\pi x}{L}\right)dx=\frac{2}{L}\left(\frac{20L\left(1-\cos(n\pi)\right)}{n\pi}\right)=\frac{40\left(1-(-1)^n\right)}{n\pi}$$

If we plug these in we get the solution,

$$u(x,t)=\sum_{n=1}^{\infty}\frac{40\left(1-(-1)^n\right)}{n\pi}\sin\left(\frac{n\pi x}{L}\right)e^{-k\left(\frac{\pi}{L}\right)^2 t}$$

That almost seems anti-climactic. This was a very short problem. Of course some of that came about because we had a really simple constant initial condition and so the integral was very simple. However, don't forget all the work that we had to put into discussing Fourier sine series, solving boundary value problems, applying separation of variables and then putting all of that together to reach this point.

While the example itself was very simple, it was only simple because of all the work that we had to put into developing the ideas that even allowed us to do this. Because of how "simple" it will often be to actually get these solutions we're not actually going to do anymore with specific initial conditions. We will instead concentrate on simply developing the formulas that we'd be required to evaluate in order to get an actual solution.

So, having said that let's move onto the next example. In this case we're going to again look at the temperature distribution in a bar with perfectly insulated boundaries. We are also no longer going to go in steps. We will do the full solution as a single example and end up with a solution that will satisfy any piecewise smooth initial condition.

Example: Find a solution to the following partial differential equation.

$$\frac{\partial u}{\partial t}=k\frac{\partial^2 u}{\partial x^2}$$

$$u(x,0)=f(x) \qquad \frac{\partial u}{\partial x}(0,t)=0 \quad \frac{\partial u}{\partial x}(0,t)=0$$

Solution: We applied separation of variables to this problem in Example of the previous section. So, after assuming that our solution is in the form,

$$u(x,t) = \varphi(x)G(t)$$

and applying separation of variables we get the following two ordinary differential equations that we need to solve.

$$\frac{dG}{dt} = --k\lambda G \qquad \frac{d^2\varphi}{dx^2} + \lambda\varphi = 0$$

$$\frac{d\varphi}{dx}(0) = 0 \quad \frac{d\varphi}{dx}(L) = 0$$

We solved the boundary value problem in Example of the Eigenvalues and Eigenfunctions section of the previous chapter for $L = 2\pi$ so as with the first example in this section we're not going to put a lot of explanation into the work here. If you need a reminder on how this works go back to the previous chapter and review the example we worked there. Let's get going on the three cases we've got to work for this problem.

$\lambda > 0$

The solution to the differential equation is,

$$\varphi(x) = c_1 \cos\left(\sqrt{\lambda} x\right) + c_2 \sin\left(\sqrt{\lambda} x\right)$$

Applying the first boundary condition gives,

$$0 = \frac{d\varphi}{dx}(0) = \sqrt{\lambda} c_2 \qquad \Rightarrow \qquad c_2 = 0$$

The second boundary condition gives,

$$0 = \frac{d\varphi}{dx}(L) = -\sqrt{\lambda} c_1 \sin\left(L\sqrt{\lambda}\right)$$

Recall that $\lambda > 0$ and so we will only get non-trivial solutions if we require that,

$$\sin\left(L\sqrt{\lambda}\right) = 0 \quad \Rightarrow \quad L\sqrt{\lambda} = n\pi \qquad n = 1, 2, 3, \ldots$$

The positive eigenvalues and their corresponding eigenfunctions of this boundary value problem are then,

$$\lambda_n = \left(\frac{n\pi}{L}\right)^2 \qquad \varphi_n(x) = \cos\left(\frac{n\pi x}{L}\right) \qquad n = 1, 2, 3, \ldots$$

$\lambda = 0$

The general solution is,

$$\varphi(x) = c_2 + c_2 x$$

Applying the first boundary condition gives,

$$0=\frac{d\varphi}{dx}(0)=c_2$$

Using this the general solution is then,

$$\varphi(x)=c_1$$

and note that this will trivially satisfy the second boundary condition. Therefore $\lambda = 0$ is an eigenvalue for this BVP and the eigenfunctions corresponding to this eigenvalue is,

$$\varphi(x)=1$$

$\lambda < 0$

The general solution here is,

$$\varphi(x)=c_1\cosh\left(\sqrt{-\lambda}\,x\right)+c_2\sinh\left(\sqrt{-\lambda}\,x\right)$$

Applying the first boundary condition gives,

$$0=\frac{d\varphi}{dx}(0)=\sqrt{-\lambda}\,c_2 \qquad \Rightarrow \qquad c_2=0$$

The second boundary condition gives,

$$0=\frac{d\varphi}{dx}(L)=\sqrt{-\lambda}\,c_1\sinh\left(L\sqrt{-\lambda}\right)$$

We know that $L\sqrt{-\lambda}\neq 0$ and so $\sinh\left(L\sqrt{-\lambda}\right)\neq 0$. Therefore we must have $c_1=0$ and so, this boundary value problem will have no negative eigenvalues.

So, the complete list of eigenvalues and eigenfunctions for this problem is then,

$$\lambda_n=\left(\frac{n\pi}{L}\right)^2 \qquad \varphi_n(x)=\cos\left(\frac{n\pi x}{L}\right) \qquad n=1,2,3,\ldots$$

$$\lambda_0=0 \qquad \varphi_0(x)=1$$

and notice that we get the $\lambda_0=0$ eigenvalue and its eigenfunction if we allow $n = 0$ in the first set and so we'll use the following as our set of eigenvalues and eigenfunctions.

$$\lambda_n=\left(\frac{n\pi}{L}\right)^2 \qquad \varphi_n(x)=\cos\left(\frac{n\pi x}{L}\right) \qquad n=0,1,2,3,\ldots$$

The time problem here is identical to the first problem we looked at so,

$$G(t)=c\,e^{-k\left(\frac{n\pi}{L}\right)^2 t}$$

Our product solutions will then be,

$$u_n(x,t) = A_n \cos\left(\frac{n\pi x}{L}\right) e^{-k\left(\frac{n\pi}{L}\right)^2 t} \qquad n = 0,1,2,3,...$$

and the solution to this partial differential equation is,

$$u(x,t) = \sum_{n=0}^{\infty} A_n \cos\left(\frac{n\pi x}{L}\right) e^{-k\left(\frac{n\pi}{L}\right)^2 t}$$

If we apply the initial condition to this we get,

$$u(x,t) = f(x) = \sum_{n=0}^{\infty} A_n \cos\left(\frac{n\pi x}{L}\right)$$

and we can see that this is nothing more than the Fourier cosine series for $f(x)$ on $0 \le x \le L$ and so again we could use the orthogonality of the cosines to derive the coefficients or we could recall that we've already done that in the previous chapter and know that the coefficients are given by,

$$A_n = \begin{cases} \frac{1}{L}\int_0^L f(x)\,dx & n = 0 \\ \frac{2}{L}\int_0^L f(x)\cos\left(\frac{n\pi x}{L}\right)dx & n \neq 0 \end{cases}$$

The last example that we're going to work in this section is a little different from the first two. We are going to consider the temperature distribution in a thin circular ring. We will consider the lateral surfaces to be perfectly insulated and we are also going to assume that the ring is thin enough so that the temperature does not vary with distance from the center of the ring.

So, what does that leave us with? Let's set $x = 0$as shown below and then let x be the arc length of the ring as measured from this point.

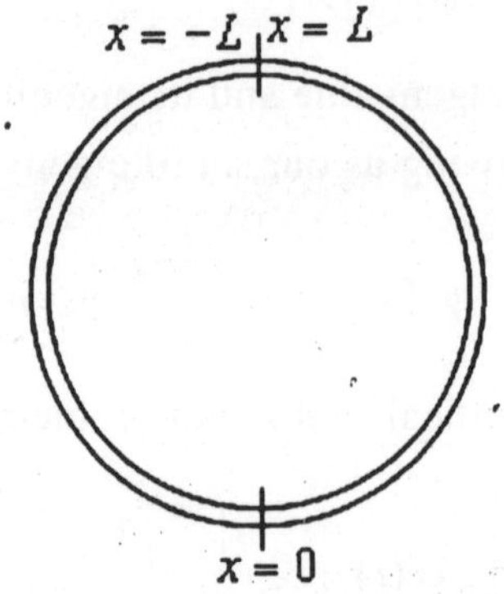

We will measure x as positive if we move to the right and negative if we move to the left of $x = 0$. This means that at the top of the ring we'll meet where $x = L$(if we move to the right) and $x = -L$(if we move to the left). By doing this we can consider this ring to be a bar of length $2L$ and the heat equation that we developed earlier in this chapter will still hold.

At the point of the ring we consider the two "ends" to be in perfect thermal contact. This means that at the two ends both the temperature and the heat flux must be equal. In other words we must have,

$$u(-L,t) = u(L,t) \qquad \frac{\partial u}{\partial x}(-L,t) = \frac{\partial u}{\partial x}(L,t)$$

If you recall from the section in which we derived the heat equation we called these periodic boundary conditions. So, the problem we need to solve to get the temperature distribution in this case is,

Example: Find a solution to the following partial differential equation.

$$\frac{\partial u}{\partial t} = k\frac{\partial^2 u}{\partial x^2}$$

$$u(x,0) = f(x) \qquad u(-L,t) = u(L,t) \qquad \frac{\partial u}{\partial x}(-L,t) = \frac{\partial u}{\partial x}(L,t)$$

Solution: We applied separation of variables to this problem in Example of the previous section. So, if we assume the solution is in the form,

$$u(x,t) = \varphi(x)G(t)$$

we get the following two ordinary differential equations that we need to solve.

$$\frac{dG}{dt} = -k\lambda G \qquad \frac{d^2\varphi}{dx^2} + \lambda\varphi = 0$$

$$\varphi(-L) = \varphi(L) \qquad \frac{d\varphi}{dx}(-L) = \frac{d\varphi}{dx}(L)$$

As we've seen with the previous two problems we've already solved a boundary value problem like this one back in the Eigenvalues and Eigenfunctions section of the previous chapter, Example to be exact with . So, if you need a little more explanation of what's going on here go back to this example and you can see a little more explanation. We again have three cases to deal with here.

$\lambda > 0$

The general solution to the **differential** equation is,

$$\varphi(x) = c_1 \cos\left(\sqrt{\lambda}x\right) + c_2 \sin\left(\sqrt{\lambda}x\right)$$

Applying the first boundary condition and recalling that cosine is an even function and sine is an odd function gives, us

$$c_1 \cos\left(-L\sqrt{\lambda}\right)+c_2 \sin\left(-L\sqrt{\lambda}\right)=c_1 \cos\left(L\sqrt{\lambda}\right)+c_2 \sin\left(L\sqrt{\lambda}\right)$$

$$-c_2 \sin\left(L\sqrt{\lambda}\right)=c_2 \sin\left(L\sqrt{\lambda}\right)$$

$$0=2c_2 \sin\left(L\sqrt{\lambda}\right)$$

At this stage we can't really say anything as either $c_2$or sine could be zero. So, let's apply the second boundary condition and see what we get.

$$-\sqrt{\lambda}c_1 \sin\left(-L\sqrt{\lambda}\right)+\sqrt{\lambda}c_2 \sin\left(-L\sqrt{\lambda}\right)=-\sqrt{\lambda}c_1 \sin\left(L\sqrt{\lambda}\right)+\sqrt{\lambda}c_2 \cos\left(L\sqrt{\lambda}\right)$$

$$\sqrt{\lambda}c_1 \sin\left(L\sqrt{\lambda}\right)=-\sqrt{\lambda}c_1 \sin\left(L\sqrt{\lambda}\right)$$

$$2\sqrt{\lambda}c_1 \sin\left(L\sqrt{\lambda}\right)=0$$

We get something similar. However notice that if $\sin\left(L\sqrt{\lambda}\right)\neq 0$ then we would be forced to have $c_1=c_2=0$ and this would give us the trivial solution which we don't want. This means therefore that we must have $\sin\left(L\sqrt{\lambda}\right)=0$ which in turn means that the positive eigenvalues for this problem are,

$$\lambda_n=\left(\frac{n\pi}{L}\right)^2 \qquad n=1,2,3,...$$

Now, there is no reason to believe that $c_1=0$ or $c_2=0$. All we know is that they both can't be zero and so that means that we in fact have two sets of eigenfunctions for this problem corresponding to positive eigenvalues. They are,

$$\varphi_n(x)=\cos\left(\frac{n\pi x}{L}\right) \qquad \varphi_n(x)=\sin\left(\frac{n\pi x}{L}\right) \qquad n=1,2,3,...$$

$\lambda = 0$

The general solution in this case is,

$$\varphi(x)=c_1+c_2x$$

Applying the first boundary condition gives,

$$c_1+c_2(-L)=c_1+c_2(L)$$

$$2Lc_2 \quad \Rightarrow \quad c_2=0$$

The general solution is then,

$$\varphi(x)=c_1$$

and this will trivially satisfy the second boundary condition. Therefore $\lambda = 0$ is an eigenvalue for this BVP and the eigenfunctions corresponding to this eigenvalue is,

$$\varphi(x) = 1$$

$\lambda < 0$

For this final case the general solution here is,

$$\varphi(x) = c_1 \cosh\left(\sqrt{-\lambda}x\right) + c_2 \sinh\left(\sqrt{-\lambda}x\right)$$

Applying the first boundary condition and using the fact that hyperbolic cosine is even and hyperbolic sine is odd gives,

$$c_1 \cosh\left(-L\sqrt{-\lambda}\right) + c_2 \sinh\left(-L\sqrt{-\lambda}\right) = -c_1 \cosh\left(L\sqrt{-\lambda}\right) + c_2 \sinh\left(L\sqrt{-\lambda}\right)$$

$$-c_2 \sinh\left(-L\sqrt{-\lambda}\right) = c_2 \sinh\left(L\sqrt{-\lambda}\right)$$

$$2c_2 \sinh\left(L\sqrt{-\lambda}\right) = 0$$

Now, in this case we are assuming that $\lambda < 0$ and so $L\sqrt{-\lambda} \neq 0$. This turn tells us that $\sinh\left(L\sqrt{-\lambda}\right) \neq 0$. We therefore must have $c_2 = 0$.

Let's now apply the second boundary condition to get,

$$\sqrt{-\lambda}c_1 \sinh\left(-L\sqrt{-\lambda}\right) = \sqrt{-\lambda}c_1 \sinh\left(L\sqrt{-\lambda}\right)$$

$$2\sqrt{-\lambda}c_1 \sinh\left(L\sqrt{-\lambda}\right) = 0 \qquad \Rightarrow \qquad c_1 = 0$$

By our assumption on we again have no choice here but to have $c_1 = 0$ and so for this boundary value problem there are negative eigenvalues.

Summarizing up then we have the following sets of eigenvalues and eigenfunctions and note that we've merged the $\lambda = 0$ case into the cosine case since it can be here to simplify things up a little.

$$\lambda_n = \left(\frac{n\pi}{L}\right)^2 \qquad \varphi_n(x) = \cos\left(\frac{n\pi x}{L}\right) \qquad n = 0, 1, 2, 3, \ldots$$

$$\lambda_n = \left(\frac{n\pi}{L}\right)^2 \qquad \varphi_n(x) = \cos\left(\frac{n\pi x}{L}\right) \qquad , n = 1, 2, 3, \ldots$$

The time problem is again identical to the two we've already worked here and so we have,

$$G(t) = ce^{-k\left(\frac{n\pi}{L}\right)^2 t}$$

Now, this example is a little different from the previous two heat problems that we've looked at. In this case we actually have two different possible product solutions that will satisfy the partial differential equation and the boundary conditions. They are,

$$u_n(x,t) = A_n \cos\left(\frac{n\pi x}{L}\right) e^{-k\left(\frac{n\pi}{L}\right)^2 t} \qquad n = 0,1,2,3,\ldots$$

$$u_n(x,t) = B_n \sin\left(\frac{n\pi x}{L}\right) e^{-k\left(\frac{n\pi}{L}\right)^2 t} \qquad n = 1,2,3,\ldots$$

The Principle of Superposition is still valid however and so a sum of any of these will also be a solution and so the solution to this partial differential equation is,

$$u_n(x,t) = \sum_{n=1}^{\infty} A_n \cos\left(\frac{n\pi x}{L}\right) e^{-k\left(\frac{n\pi}{L}\right)^2 t} + \sum_{n=1}^{\infty} B_n \sin\left(\frac{n\pi x}{L}\right) e^{-k\left(\frac{n\pi}{L}\right)^2 t}$$

If we apply the initial condition to this we get,

$$u(x,0) = f(x) = \sum_{n=1}^{\infty} A_n \cos\left(\frac{n\pi x}{L}\right) + \sum_{n=1}^{\infty} B_n \sin\left(\frac{n\pi x}{L}\right)$$

and just as we saw in the previous two examples we get a Fourier series. The difference this time is that we get the full Fourier series for a piecewise smooth initial condition on $-L \le x \le L$. As noted for the previous two examples we could either rederive formulas for the coefficients using the orthogonality of the sines and cosines or we can recall the work we've already done. There's really no reason at this point to redo work already done so the coefficients are given by,

$$A_0 = \frac{1}{2L} \int_{-L}^{L} f(x)\,dx$$

$$A_0 = \frac{1}{L} \int_{-L}^{L} f(x) \cos\left(\frac{n\pi x}{L}\right) dx \qquad n = 1,2,3,\ldots$$

$$B_0 = \frac{1}{L} \int_{-L}^{L} f(x) \sin\left(\frac{n\pi x}{L}\right) dx \qquad n = 1,2,3,\ldots$$

Note that this is the reason for setting up x as we did at the start of this problem. A full Fourier series needs an interval of $-L \le x \le L$ whereas the Fourier sine and cosines series we saw in the first two problems need $0 \le x \le L$.

You might want to go through and do the two cases where we have a zero temperature on one boundary and a perfectly insulated boundary on the other to see if you've got this process down.

HEAT EQUATION WITH NON-ZERO TEMPERATURE BOUNDARIES

In this section we want to expand one of the cases from the previous section a little bit. In the previous section we look at the following heat problem.

$$\frac{\partial u}{\partial t} = k\frac{\partial^2 u}{\partial x^2}$$
$$u(x,0) = f(x) \qquad u(0,t) = 0 \qquad n(L,t) = 0$$

Now, there is nothing inherently wrong with this problem, but the fact that **we're** fixing the temperature on both ends at zero is a little unrealistic. The other two problems we looked at, insulated boundaries and the thin ring, are a little more realistic problems, but this one just isn't all that realistic so we'd like to extend it a little.

What we'd like to do in this section is instead look at the following problem.

$$\frac{\partial u}{\partial t} = k\frac{\partial^2 u}{\partial x^2}$$
$$u(x,0) = f(x) \qquad u(0,t) = T_1 \qquad n(L,t) = T_2$$

In this case we'll allow the boundaries to be any fixed temperature, T_1or T_2. The problem here is that separation of variables will no longer work on this problem because the boundary conditions are no longer homogeneous. Recall that separation of variables will only work if both the partial differential equation and the boundary conditions are linear and homogeneous. So, we're going to need to deal with the boundary conditions in some way before we actually try and solve this.

Luckily for us there is an easy way to deal with them. Let's consider this problem a little bit. There are no sources to add/subtract heat energy anywhere in the bar. Also our boundary conditions are fixed temperatures and so can't change with time and we aren't prescribing a heat flux on the boundaries to continually add/subtract heat energy. So, what this all means is that there will not ever be any forcing of heat energy into or out of the bar and so while some heat energy may well naturally flow into our out of the bar at the end points as the temperature changes eventually the temperature distribution in the bar should stabilize out and no longer depend on time. Or, in other words it makes some sense that we should expect that as $t \to \infty$our temperature distribution, $u(x,t)$should behave as,

$$\lim_{t\to\infty} u(x,t) = u_E(x)$$

where $u_E(x)$is called the equilibrium temperature. Note as well that is should still satisfy the heat equation and boundary conditions. It won't satisfy the initial condition however because it is the temperature distribution as $t \to \infty$ whereas the initial condition is at $t = 0$. So, the equilibrium temperature distribution should satisfy,

$$0 = \frac{\partial^2 u_E}{\partial x^2} \qquad u_E(0) = T_1 \qquad u_E(L) = T_2$$

This is a really easy 2nd order ordinary differential equation to solve. If we integrate twice we get,

$$u_E(x) = c_1 x + c_2$$

and applying the boundary conditions (we'll leave this to you to verify) gives us,

$$u_E(x) = T_1 + \frac{T_2 - T_1}{L} x$$

We'll let's define the function,

$$v(x,t) = u(x,t) - u_E(x)$$

where $u(x,t)$ is the solution and $u_E(x)$ is the equilibrium temperature.

Now let's rewrite this as,

$$u(x,t) = v(x,t) + u_E(x)$$

and let's take some derivatives.

$$\frac{\partial u}{\partial t} = \frac{\partial v}{\partial t} + \frac{\partial u_E}{\partial t} = \frac{\partial v}{\partial t} \qquad \frac{\partial^2 u}{\partial x^2} = \frac{\partial^2 v}{\partial x^2} + \frac{\partial^2 u_E}{\partial x^2} = \frac{\partial^2 v}{\partial x^2}$$

In both of these derivatives we used the fact that $u_E(x)$ is the equilibrium temperature and so is independent of time t and must satisfy the differential equation.

What this tells us is that both $u(x,t)$ and $v(x,t)$ must satisfy the same partial differential equation. Let's see what the initial conditions and boundary conditions would need to be for $v(x,t)$.

$$v = (x,0) = u(x,0) - u_E(x) = f(x) - u_E(x)$$
$$v = (0,t) = u(0,t) - u_E(0) = T_1 - T_2 = 0$$
$$v = (L,t) = u(L,t) - u_E(L) = T_2 - T_2 = 0$$

So, the initial condition just gets potentially messier, but the boundary conditions are now homogeneous! The partial differential equation that $v(x,t)$ must satisfy is,

$$\frac{\partial v}{\partial t} = k \frac{\partial^2 u}{\partial x^2}$$
$$v(x,0) = f(x) - u_E(x) \qquad v(0,t) = 0 \qquad v(L,t) = 0$$

We saw how to solve this in the previous section and so we the solution is,

$$v(x,t)=\sum_{n=1}^{\infty} B_n \sin\left(\frac{n\pi x}{L}\right) e^{-k\left(\frac{n\pi}{L}\right)^2 t}$$

where the coefficients are given by,

$$B_n=\frac{2}{L}\int_0^L \left(f(x)-u_E(x)\right)\sin\left(\frac{n\pi x}{L}\right)dx \qquad n=1,2,3,\ldots$$

The solution is then,

$$u(x,t)=u_E(x)+v(x,t)$$

$$=T_1+\frac{T_2-T_1}{L}x+\sum_{n=1}^{\infty} B_n \sin\left(\frac{n\pi x}{L}\right) e^{-k\left(\frac{n\pi}{L}\right)^2 t}$$

and the coefficients are given above.

LAPLACE'S EQUATION

The next partial differential equation that we're going to solve is the 2-D Laplace's equation,

$$\nabla^2 u=\frac{\partial^2 u}{\partial x^2}+\frac{\partial^2 u}{\partial y^2}=0$$

A natural question to ask before we start learning how to solve this is does this equation come up naturally anywhere? The answer is a very resounding yes! If we consider the 2-D heat equation,

$$\frac{\partial u}{\partial t}=k\nabla^2 u+\frac{Q}{cp}$$

We can see that Laplace's equation would correspond to finding the equilibrium solution (*i.e.* time independent solution) if there were not sources. So, this is an equation that can arise from physical situations.

How we solve Laplace's equation will depend upon the geometry of the 2-D object we're solving it on. Let's start out by solving it on the rectangle given by, $0\le x\le L, 0\le y\le H$. For this geometry Laplace's equation along with the four boundary conditions will be,

$$\nabla^2 u=\frac{\partial^2 u}{\partial x^2}+\frac{\partial^2 u}{\partial y^2}=0$$

$$u(0,y)=g_1(y) \qquad u(L,y)=g_2(y)$$
$$u(x,0)=f_1(x) \qquad u(x,H)=f_2(x)$$

One of the important things to note here is that unlike the heat equation we will not have any initial conditions here. Both variables are spatial variables and each variable occurs in a 2nd order derivative and so we'll need two boundary conditions for each variable.

Next, let's notice that while the partial differential equation is both linear and homogeneous the boundary conditions are only linear and are not homogeneous. This creates a problem because separation of variables requires homogeneous boundary conditions.

To completely solve Laplace's equation we're in fact going to have to solve it four times. Each time we solve it only one of the four boundary conditions can be nonhomogeneous while the remaining three will be homogeneous. The four problems are probably best shown with a quick sketch so let's consider the following sketch.

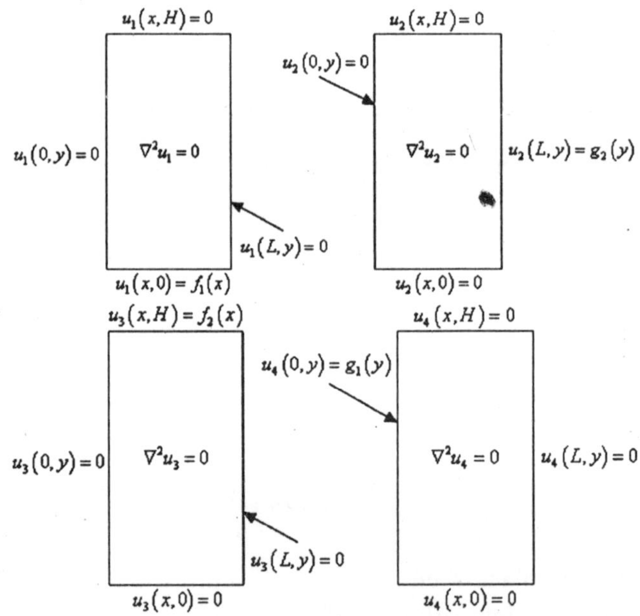

Now, once we solve all four of these problems the solution to our original system, (1), will be,

$$u(x,y)=u_1(x,y)+u_2(x,y)+u_3(x,y)+u_4(x,y)$$

Because we know that Laplace's equation is linear and homogeneous and each of the pieces is a solution to Laplace's equation then the sum will also be a solution. Also, this will satisfy each of the four original boundary conditions. We'll verify the first one and leave the rest to you to verify.

$$u(x,0)=u_1(x,0)+u_2(x,0)+u_3(x,0)+u_4(x,0)=f_1(x)+0+0+0=f_1(x)$$

In each of these cases the lone nonhomogeneous boundary condition will take the place of the initial condition in the heat equation problems that we solved a couple of sections ago. We will apply separation of variables to the each problem and find a product solution that will satisfy the differential equation and the three homogeneous boundary conditions. Using the Principle of Superposition we'll find a solution to the problem and then apply the final boundary condition to determine the value of the constant(s) that are left in the problem. The process is nearly identical in many ways to what we did when we were solving the heat equation.

We're going to do two of the cases here and we'll leave the remaining two for you to do.

Example: Find a solution to the following partial differential equation.

$$\nabla^2 u_4 = \frac{\partial^2 u_4}{\partial x^2} + \frac{\partial^2 u_4}{\partial y^2} = 0$$

$$u_4(0,y) = g_1(y) \qquad u_4(L,y) = 0$$
$$u_4(x,0) = 0 \qquad u_4(x,H) = 0$$

Solution: We'll start by assuming that our solution will be in the form,

$$u_4(x,y) = h(x)\varphi(y)$$

and then recall that we performed separation of variables on this problem (with a small change in notation) back in Example of the Separation of Variables section. So from that problem we know that separation of variables yields the following two ordinary differential equations that we'll need to solve.

$$\frac{\partial^2 h}{\partial x^2} - \lambda h = 0 \qquad \frac{d^2\varphi}{dy^2} + \lambda\varphi = 0$$
$$h(L) = 0 \qquad \varphi(0) - 0 \qquad \varphi(H) = 0$$

Note that in this case, unlike the heat equation we must solve the boundary value problem first. Without knowing what $\lambda$is there is no way that we can solve the first differential equation here with only one boundary condition since the sign of λwill affect the solution.

Let's also notice that we solved the boundary value problem in Example of Solving the Heat Equation and so there is no reason to resolve it here. Taking a change of letters into account the eigenvalues and eigenfunctions for the boundary value problem here are,

$$\lambda_n = \left(\frac{n\pi}{H}\right)^2 \qquad \varphi_n(y) = \sin\left(\frac{n\pi x}{H}\right) \qquad n = 1,2,3,\ldots$$

Now that we know what the eigenvalues are let's write down the first differential equation with λplugged in.

$$\frac{\partial^2 h}{\partial x^2} - \left(\frac{n\pi}{H}\right)^2 h = 0$$
$$h(L) = 0$$

Because the coefficient of $h(x)$in the differential equation above is positive we know that a solution to this is,

$$h(x) = c_1 \cosh\left(\frac{n\pi x}{H}\right) + c_2 \sinh\left(\frac{n\pi x}{H}\right)$$

However, this is not really suited for dealing with the $h(L) = 0$boundary condition.So, let's also notice that the following is also a solution.

$$h(x) = c_1 \cosh\left(\frac{n\pi(x-L)}{H}\right) + c_2 \sinh\left(\frac{n\pi(x-L)}{H}\right)$$

You should verify this by plugging this into the differential equation and checking that it is in fact a solution. Applying the lone boundary condition to this "shifted" solution gives,

$$0 = h(L) = c_1$$

The solution to the first differential equation is now,

$$h(x) = c_2 \sinh\left(\frac{n\pi(x-L)}{H}\right)$$

and this is all the farther we can go with this because we only had a single boundary condition. That is not really a problem however because we now have enough information to form the product solution for this partial differential equation.

A product solution for this partial differential equation is,

$$u_n(x,y) = B_n \sinh\left(\frac{n\pi(x-L)}{H}\right) \sin\left(\frac{n\pi y}{H}\right) \qquad n = 1,2,3,\ldots$$

The Principle of Superposition then tells us that a solution to the partial differential equation is,

$$u_4(x,y) = \sum_{n=1}^{\infty} B_n \sinh\left(\frac{n\pi(x-L)}{H}\right) \sin\left(\frac{n\pi y}{H}\right)$$

and this solution will satisfy the three homogeneous boundary conditions.

To determine the constants all we need to do is apply the final boundary condition.

$$u_4(0,y) = g_1(y) = \sum_{n=1}^{\infty} B_n \sinh\left(\frac{n\pi(x-L)}{H}\right)\sin\left(\frac{n\pi y}{H}\right)$$

Now, in the previous problems we've done this has clearly been a Fourier series of some kind and in fact it still is. The difference here is that the coefficients of the Fourier sine series are now,

$$B_n \sinh\left(\frac{n\pi(x-L)}{H}\right)$$

instead of just B_n. We might be a little more tempted to use the orthogonality of the sines to derive formulas for the, B_n however we can still reuse the work that we've done previously to get formulas for the coefficients here.

Remember that a Fourier sine series is just a series of coefficients (depending on n) times a sine. We still have that here, except the "coefficients" are a little messier this time that what we saw when we first dealt with Fourier series. So, the coefficients can be found using exactly the same formula from the Fourier sine series section of a function on $0 \le y \le H$ we just need to be careful with the coefficients.

$$B_n \sinh\left(\frac{n\pi(-L)}{H}\right) = \frac{2}{H}\int_0^H g_1(y)\sin\left(\frac{n\pi x}{H}\right)dy \qquad n = 1,2,3,\ldots$$

$$B_n = \frac{2}{H\sinh\left(\frac{n\pi(-L)}{H}\right)}\int_0^H g_1(y)\sin\left(\frac{n\pi x}{H}\right)dy \qquad n = 1,2,3,\ldots$$

The formulas for the are a little messy this time in comparison to the other problems we've done but they aren't really all that messy. Okay, let's do one of the other problems here so we can make a couple of points.

Example: Find a solution to the following partial differential equation.

$$\nabla^2 u_3 = \frac{\partial^2 u_3}{\partial x^2} + \frac{\partial^2 u_3}{\partial x^2} = 0$$

$$u_3(0,y) = 0 \qquad u_3(L,y) = 0$$

$$u_3(x,0) = 0 \qquad u_3(x,H) = f_2(x)$$

Solution: Okay, for the first time we've hit a problem where we haven't previous done the separation of variables so let's go through that. We'll assume the solution is in the form,

$$u_3(x,y) = h(x)\varphi(y)$$

We'll apply this to the homogeneous boundary conditions first since we'll need those once we get reach the point of choosing the separation constant. We'll let you verify that the boundary conditions become,

$$h(0)=0 \qquad h(L)=0 \qquad \varphi(0)=0$$

Next, we'll plug the product solution into the differential equation.

$$\frac{\partial^2}{\partial x^2}\big(h(x)\varphi(y)\big)+\frac{\partial^2}{\partial y^2}\big(h(x)\varphi(y)\big)=0$$

$$\varphi(y)+\frac{d^2u}{dx^2}+h(x)\frac{d^2\varphi}{dy^2}=0$$

$$\frac{1}{h}\frac{d^2h}{dx^2}=-\frac{1}{\varphi}\frac{d^2\varphi}{dy^2}$$

Now, at this point we need to choose a separation constant. We've got two homogeneous boundary conditions on h so let's choose the constant so that the differential equation for h yields a familiar boundary value problem so we don't need to redo any of that work. In this case, unlike the $u_4$case, we'll need $-\lambda$.

This is a good problem in that is clearly illustrates that sometimes you need λas a separation constant and at other times you need $-\lambda$.Not only that but sometimes all it takes is a small change in the boundary conditions it force the change.

So, after adding in the separation constant we get,

$$\frac{1}{h}\frac{d^2h}{dx^2}=-\frac{1}{\varphi}\frac{d^2\varphi}{dy^2}=-\lambda$$

and two ordinary differential equations that we get from this case (along with their boundary conditions) are,

$$\frac{d^2h}{dx^2}+\lambda h=0 \qquad \frac{d^2\varphi}{dy^2}-\lambda\varphi=0$$

$$h(0)=0 \qquad h(L)=0 \qquad \varphi(0)=0$$

Now, as we noted above when we were deciding which separation constant to work with we've already solved the first boundary value problem. So, the eigenvalues and eigenfunctions for the first boundary value problem are,

$$\lambda_n=\left(\frac{n\pi}{L}\right)^2 \qquad h_n(x)=\sin\left(\frac{n\pi x}{L}\right) \qquad n=1,2,3,\ldots$$

The second differential equation is then,

$$\frac{d^2\varphi}{dx^2} - \left(\frac{n\pi}{L}\right)^2 \varphi = 0$$

$$\varphi(0) = 0$$

Because the coefficient of the φ is positive we know that a solution to this is,

$$\varphi(y) = c_1 \cosh\left(\frac{n\pi y}{L}\right) + c_2 \sinh\left(\frac{n\pi y}{L}\right)$$

In this case, unlike the previous example, we won't need to use a shifted version of the solution because this will work just fine with the boundary condition we've got for this. So, applying the boundary condition to this gives,

$$0 = \varphi(0) = c_1$$

and this solution becomes,

$$\varphi(y) = c_1 \sinh\left(\frac{n\pi y}{L}\right)$$

The product solution for this case is then,

$$u_n(x,y) = B_n \sinh\left(\frac{n\pi y}{L}\right) \sin\left(\frac{n\pi x}{L}\right) \qquad n = 1,2,3,\ldots$$

The solution to this partial differential equation is then,

$$u_3(x,y) = \sum_{n=1}^{\infty} B_n \sinh\left(\frac{n\pi y}{L}\right) \sin\left(\frac{n\pi x}{L}\right)$$

Finally, let's apply the nonhomogeneous boundary condition to get the coefficients for this solution.

$$u_3(x,H) = f_2(x) = \sum_{n=1}^{\infty} B_n \sinh\left(\frac{n\pi H}{L}\right) \sin\left(\frac{n\pi x}{L}\right)$$

As we've come to expect this is again a Fourier sine (although it won't always be a sine) series and so using previously done work instead of using the orthogonality of the sines to we see that,

$$B_n \sinh\left(\frac{n\pi H}{L}\right) = \frac{2}{L}\int_0^L f_2(x) \sin\left(\frac{n\pi x}{L}\right) dx \qquad n = 1,2,3,\ldots$$

$$B_n = \frac{2}{L \sinh\left(\frac{n\pi H}{L}\right)} \int_0^L f_2(x) \sin\left(\frac{n\pi x}{L}\right) dx \qquad n = 1,2,3,\ldots$$

Okay, we've worked two of the four cases that would need to be solved in order to completely solve (1). As we've seen each case was very similar and yet also had some differences. We saw the use of both separation constants and that sometimes we need to use a "shifted" solution in order to deal with one of the boundary conditions.

Before moving on let's note that we used prescribed temperature boundary conditions here, but we could just have easily used prescribed flux boundary conditions or a mix of the two. No matter what kind of boundary conditions we have they will work the same.

As a final example in this section let's take a look at solving Laplace's equation on a disk of radius a and a prescribed temperature on the boundary. Because we are now on a disk it makes sense that we should probably do this problem in polar coordinates and so the first thing we need to so do is write down Laplace's equation in terms of polar coordinates. Laplace's equation in terms of polar coordinates is,

$$\nabla^2 u \frac{1}{r}\frac{\partial}{\partial r}\left(r\frac{\partial u}{\partial r}\right)+\frac{1}{r^2}\frac{\partial^2 u}{\partial \theta^2}$$

Okay, this is a lot more complicated than the Cartesian form of Laplace's equation and it will add in a few complexities to the solution process, but it isn't as bad as it looks. The main problem that we've got here really is that fact that we've got a single boundary condition. Namely,

$$u(a,\theta)=f(\theta)$$

This specifies the temperature on the boundary of the disk. We are clearly going to need three more conditions however since we've got a 2nd derivative in both r and θ.

When we solved Laplace's equation on a rectangle we used conditions at the end points of the range of each variable and so it makes some sense here that we should probably need the same kind of conditions here as well. The range on our variables here are,

$$0<r<a \qquad -\pi<\theta<\pi$$

Note that the limits on $\theta$are somewhat arbitrary here and are chosen for convenience here. Any set of limits that covers the complete disk will work, however as we'll see with these limits we will get another familiar boundary value problem arising. The best choice here is often not known until the separation of variables is done. At that point you can go back and make your choices.

Okay, we now need conditions for $r=0$and $\theta=\pm\pi$. First, note that Laplace's equation in terms of polar coordinates is singular at $r=0$(*i.e.* we get division by zero). However, we know from physical considerations that the temperature must remain finite everywhere in the disk and so let's impose the condition that,

$$|u(0,\theta)|<\infty$$

This may seem like an odd condition and it definitely doesn't conform to the other boundary conditions that we've seen to this point, but it will work out for us as we'll see.

Now, for boundary conditions for θwe'll do something similar to what we did for the 1-D head equation on a thin ring. The two limits on $\theta$are really just different sides of a line in the disk and so let's use the periodic conditions there. In other words,

$$u(-\pi,t)=u(\pi,t) \qquad \frac{\partial u}{\partial x}(-\pi,t)=\frac{\partial u}{\partial x}(\pi,t)$$

With all of this out of the way let's solve Laplace's equation on a disk of radius a.

Example: Find a solution to the following partial differential equation.

$$\nabla^2 u=\frac{1}{r}\frac{\partial}{\partial r}\left(r\frac{\partial u}{\partial r}\right)+\frac{1}{r^2}\frac{\partial^2 u}{\partial\theta^2}=0$$

$$|u(0,\theta)|<\infty \qquad u(a,\theta)=f(\theta)$$

$$u(-\pi,t)=u(\pi,t) \qquad \frac{\partial u}{\partial x}(-\pi,t)=\frac{\partial u}{\partial x}(\pi,t)$$

Solution: In this case we'll assume that the solution will be in the form,

$$u(\theta,r)=\varphi(\theta)G(r)$$

Plugging this into the periodic boundary conditions gives,

$$\varphi(-\pi)=\varphi(\pi) \qquad \frac{d\varphi}{d\theta}(-\pi)=\frac{d\varphi}{d\theta}(\pi)$$

$$|G(0)|<\infty$$

Now let's plug the product solution into the partial differential equation.

$$\frac{1}{r}\frac{\partial}{\partial r}\left(r\frac{\partial}{\partial r}(\varphi(\theta)G(r))\right)+\frac{1}{r^2}\frac{\partial^2}{\partial\theta^2}(\varphi(\theta)G(r))=0$$

$$\varphi(\theta)\frac{1}{r}\frac{d}{dr}\left(r\frac{dG}{dr}\right)+G(r)\frac{1}{r^2}\frac{d^2\varphi}{d\theta^2}=0$$

This is definitely more of a mess that we've seen to this point when it comes to separating variables. In this case simply dividing by the product solution, while still necessary, will not be sufficient to separate the variables. We are also going to have to multiply by $r^2$to completely separate variables. So, doing all that, moving each term to one side of the equal sign and introduction a separation constant gives,

$$\frac{r}{G}\frac{d}{dr}\left(r\frac{dG}{dr}\right)=-\frac{1}{\varphi}\frac{d^2\varphi}{d\theta^2}=\lambda$$

We used as the separation constant this time to get the differential equation for θto match up with one we've already done. The ordinary differential equations we get are then,

$$r\frac{d}{dr}\left(r\frac{dG}{dr}\right)-\lambda G=0 \qquad \frac{d^2\varphi}{d\theta^2}+\lambda\varphi=0$$

$$|G(0)|<\infty \qquad \theta(-\pi)=\varphi(\pi) \qquad \frac{d\varphi}{d\theta}(-\pi)=\frac{d\varphi}{d\theta}(\pi)$$

Now, we solved the boundary value problem above in Example of the Eigenvalues and Eigenfunctions section of the previous chapter and so there is no reason to redo it here. The eigenvalues and eigenfunctions for this problem are,

$$\lambda_n=n^2 \qquad \varphi_n(\theta)=\sin(n\theta) \qquad n=1,2,3,...$$

$$\lambda_n=n^2 \qquad \varphi_n(\theta)=\cos(n\theta) \qquad n=0,1,2,3,...$$

Plugging this into the first ordinary differential equation and using the product rule on the derivative we get,

$$r\frac{d}{dr}\left(r\frac{dG}{dr}\right)-n^2G=0$$

$$r\left(r\frac{d^2G}{dr^2}+\frac{dG}{dr}\right)-n^2G=0$$

$$r^2r\frac{d^2G}{dr^2}+r\frac{dG}{dr}-n^2G=0$$

This is an Euler differential equation and so we know that solutions will be in the form $G(r)=r^p$ provided p is a root of,

$$p(p-1)+p-n^2=0$$

$$p^2-n^2=0 \quad \Rightarrow \quad p=\pm n \qquad n=0,1,2,3,...$$

So, because the case will yield a double root, versus two real distinct roots if we have two cases here.They are,

$$G(r)=c_1+c_2\ln r \qquad n=0$$

$$G(r)=\overline{c_1}r^n+\overline{c_2}r^{-n} \qquad n=1,2,3,...$$

Now we need to recall the condition that $|G(0)|<\infty$ .Each of the solutions above will have $G(0)\to\infty$ as $r\to 0$ Therefore in order to meet this boundary condition we must have $c_2=\overline{c_2}=0$.

Therefore, the solution reduces to,

$$G(r) = c_1 r^n \qquad n = 0, 1, 2, 3, \ldots$$

and notice that with the second term gone we can combine the two solutions into a single solution.

So, we have two product solutions for this problem. They are,

$$u_n(\theta, r) = A_n r^n \cos(n\theta) \qquad n = 0, 1, 2, 3, \ldots$$
$$u_n(\theta, r) = B_n r^n \sin(n\theta) \qquad n = 1, 2, 3, \ldots$$

Our solution is then the sum of all these solution or,

$$u(\theta, r) = \sum_{n=1}^{\infty} A_n r^n \cos(n\theta) + \sum_{n=1}^{\infty} A_n r^n \sin(n\theta)$$

Applying our final boundary condition to this gives,

$$u(a, \theta) = f(\theta) = \sum_{n=1}^{\infty} A_n a^n \cos(n\theta) + \sum_{n=1}^{\infty} A_n a^n \sin(n\theta)$$

This is a full Fourier series for $f(\theta)$ on the interval, $-\pi \le \theta \le \pi$ *i.e.* $L = \pi$. Also note that once again the "coefficients" of the Fourier series are a little messier than normal, but not quite as messy as when we were working on a rectangle above. We could once again use the orthogonality of the sines and cosines to derive formulas for the A_n and B_n or we could just use the formulas from the Fourier series section with $L = \pi$ to get,

$$A_0 = \frac{1}{2\pi} \int_{-\pi}^{\pi} f(\theta)\, d\theta$$
$$A_n a^n = \frac{1}{\pi} \int_{-\pi}^{\pi} f(\theta) \cos(n\theta)\, d\theta \qquad n = 1, 2, 3, \ldots$$
$$B_n a^n = \frac{1}{\pi} \int_{-\pi}^{\pi} f(\theta) \sin(n\theta)\, d\theta \qquad n = 1, 2, 3, \ldots$$

Upon solving for the coefficients we get,

$$A_0 = \frac{1}{2\pi} \int_{-\pi}^{\pi} f(\theta)\, d\theta$$
$$A_n = \frac{1}{\pi a^n} \int_{-\pi}^{\pi} f(\theta) \cos(n\theta)\, d\theta \qquad n = 1, 2, 3, \ldots$$
$$B_n = \frac{1}{\pi a^n} \int_{-\pi}^{\pi} f(\theta) \sin(n\theta)\, d\theta \qquad n = 1, 2, 3, \ldots$$

Prior to this example most of the separation of variable problems tended to look very similar and it is easy to fall in to the trap of expecting everything to look like what we'd seen earlier. With this example we can see that the problems can definitely be different on occasion so don't get too locked into expecting them to always work in exactly the same way. Before we leave this section let's briefly talk about what you'd need to do on a partial disk. The periodic boundary conditions above were only there because we had a whole disk. What if we only had a disk between say $\alpha \leq \theta \leq \beta$.

When we've got a partial disk we now have two new boundaries that we not present in the whole disk and the periodic boundary conditions will no longer make sense. The periodic boundary conditions are only used when we have the two "boundaries" in contact with each other and that clearly won't be the case with a partial disk. So, if we stick with prescribed temperature boundary conditions we would then have the following conditions

$$\begin{aligned}
&|u(0,\theta)| < \infty \\
&u(a,\theta) = f(\theta) \qquad && \alpha \leq \theta \leq \beta \\
&u(r,\alpha) = g_1(r) \qquad && 0 \leq r \leq a \\
&u(r,\beta) = g_2(r) \qquad && 0 \leq r \leq a
\end{aligned}$$

Also note that in order to use separation of variables on these conditions we'd need to have $g_1(r) = g_2(r) = 0$ to make sure they are homogeneous.

As a final note we could just have easily used flux boundary conditions for the last two if we'd wanted to. The boundary value problem would be different, but outside of that the problem would work in the same manner.

We could also use a flux condition on the boundary but we haven't really talked yet about how to apply that kind of condition to our solution. Recall that this is the condition that we apply to our solution to determine the coefficients. It's not difficult to use we just haven't talked about this kind of condition yet. We'll be doing that in the next section.

VIBRATING STRING

This will be the final partial differential equation that we'll be solving in this chapter. In this section we'll be solving the 1-D wave equation to determine the displacement of a vibrating string. There really isn't much in the way of introduction to do here so let's just jump straight into the example.

Example: Find a solution to the following partial differential equation.

$$\frac{\partial^2 u}{\partial t^2} = c^2 \frac{\partial^2 u}{\partial x^2}$$

$$u(x,0) = f(x) \qquad \frac{\partial u}{\partial t}(x,0) = g(x)$$

$$u(0,t) = 0 \qquad u(L,t) = 0$$

Solution: One of the main differences here that we're going to have to deal with is the fact that we've now got two initial conditions. That is not something we've seen to this point, but will not be all that difficult to deal with when the time rolls around.

We've already done the separation of variables for this problem, but let's go ahead and redo it here so we can say we've got another problem almost completely worked out.

So, let's start off with the product solution.

$$u(x,t) = \varphi(x)h(t)$$

Plugging this into the two boundary conditions gives,

$$\varphi(0) = 0 \qquad \varphi(L) = 0$$

Plugging the product solution into the differential equation, separating and introducing a separation constant gives,

$$\frac{\partial^2}{\partial t^2}(\varphi(x)h(t)) = c^2 \frac{\partial^2}{\partial x^2}(\varphi(x)h(t))$$

$$\varphi(x)\frac{d^2 h}{dt^2} = c^2 h(t)\frac{d^2\varphi}{dx^2}$$

$$\frac{1}{c^2 h}\frac{d^2 h}{dt^2} = \frac{1}{\varphi}\frac{d^2\varphi}{dx^2} = -\lambda$$

We moved the c^2 to the left side for convenience and chose $-\lambda$ for the separation constant so the differential equation for would match a known (and solved) case.

The two ordinary differential equations we get from separation of variables are then,

$$\frac{d^2 h}{dt^2} + c^2\lambda h = 0 \qquad \frac{d^2\varphi}{dx^2} + \lambda\varphi$$

$$\varphi(0) = 0 \qquad \varphi(L) = 0$$

We solved the boundary value problem above in Example of the Solving the Heat Equation section of this chapter and so the eigenvalues and eigenfunctions for this problem are,

$$\lambda_n = \left(\frac{n\pi}{L}\right)^2 \qquad \varphi_n(x) = \sin\left(\frac{n\pi x}{L}\right) \qquad n = 1, 2, 3, \ldots$$

The first ordinary differential equation is now,

$$\frac{\partial^2 h}{\partial t^2} + \left(\frac{n\pi c}{L}\right)^2 h = 0$$

and because the coefficient of the h is clearly positive the solution to this is,

$$h(t) = c_1 cos\left(\frac{n\pi ct}{L}\right) + c_2 \sin\left(\frac{n\pi ct}{L}\right)$$

Because there is no reason to think that either of the coefficients above are zero we then get two product solutions,

$$u_n(x,t) = A_1 cos\left(\frac{n\pi ct}{L}\right)\sin\left(\frac{n\pi x}{L}\right)$$
$$u_n(x,t) = B_1 \sin\left(\frac{n\pi ct}{L}\right)\sin\left(\frac{n\pi x}{L}\right) \qquad n = 1, 2, 3, \ldots$$

The solution is then,

$$u(x,t) = \sum_{n=1}^{\infty}\left[A_1 cos\left(\frac{n\pi ct}{L}\right)\sin\left(\frac{n\pi x}{L}\right) + B_1 \sin\left(\frac{n\pi ct}{L}\right)\sin\left(\frac{n\pi x}{L}\right)\right]$$

Now, in order to apply the second initial condition we'll need to differentiate this with respect to *t* so,

$$\frac{\partial u}{\partial t} = \sum_{n=1}^{\infty}\left[-\frac{n\pi c}{L} A_n \sin\left(\frac{n\pi ct}{L}\right)\sin\left(\frac{n\pi x}{L}\right) + \frac{n\pi c}{L} B_c \cos\left(\frac{n\pi ct}{L}\right)\sin\left(\frac{n\pi x}{L}\right)\right]$$

If we now apply the initial conditions we get,

$$u(u, 0) = f(x) = \sum_{n=1}^{\infty}\left[A_n \cos(0)\sin\left(\frac{n\pi x}{L}\right) + B_c \sin(0)\sin\left(\frac{n\pi x}{L}\right)\right] = \sum_{n=1}^{\infty} A_n \sin\left(\frac{n\pi x}{L}\right)$$

$$\frac{\partial u}{\partial t}(x, 0) = g(x) = \sum_{n=1}^{\infty}\frac{n\pi x}{L} B_n \sin\left(\frac{n\pi x}{L}\right)$$

Both of these are Fourier sine series. The first is for $f(x)$ on $0 \le x \le L$ while the second is for $g(x)$ on $0 \le x \le L$ with a slightly messy coefficient.

As in the last few sections we're faced with the choice of either using the orthogonality of the sines to derive formulas for A_n and B_n or we could reuse formula from previous work. It's easier to reuse formulas so using the formulas form the Fourier sine series section we get,

$$A_n = \frac{2}{L}\int_0^L f(x)\sin\left(\frac{n\pi x}{L}\right)dx \quad n = 1, 2, 3, \ldots$$

$$\frac{n\pi c}{L} B_n = \frac{2}{L}\int_0^L g(x)\sin\left(\frac{n\pi x}{L}\right)dx \quad n = 1, 2, 3, \ldots$$

Upon solving the second one we get,

$$A_n = \frac{2}{L}\int_0^L f(x)\sin\left(\frac{n\pi x}{L}\right)dx \quad n = 1, 2, 3, \ldots$$

$$B_n = \frac{2}{L}\int_0^L g(x)\sin\left(\frac{n\pi x}{L}\right)dx \quad n = 1, 2, 3, \ldots$$

So, there is the solution to the 1-D wave equation and with that we've solved the final partial differential equation in this chapter.

SUMMARY OF SEPARATION OF VARIABLES

Throughout this chapter we've been talking about and solving partial differential equations using the method of separation of variables. However, the one thing that we've not really done is completely work an example from start to finish showing each and every step.

Each partial differential equation that we solved made use somewhere of the fact that we'd done at least part of the problem in another section and so it makes some sense to have a quick summary of the method here.

Also note that each of the partial differential equations only involved two variables. The method can often be extended out to more than two variables, but the work in those problems can be quite involved and so we didn't cover any of that here.

So with all of that out of the way here is a quick summary of the method of separation of variables for partial differential equations in two variables.

Verify that the partial differential equation is linear and homogeneous.

1. Verify that the boundary conditions are in proper form. Note that this will often depend on what is in the problem. So,
    (a) If you have initial conditions verify that all the boundary conditions are linear and homogeneous.
    (b) If there are no initial conditions (such as Laplace's equation) the verify that all but one of the boundary conditions are linear and homogeneous.
    (c) In some cases (such as we saw with Laplace's equation on a disk) a boundary condition will take the form of requiring that the solution stay finite and in these cases we just need to make sure the boundary condition is met.
2. Assume that solutions will be a product of two functions each a function in only one of the variables in the problem. This is called a product solution.
3. Plug the product solution into the partial differential equation, separate variables and introduce a separation constant. This will produce two ordinary differential equations.

4. Plug the product solution into the homogeneous boundary conditions. Note that often it will be better to do this prior to doing the differential equation so we can use these to help us chose the separation constant.
5. One of the ordinary differential equations will be a boundary value problem. Solve this to determine the eigenvalues and eigenfunctions for the problem.

   Note that this is often very difficult to do and in some cases it will be impossible to completely find all eigenvalues and eigenfunctions for the problem. These cases can be dealt with to get at least an approximation of the solution, but that is beyond the scope of this quick introduction.
6. Solve the second ordinary differential equation using any remaining homogeneous boundary conditions to simplify the solution if possible.
7. Use the Principle of Superposition and the product solutions to write down a solution to the partial differential equation that will satisfy the partial differential equation and homogeneous boundary conditions.
8. Apply the remaining conditions (these may be initial condition(s) or a single nonhomogeneous boundary condition) and use the orthogonality of the eigenfunctions to find the coefficients.

   Note that in all of our examples the eigenfunctions were sines and/or cosines however they won't always be sines and cosines. If the boundary value problem is sufficiently nice (and that's beyond the scope of this quick introduction to the method) we can always guarantee that the eigenfunctions will be orthogonal regardless of what they are.

Chapter 8

Applications of Partial Differential Equation

TANGENT PLANES AND LINEAR APPROXIMATIONS

Earlier we saw how the two partial derivatives f_x and f_y can be thought of as the slopes of traces. We want to extend this idea out a little in this section. The graph of a function $z = f(x, y)$ is a surface in $\mathbb{R}^3$ (three dimensional space) and so we can now start thinking of the plane that is "tangent" to the surface as a point.

Let's start out with a point (x_0, y_0) and let's let C_1 represent the trace to $f(x, y)$ for the plane $y = y_0$ (*i.e.* allowing x to vary with y held fixed) and we'll let C_2 represent the trace to $f(x, y)$ for the plane $x = x_0$ (*i.e.* allowing y to vary with x held fixed). Now, we know that $f_x(x_0, y_0)$ is the slope of the tangent line to the trace C_1 and $f_y(x_0, y_0)$ is the slope of the tangent line to the trace C_2. So, let L_1 be the tangent line to the trace C_1 and let L_2 be the tangent line to the trace C_2.

The tangent plane will then be the plane that contains the two lines L_1 and L_2. Geometrically this plane will serve the same purpose that a tangent line. A tangent line to a curve was a line that just touched the curve at that point and was "parallel" to the curve at the point in question. Well tangent planes to a surface are planes that just touch the surface at the point and are "parallel" to the surface at the point. Note that this gives us a point that is on the plane.

Since the tangent plane and the surface touch at (x_0, y_0) the following point will be on both the surface and the plane.

$$(x_0, y_0, z_0) = (x_0, y_0, f(x_0, y_0))$$

What we need to do now is determine the equation of the tangent plane. We know that the general equation of a plane is given by,

$$a(x - x_0) + b(y - y_0) + c(z - z_0) = 0$$

where (x_0, y_0, z_0) is a point that is on the plane, which we know already. Let's rewrite this a little. We'll move the x terms and y terms to the other side and divide both sides by c. Doing this gives,

$$z - z_0 = -\frac{a}{c}(x - x_0) - \frac{b}{c}(y - y_0)$$

Now, let's rename the constants to simplify up the notation a little. Let's rename them as follows,

$$A = -\frac{a}{c} \qquad B - \frac{b}{c}(y - y_0)$$

With this renaming the equation of the tangent plane becomes,

$$z - z_0 = A(x - x_0) + B(y - y_0)$$

and we need to determine values for A and B.

Let's first think about what happens if we hold y fixed, *i.e.* if we assume that $y = y_0$. In this case the equation of the tangent plane becomes,

$$z - z_0 = A(x - x_0)$$

This is the equation of a line and this line must be tangent to the surface at $(x - x_0)$ (since its part of the tangent plane). In addition, this line assumes that $(y - y_0)$ (*i.e.* fixed) and A is the slope of this line. But if we think about it this is exactly that the tangent toC_l is, a line tangent to the surface at $(x - x_0)$ assuming that $(y - y_0)$. In other words,

$$z - z_0 = A(x - x_0)$$

is the equation forL_l and we know that the slope of L_l is given by $f_x(x - x_0)$. Therefore we have the following,

$$A = f_x(x - x_0)$$

$$A = f_x(x - x_0)\ A = f_x(x - x_0)$$

If we hold x fixed at $x - x_0$ the equation of the tangent plane becomes,

$$z - z_0 = B(y - y_0)$$

However, by a similar argument to the one above we can see that this is nothing more than the equation for $L_l$and that it's slope is B or $f_y(x_0, y_0)$. So,

$$B = f_y(x_0, y_0)$$

The equation of the tangent plane to the surface given by $z = f(x, y)$ at (x_0, y_0) is then,

$$z - z_0 = f_x(x_0, y_0)(x - x_0) + f_y(x_0, y_0)(y - y_0)$$

Also, if we use the fact that $z_0 = f(x_0, y_0)$ we can rewrite the equation of the tangent plane as,

$$z - f(x_0, y_0) = f_x(x_0, y_0)(x - x_0) + f_y(x_0, y_0)(y - y_0)$$
$$z = f(x_0, y_0) + f_x(x_0, y_0)(x - x_0) + f_y(x_0, y_0)(y - y_0)$$

We will see an easier derivation of this formula (actually a more general formula) in the next section so if you didn't quite follow this argument hold off until then to see a better derivation.

Example: Find the equation of the tangent plane to $z = \text{In}(2x + y)$ at $(-1, 3)$.

Solution: There really isn't too much to do here other than taking a couple of derivatives and doing some quick evaluations

$$f(x, y) = \text{In}(2x + y)$$
$$f_x(x, y) = \frac{2}{2x + y}$$
$$f_x(x, y) = \frac{1}{2x + y}$$
$$z_0 = f(-1, 3) = \text{In}(1) = 0$$
$$f_x(-1, 3) = 2$$
$$f_y(-1, 3) = 1$$

The equation of the plane is then,

$$z - 0 = 2(x + 1) + (1)(y - 3)$$
$$z = 2x + y - 1$$

One nice use of tangent planes is they give us a way to approximate a surface near a point. As long as we are near to the point (x_0, y_0) then the tangent plane should nearly approximate the function at that point.Because of this we define the linear approximation to be,

$$L(x, y) = f(x_0, y_0) + f_x(x_0, y_0)(x - x_0)$$
$$+ f_y(x_0, y_0)(y - y_0)$$

and as long as we are "near" (x_0, y_0) then we should have that,

$$f(x,y) \approx L(x,y) = f(x_0,y_0) + f_x(x_0,y_0)(x-x_0) + f_y(x_0,y_0)(y-y_0)$$

Example: Find the linear approximation to $z = 3 + \frac{x^2}{16} + \frac{y^2}{9}$ at $(-4,3)$.

Solution: So, we're really asking for the tangent plane so let's find that.

$$f(x,y) = 3 + \frac{x^2}{16} + \frac{y^2}{9}$$

$$f_x(x,y) = \frac{x}{8}$$

$$f_y(x,y) = \frac{2y}{9}$$

$$f(-4,3) = 3+1+1 = 5$$

$$f_x(-4,3) = -\frac{1}{2}$$

$$f_y(-4,3) = \frac{2}{3}$$

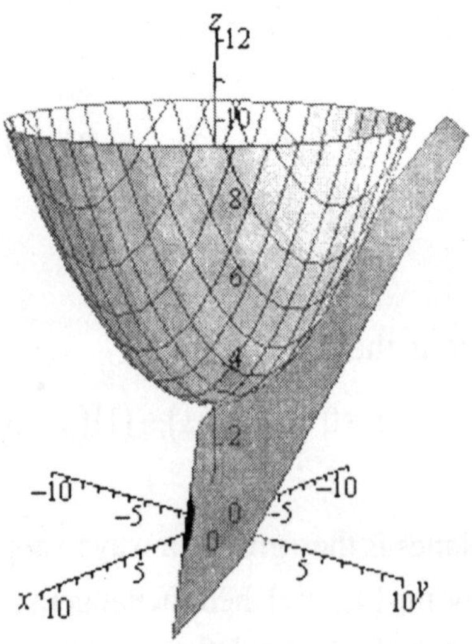

The tangent plane, or linear approximation, is then,

$$L(x,y) = 5 - \frac{1}{2}(x+4) + \frac{2}{3}(y-3)$$

For reference purposes here is a sketch of the surface and the tangent plane/linear approximation.

GRADIENT VECTOR, TANGENT PLANES AND NORMAL LINES

In this section we want to revisit tangent planes only this time we'll look at them in light of the gradient vector. In the process we will also take a look at a normal line to a surface.

Let's first recall the equation of a plane that contains the point (x_0, y_0, z_0) with normal vector $\vec{n} = \langle a,b,c \rangle$ is given by,

$$a(x-x_0)+b(y-y_0)+c(z-z_0)=0$$

When we introduced the gradient vector in the section on directional derivatives we gave the following fact.

Fact

The gradient vector $\nabla f(x_0, x_0)$ is orthogonal (or perpendicular) to the level curve $> f(x,y)=k$ at the point (x_0, y_0). Likewise, the gradient vector $\nabla f(x_0, y_0, z_0)$ is orthogonal to the level surface $f(x,y,z)=k$ at the point (x_0, y_0, z_0).

Actually, all we need here is the last part of this fact. This says that the gradient vector is always orthogonal, or *normal*, to the surface at a point. Also recall that the gradient vector is,

$$\nabla f = (f_x, f_y, f_z)$$

So, the tangent plane to the surface given by $f(x,y,z)=k$ at (x_0, y_0, z_0) has the equation,

$$f_x(x_0,y_0,z_0)(x-x_0)+f_y(x_0,y_0,z_0)(y-y_0)$$
$$+f_z(x_0,y_0,z_0)(z-z_0)=0$$

This is a much more general form of the equation of a tangent plane that the one that derived in the previous section.

Note however, that we can also get the equation from the previous section using this more general formula. To see this let's start with the equation $z = f(x,y)$ and we want to find the tangent plane to the surface given by $z = f(x,y)$ at the point (x_0, y_0, z_0) where $z_0 - f(x_{0,} y_0)$. In order to use the formula above we need to have all the variables on one side. This is easy enough to do. All we need to do is subtract a z from both sides to get,

$$f(x,y)-z=0$$

Now, if we define a new function

$$F(x, y, z) = f(x, y) - z$$

we can see that the surface given by $z = f(x, y)$ is identical to the surface given by $F(x, y, z) = 0$ and this new equivalent equation is in the correct form for the equation of the tangent plane that we derived in this section.

So, the first thing that we need to do is find the gradient vector for F.

$$\nabla F = \langle F_x, F_y, F_z \rangle = \langle f_x, f_y, -1 \rangle$$

Notice that

$$F_x = \frac{\partial}{\partial x}(f(x, y) - z) = f_x$$

$$F_y = \frac{\partial}{\partial y}(f(x, y) - z) = f_y$$

$$F_z = \frac{\partial}{\partial z}(f(x, y) - z) = -1$$

The equation of the tangent plane is then,

$$f_x(x_0, y_0)(x - x_0) + f_y(x_0, y_0)(y - y_0) - (z - z_0) = 0$$

Or, upon solving for z, we get,

$$z = f(x_0, y_0) + f_x(x_0, y_0)(x - x_0) + f_y(x_0, y_0)(y - y_0)$$

which is identical to the equation that we derived in the previous section.

We can get another nice piece of information out of the gradient vector as well. We might on occasion want a line that is orthogonal to a surface at a point, sometimes called the normal line. This is easy enough to get if we recall that the equation of a line only requires that we have a point and a parallel vector. Since we want a line that is at the point (x_0, y_0, z_0) we know that this point must also be on the line and we know that $\nabla f(x_0, y_0, z_0)$ is a vector that is normal to the surface and hence will be parallel to the line. Therefore the equation of the normal line is,

$$\vec{r}(t) = \langle x_0, y_0, z_0 \rangle + t\nabla f(x_0, y_0, z_0)$$

Example: Find the tangent plane and normal line to $x^2 + y^2 + z^2 = 30$ at the point $(1, -2, 5)$.

Solution: For this case the function that we're going to be working with is,

$$F(x, y, z) = x^2 + y^2 + z^2$$

and note that we don't have to have a zero on one side of the equal sign. All that we need is a constant. To finish this problem out we simply need the gradient evaluated at the point.

$$\nabla F = \langle 2x, 2y, 2z \rangle$$
$$\nabla F(1, -2, 5) = \langle 2, -4, 10 \rangle$$

The tangent plane is then,

$$2(x-1) - 4(y+2) + 10(z-5) = 0$$

The normal line is,

$$\begin{aligned}\vec{r}(t) &= \langle 1, -2, 5 \rangle + t\langle 2, -4, 10 \rangle \\ &= \langle 1 + 2t, -2 - 4t, 5 + 10t \rangle\end{aligned}$$

RELATIVE MINIMUMS AND MAXIMUMS

In this section we are going to extend one of the more important ideas into functions of two variables. We are going to start looking at trying to find minimums and maximums of functions. This in fact will be the topic of the following two sections as well. In this section we are going to be looking at identifying relative minimums and relative maximums. Recall as well that we will often use the word extrema to refer to both minimums and maximums.

The definition of relative extrema for functions of two variables is identical to that for functions of one variable we just need to remember now that we are working with functions of two variables. So, for the sake of completeness here is the definition of relative minimums and relative maximums for functions of two variables.

Definition:

1. A function $f(x, y)$ has a relative minimum at the point (a, b) if $f(x, y) \geq f(a, b)$ for all points (x, y) in some region around (a, b).
2. A function $f(x, y)$ has a relative maximum at the point (a, b) if $f(x, y) \leq f(a, b)$ for all points (x, y) in some region around (a, b).

Note that this definition does not say that a relative minimum is the smallest value that the function will ever take. It only says that in some region around the point (a, b) the function will always be larger than $f(a, b)$. Outside of that region it is completely possible for the function to be smaller. Likewise, a relative maximum only says that around (a,b) the function will always be smaller than $f(a,b)$. Again, outside of the region it is completely possible that the function will be larger.

Next we need to extend the idea of critical points up to functions of two variables. Recall that a critical point of the function $f(x)$ was a number $x = c$ so that either $f'(c) = 0$ or $f'(c)$ doesn't exist. We have a similar definition for critical points of functions of two variables.

Definition: The point (a,b) is a critical point (or a stationary point) of $f(a,b)$ provided one of the following is true,

1. $\nabla f(a,b) = \vec{0}$ (this is equivalent to saying that $f_x(a,b)$ and $f_y(a,b) = 0$),
2. $f_x(a,b)$ and/or $f_y(a,b)$ doesn't exist.

To see the equivalence in the first part let's start off with $\nabla f = \vec{0}$ and put in the definition of each part.

$$\nabla f(a,b) = \vec{0}$$

$$\langle f_x(a,b), f_y(a,b)\rangle = \langle 0,0\rangle$$

The only way that these two vectors can be equal is to have $f_x(a,b) = 0$ and $f_y(a,b) = 0$). In fact, we will use this definition of the critical point more than the gradient definition since it will be easier to find the critical points if we start with the partial derivative definition.

Note as well that BOTH of the first order partial derivatives must be zero at (a,b). If only one of the first order partial derivatives are zero at the point then the point will NOT be a critical point. We now have the following fact that, at least partially, relates critical points to relative extrema.

Fact

If the point (a,b) is a relative extrema of the function $f(x,y)$ then (a,b) is also a critical point of $f(x,y)$ and in fact we'll have $\nabla f(a,b) = \vec{0}$.

Proof: This is a really simple proof that relies on the single variable version, often called Fermat's Theorem.

Let's start off by defining $g(x) = f(x,b)$ and suppose that $f(x,b)$ has a relative extrema at (a,b). However, this also means that $g(x)$ also has a relative extrema (of the same kind as $f(x,y)$) at $x=a$. By Fermat's Theorem we then know that $g'(x)=0$. But we also know that $g'(a)=f_x(a,b)$ and so we have that $f_x(a,b)=0$.

If we now define $h(y)=f(a,y)$ and going through exactly the same process as above we will see that $f_y(a,y)=0$.

So, putting all this together means that $\nabla f(a,y)=\vec{0}$ and so $f(x,y)$ has a critical point at (a,y).

Note that this does NOT say that all critical points are relative extrema. It only says that relative extrema will be critical points of the function. To see this let's consider the function

$$f(x,y)=xy$$

The two first order partial derivatives are,

$$f_x(x,y)=y$$
$$f_y(x,y)=x$$

The only point that will make both of these derivatives zero at the same time is $(0,0)$ and so $(0,0)$ is a critical point for the function. Here is a graph of the function.

Note that the axes are not in the standard orientation here so that we can see more clearly what is happening at the origin, *i.e.* at $(0,0)$.

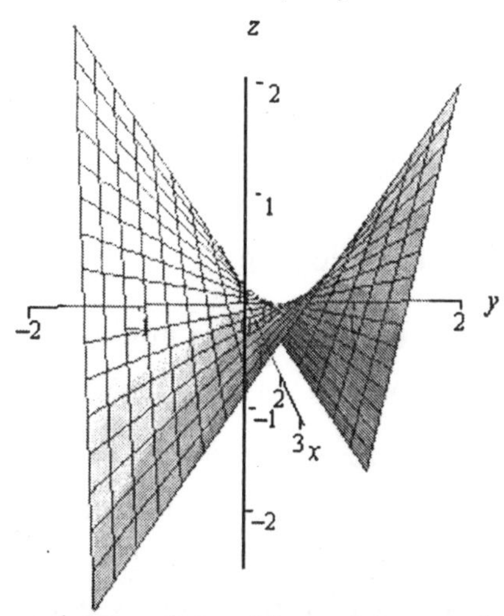

If we start at the origin and move into either of the quadrants where both x and y are the same sign the function increases. However, if we start at the origin and move into either of the quadrants where x and y have the opposite sign then the function decreases. In other words, no matter what region you take about the origin there will be points larger than $f(0,0)=0$ and points smaller than $f(0,0)=0$. Therefore, there is no way that $(0,0)$ can be a relative extrema.

Critical points that exhibit this kind of behavior are called saddle points.

While we have to be careful to not misinterpret the results of this fact it is very useful in helping us to identify relative extrema. Because of this fact we know that if we have all the critical points of a function then we also have every possible relative extrema for the function. The fact tells us that all relative extrema must be critical points so we know that if the function does have relative extrema then they must be in the collection of all the critical points. Remember however, that it will be completely possible that at least one of the critical points won't be a relative extrema. So, once we have all the critical points in hand all we will need to do is test these points to see if they are relative extrema or not. To determine if a critical point is a relative extrema (and in fact to determine if it is a minimum or a maximum) we can use the following fact.

Fact

Suppose that (a,b) is a critical point of $f(x,y)$ and that the second order partial derivatives are continuous in some region that contains (a,b). Next define,

$$D=D(a,b)=-f_{xx}(a,b)f_{yy}(a,b)-\left[f_{xx}(a,b)\right]^2$$

We then have the following classifications of the critical point.

1. If $D>0$ and $f_{xx}(a,b)>0$ then (a,b) is a relative minimum.
2. If $D>0$ and $f_{xx}(a,b)<0$ then (a,b) is a relative maximum.
3. If $D<0$ then (a,b) is a saddle point.
4. If $D=0$ then (a,b) may be a relative minimum, relative maximum or a saddle point. Other techniques would need to be used to classify the critical point.

Note that if $D>0$ then both $f_{xx}(a,b)$ and $f_{yy}(a,b)$ will have the same sign and so in the first two cases above we could just as easily replace $f_{xx}(a,b)$ with $f_{yy}(a,b)$. Also note that we aren't going to be seeing any cases in this class where $D=0$. We will be able to classify all the critical points that we find. Let's see a couple of examples.

Example: Find and classify all the critical points of $f(x,y)=4+x^3+y^3-3xy$.

Solution: We first need all the first order (to find the critical points) and second order (to classify the critical points) partial derivatives so let's get those.

$$f_x = 3x^2 - 3y$$
$$f_y = 3y^2 - 3x$$
$$f_{xx} = 6x$$
$$f_{yy} = 6y$$
$$f_{xy} = -3$$

Let's first find the critical points. Critical points will be solutions to the system of equations,

$$f_x = 3x^2 - 3y = 0$$
$$f_y = 3y^2 - 3x = 0$$

This is a non-linear system of equations and these can, on occasion, be difficult to solve. However, in this case it's not too bad. We can solve the first equation for y as follows,

$$3x^2 - 3y = 0 \quad \Rightarrow \quad y = x^2$$

Plugging this into the second equation gives,

$$3\left(x^2\right)^2 - 3x = 3x\left(x^3 - 1\right) = 0$$

From this we can see that we must have $x = 0$or.$x = 1$ Now use the fact that $y = x^2$to get the critical points.

$$x = 1: \; y = 0^2 = 0 \quad \Rightarrow \quad (0,0)$$
$$x = 1: \; y = 1^2 = 1 \quad \Rightarrow \quad (1,1)$$

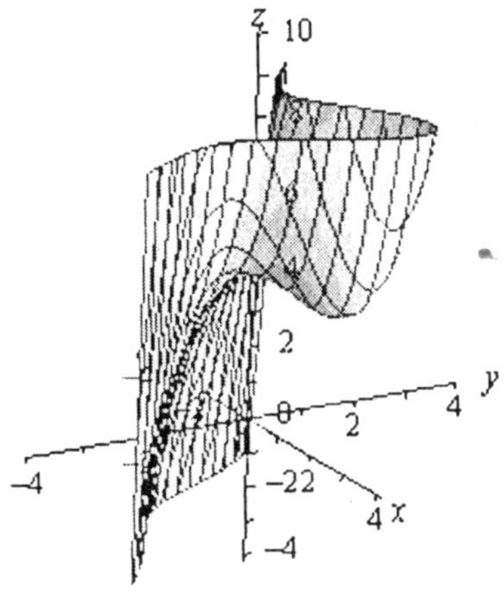

So, we get two critical points. All we need to do now is classify them. To do this we will need *D*. Here is the general formula for *D*.

$$D(x,y) = f_{xx}(x,y)f_{yy}(x,y) - \left[f_{xy}(x,y)\right]^2$$
$$= (6x)(6y) - (-3)^2$$
$$= 36xy - 9$$

To classify the critical points all that we need to do is plug in the critical points and use the fact above to classify them. $(0,0)$:

$$D = D(0,0) = -9 < 0$$

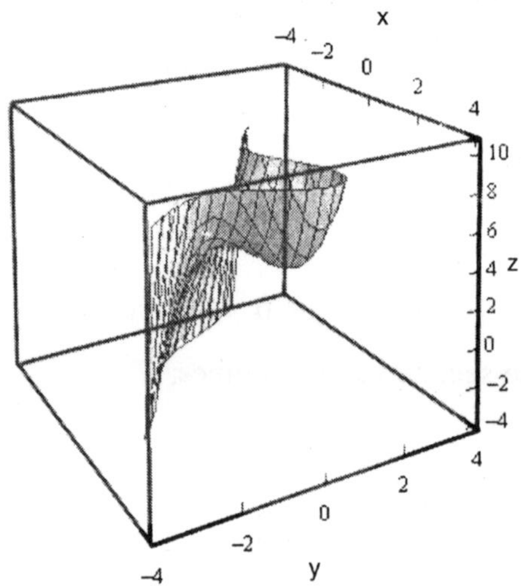

So, for $(0,0)$ *D* is negative and so this must be a saddle point. $(1,1)$:

$$D = D(1,1) = 36 - 9 = 27 > 0$$
$$f_{xx}(1,1) = 6 > 0$$

For $(1,1)$ *D* is positive and f_{xx} is positive and so we must have a relative minimum.

For the sake of completeness here is a graph of this function.

Notice that in order to get a better visual we used a somewhat nonstandard orientation. We can see that there is a relative minimum at $(1,1)$ and (hopefully) it's clear that at $(0,0)$ we do get a saddle point.

Example: Find and classify all the critical points for $f(x,y) = 3x^2y + y^3 - 3x^2 - 3y^2 + 2$

Solution: As with the first example we will first need to get all the first and second order derivatives.

$$f_x = 6xy - 6x$$
$$f_y = 3x^2 + 3y^2 - 6y$$
$$f_{xx} = 6y - 6$$
$$f_{yy} = 6y - 6$$
$$f_{xy} = 6x$$

We'll first need the critical points. The equations that we'll need to solve this time are,

$$6xy - 6x = 0$$
$$3x^2 + 3y^2 - 6y = 0$$

These equations are a little trickier to solve than the first set, but once you see what to do they really aren't terribly bad. First, let's notice that we can factor out a $6x$ from the first equation to get,

$$6x(y-1) = 0$$

So, we can see that the first equation will be zero if $x = 0$ or $y = 1$. Be careful to not just cancel the x from both sides. If we had done that we would have missed $x = 0$.

To find the critical points we can plug these (individually) into the second equation and solve for the remaining variable.

$$x = 0 : 3y^2 - 6y = 3y(y-2) = 0 \quad \Rightarrow \quad y = 0, y = 2$$
$$x = 1 : 3x^2 - 3 = 3\left(x^2 - 1\right) = 0 \quad \Rightarrow \quad x = -1, x = 1$$

So, if $x = 0$ we have the following critical points,

$$(0,0) \ (0,2)$$

and if $y = 1$ the critical points are,

$$(1,1) \ (-1,1)$$

Now all we need to do is classify the critical points. To do this we'll need the general formula for D.

$$D(x,y) = (6y-6)(6y-6) - (6x)^2 = (6y-6)^2 - 36x^2$$
$$(0,0) : D = D(0,0) = 36 > 0 \quad f_{xx}(0,0) = -6 < 0$$

$$(0,2): D = D(0,2) = 36 > 0 \quad f_{xx}(0,0) = -6 > 0$$

$$(1,1): D = D(1,1) = -36 < 0$$

$$(-1,1): D = D(-1,1) = -36 < 0$$

So, it looks like we have the following classification of each of these critical points.

$$(0,0): \text{ Relative Maximum}$$

$$(0,2): \text{ Relative Maximum}$$

$$(1,1): \text{ Saddle Point}$$

$$(-1,1): \text{ Saddle Point}$$

Here is a graph of the surface for the sake of completeness.

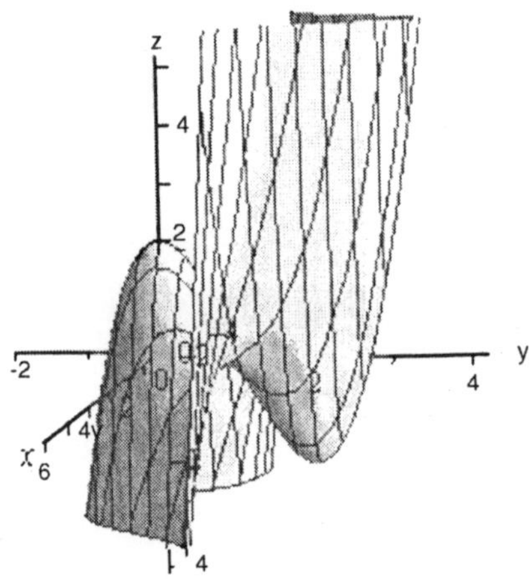

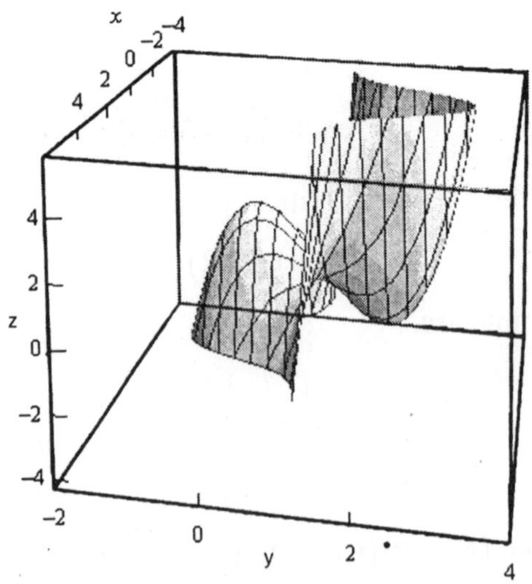

Let's do one more example that is a little different from the first two.

Example: Determine the point on the plane $4x-2y+z=1$ that is closest to the point $(-2,-1,5)$.

Solution: Note that we are NOT asking for the critical points of the plane. In order to do this example we are going to need to first come up with the equation that we are going to have to work with.

First, let's suppose that (x,y,z) is any point on the plane. The distance between this point and the point in question, $(-2,-1,5)$, is given by the formula,

$$d=\sqrt{(x+2)^2+(y+1)^2+(z-5)^2}$$

What are then asking is to find the minimum value of this equation. The point (x,y,z) that gives the minimum value of this equation will be the point on the plane that is closest to $(-2,-1,5)$.

There are a couple of issues with this equation. First, it is a function of x, y and z and we can only deal with functions of x and y at this point. This is easy to fix however. We can solve the equation of the plane to see that,

$$z=1-4x+2y$$

Plugging this into the distance formula gives,

$$\begin{aligned} d &= \sqrt{(x+2)^2+(y+1)^2+(1-4x+2y-5)^2} \\ &= \sqrt{(x+2)^2+(y+1)^2+(-4-4x+2y)^2} \end{aligned}$$

Now, the next issue is that there is a square root in this formula and we know that we're going to be differentiating this eventually. So, in order to make our life a little easier let's notice that finding the minimum value of d will be equivalent to finding the minimum value of d^2.

So, let's instead find the minimum value of

$$f(x,y)=d^2=(x+2)^2+(y+1)^2+(-4-4x+2y)^2$$

Now, we need to be a little careful here. We are being asked to find the closest point on the plane to $(-2,-1,5)$ and that is not really the same thing as what we've been doing in this section. In this section we've been finding and classifying critical points as relative minimums or maximums and what we are really asking is to find the smallest value the function will take, or the absolute minimum. Hopefully, it does make sense from a physical standpoint that there will be a closest point on the plane to $(-2,-1,5)$. Also, this point should be a relative minimum.

So, let's go through the process from the first and second example and see what we get as far as relative minimums go. If we only get a single relative minimum then we will be done since that point will also need to be the absolute minimum of the function and hence the point on the plane that is closest to $(-2,-1,5)$.

We'll need the derivatives first.

$$f_x = 2(x+2)+2(-4)+(-4-4x+2y) = 36+34x-16y$$

$$f_y = 2(y+1)+2(2)+(-4-4x+2y) = -14-16x+10y$$

$$f_{xx} = 34$$

$$f_{yy} = 10$$

$$f_{xy} - 16$$

Now, before we get into finding the critical point let's compute D quickly.

$$D = 34(10)-(-16)^2 = 84 > 0$$

So, in this case D will always be positive and also notice that is always positive and so any critical points that we get will be guaranteed to be relative minimums.

Now let's find the critical point(s). This will mean solving the system.

$$36+34x-16y = 0$$

$$-14-16x+10y = 0$$

To do this we can solve the first equation for x.

$$x = \frac{1}{34}(16y-36) = \frac{1}{17}(8y-18)$$

Now, plug this into the second equation and solve for y.

$$-14-\frac{16}{17}(8y-18)+10y = 0 \quad \Rightarrow \quad y = -\frac{25}{21}$$

Back substituting this into the equation for x gives $x=-\frac{34}{21}$.

So, it looks like we get a single critical point : $\left(-\frac{34}{21},-x=-\frac{25}{21}\right)$. Also, since we know this will be a relative minimum and it is the only critical point we know that this is also the x

and y coordinates of the point on the plane that we're after. We can find the z coordinate by plugging into the equation of the plane as follows,

$$z = 1 - 4\left(-\frac{34}{21}\right) + 2\left(-\frac{25}{21}\right) = \frac{107}{21}$$

So, the point on the plane that is closest to $(-2,-1,5)$ is $\left(-\frac{34}{21},-\frac{25}{21},\frac{107}{21}\right)$

ABSOLUTE MINIMUMS AND MAXIMUMS

In this section we are going to extend the work from the previous section. In the previous section we were asked to find and classify all critical points as relative minimums, relative maximums and/or saddle points. In this section we are want to optimize a function, that is identify the absolute minimum and/or the absolute maximum of the function, on a given region in $\mathbb{R}^2$.

Note that when we say we are going to be working on a region in $\mathbb{R}^2$ we mean that we're going to be looking at some region in the xy-plane. In order to optimize a function in a region we are going to need to get a couple of definitions out of the way and a fact. Let's first get the definitions out of the way.

DEFINITIONS

1. A region in $\mathbb{R}^2$ is called closed if it includes its boundary. A region is called open if it doesn't include any of its boundary points.
2. A region in $\mathbb{R}^2$ is called bounded if it can be completely contained in a disk. In other words, a region will be bounded if it is finite.

Let's think a little more about the definition of closed. We said a region is closed if it includes its boundary. Just what does this mean? Let's think of a rectangle. Below are two definitions of a rectangle, one is closed and the other is open.

Open	Closed
$-5 < x < 3$	$-5 \leq y \leq 3$
$1 < y < 6$	$1 \leq y \leq 6$

In this first case we don't allow the ranges to include the endpoints (*i.e.* we aren't including the edges of the rectangle) and so we aren't allowing the region to include any points on the edge of the rectangle. In other words, we aren't allowing the region to include its boundary and so it's open. In the second case we are allowing the region to contain points on the edges and so will contain its entire boundary and hence will be close. This is an important idea because of the following fact.

EXTREME VALUE THEOREM

If $f(x, y)$ is continuous in some closed, bounded set D in $\mathbb{R}^2$ then there are points in D, (x_1, y_1) and (x_2, y_2) so that is the absolute maximum and is the absolute minimum of the function in D.

Note that this theorem does NOT tell us where the absolute minimum or absolute maximum will occur. It only tells us that they will exist. Note as well that the absolute minimum and/or absolute maximum may occur in the interior of the region or it may occur on the boundary of the region. The basic process for finding absolute maximums is pretty much identical to the process when we looked at finding absolute extrema of functions of single variables. There will however, be some procedural changes to account for the fact that we now are dealing with functions of two variables. Here is the process.

FINDING ABSOLUTE EXTREMA

1. Find all the critical points of the function that lie in the region D and determine the function value at each of these points.
2. Find all extrema of the function on the boundary.
3. The largest and smallest values found in the first two steps are the absolute minimum and the absolute maximum of the function.

Let's take a look at an example or two.

Example: Find the absolute minimum and absolute maximum of $f(x,y) = x^2 + 4y^2 - 2x^2y + 4$ on the rectangle given by $-1 \le x \le 1$ and $-1 \le y \le 1$.

Solution: Let's first get a quick picture of the rectangle for reference purposes.

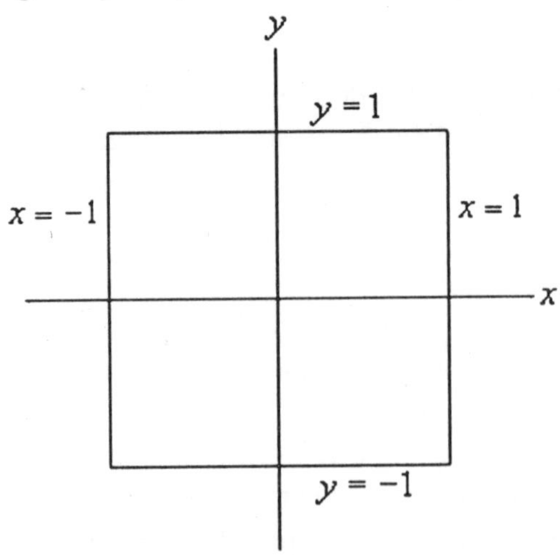

The boundary of this rectangle is given by the following conditions.

$$\text{Right side:} \quad x = 1, -1 \le y \le 1$$

$$\text{Left side:} \quad x = -1, -1 \le y \le 1$$

$$\text{Upper side:} \quad y = 1, -1 \le x \le 1$$

$$\text{Lower side:} \quad y = -1, -1 \le x \le 1$$

These will be important in the second step of our process.

We'll start this off by finding all the critical points that lie inside the given rectangle. To do this we'll need the two first order derivatives.

$$f_x = 2x - 4xy$$

$$f_y = 8y - 2x^2$$

Note that since we aren't going to be classifying the critical points we don't need the second order derivatives. To find the critical points we will need to solve the system,

$$2x - 4xy = 0$$

$$8y - 2x^2 = 0$$

We can solve the second equation for y to get,

$$y = \frac{x^2}{4}$$

Plugging this into the first equation gives us,

$$2x - 4x\left(\frac{x^2}{4}\right) = 2x - x^3 = x\left(2 - x^2\right) = 0$$

This tell us that must have $x = 0$ or $z = \pm\sqrt{2} = \pm 1.414...$. Now, recall that we only want critical points in the region that we're given. That means that we only want critical points for which $-1 \le x \le 1$.

The only value of x that will satisfy this is the first one so we can ignore the last two for this problem. Note however that a simple change to the boundary would include these two so don't forget to always check if the critical points are in the region (or on the boundary since that can also happen).

Plugging $x = 0$ into the equation for y gives us,

$$y = \frac{0^2}{4} = 0$$

The single critical point, in the region (and again, that's important), is $(0,0)$. We now need to get the value of the function at the critical point.

$$f(0,0) = 4$$

Eventually we will compare this to values of the function found in the next step and take the largest and smallest as the absolute extrema of the function in the rectangle.

Now we have reached the long part of this problem. We need to find the absolute extrema of the function along the boundary of the rectangle. What this means is that we're going to need to look at what the function is doing along each of the sides of the rectangle listed above. Let's first take a look at the right side. As noted above the right side is defined by

$$x = 1, -1 \le y \le 1$$

Notice that along the right side we know that $x = 1$. Let's take advantage of this by defining a new function as follows,

$$g(y) = f(1, y) = 1^2 + 4y^2 - 2(1^2)y + 4 = 5 + 4y^2 - 2y$$

Now, finding the absolute extrema of $f(x, y)$ along the right side will be equivalent to finding the absolute extrema of $g(y)$ in the range $-1 \le y \le 1$. We find the critical points of $g(y)$ in the range $-1 \le y \le 1$ and then evaluate $g(y)$ at the critical points and the end points of the range of y's.

Let's do that for this problem.

$$g'(y) = 8y - 2 \quad \Rightarrow \quad y = \frac{1}{4}$$

This is in the range and so we will need the following function evaluations.

$$g(-1) = 11$$
$$g(1) = 7$$
$$g\left(\frac{1}{4}\right) = \frac{19}{4} = 4.75$$

Notice that, using the definition of $g(-1)$ these are also function values for $f(x, y)$.

$$g(-1) = f(1, -1) = 11$$
$$g(1) = f(1, 1) = 7$$
$$g\left(\frac{1}{4}\right) = f\left(1, \frac{1}{4}\right) = \frac{19}{4} = 4.75$$

We can now do the left side of the rectangle which is defined by,

$$x = -1, -1 \le y \le 1$$

Again, we'll define a new function as follows,

$$g(y) = f(-1, y) = (-1)^2 + 4y^2 - 2(-1)^2 y + 4 = 5 + 4y^2 - 2y$$

Notice however that, for this boundary, this is the same function as we looked at for the right side. This will not always happen, but since it has let's take advantage of the fact that we've already done the work for this function. We know that the critical point is $y = \frac{1}{4}$ and we know that the function value at the critical point and the end points are,

$$g(-1) = 11$$
$$g(1) = 7$$
$$g\left(\frac{1}{4}\right) = \frac{19}{4} = 4.75$$

The only real difference here is that these will correspond to value of $f(x, y)$ at different points than for the right side. In this case these will correspond to the following function values for $f(x, y)$.

$$g(-1) = f(-1, -1) = 11$$
$$g(1) = f(-1, 1) = 7$$
$$g\left(\frac{1}{4}\right) = f\left(-1, \frac{1}{4}\right) = \frac{19}{4} = 4.75$$

We can now look at the upper side defined by, $y = 1, -1 \le x \le 1$. We'll again define a new function except this time it will be a function of x.

$$f(x) = f(x, 1) = x^2 + 4(1^2) - 2x^2(1) + 4 = 8 - x^2$$

We need to find the absolute extrema of $h(x)$ on the range $-1 \le x \le 1$. First find the critical point(s).

$$h'(x) = -2x \quad \Rightarrow \quad x = 0$$

The value of this function at the critical point and the end points is,

$$h(1) = 7 \qquad h(1) = 7 \qquad h(0) = 8$$

and these in turn correspond to the following function values for $f(x, y)$

$$h(-1) = f(-1, 1) = 7$$
$$h(1) = f(1, 1) = 7$$
$$h(0) = f(0, 1) = 8$$

Note that there are several "repeats" here. The first two function values have already been computed when we looked at the right and left side. This will often happen.

Finally, we need to take care of the lower side. This side is defined by,

$$y = -1, -1 \leq x \leq 1$$

The new function we'll define in this case is,

$$h(x) = f(x,-1) = x^2 + 4(-1)^2 - 2x^2(-1) + 4 = 8 + 3x^2$$

The critical point for this function is,

$$h'(x) = 6x \quad \Rightarrow \quad x = 0$$

The function values at the critical point and the endpoint are,

$$h(-1) = 11 \quad h(1) = 11 \quad h(0) = 8$$

and the corresponding values for $f(x, y)$ are,

$$h(-1) = f(-1,-1) = 11$$
$$h(1) = f(1,-1) = 11$$
$$h(0) = f(0,1) = 8$$

The final step to this (long...) process is to collect up all the function values for $f(x, y)$ that we've computed in this problem. Here they are,

$$f(0,0) = 4 \qquad f(1,-1) = 11 \qquad f(1,1) = 7$$

$$f\left(1,\frac{1}{4}\right) = 4.75 \qquad f(-1,1) = 7 \qquad f(-1,-1) = 11$$

$$f\left(-1,\frac{1}{4}\right) = 4.75 \qquad f(0,1) = 8 \qquad f(0,-1) = 8$$

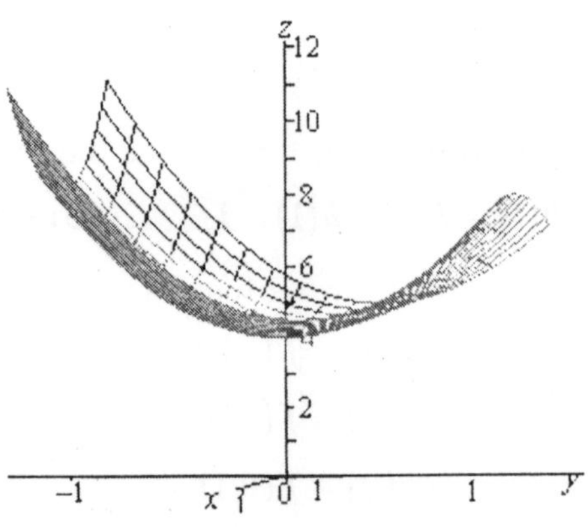

The absolute minimum is at $(0,0)$ since gives the smallest function value and the absolute maximum occurs at $(1,-1)$ and $(1,-1)$ since these two points give the largest function value.

Here is a sketch of the function on the rectangle for reference purposes.

As this example has shown these can be very long problems. Let's take a look at an easier problem with a different kind of boundary.

Example: Find the absolute minimum and absolute maximum of $f(x,y) = 2x^2 - y^2 + 6y$ on the disk of radius 4, $x^2 + y^2 \le 16$.

Solution: First note that a disk of radius 4 is given by the inequality in the problem statement.

The "less than" inequality is included to get the interior of the disk and the equal sign is included to get the boundary. Of course, this also means that the boundary of the disk is a circle of radius 4.

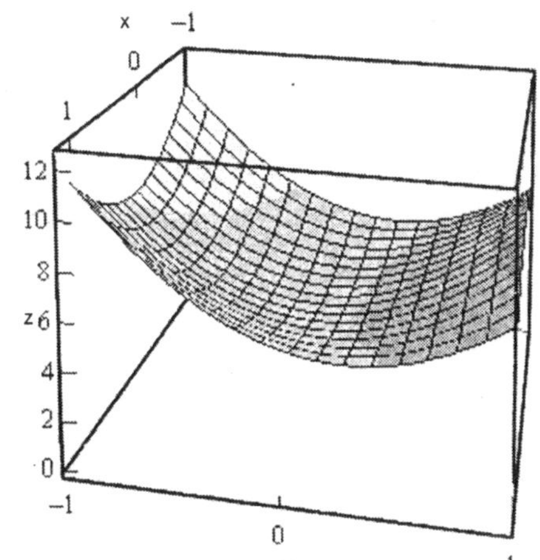

Let's first find the critical points of the function that lie inside the disk. This will require the following two first order partial derivatives.

$$f_x = 4x$$
$$f_y = -2y + 6$$

To find the critical points we'll need to solve the following system.

$$4x = 0$$
$$-2y + 6 = 0$$

This is actually a fairly simple system to solve however. The first equation tells us that $x = 0$ and the second tells us that $y = 3$. So the only critical point for this function is $(0,3)$ and this is inside the disk of radius 4. The function value at this critical point is,

$$f(0,3) = 9$$

Now we need to look at the boundary. This one will be somewhat different from the previous example. In this case we don't have fixed values of x and y on the boundary. Instead we have,

$$x^2 + y^2 = 16$$

We can solve this for x^2 and plug this into the x^2 in $f(x,y)$ to get a function of y as follows.

$$x^2 = 16 - y^2$$

$$\begin{aligned} g(y) &= 2(16 - y^2) - y^2 + 6y \\ &= 32 - 3y^2 + 6y \end{aligned}$$

We will need to find the absolute extrema of this function on the range $-4 \le y \le 4$ (this is the range of y's for the disk....). We'll first need the critical points of this function.

$$g'(y) = -6y + 6 \quad \Rightarrow \quad y = 1$$

The value of this function at the critical point and the endpoints are,

$$g(-4) = -40 \quad g(4) = 8 \quad g(1) = 35$$

Unlike the first example we will still need to find the values of x that correspond to these. We can do this by plugging the value of y into our equation for the circle and solving for y.

$$\begin{aligned} y = -4: &\quad x^2 = 16 - 16 = 0 \quad \Rightarrow \quad x = 0 \\ y = 4: &\quad x^2 = 16 - 16 = 0 \quad \Rightarrow \quad x = 0 \\ y = 1: &\quad x^2 = 16 - 1 = 15 \quad \Rightarrow \quad x = \pm\sqrt{15} = \pm 3.87 \end{aligned}$$

The function values for $g(y)$ then correspond to the following function values for $f(x,y)$.

$$\begin{aligned} g(-4) = -40 &\quad \Rightarrow \quad f(0,-4) = -40 \\ g(4) = 8 &\quad \Rightarrow \quad f(0,4) = 8 \\ g(1) = 35 &\quad \Rightarrow \quad f(-\sqrt{15},1) = 35 \text{ and } f(\sqrt{15},1) = 35 \end{aligned}$$

Note that the third one actually corresponded to two different values for $f(x,y)$ since that y also produced two different values of x.

So, comparing these values to the value of the function at the critical point of $f(x,y)$ that we found earlier we can see that the absolute minimum occurs at $(0,-4)$ while the absolute maximum occurs twice at $\left(-\sqrt{15},1\right)$ and $\left(-\sqrt{15},1\right)$.

Here is a sketch of the region for reference purposes.

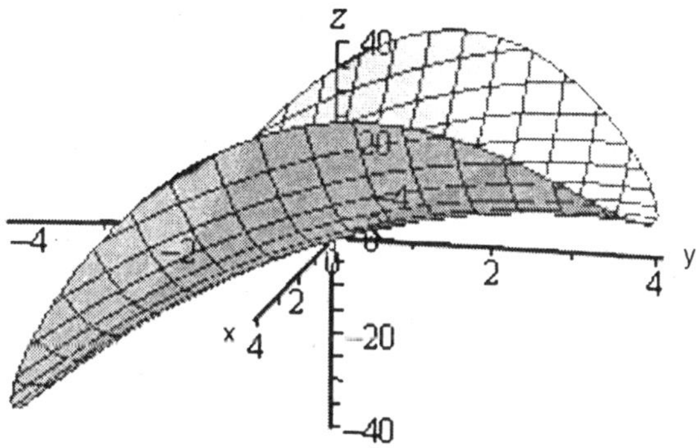

In both of these examples one of the absolute extrema actually occurred at more than one place. Sometimes this will happen and sometimes it won't so don't read too much into the fact that it happened in both examples given here.

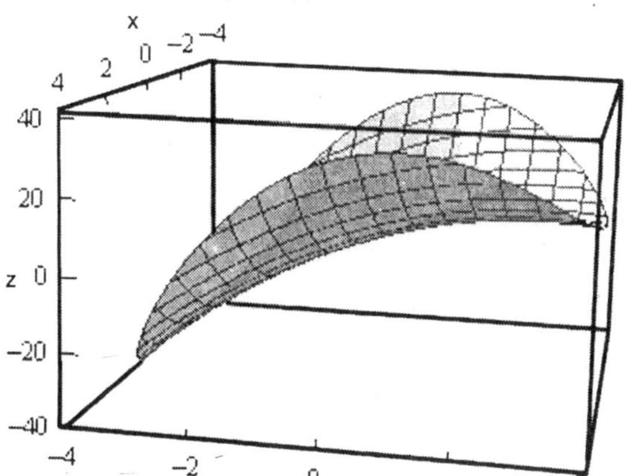

Also note that, as we've seen, absolute extrema will often occur on the boundaries of these regions, although they don't have to occur at the boundaries.

Had we given much more complicated examples with multiple critical points we would have seen examples where the absolute extrema occurred interior to the region and not on the boundary.

LAGRANGE MULTIPLIERS

In the previous section we optimized (*i.e.* found the absolute extrema) a function on a region that contained its boundary. Finding potential optimal points in the interior of the region isn't too bad in general, all that we needed to do was find the critical points and plug them into the function. However, as we saw in the examples finding potential optimal points on the boundary was often a fairly long and messy process.

In this section we are going to take a look at another way of optimizing a function subject to given constraint(s). The constraint(s) may be the equation(s) that describe the boundary of a region although in this section we won't concentrate on those types of problems since this method just requires a general constraint and doesn't really care where the constraint came from. So, let's get things set up. We want to optimize (find the minimum and maximum) of a function, $f(x,y,z)$, subject to the constraint $g(x,y,z)=k$.

Again, the constraint may be the equation that describes the boundary of a region or it may not be. The process is actually fairly simple, although the work can still be a little overwhelming at times.

METHOD OF LAGRANGE MULTIPLIERS

Solve the following system of equations.

$$\nabla f(x,y,z)=\lambda\nabla g(x,y,z)$$
$$g(x,y,z)=c$$

Plug in all solutions, (x,y,z), from the first step into $f(x,y,z)$ and identify the minimum and maximum values, provided they exist.

The constant, λ, is called the Lagrange Multiplier.

Notice that the system of equations actually has four equations, we just wrote the system in a simpler form. To see this let's take the first equation and put in the definition of the gradient vector to see what we get.

$$\left\langle f_x \cdot f_y, f_z\right\rangle=\lambda\left\langle g_x, g_y, g_z\right\rangle=\left\langle \lambda g_x, \lambda g_y, \lambda g_z\right\rangle$$

In order for these two vectors to be equal the individual components must also be equal. So, we actually have three equations here.

$$f_x=\lambda g_x$$
$$f_y=\lambda g_y$$
$$f_z=\lambda g_z$$

These three equations along with the constraint, $g(x,y,z)=c$, give four equations with four unknowns x, y, z, and λ.

Note as well that if we only have functions of two variables then we won't have the third component of the gradient and so will only have three equations in three unknowns x, y, and λ. As a final note we also need to be careful with the fact that in some cases minimums and maximums won't exist even though the method will seem to imply that they do. In every problem we'll need to go back and make sure that our answers make sense.

Let's work a couple of examples.

Example: Find the dimensions of the box with largest volume if the total surface area is 64 cm^2.

Solution: Before we start the process here note that we also saw a way to solve this kind of problem, except in those problems we required a condition that related one of the sides of the box to the other sides so that we could get down to a volume and surface area function that only involved two variables. We no longer need this condition for these problems.

Now, let's get on to solving the problem. We first need to identify the function that we're going to optimize as well as the constraint.

Let's set the length of the box to be x, the width of the box to be y and the height of the box to be z.

Let's also note that because we're dealing with the dimensions of a box it is safe to assume that x, y, and z are all positive quantities. We want to find the largest volume and so the function that we want to optimize is given by,

$$f(x, y, z) = xyz$$

Next we know that the surface area of the box must be a constant 64. So this is the constraint. The surface area of a box is simply the sum of the areas of each of the sides so the constraint is given by,

$$2xy + 2xz + 2yz = 64 \quad \Rightarrow \quad xy + xz + yz = 32$$

Note that we divided the constraint by 2 to simplify the equation a little. Also, we get the function $g(x, y, z)$ from this.

$$g(x, y, z) = xy + xz + yz$$

Here are the four equations that we need to solve.

$$yz = \lambda(y + z) \quad (f_x = \lambda g_x)$$

$$xz = \lambda(x + z) \quad (f_y = \lambda g_y)$$

$$xy = \lambda(x + y) \quad (f_z = \lambda g_z)$$

$$xy + xz + yz = 32 \quad (g(x, y, z) = 32)$$

There are many ways to solve this system. We'll solve it in the following way. This gives,

$$xyz = \lambda x(y+z)$$
$$xyz = \lambda y(x+z)$$
$$xyz = \lambda z(x+y)$$

Doing this gives,

$$\lambda x(y+z) = \lambda y(x+z)$$
$$\lambda(xy+xz) = \lambda(yx+yz) = 0$$
$$\lambda(xz-yz) = 0 \quad \Rightarrow \quad \lambda = 0 \text{ or } xz = yz$$

This gave two possibilities. The first, $\lambda = 0$ is not possible since if this was the case equation would reduce to $yz = 0 \quad \Rightarrow \quad y = 0 \text{ or } z = 0$

Since we are talking about the dimensions of a box neither of these are possible so we can discount $\lambda = 0$. This leaves the second possibility.

$$xz = yz$$

Since we know that $z \neq 0$ (again since we are talking about the dimensions of a box) we can cancel the z from both sides. This gives,

$$x = y$$

Doing this gives,

$$\lambda y(x+z) = \lambda z(x+y)$$
$$\lambda(yx+yz-zx-zy) = 0$$
$$\lambda(yx-zx) = 0 \quad \Rightarrow \quad \lambda = 0 \text{ or } yx = zx$$

As already discussed we know that $\lambda = 0$ won't work and so this leaves,

$$yx = zx$$

We can also say that since we are dealing with the dimensions of a box so we must have,

Plugging equations we get,

$$y^2 + y^2 + y^2 = 3y^2 = 32$$
$$y = \pm\sqrt{\frac{32}{3}} = \pm 3.266$$

However, we know that y must be positive since we are talking about the dimensions of a box. Therefore the only solution that makes physical sense here is

$$x = y = z = \pm 3.266$$

So, it looks like we've got a cube here.

We should be a little careful here. Since we've only got one solution we might be tempted to assume that this is dimensions that will give the largest volume.

The method of Lagrange Multipliers will give a set of points that will either maximize or minimize a given function subject to the constraint, provided there actually are minimums or maximums. The function itself, $f(x, y, z) = xyz$ will clearly have neither minimums or maximums unless we put some restrictions on the variables.

The only real restriction that we've got is that all the variables must be positive. This, of course, instantly means that the function does have a minimum, zero. The function will not have a maximum if all the variables are allowed to increase without bound. That however, can't happen because of the constraint,

$$xy + xz + yz = 32$$

Here we've got the sum of three positive numbers (because x, y, and z are positive) and the sum must equal 32. So, if one of the variables gets very large, say x, then because each of the products must be less than 32 both y and z must be very small to make sure the first two terms are less than 32.

So, there is no way for all the variables to increase without bound and so it should make some sense that the function, $f(x, y, z) = xyz$, will have a maximum.

This isn't a rigorous proof that the function will have a maximum, but it should help to visualize that in fact it should have a maximum and so we can say that we will get a maximum volume if the dimensions are : $x = y = z = 3.266$. Notice that we never actually found values for λin the above example. This is fairly standard for these kinds of problems. The value of λisn't really important to determining if the point is a maximum or a minimum so often we will not bother with finding a value for it. On occasion we will need its value to help solve the system, but even in those cases we won't use it past finding the point.

Example: Find the maximum and minimum of $f(x, y) = 5x - 3y$ subject to the constraint $x^2 + y^2 = 136$.

Solution: This one is going to be a little easier than the previous one since it only has two variables. Also, note that it's clear from the constraint that region of possible solutions lies on a disk of radius $\sqrt{136}$ with is a closed and bounded region and hence by the Extreme Value Theorem we know that a minimum and maximum value must exist.

Here is the system that we need to solve.

$$5 = 2\lambda x$$
$$-3 = 2\lambda y$$
$$x^2 + y^2 = 136$$

Notice that, as with the last example, we can't have $\lambda = 0$ since that would not satisfy the first two equations. So, since we know that $\lambda \neq 0$ we can solve the first two equations for x and y respectively. This gives,

$$x = \frac{5}{2\lambda}$$
$$y = -\frac{3}{2\lambda}$$

Plugging these into the constraint gives,

$$\frac{25}{4\lambda^2} + \frac{9}{4\lambda^2} = \frac{17}{4\lambda^2} = 136$$

We can solve this forλ.

$$\lambda^2 = \frac{1}{16} \quad \Rightarrow \quad \lambda = \pm\frac{1}{4}$$

Now, that we know lwe can find the points that will be potential maximums and/or minimums.

If $\lambda = -\frac{1}{4}$ we get,

$$x = -10 \quad y = 6$$

and if $\lambda = \frac{1}{4}$ we get,

$$x = 10$$
$$y = -6$$

To determine if we have maximums or minimums we just need to plug these into the function. Also recall from the discussion at the start of this solution that we know these will be the minimum and maximums because the Extreme Value Theorem tells us that minimums and maximums will exist for this problem.

Here are the minimum and maximum values of the function.

$$f(-10,6) = -68 \quad \text{Minimum at } (-10,6)$$
$$f(-10,-6) = 68 \quad \text{Maximum at } (10,-6)$$

In the first two examples we've excluded $\lambda = 0$ either for physical reasons or because it wouldn't solve one or more of the equations. Do not always expect this to happen. Sometimes we will be able to automatically exclude a value of λ and sometimes we won't.

Let's take a look at another example.

Example: Find the maximum and minimum values of $f(x, y, z) = xyz$ subject to the constraint $x + y + z = 1$. Assume that $x, y, z \geq 0$.

Solution: First note that our constraint is a sum of three positive or zero number and it must be 1. Therefore it is clear that our solution will fall in the range $0 \leq x, y, z \leq 1$.

Therefore the solution must lie in a closed and bounded region and so by the Extreme Value Theorem we know that a minimum and maximum value must exist.

Here is the system of equation that we need to solve.

$$yz = \lambda$$
$$xz = \lambda$$
$$xy = \lambda$$
$$x + y + z = 1$$

Let's start this solution process off by noticing that since the first three equations all have λ they are all equal. So, let's start off by setting equations.

$$yz = xz \quad \Rightarrow \quad z(y - x) = 0 \quad \Rightarrow \quad z = 0 \text{ or } y = x$$

So, we've got two possibilities here. Let's start off with by assuming that $z = 0$. In this case we can see from either equation that we must then have $\lambda = 0$. From equation we see that this means that $xy = 0$. This in turn means that either $x = 0$ or $y = 0$. So, we've got two possible cases to deal with there. In each case two of the variables must be zero. Once we know this we can plug into the constraint, equation, to find the remaining value.

$$z = 0, x = 0: \quad \Rightarrow \quad y = 1$$
$$z = 0, y = 0: \quad \Rightarrow \quad x = 1$$

So, we've got two possible solutions $(0,1,0)$ and $(1,0,0)$.

Now let's go back and take a look at the other possibility, $y = x$. We also have two possible cases to look at here as well. This first case is $x = y = 0$. In this case we can see from the constraint that we must have $z = 1$ and so we now have a third solution $(0,0,1)$.

The second case is $x = y \neq 0$. Let's set equations.

$$xz = xy \quad \Rightarrow \quad x(z - y) = 0 \quad \Rightarrow \quad x = 0 \text{ or } z = y$$

Now, we've already assumed that $x \neq 0$ and so the only possibility is that $z = y$. However, this also means that,

$$x = y = z$$

Using this in the constraint gives,

$$3x = 1 \quad \Rightarrow \quad x = \frac{1}{3}$$

So, the next solution is $\left(\frac{1}{3}, \frac{1}{3}, \frac{1}{3}\right)$.

We got four solutions by setting the first two equations equal.

To completely finish this problem out we should probably set equations equal as well as setting equations equal to see what we get. Doing this gives,

$$yz = xy \quad \Rightarrow \quad y(z-x) = 0 \quad \Rightarrow \quad y = 0 \text{ or } z = x$$

$$xz = xy \quad \Rightarrow \quad x(z-y) = 0 \quad \Rightarrow \quad x = 0 \text{ or } z = y$$

Both of these are very similar to the first situation that we looked at and we'll leave it up to you to show that in each of these cases we arrive back at the four solutions that we already found. So, we have four solutions that we need to check in the function to see whether we have minimums or maximums.

$$f(0,0,1) = 0 \quad f(0,1,0) = 0 \quad f(1,0,0) = 0 \qquad \text{All Minimums}$$

$$f\left(\frac{1}{3}, \frac{1}{3}, \frac{1}{3}\right) = \frac{1}{27} \qquad \text{Maximum}$$

So, in this case the maximum occurs only once while the minimum occurs three times.

Note as well that we never really used the assumption that $x, y, z \geq 0$ in this problem. This assumption is here mostly to make sure that we really do have a maximum and a minimum of the function.

Without this assumption it wouldn't be too difficult to find points that give both larger and smaller values of the functions. For example.

$$x = -100,\ y = 100,\ z = 1:\ -100 + 100$$
$$= 1\ f(-100, 100, 1) = -10000\ .$$
$$x = -50,\ y = -50,\ z = 101:\ -50 - 50 + 101$$
$$= 1\ f(-50, -50, 101) = 252500$$

With these examples you can clearly see that it's not too hard to find points that will give larger and smaller function values. However, all of these examples required negative values of x, y and/or z to make sure we satisfy the constraint.

By eliminating these we will know that we've got minimum and maximum values by the Extreme Value Theorem. To this point we've only looked at constraints that were equations. We can also have constraints that are inequalities.

The process for these types of problems is nearly identical to what we've been doing in this section to this point.

The main difference between the two types of problems is that we will also need to find all the critical points that satisfy the inequality in the constraint and check these in the function when we check the values we found using Lagrange Multipliers. Let's work an example to see how these kinds of problems work.

Example: Find the maximum and minimum values of $f(x,y)=4x^2+10y^2$ on the disk $x^2+y^2\le 4$.

Solution: Note that the constraint here is the inequality for the disk. Because this is a closed and bounded region the Extreme Value Theorem tells us that a minimum and maximum value must exist.

The first step is to find all the critical points that are in the disk (*i.e.* satisfy the constraint). This is easy enough to do for this problem. Here are the two first order partial derivatives.

$$f_x = 8x \quad \Rightarrow \quad 8x = 0 \quad \Rightarrow x = 0$$
$$f_y = 20y \quad \Rightarrow \quad 20y = 0 \quad \Rightarrow y = 0$$

So, the only critical point is $(0,0)$ and it does satisfy the inequality.

At this point we proceed with Lagrange Multipliers and we treat the constraint as an equality instead of the inequality. We only need to do deal with the inequality when finding the critical points.

So, here is the system of equations that we need to solve.

$$8x = 2\lambda x$$
$$20y = 2\lambda y$$
$$x^2 + y^2 = 4$$

From the first equation we get,

$$2x(4-\lambda) = 0 \quad \Rightarrow \quad x = 0 \text{ or } \lambda = 4$$

If we have $x=0$ then the constraint gives us $y=\pm 2$.

If we have $\lambda = 4$ the second equation gives us,

$$20y = 8y \quad \Rightarrow \quad y = 0$$

The constraint then tells us that $x = \pm 2$.

If we'd performed a simiiar analysis on the second equation we would arrive at the same points.

So, Lagrange Multipliers gives us four points to check: (0, 2), (0, –2), (2, 0) and (–2, 0).

To find the maximum and minimum we need to simply plug these four points along with the critical point in the function.

$$f(0,0) = 0 \qquad \text{Minimum}$$
$$f(2,0) = f(-2,0) = 16$$
$$f(0,2) = f(0,-2) = 40 \qquad \text{Maximum}$$

In this case, the minimum was interior to the disk and the maximum was on the boundary of the disk.

The final topic that we need to discuss in this section is what to do if we have more than one constraint. We will look only at two constraints, but we can naturally extend the work here to more than two constraints.

We want to optimize $f(x,y,z)$ subject to the constraints $g(x,y,z) = c$ and. $h(x,y,z) = k$ The system that we need to solve in this case is,

$$\nabla f(x,y,z) = \lambda \nabla g(x,y,z) + \mu \nabla h(x,y,z)$$
$$g(x,y,z) = c$$
$$h(x,y,z) = k$$

So, in this case we get two Lagrange Multipliers. Also, note that the first equation really is three equations as we saw in the previous examples. Let's see an example of this kind of yoptimization problem.

Example: Find the maximum and minimum of $f(x,y,z) = 4y - 2z$ subject to the constraints $2x - y - z = 2$ and $x^2 + y^2 = 1$.

Solution: Verifying that we will have a minimum and maximum value here is a little trickier. Clearly, because of the second constraint we've got to have $0 \le x, y \le 1$.

With this in mind there must also be a set of limits on z in order to make sure that the first constraint is met. If one really wanted to determine that range you could find the minimum and maximum values of $2x - y$ subject to $x^2 + y^2 = 1$ and you could then use this to determine the minimum and maximum values of z. We won't do that here.

The point is only to acknowledge that once again the possible solutions must lie in a closed and bounded region and so minimum and maximum values must exist by the Extreme Value Theorem.

Here is the system of equations that we need to solve.

$$0 = 2\lambda + 2\mu x \quad (f_x = \lambda g_x + \mu h_x)$$

$$4 = -\lambda + 2\mu y \quad (f_y = \lambda g_y + \mu h_y)$$

$$-2 = -\lambda \quad (f_z = \lambda g_z + \mu h_z)$$

$$2x - y - z = 2$$

$$x^2 + y^2 = 1$$

First, let's notice that from equation we get $\lambda = 2$. Plugging this into equation and solving for *x* and *y* respectively gives,

$$0 = 4 + 2\mu x \quad \Rightarrow \quad x = -\frac{2}{\mu}$$

$$4 = -2 + 2\mu y \quad \Rightarrow \quad y = \frac{3}{\mu}$$

Now,

$$\frac{4}{\mu^2} + \frac{9}{\mu^2} = \frac{13}{\mu^2} = 1 \quad \Rightarrow \quad \mu = \pm\sqrt{13}$$

So, we have two cases to look at here. First, let's see what we get when. In this case we know that,

$$x = -\frac{2}{\sqrt{13}} \quad y = \frac{3}{\sqrt{13}}$$

Plugging these into equation gives,

$$-\frac{4}{\sqrt{13}} - \frac{3}{\sqrt{13}} - z = 2 \quad \Rightarrow \quad z = -2 - \frac{7}{\sqrt{13}}$$

So, we've got one solution.

Let's now see what we get if we take $\mu = -\sqrt{13}$. Here we have,

$$x = \frac{2}{\sqrt{13}} \quad y = -\frac{3}{\sqrt{13}}$$

Plugging these into equation gives,

$$\frac{4}{\sqrt{13}}+\frac{3}{\sqrt{13}}-z=2 \quad\Rightarrow\quad z=-2+\frac{7}{\sqrt{13}}$$

and there's a second solution.

Now all that we need to is check the two solutions in the function to see which is the maximum and which is the minimum.

$$f\left(-\frac{2}{\sqrt{13}},\frac{3}{\sqrt{13}},-2-\frac{7}{\sqrt{13}}\right)=4+\frac{26}{\sqrt{13}}=11.2111$$

$$f\left(\frac{2}{\sqrt{13}},-\frac{3}{\sqrt{13}},-2+\frac{7}{\sqrt{13}}\right)=4-\frac{26}{\sqrt{13}}=-3.2111$$

So, we have a maximum at $f\left(\frac{2}{\sqrt{13}},-\frac{3}{\sqrt{13}},-2+\frac{7}{\sqrt{13}}\right)$ and a minimum at $\left(\frac{2}{\sqrt{13}},-\frac{3}{\sqrt{13}},-2+\frac{7}{\sqrt{13}}\right)$.

Chapter 9

Fourier Series

There is another class of infinite series which has a particularly important role both in pure and applied mathematics. These are the Fourier series, in which the individual terms are trigonometric functions and the sum is a periodic function.

PERIODIC FUNCTIONS

Periodic functions of the time, i.e., functions which repeat their course after a definite interval of time, are encountered in many.applications. In most machines, a periodic process occurs in rhythm with the rotation of a flywheel, e.g., the alternating current developed by a dynamo.

Periodic functions are also associated with all vibration phenomena. A periodic function with period $2l$ is represented by the equation $f(x + 2l) = f(x)$, true for all values of x. We call especially attention to the fact that $2l$ is called the period.

In representing periodic functions, it is often convenient to denote the independent variable x by a point on the circumference of a circle instead of the usual point on a straight line. If a function $f(x)$ has the period $2p$, i.e., if the equation $f(x + 2\pi) = f(x)$ holds for all values of x and we denote by x the angle at the centre of a circle of unit radius, which is included between an arbitrary initial radius and the radius to a variable point on the circumference, then the periodicity of the function $f(x)$ is expressed simply by the fact that there corresponds to each point on the circumference just one value of the function. For example, in the case of a machine, periodicity may be expressed in terms of the position of a point on the flywheel.

It is worth noting that in addition to the period $2l$, the function $f\{x)$ necessarily has also the period $4l$, since $f(x+ 4l) = f(x+ 2l) = f(x)$; similarly, $f(x)$ has the periods $6l$, $8l$,...; and it is also possible (though not necessarily true) that $f(x)$ may have shorter periods such as l or $l/5$. Graphically speaking, the graph of the function has in any two consecutive intervals of length $2l$ exactly the same shape.

In order to have available a second interpretation, which some readers may prefer, we may think of the variable x as the time and write, accordingly, sometimes t instead of x; the function $f(x)$ then represents a periodic process or, as we shall also say, a vibration (or oscillation).

The period $2l = T$ is then called the period of vibration (or period of oscillation).

If any arbitrary function $f(x)$ is given in a definite interval, say. -l ύ□ x ύ□ l, it can always be extended as a periodic function; we have only to define $f(x)$ outside the interval by the equation $f(x + 2nl) = f(x)$, where n is an arbitrary positive or negative integer.

We must point out here that, if $f(x)$ is continuous in the interval -l ⁒ x ⁒ l, but $f(-l)$ ⁒ $f(l)$, our extended periodic function will be discontinuous at the points l, $3l$..., where $l = p$. Moreover, in this case, the extension fails to yield a single-valued function $f(x)$ at the points $x = l$, $3l$, since, for example, we have defined $f(3l)$ as $f(l+2l)$, which gives $f(3l) = f(l)$, and we also have defined it as/(-l+ $4l$), which yields $f(3l) = f(-l)$.

We avoid this difficulty by not extending the function as defined for -l ύ□ x ύ□ l, but as defined either for $-l < x < l$ or $-l < x < l$, i.e., we either discard the original value $f(-l)$ or the original value $f(+l)$.

We should point out here a general fact,, which relates to periodic functions and is expressed by the equation

$$\int_{-l-a}^{l-a} f(x)dx = \int_{-l}^{l} f(x)dx$$

or, in words: The integral of a periodic function over an interval, the length of which is one period $T == 2l$, has always the same value, no matter where the interval lies.

In order to prove this, we need only note that, by virtue of the equation $f(\xi - 2l) = f(\xi)$, the substitution $x = \xi - 2l$ yields

$$\int_{\varsigma}^{\beta} f(x)\,dx = \int_{\alpha+2l}^{\beta+2l} f(\xi)\,d\xi = \int_{\alpha+2l}^{\beta+2l} f(x)\,dx$$

In particular, for $\alpha = -l - \alpha$ and $\beta = -l$, it follows that

$$\int_{-l-a}^{-l} f(x)\,dx = \int_{-l-a}^{-l} f(x)\,dx,$$

whence

$$\int_{-l-a}^{-l} f(x)\,dx = \int_{-l-a}^{-l} f(x)\,dx + \int_{-l}^{l-a} f(x)\,dx$$

$$= \int_{-l-a}^{-l} f(x)\,dx + \int_{-l}^{l-a} f(x)\,dx = \int_{-l}^{l} f(x)dx,$$

which proves our statement. If we recall the geometrical meaning of the integral, the statement is made obvious by figure.

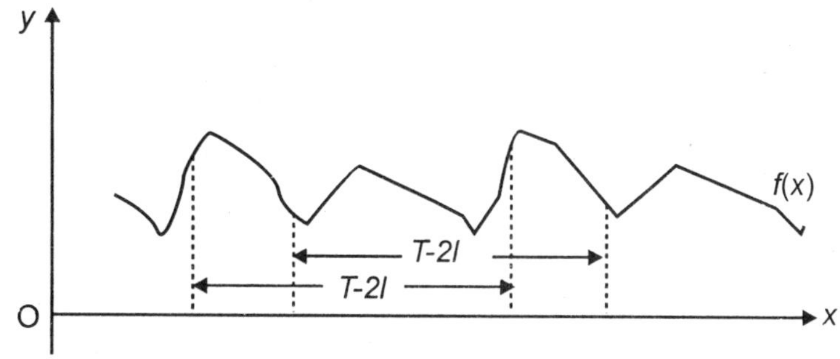

Fig. To Illustrate the Integral over a whole Period

The simplest periodic functions, from which we shall later compose the most general periodic functions, are $a \sin\omega x$ and $a \cos\omega x$, or, more generally, $a\sin\omega(x - \xi)$ and $a\cos w(x - \xi)$, where $a(0)$, $\omega(0)$ and x are constants. We call the processes represented by these functions* sinusoidal vibrations or simple harmonic vibrations (or oscillations). The period of vibration is $T=2\pi/\omega$. The number ω is called the circular frequency of the vibration; since $1/T$ is the number of vibrations in unit time or frequency, ω is the number of vibrations in time 2π. The number a is called the amplitude of the vibration; it represents the maximum value of the function $a\sin\omega(x - \xi)$ or $a\cos w(x - \xi)$, since both sine and cosine have the maximum value 1. The number $\omega(x - \xi)$ is called the phase and the number wx the phase displacement.

We obtain these functions graphically by stretching the sine curve in the ratios $1: \omega$ along the x-axis and a: 1 along the y-axis and then translating the curve by a distance ξ along the x-axis in the positive direction.

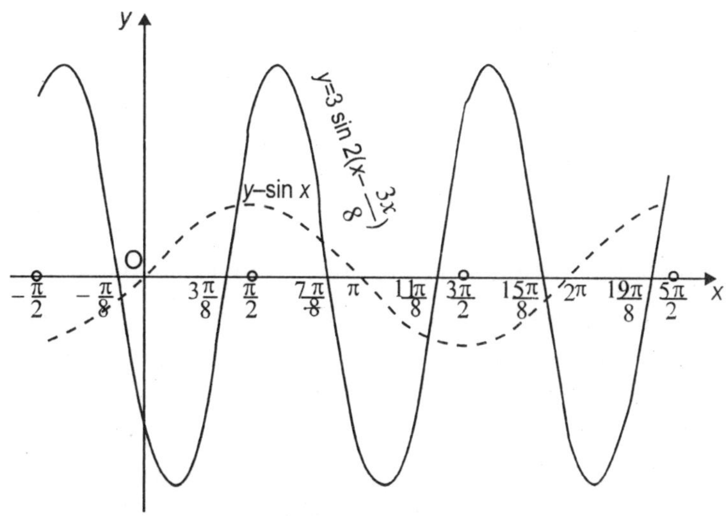

Fig. Sinusoidal Vibrations

We can also repress sinusoidal vibrations by the addition formula for the trigonometric functions in the form $\alpha \cos\omega x + \beta \sin\omega x$ *and* $\beta \cos\omega x - a \sin\omega x$ respectively, where $\alpha = -$

$a\sin\omega\xi$ and $\beta = a\cos\omega\xi$. Conversely, every function of the form $\alpha\cos\omega x + \beta\sin\omega x$ represents a sinusoidal vibration $a\sin\omega(x - \xi)$ with the amplitude $a = \sqrt{(a^2 + \beta^2)}$ and the phase displacement $\omega\xi$ given by the equations $\alpha = -a\sin\omega\xi$ and $\beta = a\cos\omega\xi$. By using the expression $a\sin wx + b\cos wx$, we see that the sum of two or more such functions with the same circular frequency w always represent another sinusoidal vibration with the circular frequency w.

SUPERPOSITION OF SINUSOIDAL VIBRATIONS. HARMONICS. BEATS

Although many vibrations are found to be sinusoida, it is nevertheless true that most periodic motions have a more complicated character, being obtained by superposition of sinusoidal vibrations. Mathematically speaking, this simply means that the motion, e.g., the distance of a point from its initial position as a function of the time, is given by a function which is the sum of a number of pure periodic functions of the above type. The sine waves of the function are then piled up on top of one another (that is, their ordinates are added), or, as we say, they are superposed. In this superposition, we assume that all the circular frequencies (and, of course, the periods) of the superposed vibrations differ, because the superposition of two sinusoidal vibrations with the same circular frequency gives us another sinusoidal vibration with the same circular frequency (but with a different amplitude and phase displacement), as shown above.

If we consider the simplest case, the superposition of two sinusoidal vibrations with the circular frequencies ω_1 and ω_2, we find that there are two fundamentally different cases, depending on whether the two circular frequencies have a rational ratio or not, or, as we say, whether they are commensurable or incommensurable.

We begin with the first case and, by way of an example, take the second circular frequency to be twice the first: $\omega_2 = 2\omega_1$. The period of the second vibration will then be half the period of the first vibration, $2\pi/2\omega_1 = T_2 = T_1/2$, whence it will necessarily have not only the period T_2, but also the doubled period T_1, since the function repeats itself after this double period; and the function after their superposition will also have the period T_1. The second vibration with twice the circular frequency and half the period of the first vibration is called the first harmonic of the first vibration (the fundamental).

Corresponding statements hold, if we introduce a further vibration with the circular frequency $\omega_3 = 3\omega_1$. Here again, the vibration function $\sin 3\omega_1 x$ will necessarily repeat itself with the period $2\pi/\omega_1 = T_1$. Such a vibration is called a second harmonic of the given vibration.

Similarly, we can consider third, fourth, (n — l)-th harmonics with the circular frequencies $\omega_4 = 4\omega_1$, $\omega_5 = 5\omega_1$, $\omega_n = nw_1$, and, moreover, with any phase displacements of your choice. Every such harmonic will necessarily repeat itself after the period $T_1 = 2\pi/\omega_1$ and, consequently, every function obtained by superposition of a number of vibrations, each of which is a harmonic of a given fundamental circular frequency ω_1, will itself be a periodic function with the period $2\pi/\omega_1=T_1$. By superposing vibrations with circular frequencies ranging from that of the fundamental to that of the $(n-1)$-th harmonic, we obtain a periodic function of the form

$$S(x)= a\sum_{n=1}^{n}(a_n \cos nwx+b_n \sin nwx)$$

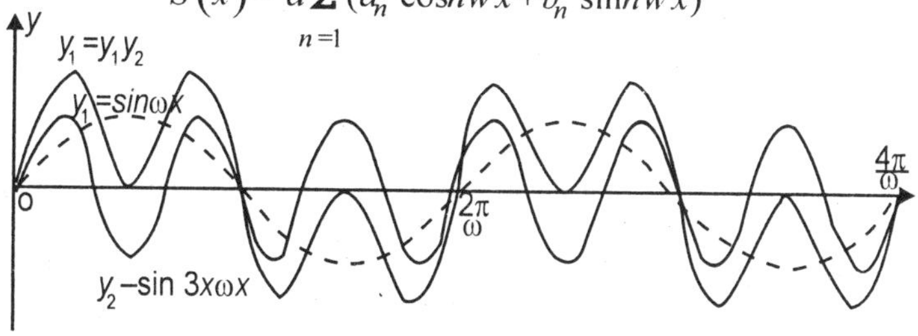

Fig. Combination of vibrations

The proportions of figure correspond to the assumption $w=1$.

Since this function contains $2n+1$ constants, which we can choose arbitrarily, we are thus able to generate very complicated curves which are not at all like the original curves. The term harmonic originated in acoustics, where we find that there corresponds to a fundamental vibration with circular frequency a note of a certain pitch, then the first, second, third, etc., harmonics correspond to the sequence of harmonics of the fundamental, i.e., to the octave, octave + fifth, double octave, etc. (In acoustics, also the terms overtone, (upper) partial are used.

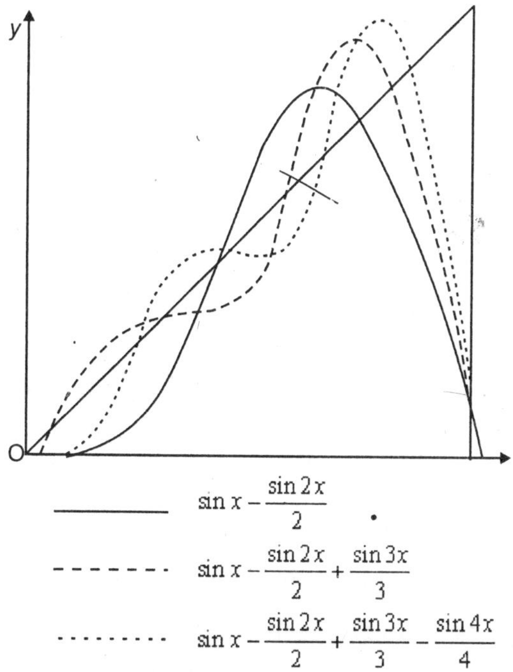

Fig. Combination of vibrations

In general, in the case of superposition of vibrations in which the circular frequencies have rational ratios, these circular frequencies can all be represented as integral multiples of a common fundamental circular frequency. However, the superposition of two vibrations with incommensurable circular frequencies ω_1 and ω_2 represents an intrinsically different type of phenomenon. In that case, the process resulting from the superposition of sinusoidal vibrations will no longer be periodic. We cannot go here into the mathematical discussions that arise from this, but will merely remark that such functions always have an approximately periodic character or, as we say, are almost periodic.

A final remark on the superposition of sinusoidal vibrations concerns the phenomenon of beats. If we superpose two vibrations, both with unit amplitude but different circular frequencies ω_1 and ω_2, and, for the sake of simplicity, take the same value of ξ for both (the generalization to arbitrary phase can be left to the reader), then we are merely concerned with the behaviour of the function

$$y = \sin\omega_1 x + \sin\omega_2 x \ (\omega_1 > \omega_1 > 0)$$

A well-known trigonometrical formula yields

$$y = 2\cos\frac{1}{2}(\omega_1 - \omega_2)x \sin\frac{1}{2}(\omega_1 + \omega_2)x$$

This equation represents a phenomenon which we may think of as follows: We have a vibration with the circular frequency $(\omega_1 + \omega_2)$ and the period $4\pi/(\omega_1+\omega_2)$. However, this vibration does not have a constant amplitude; on the contrary, its amplitude is given by $2\cos(\omega_1 + \omega_2)x$, which varies with the longer period $4\pi/(\omega_1-\omega_2)$). This point of view is particularly useful and easy to interpret when the two circular frequencies $\omega_1$and ω_2 are relatively large, while their difference $\omega_1-\omega_2$ is small in comparison with them. Then the amplitude

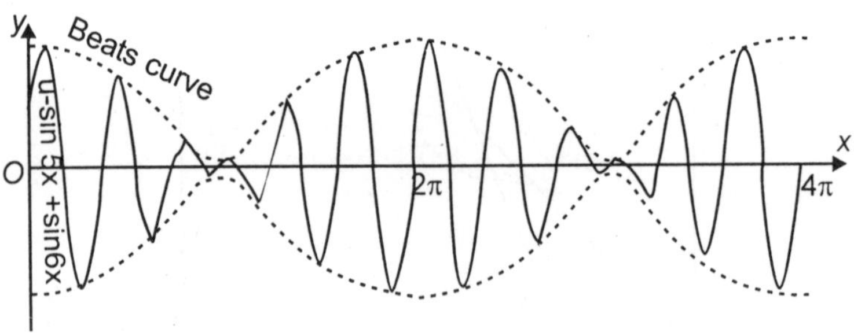

Fig. Beats

of the vibration $2\cos\frac{1}{2}(\omega_1 - \omega_2)x$ with period $4\pi/(\omega_1 - \omega_2)$ will vary only slowly compared with the period of vibration and this change of amplitude will repeat itself periodically with the long period $4\pi/(\omega_1 - \omega_2)$. These rhythmic changes of amplitude are called beats. Everyone is acquainted with this phenomenon in acoustics, and perhaps also in wireless telegraphy, in which the circular frequencies $\omega_1$and ω_2 are, as a rule, far above those which the ear can detect, while the difference $\omega_1 - \omega_2$ falls in the range of audible notes. The beats can then be heard, while the original vibrations remain imperceptible to the ear.

Use of complex notation

General Remarks: An investigation of vibration phenomena and periodic functions gains in formal simplicity if we employ complex numbers, combining each pair of trigonometric functions $\cos w\, x$ and $\sin w\, x$ to form an expression of the type $\cos w\, x + i \sin \omega\, x = e^{i\omega x}$. We must here bear in mind that one equation between complex quantities is equivalent to two equations between real quantities and that our results must always be interpreted and made intelligible in the real domain.

If we replace everywhere the trigonometric functions by exponential functions in accordance with the formulae

$$2\cos\theta = e^{i\theta} + e^{i\theta},$$

$$2i\sin\theta = e^{i\theta} - e^{i\theta}$$

we express sinusoidal vibrations in terms of the complex quantities $e^{i\omega x}$, $e^{-i\omega x}$ or $ae^{i\omega(x-\xi)}$, $ae^{-i\omega(x-\xi)}$, respectively, where a, ω and $\omega\xi$ are the real quantities: Amplitude, circular frequency and phase displacement. The real vibrations are obtained from this complex expression simply by taking real and imaginary parts.

The convenience of this mode of representation for many purposes is due to the fact that the derivatives of the real vibrations with respect to the time are obtained by differentiating the complex exponential function just as if it were a real constant, as is expressed by the formula

$$\frac{d}{dx} a\{\cos\omega(x-\xi) + i\sin\omega(x-\xi)\}$$

$$= a\omega\{-\sin\omega(x-\xi) + i\cos\omega(x-\xi)\}$$

$$= ia\omega\{\cos\omega(x-\xi) + i\sin\omega(x-\xi)\},$$

or

$$\frac{d}{dx} ae^{i\omega(x-\xi)} = ia\omega e^{i\omega(x-\xi)}$$

Application to the Study of Alternating Currents: We shall now illustrate these matters by means of an important example and denote here the independent variable, the time, by t instead of x.

We consider an electric circuit with resistance R and inductance L, on which an external electromotive force (voltage) E is impressed. In the case of a direct current, E is constant and the current I is given by Ohm's law $E = RI$.

However, if we are dealing with an alternating current, E is a function of the time t, and consequently so is I; Ohm's law then becomes

$$E - L\frac{dI}{dt} = RI$$

In the simplest case, to which we will restrict ourselves here, the external electromotive force E is sinusoidal with circular frequency ω. Now, instead of taking this oscillation in the form *accost t* or *a*sin*w t*, we combine both possibilities formally in the complex form

$$E = \varepsilon e^{i\omega t} = \varepsilon \cos \omega t + i\varepsilon \sin \omega t,$$

where ε (>0) represents the amplitude. We shall operate with this complex voltage as if it were a real parameter and thus obtain a complex current I. Then, the significance of the relation thus found between the complex qnantities E and I is that the current corresponding to an electromotive force $\varepsilon\cos\omega t$ is the real part of I, while the current corresponding to an electromotive force $\varepsilon\sin\omega t$ is the imaginary part of I. The complex current can be calculated immediately, if we write down for I an expression of the form $I = \alpha e^{i\omega t} = \alpha (\cos\omega t + i\sin\omega t)$, i.e., if we assume that I is also sinusoidal with circular frequency ω. The derivative of I is then given formally by the expression

$$\frac{dI}{dt} = i\alpha\omega e^{i\omega t} = \alpha\omega\left(-\sin \omega t + i\cos \omega t\right)$$

On substitution of these quantities into the generalized form of Ohm's law and dividing by the factor $e^{i\omega t}$, we obtain the equation $\varepsilon - \alpha Lic\omega = R\alpha$, or

$$\alpha = \frac{\varepsilon}{R + i\omega L},$$

so that $E = (R + i\omega L)I = WI$.

We may regard this last equation as Ohm's law for alternating currents in complex form, if we call the quantity $W = R + i\omega L$ the complex resistance of the circuit. Ohm's law is then the same as for direct current: The current is equal to the voltage divided by the resistance.

If we write the complex resistance in the form

$$W = \omega e^{i\delta} = \omega \cos\delta + i\omega \sin\delta,$$

wheref

$$\omega = \sqrt{\left(R^2 + L^2\omega^2\right)}, \quad \tan\delta = \frac{\omega L}{R}$$

we obtain $$I = \frac{\varepsilon}{\omega}e^{i(\omega t - \delta)},$$

According to this formula, the current has the same period (and circular frequency) as the voltage; the amplitude a of the current is connected with the amplitude e of the electromotive force by the equation

$$a = \frac{\varepsilon}{\omega},$$

and, in addition, there is a phase difference between the current and the voltage. The current does not reach its maximum at the same time as the voltage, but at a time δ/ω later and the same is of course true for the minimum.

In electrical engineering, the quantity $\omega = \sqrt{(R^2 + L^2\omega^2)}$ is frequently called the impedance or the alternating current resistance of the circuit for the circular frequency ω; the phase displacement, usually stated in degrees, is called the lag.

Complex Representation of the Superposition of Sinusoidal Vibrations

So far, the complex notation has been used to denote the combination of two sinusoidal vibrations. But a single vibration or a compound vibration of the type

$$S(x) = a + \sum_{\nu=1}^{n} (a_\nu \cos\nu x + b_\nu \sin\nu x)$$

(for the sake of simplicity, we have taken $\omega = 1$) can also be reduced to complex form by the substitution

$$\cos\nu x = \frac{1}{2}\left(e^{i\nu x} + e^{-i\nu x}\right) \quad \sin\nu x = \frac{1}{2}\left(e^{i\nu x} - e^{-i\nu x}\right)$$

The above expression then assumes the form

$$S(x) = \sum_{\nu=-\eta}^{n} a_\nu e^{i\nu x},$$

where the complex numbers a_ν are linked to the real numbers a, a_ν and b_ν by the equations

$$a_\nu = a_\nu + a_{-\nu}, \; a = a_0, \; b_\nu = i(a_\nu - a_{-\nu})$$

In order that the equation $a_\nu = a_\nu + a_{-\nu}$ shall formally include the case $n = 0$, we often put $a = a_0 = a_0/2$.

Conversely, we may regard any arbitrary expression of the form $\sum_{\nu=-n}^{n} a_\nu e^{i\nu n}$ as a function which represents the superposition of vibrations written in complex form. In order that the result of this superposition may be real, it is only necessary that $a_\nu + a_{-\nu}$ should be real and $a_\nu - a_{-\nu}$ pure imaginary, i.e., that a_ν and $a_{-\nu}$ are conjugate complex numbers.

Deduction of a trigonometric Formula

By using complex notation, we obtain a very simple proof of a formula which we shall require below. This is the trigonometric summation formula

$$\alpha_n(\alpha) = \frac{1}{2} + \cos\alpha + \cos 2\alpha + ... + \cos 2n\alpha = \frac{\sin\left(n+\frac{1}{2}\right)\alpha}{2\sin\frac{1}{2}\alpha}$$

which holds for all values of *a* except the values 0, 2*p*, 4*p*,.... In order to prove this, we replace the cosine function by its exponential expression and thus write the sum $\sigma_n(\alpha)$ in the form

$$\sigma_n(\alpha) = \frac{1}{2}\sum_{v=-n}^{n} e^{iva}$$

On the right hand side, we have a geometric progression with the common ratio $q = e^{i\alpha} \} 1$. Using the ordinary formula for the sum, we have

$$\sigma_n(\alpha) = \frac{1}{2}e_{-}^{\ in\alpha}\frac{1-q^{2n+1}}{1-q} = \frac{1}{2}\frac{e^{-ina} - e^{(n+1)i\alpha}}{1-e^{i\alpha}}$$

On multiplying the numerator and denominator by $e^{-ia/2}$, we obtain

$$\sigma_n(\alpha) = \frac{\sin\left(n+\frac{1}{2}\right)\alpha}{2\sin\frac{1}{2}\alpha}$$

as has been stated above.

Fourier Series

The function

$$S(x) = a + \sum_{v=1}^{n}(a_v \cos vx + b_v \sin vx),$$

resulting from the superposition of sinusoidal vibrations, contains $2n + 1$ arbitrary constants a, a_v, b_v. There arises now the question whether these constants can be chosen such that in the interval $-\pi < x < +p$ the sum $S(x)$ shall approximate to a given function $f(x)$, and if so, how they are to be found. More precisely, we ask whether the given function $f(x\}$ can be expanded in an infinite series

$$f(x) = a + \sum_{v=1}^{\infty}(a_v \cos vx + b_v \sin vx),$$

If we assume for the moment that this expansion of the function $f(x)$ is actually possible and that the series converges uniformly in the interval $-\pi < x < +\pi$, we readily obtain a simple relation between the function $f(x)$ and the coefficients $a = a_0$, a_v and b_v. We multiply the above hypothetical expansion by cos*nx* and integrate term by term, which is permissible on account of its uniform convergence. By virtue of the orthogonality relations

$$\int_{-\pi}^{+\pi} \sin mx \sin nx\, dx = \begin{cases} 0, & if\ m \neq n, \\ \pi, if\ m \neq n \neq 0, \end{cases}$$

$$\int_{-\pi}^{+\pi} \sin mx \ \cos nx dx = 0,$$

$$\int_{-\pi}^{+\pi} \cos mx \ \cos nx dx = \begin{cases} 0, & \textit{if } m \neq n, \\ \pi, \textit{if } m \neq n \neq 0, \end{cases}$$

proved at the end of 4.3, we obtain at once for the coefficients the formula

$$a_\nu = \frac{1}{\pi}\int_{-\pi}^{+\pi} f(x)\cos \nu x \, dx$$

Similarly, multiplying the series by sin νx and integrating, we find

$$b_\nu = \frac{1}{\pi}\int_{-\pi}^{+\pi} f(x)\sin \nu x dx$$

These formulae assign a definite sequence of coefficients a_ν and b_ν, usually called Fourier coefficients, to every function $f(x)$ which is defined and continuous in the interval $-\pi < x < +\pi$, or has only a finite number of jump discontinuities there. If the function $f(x)$ is given, we can employ these quantities a_ν, b_ν to form the Fourier partial sum

$$S_n(x) = \frac{1}{2}a_0 + \sum_{\nu=1}^{n}(a_\nu \cos \nu x + b_\nu \sin \nu x),$$

and we may also write down formally the corresponding infinite Fourier series. Our problem is now to distinguish simple classes of functions $f(x)$ for which these Fourier series actually converge and represent them.

In order to formulate the result which we wish to prove, we introduce the definition:

A function $f(x)$ is said to be sectionally continuous in an interval, if it is itself sectionally continuous (i.e., is continuous in the interval except for a finite number of jump discontinuities) and if, in addition, its first derivative $f(x)$ is sectionally continuous.

We shall imagine that the function $f(x)$, originally defined in the interval $-\pi \rangle x \rangle \pi$, is periodically extended. At each point at which the function $f(x)$ has a jump discontinuity, we shall, if necessary, alter the function and assign to it the value which is the arithmetic mean of the left hand and right-hand limits of $f(x)$, i.e., we shall write

$$f(x) = \frac{1}{2}(f(x-0) + f(x+0)),$$

where $f(x-0)$ and $f(x+0)$ are simply the limits of $f(x)$ as x approaches from the left and from the right hand side, respectively. This equation is obviously true for every point x at which $f(x)$ is continuous.

Our goal is now the theorem: If the function $f(x)$ if sectionally smooth and satisfies the above equation, then its Fourier series converges at any point x and represents the function. Note that this theorem can be proved for more general classes of functions. However, the result formulated here is sufficient for all applications.

Moreover, we shall prove the theorem: In every closed interval in which the function $f(x)$ (imagined to be periodically extended) is continuous as well as sectionally smooth, the Fourier series converges uniformly.

Finally: If the function $f(x)$ is sectionally smooth and has no discontinuities, the Fourier series converges absolutely.

The proofs of these theorems will be postponed. We merely wish to emphasize here that the functions, which can be expanded according to these theorems, have a very high degree of arbitrariness; it is by no means necessary that the function should be given by a single analytical expression. In the next section, we shall display the extraordinary fertility of Fourier expansions by discussing a number of examples.

Examples of Fourier Series

We shall assume that our functions $f(x)$ have the period 2π and are defined in the interval $-\pi <\square\, x < +\pi$,. Beyond this interval, to the left and right hand sides, they are to be extended periodically.

Before going into details, we note that if $f(x)$ is an even function, then clearly $f(a)$ sin nx is odd and $f(x)$ cos νx is even, so that

$$b_\nu = \frac{1}{\pi}\int_{-\pi}^{+\pi} f(x)\sin\nu x dx = 0;$$

$$a_\nu = \frac{2}{\pi}\int_{-\pi}^{\pi} f(x)\cos\nu x dx .$$

We thus obtain a cosine series. On the other hand, if $f(x)$ is an odd function, then

$$a_\nu = \frac{1}{\pi}\int_{-\pi}^{+\pi} f(x)\cos\nu x dx = 0;$$

$$b_\nu = \frac{2}{\pi}\int_{0}^{\pi} f(x)\sin\nu x dx,$$

and we obtain a sine series.

Consequently, if the function $f(x)$ is initially given only in the interval $0 < x < \pi$, we can extend it in the interval $-\pi < x < 0$ either as an odd function or as an even function, and, correspondingly, expand it in the interval $0 < x < \pi$ in a sine series or in a cosine series.

Expansion of the Functions $\psi(x) = x$ and $\psi(x) = x$: For the odd function x, we have

$$b_\nu = \frac{2}{\pi}\int_{0}^{\pi} f(x)\sin\nu x dx,$$

and, on integration by parts,

$$\frac{\pi}{2}b_\nu = \frac{-x\cos\nu x}{\nu}\bigg|_0^\pi + \frac{1}{\nu}\int_0^\pi \cos\nu x dx = (-1)^{\nu+1}\frac{\pi}{\nu}$$

Hence, we obtain for the periodic function $\psi(x)$, which is equal to x in the interval $-\pi < x < \pi$

$$\psi(x) = 2\left(\frac{\sin x}{1} - \frac{\sin 2x}{2} + \frac{\sin 3x}{3} - + \ldots\right)$$

If we set $x = \pi/2$, we obtain Gregory's series

$$\frac{\pi}{4} = 1 - \frac{1}{3} + \frac{1}{5} - + \ldots,$$

with which we are already familiar. The function $y(x)$, represented by this series, is not a continuous function; on the contrary, it jumps by 2π at the points $x = k\pi$, $k = 1, 3. 5, \ldots$. At these points of discontinuity, that is, at the points $x = k\pi$, $k = 1, 3, 5, \ldots$, each term of the series is zero, whence the function itself is zero. Hence, at the points of discontinuity, the series represents the arithmetic mean of the left hand and right hand limits.

If x is any fixed number between $-\pi$ and π and we replace x in the above series by $(x - \xi)$, we obtain the series

$$\psi(x - \xi) = 2\left(\frac{\sin(x-\xi)}{1} - \frac{\sin 2(x-\xi)}{2} + \frac{\sin 3(x-\xi)}{3} - + \ldots\right)$$

$$= -\frac{2}{1}\sin\xi\cos x + \frac{2}{1}\cos\xi\sin x + \frac{2}{2}\sin 2\xi\cos 2x$$

$$-\frac{2}{2}\cos 2\xi\sin 2x - \frac{2}{3}\sin 3\xi\cos 3x + \frac{2}{3}\cos 3\xi\sin 3x + \ldots$$

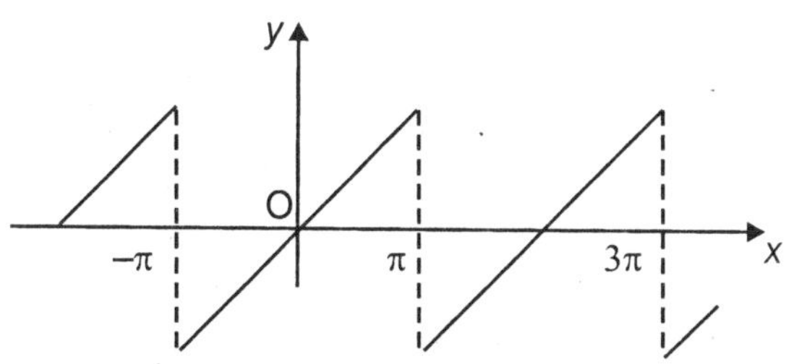

This series may also be written in the form of a Fourier series with the coefficients

$$a_0 = 0, a_n = 2\frac{(-1)^n}{n}\sin n\xi,\ b_n = 2\frac{(-1)^{n-1}}{n}\cos n\xi$$

which tend to zero as n increases; this series represents a function with the discontinuities, described above, at the points $x = \xi\ \pi,, \xi\ 3\pi, \ldots$.

For the even function $\varphi(x) = x$, on integrating twice by parts, we find

$$a_\nu = \frac{2}{\pi}x^2\cos\nu x dx = (-1)^\nu\frac{4}{\nu^2}\ (\nu > 0),\ a_0 = \frac{2\pi^2}{3},$$

Differentiating this series term by term and dividing by 2, we recover formally the series for $\psi(x) = x$.

Expansion of the Function $x\cos x$:. For this odd function, we find

$$a_\nu = 0,\ b_\nu = \frac{2}{\pi}\int_0^\pi x\cos x \sin \nu x dx$$

Using the formula

$$\int_0^\pi x\sin \mu x\, dx = (-1)^{\mu-1}\frac{\pi}{\mu} \quad (\mu = 1, 2, 3,...)$$

above, we find

$$b_\nu = \frac{2}{\pi}\int_0^\pi x\cos x\sin \nu x\, dx = \frac{1}{\pi}\int_0^\pi x\big(\sin(\nu+1)x + \sin(\nu-1)x\big)dx$$

$$= \frac{(-1)^{\nu+2}}{\nu+1} + \frac{(-1)^\nu}{\nu-1} = (-1)^\nu \frac{2\nu}{\nu^2-1} \quad (\nu = 2, 3,...),$$

$$b_1 = -\frac{1}{2}$$

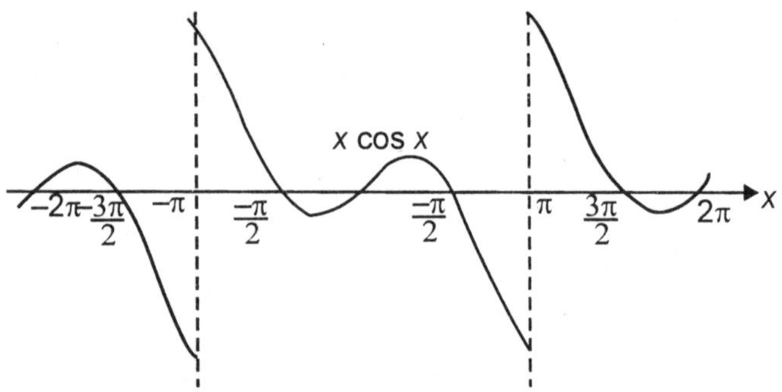

Hence we obtain the series

$$x\cos x = -\frac{1}{2}\sin x + 2\sum_{\nu=2}^{\infty}\frac{(-1)^\nu \nu}{\nu^2-1}\sin \nu x,$$

and, if we add the series in 9.4.2, the series

$$x\cos x = -\frac{1}{2}\sin x + 2\sum_{\nu=2}^{\infty}\frac{(-1)^\nu \nu}{\nu^2-1}\sin \nu x,$$

When the function, which is equal to $x\cos x$ in the interval $-\pi < x < \pi$, is extended periodically beyond this interval, the same discontinuities appear as are exhibited by the function $\psi(x)$. On the other head, if the function $x(1 + \cos x)$ is periodically extended, it remains continuous at the end-points of the interval and, in fact, its derivative also remains continuous, since the discontinuities are eliminated by the factor $1+\cos x$ which vanishes together with its derivative at the end-points.

The Function $f(x) = |x|$: This function is even, whence $b_\nu = 0$ and

$$a_\nu = \frac{2}{\pi}\int_0^\pi x\cos\nu x\,dx,$$

and, on integration by parts, we obtain

$$\int_0^\pi x\cos\nu x\,dx = \frac{1}{\nu}x\sin\nu x\Big|_0^\pi - \frac{1}{\nu}\int_0^\pi \sin\nu x\,dx$$

$$= \begin{cases} 0, \text{if } \nu \text{ is even and } \neq 0, \\ -\dfrac{2}{\nu^2}, \text{if } \nu \text{ is odd}, \end{cases}$$

whence

$$f(x) = \frac{\pi}{2} - \frac{4}{\pi}\left(\cos x + \frac{\cos 3x}{3^2} + \frac{\cos 5x}{5^2} + \ldots\right)$$

Setting $x = 0$, we obtain the remarkable formula

$$\frac{\pi^2}{8} = 1 + \frac{1}{3^2} + \frac{1}{5^2} + \ldots$$

The Step Function: The function defined by the equations

$$f(x) = \begin{cases} -1, \text{for } -\pi < x < 0, \\ 0, \text{for } x = 0, \\ +1, \text{for } 0 < x < \pi, \end{cases}$$

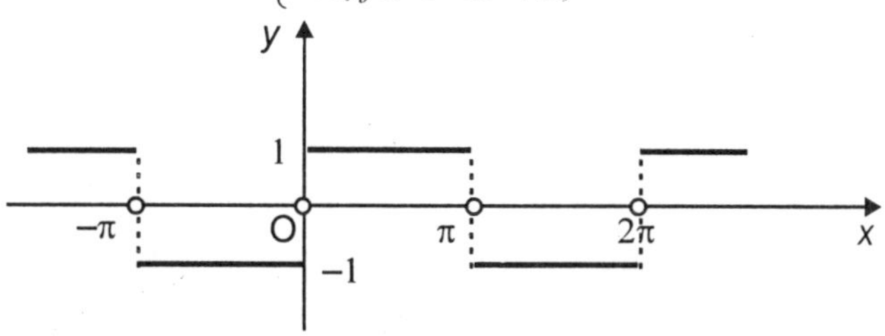

as, indicated by Fig., is an odd function, whence $a_\nu = 0$ and

$$b_\nu = \frac{2}{\pi}\int_0^\pi \sin\nu x\,dx = \begin{cases} 0 \text{ if } \nu \text{ is even}, \\ \dfrac{4}{\pi\nu} \text{ if } \nu \text{ is odd}, \end{cases}$$

so that its Fourier series is

$$f(x) = \frac{4}{\pi}\left(\frac{\sin x}{1} + \frac{\sin 3x}{3} + \ldots\right)$$

For $x = \pi/2$, in particular, this again yields Gregory's series.

Note that this series can be derived *formally* from that for $|x|$ by term by term differentiation.

The Function $f(x) = |\sin x|$: The even function $f(x) = |\sin x|$ can be expanded in a sine series with the coefficients a_n given by

$$\frac{\pi}{2}a_\nu = \int_0^\pi \sin x \cos \nu x\,dx = \frac{1}{2}\int_0^\pi \{\sin(\nu+1)x - \sin(\nu-1)x\}\,dx$$

$$= \begin{cases} 0 \text{ if v is odd,} \\ \dfrac{-2}{\nu^2-1} \text{ if } \nu \text{ is even,} \end{cases}$$

We thus obtain

$$f(x) = |\sin x| = \frac{2}{\pi} - \frac{4}{\pi}\sum_{\mu=1}^{\infty}\frac{\cos 2\mu x}{4\mu^2-1}$$

Expansion of the Function cos μx. Resolution of cotan into Partial Fractions. The Infinite Product for sin x:. Let $f(x) = \cos \mu x$ for $-\pi < x < \pi$, where m is not an integer. Since $f(x)$ is even, we obtain $b_\nu = 0$, while

$$\frac{\pi}{2}a_\nu = \int_0^\pi \cos \mu x \cos \nu x\,dx = \frac{1}{2}\int_0^\pi \{\cos(\mu+\nu)x + \cos(\mu-\nu)x\}\,dx$$

$$= \frac{1}{2}\left\{\frac{\sin(\mu+\nu)\pi}{\mu+\nu} + \frac{\sin(\mu-\nu)\pi}{\mu-\nu}\right\}$$

$$= \frac{\mu(-1)^\nu}{\mu^2-\nu^2}\sin \mu\pi$$

We thus find

$$\cos \mu x = \frac{2\mu \sin \mu\pi}{\pi}\left(\frac{1}{2\mu^2} - \frac{\cos x}{\mu^2-1^2} + \frac{\cos 2x}{\mu^2-2^2} - + \ldots\right)$$

This function remains continuous at the points $x = \} \pi$. If we set $x = \pi$, divide both sides of the equation by sin μx and then write x instead of μ, we obtain

$$\cos \pi x = \frac{2x}{\pi}\left(\frac{1}{2x^2} + \frac{1}{x^2-1^2} + \frac{1}{x^2-2^2} + \ldots\right)$$

This is the so-called resolution of cotan x into partial fractions, a very important formula which is frequently discussed in analysis. We now rewrite this series in the form

$$\cos \pi x - \frac{1}{\pi x} = \frac{2x}{\pi}\left\{\frac{1}{1^2-x^2} + \frac{1}{2^2-x^2} + \ldots\right\}$$

If x lies in an interval $0 \} x \} q < 1$, the absolute value of the n-th term on the right hand side is less than $\frac{2}{\pi}\frac{1}{n^2-q^2}$ whence the series converges uniformly in this interval and can be integrated term by term. We thus obtain

$$\pi\int_0^x\left(\cot\pi t-\frac{1}{\pi t}\right)dt = \lim_{a\to 0}\log\frac{\sin\pi a}{\pi a}=\log\frac{\sin\pi x}{\pi x}$$

on the left hand side and

$$\log\left(1-\frac{x^2}{1^2}\right)+\log\left(1-\frac{x^2}{2^2}\right)+\ldots = \lim_{n\to\infty}\sum_{\nu=1}^{n}\log\left(1-\frac{x^2}{\nu^2}\right)$$

on the right hand side after multiplying both sides by π. If we step over from the logarithm to the exponential function, we find

$$\frac{\sin\pi x}{\pi x} = e^{\lim\limits_{n\to\infty}\sum\limits_{\nu=1}^{n}\log\left(1-x^2/\nu^2\right)} = \lim_{n\to\infty} e^{\lim\limits_{n\to\infty}\sum\limits_{\nu=1}^{n}\log\left(1-x^2/\nu^2\right)}$$

$$= \lim_{n\to\infty}\prod_{\nu=1}^{n}\left(1-\frac{x^2}{\nu^2}\right)$$

Hence

$$\sin\pi x = \pi x\left(1-\frac{x^2}{1^2}\right)\left(1-\frac{x^2}{2^2}\right)\left(1-\frac{x^2}{3^3}\right)\ldots$$

Thus, we have obtained the famous expression for $\sin x$ as an infinite product. Setting x = ύ□, we obtain Wallis' product

$$\frac{\pi}{2} = \prod_{\nu=1}^{\infty}\frac{2\nu}{2\nu-1}\cdot\frac{2\nu}{2\nu+1} = \frac{2}{1}\cdot\frac{2}{3}\cdot\frac{4}{3}\cdot\frac{4}{5}\ldots,$$

The formula for $\sin\pi x$ is particularly interesting, because it shows directly that the function $\sin\pi x$ vanishes at the points $x = 0, 1, 2, \ldots$. In this respect, it corresponds to the factorization of a polynomial when its zeroes are known.

The Convergence of Fourier Series

The Convergence of the Fourier Series of a Sectionally smooth Function:. We first recall that, if $f(x)$ is any function which is defined and sectionally continuous (i.e., continuous, except at most at a finite number of jump discontinuities) in the interval $-\pi < x < \pi$, we can form its Fourier coefficients according to the formulae

$$a_\nu = \frac{1}{\pi}\int_{-\pi}^{+\pi} f(t)\cos\nu t\,dt,$$

$$b_\nu = \frac{1}{\pi}\int_{-\pi}^{+\pi} f(t)\sin\nu t\,dt,$$

and write down formally the series $\frac{1}{2}a_0 + \sum_{\nu=1}^{\infty}(a_\nu \cos\nu x + b_\nu \sin\nu x)$

This series is called the Fourier series corresponding to $f(x)$, irrespectively of whether it converges or not. We will now find the conditions, which must be imposed on $f(x)$, in order to ensure that the Fourier series corresponding to $f(x)$ converges and represents $f(x)$. We will assume that $f(x)$ is extended periodically beyond the interval $-\pi < x < \pi$.

We shall now prove the theorem: If the function $f(x)$ is sectionally smooth, i.e. $f(x)$ and its derivative $f'(x)$ are sectionally continuous, and satisfies at each point of discontinuity (s) the condition $f(x)=\{f(x-0)+f(x+0)\}$, then the Fourier series corresponding to $f(x)$ converges at every point and represents the function $f(x)$.

In order to prove this theorem, we consider the partial sums

$$S_n(x) = \frac{1}{2}a_0 + \sum_{\nu=1}^{n}(a_\nu \cos\nu x + b_\nu \sin\nu x)$$

If we substitute for the coefficients the above integral expressions and then interchange the order of integration and summation, we obtain

$$S_n(x) = \frac{1}{\pi}\int_{-\pi}^{+\pi}\sum_{\nu=1}^{n}(\cos\nu t\cos\nu x + \sin\nu t\sin\nu x)\,dt,$$

or, by the addition theorem for the cosine,

$$S_n(x) = \frac{1}{\pi}\int_{-\pi}^{+\pi} f(t)\left\{\frac{1}{2} + \sum_{\nu=1}^{n}\cos\nu(t-x)\right\}dt$$

If we now apply the summation formula obtained above, this becomes

$$S_n(x) = \frac{1}{2\pi}\int_{-\pi}^{+\pi} f(t)\frac{\sin\left(n+\frac{1}{2}\right)(t-x)}{\sin\frac{1}{2}(t-x)}dt$$

Finally, applying the transformation $t = (\tau - x)$ and noting the periodicity of the integrand, we obtain

$$S_n(x) = \frac{1}{2\pi}\int_{-\pi}^{+\pi} f(x+\tau)\frac{\sin\left(n+\frac{1}{2}\right)\tau}{\sin\frac{1}{2}\tau}d\tau$$

Starting with this form of the partial sum $S_n(x)$, we can prove that it tends to $f(x)$, by means of the lemma: If the function $s(x)$ is sectionally continuous in the interval $a \} x \} b$, then the integral

$$I = \int_a^b s(t)\sin\lambda t\,dt$$

tends to 0 as λ increases.

In the proof, we may assume that $s(x)$ is continuous in the entire interval, since otherwise we merely need carry out the argument for each sub-interval in which $s(x)$ is continuous

As in the similar argument, we note that, if λ is positive, the function $\sin\lambda t$ is alternately positive and negative in successive intervals of length π/λ. For large values of λ, the contributions to the integral from adjacent intervals almost cancel one another, since, on account of the continuity, the values of $s(x)$ in two such adjacent intervals differ only slightly from each another. We make use of this circumstance by transforming the integral I by the substitution $t = \tau + h$, where $h = \pi/\lambda$; then, since $\sin \lambda t = -\sin \lambda\tau$, we obtain

$$I = -\int_{a-h}^{b-h} s(\tau + h)\sin \lambda\tau \, d\tau$$

If we again replace the letter τ by t and then add the two expressions for I, we find

$$2I = -\int_{a-h}^{a} s(t+h)\sin \lambda t \, dt + \int_{a}^{b-h} \{s(t) - s(t+h)\}\sin \lambda t \, dt$$

$$+ \int_{b-h}^{b} s(t)\sin \lambda t \, dt$$

If M is an upper bound for the absolute value of $s(x)$, i.e., if for all values of x in the interval under consideration $|s(x)| \leq M$, then there follows at once from this expression for I the inequality $2|I| \leq 2Mh + \int_{a}^{b-h} |s(t) - s(t+h)| dt$.

Now, let ε be any positive number; if we choose λ so large that in the entire interval $a < t < b - h$ the expression $|s(t) - s(t+h)|$ remains less than $\varepsilon/(b-a)$ and also $Mh = M\pi/\lambda < \varepsilon/2$, then $|I|] < \varepsilon$; consequently, since ε can be chosen as small as we please, $\lim_{\lambda\to\infty} I = 0$

If we assume that $s(x)$, besides being continuous, has a sectionally continuous derivative $s'(x)$, the proof of this lemma follows simply on integration by parts. In fact,

$$\int_a^b s(t)\sin \lambda t \, dt = \frac{1}{\lambda}\left\{s(a)\cos \lambda a - s(b)\cos \lambda b + \int_a^b \theta'(t)\cos \lambda t \, dt\right\}$$

We see here at once that, as λ increases, the right hand side tends to zero.

Besides this lemma, we need the integration formula

$$\int_0^\pi \frac{\sin\left(n + \frac{1}{2}\right)t}{2\sin\frac{1}{2}t} dt = \frac{\pi}{2},$$

which is true for every positive integer n. We establish this readily by using our summation formula for the cosine, since

$$\int_0^\pi \frac{\sin\left(n + \frac{1}{2}\right)t}{2\sin\frac{1}{2}t} dt = \int_0^\pi \left(\frac{1}{2} + \sum_1^n \cos \nu t\right) dt = \frac{\pi}{2}$$

Proof of the Main Theorem

By means of the lemma, the main theorem is readily proved, i.e., the formula

$$\lim_{n\to\infty} S_n(x) = \lim_{n\to\infty} \frac{1}{2\pi}\int_{-\pi}^{+\pi} f(x+t)\frac{\sin\left(n+\frac{1}{2}\right)t}{2\sin\frac{1}{2}t}dt = f(x)$$

We begin by subdividing the interval of integration at the origin. For fixed values of x, the function

$$s(t) = \frac{f(x+t)-f(x+0)}{2\sin\frac{1}{2}t}$$

is sectionally continuous in the interval 0 } t } π. In fact, this is obvious when 0 } t } π, while the continuity at $t = 0$ follows from the assumed existence of the right hand derivative

$$\lim_{t\to 0, t>0} \frac{f(x+t)-f(x+0)}{t} = \lim_{t\to+0} \frac{f(x+t)-f(x+0)}{2\sin\frac{1}{2}t}\cdot\frac{2som\frac{1}{2}t}{t}$$

$$= \lim_{t\to+0} \frac{f(x+t)-f(x+0)}{2\sin\frac{1}{2}t}$$

Hence, as $\lambda = n +$ increases, the integral

$$\frac{1}{\pi}\int_0^{\pi} s(t)\sin\lambda t\,dt = \frac{1}{2\pi}\int_0^{\pi} f(x+t)\frac{\sin\lambda t}{\sin\frac{1}{2}t}dt - \frac{1}{2\pi}\int_0^{\pi} f(x+0)\frac{\sin\lambda t}{\sin\frac{1}{2}t}dt$$

tends to zero.

However, since the factor $f(x + 0)$ can be taken out of the second integral on the right hand side and, for $\lambda = n +$, the integral $\int_0^{\pi}\frac{\sin\lambda t}{2\sin\frac{1}{2}t}$ is equal to $\pi/2$, we obtain immediately

$$\lim_{\lambda\to\infty}\frac{1}{2\pi}\int_0^{\pi} f(x+t)\frac{\sin\lambda t}{\sin\frac{1}{2}t}dt = \frac{1}{2}f(x+0)$$

By setting in this equation $x = 0$, $f(t) = (\sin t)/t$ and then replacing t by u/λ, we obtain the important relation $\lim_{\lambda\to\infty}\int_0^{\lambda\pi}\frac{\sin u}{u}du = \frac{\pi}{2}$.

In the same way, we obtain for the interval $-\pi < t < 0$

$$\lim_{\lambda\to\infty}\frac{1}{2\pi}\int_{-\pi}^{0} f(x+t)\frac{\sin\lambda t}{\sin\frac{1}{2}t}dt = \frac{1}{2}f(x-0)$$

and by addition

$$\lim_{\lambda\to\infty}\frac{1}{2\pi}\int_{-\pi}^{+\pi} f(x+t)\frac{\sin\lambda t}{\sin\frac{1}{2}t}dt = f(x)$$

Further Investigation of Convergence: In the neighbourhood of those points, where the function $f(x)$ is discontinuous, the Fourier series does not converge uniformly; a uniformly convergent series of continuous functions possesses a continuous sum. Nevertheless, we have the important theorem:

If a sectionally smooth periodic function has no discontinuities, its Fourier series converges absolutely and uniformly. The convergence of the Fourier series for any sectionally smooth function whatsoever is uniform in every closed interval which contains no point of discontinuity of the function.

In order to prove this theorem, we start from a fundamental inequality satisfied by the Fourier coefficients of any function $f(x)$ which is sectionally continuous. This so-called Bessel inequality states that for all values of $n\,\frac{1}{2}a_0^2+\sum_{\nu=1}^{n}\left(a_\nu^2+b_\nu^2\right)\leq\frac{1}{\pi}\int_{-\pi}^{+\pi}\{f(x)\}^2\,dx$.

The proof follows from the fact that the expression

$$\int_{-\pi}^{+\pi}\left\{f(x)-\frac{1}{2}a_0-\sum_{\nu-1}^{n}\left(a_\nu\cos\nu x+b_\nu\sin\nu x\right)\right\}^2 dx$$

is always positive or zero. If we evaluate the integral by expanding the bracket under the integral sign and recall the orthogonality relations and definitions of the Fourier coefficients, we obtain at once Bessel's inequality in the form $\int_{-\pi}^{+\pi}\{f(x)\}^2\,dx-\pi\left\{\frac{1}{2}a_0^2+\sum_{\nu=1}^{n}\left(a_\nu^2+b_\nu^2\right)\right\}\geq 0$

In addition to Bessel's inequality, we employ Schwarz's inequality: If $u_1,u_2,\ldots,u_n$ and $v_1,v_2,\ldots,v_n$ are arbitrary real numbers, it is always true that $\left(\sum_{\nu=1}^{n}u_\nu v_\nu\right)^2\leqq\sum_{\nu=1}^{n}u_\nu^2\cdot\sum_{\nu=1}^{n}v_\nu^2$ the equality sign occurring only when the sequence u is proportional to the sequence v.

We now assume that the periodic function $f(x)$ is sectionally smooth as well as continuous. The derivative $g(x) = f'(x)$ is sectionally continuous and we easily show that c_ν and d_ν, the Fourier coefficients of $g(x)$, satisfy the relations

$$c_0 = 0,$$

$$c_\nu = \nu b_\nu \ (\nu \geqq 1);$$

$$d_\nu = -\nu a_\nu \ (\nu \geqq 1);$$

in fact, on integration by parts, we have

$$c_\nu = \frac{1}{\pi}\int_{-\pi}^{+\pi} g(x)\cos\nu x\,dx$$

$$= \frac{1}{\pi} f(x)\cos\nu x\Big|_{-\pi}^{+\pi} + \frac{\nu}{\pi}\int_{-\pi}^{+\pi} f(x)\sin\nu x\,dx = \nu b_\nu$$

similar proofs holding for the other statements.

Hence, Bessel's inequality applied to the function $g(x)$ yields

$$\sum_{\nu=1}^{n}\nu^2\left(a_\nu^2 + b_\nu^2\right) = \sum_{\nu=1}^{n}\left(c_\nu^2 + d_\nu^2\right) \leqq \frac{1}{\pi}\int_{-\pi}^{+\pi}\{g(x)\}^2 dx.$$

If, for the sake of brevity, we denote the right hand side of this inequality by M and apply Schwarz's inequality, we find that for $m > n$

$$\sum_{\nu=n+1}^{m}\left|a_\nu\cos\nu x + b_\nu\sin\nu x\right| \leqq \sum_{\nu=n+1}^{m}\sqrt{\left(a_\nu^2 + b_\nu^2\right)} = \sum_{\nu=n+1}^{m}\left\{\frac{1}{\nu}\left(\nu\sqrt{\left(a_\nu^2 + b_\nu^2\right)}\right)\right\} \leqq M\sqrt{\left(\sum_{\nu=n+1}^{m}\frac{1}{\nu^2}\right)},$$

since $\sqrt{\left(a_\nu^2 + b_\nu^2\right)}$ is the amplitude of the periodic function $a_\nu\cos\nu x + b_\nu\sin\nu x$

However, owing to the $\sum_{\nu=1}^{\infty}\frac{1}{\nu^2}$, convergence of, the right hand side, which is independent of x, can be made as small as we please by choosing n and m large enough, which proves the absolute and uniform convergence of the series.

Incidentally, the same considerations show that for periodic functions with continuous derivatives of the (h- l)-th order and derivatives of the (h-1)-th order, which are at least sectionally continuous, the sum $\sum^{n}\nu^{2h}\left(a_\nu^2 + b_\nu^2\right)$ remains below a fixed bound.

This gives us a definite statement about the order to which the Fourier coefficients vanish. For such a function, the Fourier series of the derivatives up to the order (h - 1) converge absolutely and uniformly.

In order to prove the above theorem for sectionally smooth functions which are discontinuous, we first consider a special function $\psi(x)$ of this type.

In the interval $-\pi < x < \pi$, we define $y(x)$ as equal to x, outside this interval, $y(x)$ is extended periodically. According to equation, its Fourier series is $2\left(\frac{\sin x}{1}-\frac{\sin 2x}{2}+\frac{\sin 3x}{3}-+\ldots\right)$

This series cannot be uniformly convergent, because its sum is the discontinuous function $y(x)$. However, we shall show that the convergence is uniform in every interval $-l \leq x \leq l$ for which $0 < l < \pi$.

The proof is based on a special artifice*. We observe that in the interval $-l \leq x \leq l$ the function $\cos x/2$ is never less than the positive quantity $\cos l/2 = \kappa$. If we multiply the absolute value of the difference between the m-th and n-th partial sums of the above series $(m > n)$, i.e., the expression

$$\left|S_m(x)-S_n(x)\right| = 2\left|\frac{\sin(n+1)x}{n+1}-\frac{\sin(n+2)x}{n+2}+-\ldots\frac{\sin mx}{m}\right|,$$

by the function $\cos x/2$, then, in accordance with the well-known trigonometric formula

$2\sin u\cos v = \sin(u+v)+\sin(u-v)$, we obtain the absolute value of the expression

If we combine the terms on the right hand side with the same numerators, we obtain

$$\frac{\sin\left(n+\frac{1}{2}\right)x}{n+1}\pm\frac{\sin\left(m+\frac{1}{2}\right)x}{m}+\frac{\sin\left(n+\frac{2}{2}\right)x}{(n+1)(n+2)}-\frac{\sin\left(m+\frac{5}{2}\right)x}{(n+2)(m+3)}+-\ldots\mp\frac{\sin\left(m-\frac{1}{2}\right)x}{(m-1)m},$$

and, since $\cos x/2 < \kappa$ and $|\sin u| < 1$, the estimate

$$\left|S_m(x)-S_n(x)\right| \leqq \frac{1}{k}\left[\frac{1}{n+1}+\frac{1}{m}+\frac{1}{(n+1)(n+2)}+\ldots+\frac{1}{(m-1)m}\right]$$

But the expression on the right hand side does not depend on x and, by virtue of the convergence of the series $\sum_{\nu=1}^{\infty}\frac{1}{\nu(\nu+1)}$, it can be made as small as we please by choosing n and m large enough. This implies the uniform convergence of the Fourier series, as we have asserted.

Now that we have obtained the expression for a particular discontinuous function, we can transfer the discontinuity to any arbitrary point in the interval by translation of the curve or of the co-ordinate system. In fact, the function

$$\psi(x-\xi) = 2\left(\frac{\sin(x-\xi)}{1}-\frac{\sin 2(x-\xi)}{2}+\frac{\sin 3(x-\xi)}{3}-+\ldots\right)$$

is continuous except at the points $(2k+1)\pi+\xi$, where k is an integer. However, on passing these points, the function jumps by an amount -2π from the value π to the value $-\pi$, while at these points themselves the value of the function is zero.

If now $f(x)$ is any sectionally smooth function, which in the interval $-\pi < x < \pi$ is discontinuous only at the points $\xi_1, \xi_2, \ldots, \xi_m$ and if on passing these points from the left to the right hand side the function jumps by the amounts $\delta_1, \delta_2, \ldots, \delta_m$, respectively, then the function

$$f(x)+\frac{\delta_1}{2\pi}\psi(x+\pi-\xi_1)+\frac{\delta_2}{2\pi}\psi(x+\pi-\xi_2)+\ldots+\frac{\delta_m}{2\pi}\psi(x+\pi-\xi_m)$$

will be continuous and sectionally smooth, whence, by the previous proof, it can be expanded in a uniformly convergent Fourier series. We now obtain the Fourier series of the function $f(x)$ by adding term by term the finite number of Fourier series corresponding to the functions

$$-\frac{\delta_1}{2\pi}\psi(x+\pi-\xi_1),\ldots,-\frac{\delta_m}{2\pi}\psi(x+\pi-\xi_m)$$

Hence the theorem is proved.

This result is quite adequate for most mathematical investigations and applications. However, we point out that the investigation of Fourier series has been advanced much further. The conditions for an expansion in Fourier series, which we have found here to be sufficient, are by no means necessary. Functions with far fewer continuity properties than those discussed here can be represented by Fourier series. There is an extensive literature devoted to these questions and to the general problem of the expandability of a function in a Fourier series.

As a remarkable result of such investigations, we mention thee fact that there are continuous functions, the Fourier series of which do not converge in any interval, no matter how small. Such a result does not in any impugnway reduce the usefulness of Fourier series; on the contrary, it must be regarded as evidence that the concept of a continuous function involves fairly complicated possibilities, as has already been shown by the example of continuous functions which nowhere have a derivative.

A Sketch of the Theory of Functions of Several Variables

Up to this point, we have been concerned exclusively with functions of a single independent variable. We must now go on to consider functions of several independent variables. Even the applications of the calculus force us to take this step. In almost all the relationBhips which occur in nature, in fact, the functions in question do not depend on a single independent variable; on the contrary, the dependent variable is usually determined by two, three, or more independent variables. Thus, for example, the volume of an ideal gas is a function of a single variable, the pressure, if we keep the temperature constant, but not otherwise. As a rule, the temperature also varies and the volume depends upon a pair of values, namely, the value of the pressure and that of the temperature, whence it is a function of two independent variables.

Also from the point of view of pure mathematics, the need for a detailed study of functions of several independent variables is urgent. Here we shall be able to take advantage of what we have learned previously, so that in many cases we have only to make simple extensions of our arguments.

It is usually sufficient to consider the case of only two independent variables x and y, as long as no essentially new considerations are required for an extension to functions of three or more variables. Hence, in order to keep our statements and notation simple, we shall, as a rule, consider only two independent variables.

THE CONCEPT OF FUNCTION IN THE CASE OF SEVERAL VARIABLES

Functions and their Ranges of Definition: Equâtions of the form

$$u = x^2 + y^2, \qquad u = x - y,$$

$$u = xy, \text{ or} \qquad u = \sqrt{\left(1 - x^2 - y^2\right)}$$

assign a functional value u to each pair of values (x, y). In the first three of our example, this correspondence holds for every system of values (x, y), while in the last case the correspondence has a meaning only for those pairs of values (x, y) for which the inequality $x + y < 1$ is true.

In these cases we say that u is a function of the independent variables x and y. In general, we use this expression whenever some law assigns a value of u as dependent variable, corresponding to each pair of values (x,y) belonging to a certain specified set. The relation between x, y and u may be stated in terms of a functional equation, as above, or by means of a verbal description such as u is the area of the rectangle with sides x and y, or it may follow from physical observations as for instance in the case of the magnetic declination at different latitudes and longitudes. The essential thing is that there exists a correspondence. Similarly, u is said to be a function of the three independent variables x, y, z, if for each triad of values (x, y, u) of a certain set there exists a corresponding value of u given by some definite law; it is similar for the general case of functions of n independent variables $x_1, x_2, \ldots, x_n$.

The set of values which the pair (x, y) can assume is called the range of definition of the function $u = f(x, y)$. We shall restrict our attention to the simplest types of range of definition. We shall consider that (x, y) is limited either to a so-called rectangular region (domain) $a \leqq x \leqq b$, $c \leqq y \leqq d$, or else to a circle, determined by an inequality of the form $(x - a)^2 + (y - b)^2 \leqq r^2$

In the case of functions of three variables x, y, z, we shall again consider only rectangular regions $a \leqq x \leqq b,\ c \leqq y \leqq d,\ e \leqq z \leqq f$, and spherical regions $(x - a)^2 + (y - b)^2 (z - c)^2 \leqq r^2$

In dealing with more than three independent variables, geometrical intuition fails us, but it is often convenient to also extend geometrical terminology to this case. Thus, for functions of n variables $x_1, x_2, \ldots, x_n$, we shall consider regions $a_1 \leqq x_1 \leqq b_1,\ a_2 \leqq x_2 \leqq b_2, \ldots,$ $a_n \leqq x_n \leqq b_n$, and also regions $(x_1 - a_1)^2 + (x_2 - a_2)^2 + \ldots + (x_n - a_n)^2 \leqq r^2$, which we will call rectangular and spherical regions, respectively.

The Simplest Types of Functions. Just as in the case of functions of one variable, the simplest functions are the rational integral functions or polynomials. The most general polynomial of the first degree (linear function) has the form $u = ax + by + c$, where a, b and c are constants. The general polynomial of the second degree has the form $u = ax^2 + bxy + cy^2 + dx + ey + f$.

The general polynomial is a sum of terms of the form $a_{mn}x^m y^n$, where the quantities a_{mn} are arbitrary constants.

Rational fractional functions are quotients of polynomials; for example, there belongs to this class the linear fractional function

$$u = \frac{ax + by + c}{a'x + b'y + c'}$$

By extraction of roots, we pass on from the rational functions to certain algebraic functions, for example,

$$u = \sqrt{\left(\frac{x - y}{x + y}\right)} + \sqrt[3]{\left\{\frac{(x + y)^2}{x^3 + xy}\right\}}$$

In the construction of more complicated functions of several variables, we almost always fall back on the well-known functions of a single variable, for example,

$$u = \sin(xy) \text{ or } \qquad u = \log(y^2 + \cos\frac{1}{2}x)$$

GEOMETRICAL REPRESENTATION OF FUNCTIONS

Just as we represent functions of one variable by means of curves, we seek to represent functions of two variables geometrically by means of surfaces; we shall consider hereafter only those functions which can actually be represented in this way.

We achieve this representation very simply by considering a rectangular co-ordinate system in space with co-ordinates x, y and u, and marking off above each point (x, y) of the range (R) of definition of the function the point P with the third co-ordinate $u = f(x, y)$. As the point (x, y) ranges over the region R, the point P describes a surface in space. We take this surface as the geometrical representation of the function.

Conversely, in analytical geometry, surfaces in space are represented by functions of two variables, so that there is between such surfaces and functions of two variables a reciprocal relationship.

For example, there corresponds to the function

$$u = \sqrt{(1 - x^2 - y^2)}$$

the hemi-sphere above the x, y plane with unit radius and centre at the origin, to the function

$$u = x^2 + y^2,$$

the so-called parabaloid of revolution, obtained by rotating the parabola $w=x$ about the u-axis, to the functions

$$u = x^2 - y^2, \text{ and } u = xy$$

the so-called hyperbolic paraboloids. The linear function

$$u = ax + by + \text{c}$$

has for its graph a plane in space.

If in the function $u = f(x, y)$ one of the independent variables, say y, does not occur, so that u depends on x only, say $u = g(x)$, the function is represented in the xyu-space by a cylindrical surface, obtained by erecting the perpendiculars to the ux-plane at the points of the curve $u=g(x)$.

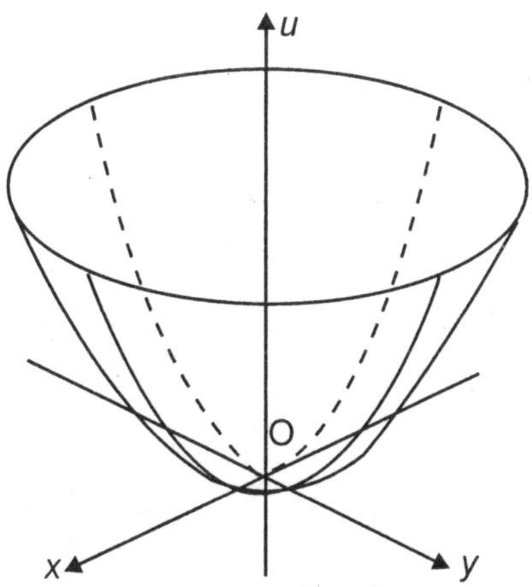

Fig. $u = x^2 + y^2$

However, this representation by means of rectangular co-ordinates has two disadvantages. Firstly, intuition fails us whenever we have to deal with three or more independent variables. Secondly, even in the case of two independent variables, it is often more convenient to confine the discussion to the xy-plane, since in the plane we can sketch and construct geometrically without difficulty. From this point of view, another geometrical representation of the function by means of contour lines is to be preferred. In the xy-plane, we take all the points for which $u = f(x, y)$ has a constant value, say $u = k$. These points will usually lie on a curve or curves, the so-called contour line for the given constant value of the function. We can also obtain these curves by cutting the surface $u = f(x, y)$ by the plane $u = k$ parallel to the xy-plane and projecting the curves of intersection perpendicularly onto the xy-plane.

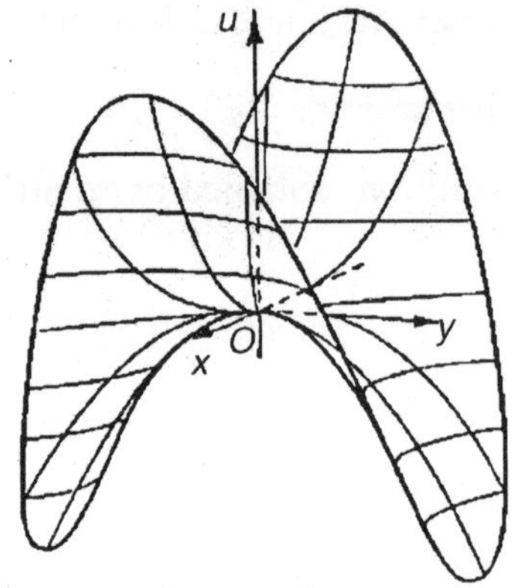

Fig. $u = x^2 - y^2$

The system of these contour lines, marked with the corresponding values k_1, k_2, ... of the height k, gives us a representation of the function. As a rule, k is assigned values in arithmetic progression, say $k = n\,h$, where $n = 1, 2, \ldots$ The distance between the contour lines then gives us a measure of the steepness of the surface $u=f(x,y)$; in fact, between every two neighbouring lines, the value of the function changes by the same amount.

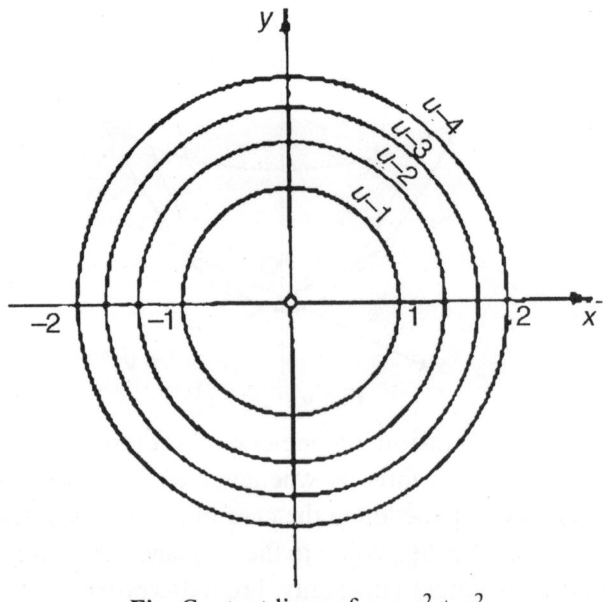

Fig. Content lines of $u = x^2 + y^2$

Where the contour lines are close together, the function rises or falls steeply, where they are far apart, the surface is flattish. This is the principle on which contour maps such as those of the Ordnance Surrey and the U.S. Geological Survey are constructed.

In this method, the linear function $u = ax + by + c$ *is* represented by a system of parallel straight lines $ax + by + c = k$. The function $u = x + y$ is represented by a system of concentric circles. The function $u = x - y$, the surface of which has a saddle point at the origin is represented by the system of hyperbolas.

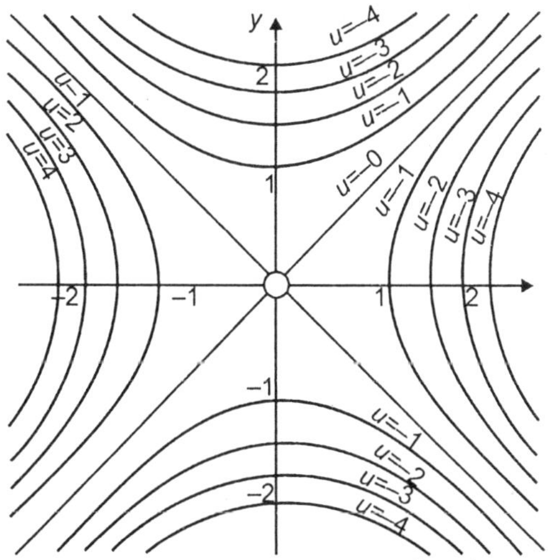

Fig. Content Lines of $u = x^2 - y^2$

The method of representing the function $u = f(x,y)$ by contour lines has the advantage of it being capable of extension to functions of three independent variables. Instead of contour lines, we then have level surfaces $f(x,y,z) = k$, where k is a constant to which we can assign any suitable sequence of values.

For example, the level surfaces for the function $u = x + y + z$ are concentric spheres about the origin of the co-ordinate system.

CONTINUITY

As in the case of functions of a single variable, the basic requirement, by which functions should he capable of being represented geometrically leads to the analytic condition of continuity. Here again, the concept of continuity is given by the definition:

A function $u = f(x, y)$, defined in a region R, is said to be continuous at a point (ξ, η) of R, if for all points (x,y) near (x,h) the value of the function $f(x, y)$ differs but little from $f(\xi, \eta)$, the difference being arbitrarily small if only (x, y) is near enough to (ξ, η).

More precisely: The function $f(x, y)$, defined in the region R, is continuous at the point (ξ, η) of R, provided that it is possible for every positive number ε to find a positive distance $\delta = \delta(\varepsilon)$ (in general, depending on ε and tending to 0 with ε) such that for all points of the region, the distance of which from (ξ,η) is less than δ, that is, for which the inequality $(x-\xi)^2 + (y-\eta)^2 + \leqq \delta^2$ holds, there is satisfied the relation

$$|f(x,y) - f(\xi,\eta)| \leqq \in$$

or, in other words, the relation

$$|f(\xi+h,\eta+k) - f(\xi,\eta)| \leqq \in$$

is to hold for all pairs of values (h, k) such that $h + k \rangle \delta$ and $(\xi + h, \eta + k)$ belongs to the region R.

If a function is continuous at every point of a region B, we say that it is continuous in R.

In the definition of continuity, we can replace the distance condition $h+k\rangle\delta\rangle$ by the equivalent condition:

There shall correspond to every $\varepsilon > 0$ two positive numbers $\delta_1$and δ_2 such that

$$|f(\xi+h,\eta+k) - f(\xi,\eta)| \leqq \in$$

whenever $|h| \leqq \delta_1$ and $|k| \leqq \delta_2$

The two conditions are equivalent. In fact, if the original condition is fulfilled, so is the second one, if we take $\delta_1 = \delta_2 = \delta/2$; and conversely, if the second condition is fulfilled, so is the first, if we take for δ the smaller one of the two numbers $\delta_1$and δ_2.

The following facts are almost obvious: The sum, difference and product of continuous functions are also continuous. The quotient of continuous functions is continuous except when the denominator vanishes. Continuous functions of continuous functions are themselves continuous. In particular, all polynomials are continuous and so are all rational fractional functions except when the denominator vanishes.

Another obvious fact, which, however, is worth stating, is:

If a function $f(x, y)$ is continuous in a region R and differs from zero at an interior point P of the region, it is possible to mark off about P a neighbourhood, say a circle, belonging entirely to R, in which $f(x,y)$ is nowhere equal to zero. In fact, if the value of the function at P is a, we can mark off about P a circle so small that the value of the function within the circle differs from a by less than $a/2$, and therefore it is certainly not zero.

Examples of Discontinuities

In the case of functions of one variable, we encounter three kinds of discontinuities: Infinite and jump discontinuities, and discontinuities at which no limit is approached from one or both sides. With functions of two or more variables, no such simple classification is possible. In particular, the situation is made more complicated by the fact that discontinuities may occur not merely at isolated points, but also along entire curves.

Thus, for the function $u = 1/(x - y)$, the line $x = y$ is a line of infinite discontinuity. As we approach the line from one side or the other, the values of u increase numerically beyond all bounds through positive or negative values. The function $u = 1/(x - y)$ has the same line of discontinuity, but tends to + } as we approach the line from either side. The function $u = 1/(x + y)$ has the single point of discontinuity $x = 0, y = 0$. The function $u = \sin \frac{1}{\sqrt{(x^2 + y^2)}}$ tends to no limit as we approach the origin; the surface which it represents is obtained by rotating the graph of the function $u = \sin 1/x$ about the u-axis.

Another instructive example of a discontinuous function is given by the rational function $u = 2xy/(x + y)$. In the first instance, the function is undefined at $x = 0$, $y = 0$, and we supplement the definition by assuming that $u(0,0) = 0$. This function has a peculiar type of discontinuity at the origin. If we put $x = 0$, i.e., if we move along the y-axis, the function becomes $u(0,y) = 0$, which has the constant value 0 for all values of y. Along the x-axis, we likewise have $u(x, 0) = 0$. Thus, at the origin, the function $u(x,y)$ is continuous in x, if we keep y at the constant value 0, and is continuous in y, if we keep x at the constant value 0. Nevertheless, the function is discontinuous when considered as a function of the two variables x and y. In fact, at every point of the line $y=x$, we find that $u = 1$, so that arbitrarily near to the origin we can find points at which u assumes the value 1. The function is therefore discontinuous at the origin and cannot be defined at the origin in such a way as to make it continuous.

More generally, on the straight line $y = x \tan \alpha$, inclined at the angle a to the x-axis, we have $u = 2\text{Tan}a/(1+\text{Tan}a) = 2\text{Sin}a\ \text{Cos}a = \text{Sin}2a$.

The surface corresponding to the $u = 2\,xy/(x+y)$ is therefore formed by rotating a straight line at right angle to the u-axis about that axis until it coincides with the x-axis and simultaneously raising or lowering it so that the height $\sin 2\alpha$ is associated with the angle a. As a increases to 45, the straight line rises to the height 1, and consequently falls to the level of the y-axis and below to the depth 1, thereafter rising again to the level of the x-axis. The surface enveloped by the moving straight line is known as the cylindroid and is important in mechanics.

The above example shows that a function can he continuous in x for every fixed value of y and continuous in y for every fixed value of x and yet be discontinuous when considered as a function of the two variables. The essential point in this definition of continuity is that the value of the function at a point P must be arbitrarily close to the value of the function at a point Q, provided only that Q is near enough to P; it is not permissible to restrict the position of Q relative to P in any other way.

Multiple and Repeated Integrals

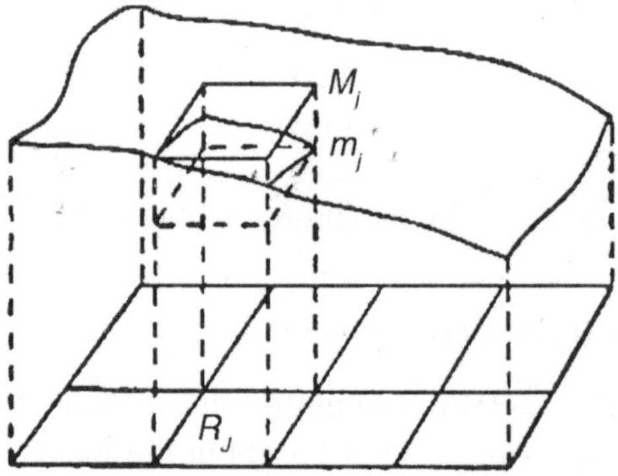

Multiple Integrals: Consider a function $u = f(x, y)$ which is defined and continuous in the rectangle *R*(*a* ú□ *x* ú□ *b*, *c* ú□ *y* ú□ *d*) and which assumes only positive values. We wish to assign a volume to the portion of three-dimensional space bounded by the rectangle *R*, the surface $u = f(x, y, z)$, and the four planes $x = a$, $x = b$, $y = c$, $y = d$ perpendicular to the *xy*-plane. Moreover, the volume should be defined so as to satisfy certain elementary conditions:

1. If the three-dimensional region is a prism, i.e., if the function *u* is a constant *k*, the volume should be the product of the base by the height, $V=(b-a)(d-c)k$;
2. If we divide the rectangle *R* into smaller rectangles R_1 and R_2 by drawing straight lines, then the volume over R_1 should be equal to the volume over R_1 plus the volume over R_2;
3. If the three-dimensional region R_1 completely includes R_2, the volume of R_1 should be at least as large as that of R_2.

These considerations lead us to a method of defining *V* which is an immediate extension of the method of defining area. By constructing lines parallel to the sides, we subdivide the rectangle *R* into smaller rectangles $R_1, R_2, \ldots, R_{n,}$ the areas of which will be denoted by DR_1, DR_2, ..., DR_n. In each rectangle R_j, the fonction has a least value m_j and a larger value M_j, whence a prism with base *R* and height M_j completely includes the portion of the region over R_j, while this portion of the region contains the prism with base R_j and height m_j. Hence, we see that the vohume of the portion in question lies between m_jR_j and M_jR_j.

Thus, the total volume *V* should be such that $\sum_{i=1}^{n} m_i \Delta R_i \leqq V \leqq \sum_{i=1}^{n} M_i \Delta R_i$

Now let the number *n* of rectangles increase beyond all bounds in such a way that the length of the longest diagonal tends to zero. Intuition leads us to expect that both the sums $\sum m_i \, \Delta R_i$ and $\sum M_i \, \Delta R_i$ will converge and tend to the same limit, whence we call this limit the volume *V*.

The reader will have observed that we have carried out an immediate generalization of the discussion. We call the common limit of the sums $\sum m_i \Delta R_i$ and $\sum M_i \Delta R_i$ the integral of the function $n = f(x, y)$ over the rectangle R and denote it by the symbol $\iint_R f(x,y)dr$

It is at once clear that, if we choose in each rectangle R_j a point (ξ_j, η_j) and find the corresponding value of the function $f(\xi_j, \eta_j)$, then there must hold the limiting relation

$$\lim_{n\to\infty} \sum f(\xi_i, \eta_i)\Delta R_i = \iint_R f(x,y)dr ;$$

in fact, the sum $\sum f(\xi_i, \eta_i)\Delta R_i$ lies between $\sum m_i \Delta R_i$ and $\sum M_i \Delta R_i$, both of which approach the integral as a limit.

As a particular method of subdividing r into smaller rectangles, we may subdivide the side $a \} x \} b$ into n intervals of length $\Delta x = (b - a)/n$ and the side $c \} y \} d$ into m intervals of length $\Delta y = (d - c)/m$, and then draw parallels to the axes through the points of division thus marked. The area of each rectangle R_j is then $\Delta R_j = \Delta x \Delta y$. Choosing a point (ξ_j, η_j) arbitrarily in each rectangle R_j, we form the sum

$$\sum f(\xi_i, \eta_i)\Delta R_i = \sum_i f(\xi_i, \eta_i)\Delta x \Delta y$$

As both n and m increase without limit, this sum approaches the integral as a limit. This type of subdivision suggests a second notation for the integral, which has been in common use since the time of Leibnitz, namely $\iint_R f(x,y)dxdy$.

The proof that such a limit exists if $u = f(x, y)$ is continuous can be carried out. However. we shall assume without proof the even stronger statement:

The function $f(x, y)$ is continuous except along a finite number of smooth curves (curves with continuous derivatives) $y = f(x)$ or $x = \phi(y)$ along which $f(x,y)$ has jump discontinuities, then there exists double integral $\iint_R f(x,y)dxdy$.

We leave its proof to Volume II. It depends essentially on the fact that, as the number of rectangles increases, the total area of the rectangles having points in common with the curves of discontinmty tends to zero. Thus, even though M_j and m_j may differ considerably for such rectangles, they give rise to a little difference between the sums $\sum m_i \Delta R_i$ and $\sum M_i \Delta R_i$.

With this assumption, we can find the area under surfaces $u = f(x, y)$ for which (x, y) ranges over quite complicated regions R. In fact, let the region R be bounded by a finite number of curves $x = \phi(y)$ or $y = \psi(x)$ with continuous derivatives and $f(x, y)$ be continuous in R. We enclose R in a rectangle R' and assign the points of R', which $\iint_R f(x,y)dr$ taken over the region R', as the volume under the surface $u = f(x, y)$, where (x, y) is in R. This integral is usually denoted by $\iint_R f(x,y)dxdy$.

Certain simple, but important theorems relating to these double integrals follow directly from the definition. We shall simply state the theorems, as the reader will be able to prove them without any trouble.

If $f(x, y)$ and $g(x, y)$ are integrable over a rectangle, then so are fg and cf, where c is a constant:

$$\iint_R \{f(x,y) \pm g(x,y)\}\, dr = \iint_R f(x,y)\, dr \pm \iint_R g(x,y)\, dr$$

If $f(x, y) \geq g(x, y)$ in R, then $\iint_R f(x,y)\, dr \geq \iint_R g(x,y)\, dr$

If $I_j R$ is the sum of two regions R_1 and R_2, then

$$\iint_R f(x,y)\, dr = \iint_{R_1} f(x,y)\, dr + \iint_{R_1} g(x,y)\, dr$$

Reduction of Double Integrals to Repeated Single Integrals

We now have a definition of the double integral with its interpretation as a volume and with the many possibilities of usefulness which our experience with the single integral suggests; however, as yet, we do not possess a method for evaluating such integrals. In this section, we shall see how the calculation of a double integral can be reduced to that of two single integrals.

Let $u = f(x, y)$ be a function which is defined and continuous in a rectangle R,

$a \leq x \leq b$, $c \leq y \leq d$. If we fix upon any value x_0 in the interval $a \leq x \leq b$, the function $f(x_0, y)$ is a continuous function of the remaining variable y, whence there exists the integral $\int_c^d f(x_0, y)\, dy$ and it can be evaluated by the methods. This integral has a definite value for each value of x_0 which we may select; in other words, the integral is a function $\phi(x_0)$ of the quantity x_0:

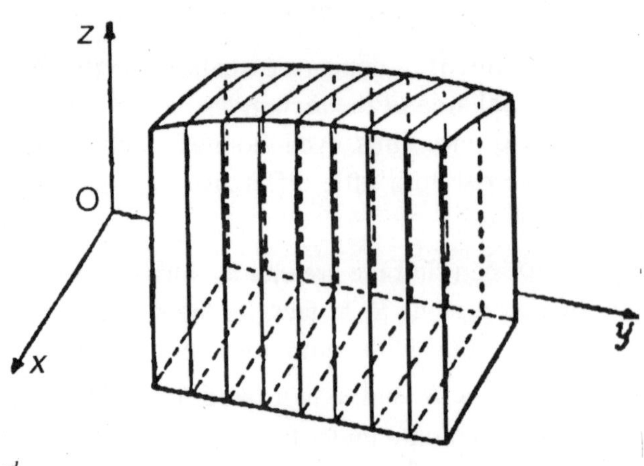

$$\int_c^d f(x_0, y)\, dy = \phi(x)$$

For example, let u = $f(x, y) = xy$, 0 x 1, 0 y 3. For each fixed x in the interval $\int_0^3 x^2 y^3 dy$ 0 x 1, the integral can be evaluated and, in fact, is $81x/4$, i.e., it is a function of x. Or if $f(x, x) = e^{xy}$, 1 x 2, 1 y 4, we have $\int_1^4 e^{xv} dy = \frac{1}{x}\left(e^{4x} - e^x\right)$

Having thus found the function $\phi(x)$, we can prove that it is continuous; this is a simple consequence of the uniform continuity of $f(x, y)$.It is therefore possible to integrate $\phi(x)$ between the limits a and b, thus obtaining the repeated integral

$$\int_a^b \phi(x)dx = \int_a^b \left(\int_c^d f(x,y)dy \right) dx$$

By reversing the order of the process, first calculating the function of y defined by $\int_a^b f(x, y)dx$ and then integrating from c to d, *we* obtain the other repeated integral $\int_c^b \left(\int_a^b f(x,y)dx \right) dy$.

These integrals, as we have seen, are obtained by a double application of the ordinaly simple integration which we have studied before. Their importance lies in the following fact:

For continuous functions $f(x, y)$ and for functions $f(x, y)$ with at most jump discontinuities on a finite number of smooth curves, the repeated integrals are equal to the double integral:

$$\int\int_R f(x,y)dr = \int_a^b \left(\int_c^d f(x,y)dy \right) dx$$
$$= \int_c^d \left(\int_a^b f(x,y)dx \right) dy$$

We shall content ourselves with an intuitive discussion of the case where $f(x,y)$ is continuous. In our original discussion of the double integral regarded as the volume lying above the and below the surface

$u = f(x, y)$, we obtained this volume by subdividing the solid into vertical colunms and then letting the diagonals of the bases of these columns approach zero. Instead of this, we can divide the solid into slices of thickness $k=(d\text{-}c)/n$ by drawing the lines $y = c + nk$ (n = 0, 1, ..., n) parallel to the x-axis and then constructing a plane perpendicular to the xy-plane through each line.

These planes cut the solid into n slices which grow thinner as n increases and the total volume of which is equal to the double integral. We now see that the volume of each slice is approximately (but, of course, nct as a rule exactly) equal to the product of the thickness k by the area of the left-hand face, i.e., equal to $k\int_a^b f(x, c+vk)dx$.

Hence, if we write

$$\phi(y) = \int_a^b f(x,y)dx$$

the desired volume is represented approximately by $\sum_{\nu=0}^{n-1} k\phi(c+\nu k)$

As *n*, these sums tend to $\int_a^b \phi(y)dy$

It is therefore reasonable to expect that the volume or doube integral is exactly equal to

$$\int_a^b \phi(y)dy = \int_c^d \left(\int_a^b f(x,y)\,dx \right) dy,$$

which is the statement made ahove. A similar discussioa makes it equally plausible that the statement

$$\int_a^b \left(\int_c^d f(x,y)\,dy \right) dx = \int\int_R f(x,y)\,dr \text{ is also true.}$$

MULTIPLE INTEGRALS

Integrating functions of several variables

Recall that we think of the integral in two different ways. In one way we interpret it as the area under the graph $y = f(x)$, while the fundamental theorem of the calculus enables us to compute this using the process of "anti-differentiation" — undoing the differentiation process. We think of the area as

$$\sum f(x_i)\,dx_i = \int f(x)\,dx,$$

where the first sum is thought of as a limiting case, adding up the areas of a number of rectangles each of height $f(x_i)$, and width dx_i. This leads to the natural generalisation to several variables: we think of the function $z = f(x, y)$ as representing the height of f at the point (x, y) in the plane, and interpret the integral as the sum of the volumes of a number of small boxes of height $z = f(x, y)$ and area $dx_i dy_j$. Thus the volume of the solid of height $z = f(x, y)$ lying above a certain region R in the plane leads to integrals of the form $\int\int_R = \sum_{i=1}^{n}\sum_{j=1}^{n} f(x_i, y_j)dx_i dy_j = \lim S_{mn}$.

We write such a double integral as $\int\int_R f(x, y)\,dA$.

Repeated Integrals and Fubini's Theorem

As might be expected from the form, in which we can sum over the elementary rectangles *dx dy* in any order, the order does not matter when calculating the answer. There are two important orders — where we first keep x constant and vary y, and then vary x; and the opposite way round. This gives rise to the concept of the repeated integral, which we write as

$$\int\left(\int_R f(x,y)\,dx\right)dy \qquad or \qquad \int\left(\int_R f(x,y)\,dy\right)dx.$$

Our result that the order in which we add up the volume of the small boxes doesn't matter is the following, which also formally shows that we evaluate a double integral as any of the possible repeated integrals. Theorem (Fubini's theorem for Rectangles) Let $f(x, y)$ be continuous on the rectangular region R: *axb;cyd*. Then

$$\iint_R f(x,y)\,dA = \int_c^d\left(\int_a^b f(x,y)\,dx\right)dy = \int_a^b\left(\int_c^d f(x,y)\,dy\right)dx$$

Note that this is something like an inverse of partial differentiation. In doing the first inner (or repeated) integral, we keep y constant, and integrate with respect to x. Then we integrate with respect to y. Of course if f is a particularly simply function, say $f(x, y) = g(x)h(y)$, then it doesn't matter which order we do the integration, since

$$\iint_R f(x,y)\,dA = \int_a^b g(x)\,dx \int_c^d h(y)\,dy$$

We use the Fubini theorem to actually evaluate integrals, since we have no direct way of calculating a double (as opposed to a repeated) integral.

Example: Integrate $z = 4 - x - y$ over the region $0 \leqq x \leqq 2$ and $0 \leqq y \leqq 1$. Hence calculate the volume under the plane $z = 4 - x - y$ above the given region.

Solution: We calculate the integral as a repeated integral, using Fubini's theorem.

$$V = \int_{x=0}^{2}\int_{y=0}^{1}(4 - x - y)\,dy\,dx = \int_{x=0}^{2}\left[\left(4y - xy - y^2/2\right)\right]_0^1$$

$$dx = \int_0^2 (4 - x - 1/2)\,dx$$

From our interpretation of the integral as a volume, we recognise V as volume under the plane $z = 4 - x - y$ which lies above $\{(x, y) \mid 0 \leqq x \leqq 2; 0 \leqq y \leqq 1\}$. In fact Fubini's theorem is valid for more general regions than rectangles. Here is a pair of statements which extend its validity.

Fubini's Theorem

Let $f(x, y)$ be continuous on a region R

1. If R is defined as $a \leqq x \leqq b; g_1(x) \leqq y \leqq g_2(x)$. Then

$$\iint_R f(x,y)\,dA = \int_a^b\left(\int_{g_1(x)}^{g_2(x)} f(x,y)\,dy\right)dx$$

2. If R is defined as $c \leqq y \leqq d; h_1(y) \leqq x \leqq h_2(y)$. Then

$$\iint_R f(x,y)dA = \int_c^d \left(\int_{h_1(y)}^{h_2(y)} f(x,y)\,dx \right) dy$$

Proof: We give no proof, but the reduction to the earlier case is in principle simple; we just extend the function to be defined on a rectangle by making it zero on the extra bits. The problem with this as it stands is that the extended function is not continuous. However, the difficulty can be fixed. This last form enables us to evaluate double integrals over more complicated regions by passing to one of the repeated integrals.

Evaluate the Integral

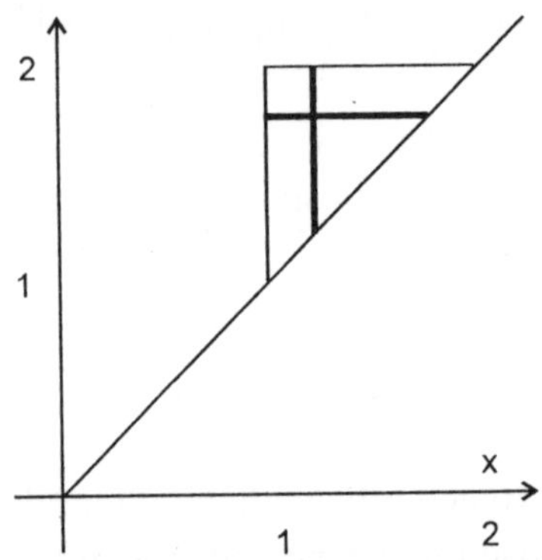

Fig. Area of Integration $\int_1^2 \left(\int_x^2 \frac{y^2}{y^2} dy \right) dx$ as it Stands, and Sketch the Region of Integration.

Reverse the order of integration, and verify that the same answer is obtained.

Solution: The diagram shows the area of integration. We first integrate in the given order.

$$\int_1^2 \left(\int_x^2 \frac{y^2}{x^2} dy \right) dx = \int_1^2 \left[\frac{y^3}{3x^2} \right]_x^2 dx \int_1^2 \left(\frac{8}{3x^2} - \frac{x}{3} \right) = \left[\frac{8}{3x} - \frac{x^2}{6} \right]_1^2 = \left(\frac{4}{3} - \frac{2}{3} \right)\left(\frac{8}{3} - \frac{1}{3} \right) = \frac{5}{6}$$

Reversing the order, using the diagram, gives

$$\int_1^2 \left(\int_1^y \frac{y^2}{x^2} dx \right) dy = \int_1^2 \left[-\frac{y^2}{x} \right]_x^y dy = \int_1^2 \left(-y + y^2 \right) dy$$

$$= \left[-\frac{y^2}{2} - \frac{y^3}{3} \right]_1^2 = \left(-2 + \frac{8}{3} \right) - \left(-\frac{1}{2} + \frac{1}{3} \right) = \frac{5}{6}$$

Thus the two orders of integration give the same answer. Another use for the ideas of double integration just automates a procedure you would have used anyway, simply from your knowledge of 1-variable results.

Chapter 10

Properties of Fourier Series

As mentioned earlier that it should not be assumed that the existence of the coefficients a_k and b_k in itself implies that the associated series converges to the function $f(x)$. In order that the function have a convergent Fourier series expansion it must satisfy a certain set of conditions. To prove a theorem giving the sufficient conditions for convergence of Fourier series we shall start with some auxiliary results.

RIEMANN LEMMA

If $f(x)$ is a square integrable function on the interval $[-\pi, \pi]$ then from Bessel's inequality it follows that $a_k = \lim b_k = 0$. By the formulae for finding the Fourier coefficients a_k and b_k we have that

$$\lim_{k\to\infty}\int_{-\pi}^{\pi} f(x)\cos kx\,dx = \lim_{k\to\infty}\int_{-\pi}^{\pi} f(x)\sin kx\,dx = 0$$

It is possible to prove that in the last condition the interval $[-\pi, \pi]$ can be replaced with any interval $[a; b]$ i.e. for each square integrable function $f(x)$ on an interval $[a; b]$

$$\lim_{k\to\infty}\int_{b}^{a} f(x)\cos kx\,dx = \lim_{k\to\infty}\int_{a}^{b} f(x)\sin kx\,dx = 0$$

A more general result called Riemann lemma holds.

Lemma: If the function f(x) is continuous in the interval [a; b] with the exception of a finite number of points and absolutely integrable in this interval then

$$\lim_{v\to\infty}\int_{b}^{a} f(x)\sin vx\,dx = \lim_{v\to\infty}\int_{a}^{b} f(x)\cos vx\,dx = 0$$

Here v represents not necessarily an integer but a real variable.

Sum of Cosines

We need a formula for calculating the sum

$$D_N(u) = \frac{1}{2} + \sum_{k=1}^{N}\cos ku$$

Obviously, multiplying both sides of this equality by $2\sin\frac{u}{2}$, we have

$$2D_N(u)\sin\frac{u}{2} = \sin\frac{u}{2} + 2\cos u\sin\frac{u}{2} + 2\cos 2u\sin\frac{u}{2} + \ldots + 2\cos Nu\sin\frac{u}{2}$$

and after using the formula

$$2\cos\alpha\sin\beta = \sin(\alpha+\beta) - \sin(\alpha-\beta)$$

we find
$$D_n(u)\sin\frac{u}{2} = \sin\frac{u}{2} + \left(\sin\frac{3u}{2} - \sin\frac{u}{2}\right) + \left(\sin\frac{5u}{2} - \sin\frac{3u}{2}\right) +$$

$$\ldots + \left(\sin\frac{(2N+1)u}{2} - \sin\frac{(2N-1)u}{2}\right)$$

or

$$2D_n(u)\sin\frac{u}{2} = \sin\frac{(2N+1)u}{2}$$

that is

$$D_n(u) = \frac{\sin\frac{(2N+1)u}{2}}{2\sin\frac{u}{2}}$$

So we have the formula

$$\frac{1}{2} + \sum_{k=1}^{N}\cos ku = \frac{\sin\frac{(2N+1)u}{2}}{2\sin\frac{u}{2}}$$

Integrating the last equality over the interval $[-\pi;\ \pi]$ we obtain

$$1 = \frac{1}{\pi}\int_{-\pi}^{\pi}\frac{\sin\frac{(2N+1)u}{2}}{2\sin\frac{u}{2}}du$$

Integral Formula of Partial Sums

Next we need an integral formula for the partial sums of a Fourier series. Let $f(x)$ be a 2π-periodic function with its Fourier series expansion

$$f(x) \approx \frac{a_0}{2} + \sum_{k=1}^{\infty} a_k\cos kx + b_k\sin kx$$

Let $S_N(x)$ designates the N th partial sum of this expansion, i.e.

$$S_N(x) = \frac{a_0}{2} + \sum_{k=1}^{N} a_k\cos kx + b_k\sin kx$$

Replacing the coefficients a_0, a_k and b_k by the formulas (13) – (15), we obtain that

$$S_N(x) = \frac{1}{2\pi}\int_{-\pi}^{\pi} f(t)\,dt + \frac{1}{\pi}\sum_{k=1}^{N}\left[\int_{-\pi}^{\pi} f(t)\cos k\,t\,dt.\cos k\,x + \int_{-\pi}^{\pi} f(t)\sin k\,t\,dt.\sin k\,x\right]$$

$$= \frac{1}{\pi}\int_{-\pi}^{\pi} f(t)\left[\frac{1}{2} + \sum_{k=1}^{N}(\cos k\,t\,d\,t.\cos k\,x)\right]dt$$

$$= \frac{1}{\pi}\int_{-\pi}^{\pi} f(t)\left[\frac{1}{2} + \sum_{k=1}^{N}(\cos k(t-x)\right]dt$$

By the formula for sum of cosines (1)

$$S_N(x) = \frac{1}{\pi}\int_{-\pi}^{\pi} f(t)\,\frac{\sin\dfrac{(2N+1)(f-x)}{2}}{2\sin\dfrac{f-x}{2}}\,dt$$

which, after changing the variable $u = t - x$, gives

$$S_N(x) = \frac{1}{\pi}\int_{-\pi-x}^{\pi-x} f(x+u)\,\frac{\sin\dfrac{(2N+1)u}{2}}{2\sin\dfrac{u}{2}}\,du$$

or, since the integrand is a periodic function with period 2π,

$$S_N(x) = \frac{1}{\pi}\int_{-\pi}^{\pi} f(x+u)\,\frac{\sin\dfrac{(2N+1)u}{2}}{2\sin\dfrac{u}{2}}\,du$$

Convergence of Fourier Series

Now we are ready to prove the theorems giving the sufficient conditions for convergence of Fourier series. Note that there exist various sets of conditions that the given function must satisfy in order that it have a convergent Fourier series expansion. We shall prove one, maybe the simplest, theorem and state some others.

Dirichlet's Theorem

Theorem: If f(x) is an absolutely integrable periodic function with period 2π in the interval $[-\pi, \pi]$ that has a finite number of points of discontinuity, then at any point where f(x) is differentiable its Fourier series expansion converges to the function f(x).

Proof: We have to prove that under these conditions

$$\lim_{N\to\infty} S_N(x) = f(x).$$

Let $f(x)$ be differentiable at the point x. Multiplying (3.2) by $f(x)$ we have

$$f(x)=\frac{1}{\pi}\int_{-\pi}^{\pi} f(x)\frac{\sin\frac{(2N+1)u}{2}}{2\sin\frac{u}{2}}du$$

and subtracting the last equality from (3)

$$S_N(x)-f(x)=\frac{1}{\pi}\int_{-\pi}^{\pi}[f(x+u)-f(x)]\frac{\sin\frac{(2N+1)u}{2}}{2\sin\frac{u}{2}}du$$

$$S_N(x)-f(x)=\frac{1}{\pi}\int_{-\pi}^{\pi}\varphi(u)\sin\left(N+\frac{1}{2}\right)u\,du$$

where
$$\varphi(u)=\frac{f(x+u)-f(x)}{2\sin\frac{u}{2}}.$$

It remains to show that the function $\varphi(u)$ satisfies the conditions of the Riemann lemma. We have

$$\varphi(u)=\frac{f(x+u)-f(x)}{u}\frac{u}{2\sin\frac{u}{2}}.$$

Since the function $f(x)$ is differentiable at the point x, there exists

$$f'(x)=\lim_{u\to 0}\frac{f(x+u)-f(x)}{u}.$$

Thus, for $u\to 0$

$$\varphi(u)=\frac{f(x+u)-f(x)}{u}\frac{\frac{u}{2}}{\sin\frac{u}{2}}\to f'(x).1=f'(x)$$

i.e. the function $\varphi(u)$ has at the point $u=0$ a removable discontinuity. All the other points of discontinuity of $\varphi(u)$ are the same as the points of discontinuity of $f(x+u)$. Therefore, if the conditions of the theorem are satisfied, the function $\varphi(u)$ satisfies the conditions of Riemann lemma and

$$\lim_{N\to\infty}\frac{1}{\pi}\int_{-\pi}^{\pi}\varphi(u)\sin\left(N+\frac{1}{2}\right)u\,du=0,$$

that is
$$\lim_{N\to\infty} S_N(x)=f(x).$$

The theorem does not state anything about the convergence at the points of discontinuity. This leads us to state the more general theorem.

The left- and right-hand derivatives of the function $f(x)$ at a point x are defined as the one-sided limits

$$f'(x-) \lim_{h\to 0-} \frac{f(x+h)-f(x)}{h}$$

and
$$f'(x+) \lim_{h\to 0+} \frac{f(x+h)-f(x)}{h}$$

respectively.

Theorem: If f(x) is an absolutely integrable periodic function with period in the interval $[-\pi, \pi]$, that has a finite number of points of discontinuity, then at any point where f(x) has the left- and right-hand derivatives $f'(x-)$ and $f'(x+)$ its Fourier series expansion converges to the function f(x) at all points where f(x) is continuous and to the average of the left- and right-hand limits of f(x), i.e. to $\frac{f(x-)+f(x+)}{2}$, at points where f(x) is discontinuous.

Such a theorem also holds for some different sets of conditions.

Theorem: If f(x) is a bounded periodic function that in the interval $[-\pi, \pi]$ has a finite number maxima and minima and a finite number of points of finite discontinuity then the Fourier series expansion of f(x) converges to f(x) at all points where f(x) is continuous and to the average of the left- and right-hand limits of f(x), i.e. to $\frac{f(x-)+f(x+)}{2}$, at points where f(x) is discontinuous.

Every function is differentiable at any interior point of domain. This fact is also easily seen taking in the Fourier series expansion

$$f(x) \approx 2\sum_{k=1}^{\infty}(-1)^{k+1}\frac{\sin kx}{k}$$

$x = -\pi$ or $x = \pi$. In both cases the Fourier series consists of zeros and, of course, equals.

By the Theorem the Fourier series expansion of the function converges to the some given function everywhere and thus we can replace the sign '~' with the sign of equality

$$f(x) = \frac{\pi}{2} - \frac{4}{\pi}\sum_{k=1}^{\infty}\frac{\cos(2k-1)x}{(2k-1)^2}.$$

If $x \in [-\pi;\pi]$ then $f(x) = |x|$ and taking in (4) $x = 0$ we have

$$0 = \frac{\pi}{2} - \frac{4}{\pi}\sum_{k=1}^{\infty}\frac{1}{(2k-1)^2}$$

or

$$1 + \frac{1}{9} + \frac{1}{25} + \ldots \frac{1}{(2k-1)^2} + \ldots = \frac{\pi^2}{8}$$

We can also calculate the sum

$$S = \sum_{k=1}^{\infty}\frac{1}{k^2}$$

because

$$S = 1 + \frac{1}{2^2} + \frac{1}{3^2} + \frac{1}{4^2} + \frac{1}{5^2} + \frac{1}{6^2} + \ldots = 1 + \frac{1}{3^2} + \frac{1}{5^2} + \ldots \frac{1}{4}\left(1 + \frac{1}{2^2} + \frac{1}{3^2} + \ldots\right).$$

Therefore,

$$S = \frac{\pi^2}{8} + \frac{1}{4}S$$

and it follows that

$$S = \frac{\pi^2}{6}.$$

Integration of a Fourier Series

Let $f(x)$ be absolutely integrable 2π-periodic function with its Fourier series expansion

$$f(x) \approx \frac{a_0}{2} + \sum_{k=1}^{\infty} a_k \cos kx + b_k \sin kx .$$

Consider the function

$$F(x) = \int_0^x \left[f(x) - \frac{a_0}{2}\right] dx$$

in the interval $[-\pi ; \pi]$. The function $F(x)$ is continuous and has an absolutely integrable derivative, moreover

$$F(x+2\pi) = \int_0^{x+2\pi}\left[f(x) - \frac{a_0}{2}\right] dx = \int_0^{x}\left[f(x) - \frac{a_0}{2}\right] dx + \int_0^{x+2\pi}\left[f(x) - \frac{a_0}{2}\right] dx$$

$$= F(x) = \int_{-\pi}^{\pi} f(x)\, dx - \frac{a_0}{2}\int_0^x dx = F(x).$$

Thus, $F(x)$ is a 2π-periodic, absolutely integrable and differentiable function, i.e. it satisfies the conditions of the the Theorem and, therefore, is representable by its Fourier series

$$F(x)=\frac{A_0}{2}+\sum_{k=1}^{\infty}A_k\cos kx+B_k\sin kx$$

Integration by parts gives the coefficients

$$A_k=\frac{1}{\pi}\int_{-\pi}^{\pi}F(x)\cos kx\,dx=\frac{1}{\pi}\frac{F(x)\sin kx}{k}\bigg|_{-\pi}^{\pi}-\frac{1}{\pi k}\int_{-\pi}^{\pi}\left[f(x)-\frac{a_0}{2}\right]\sin kx\,dx$$

$$=-\frac{1}{\pi k}\int_{-\pi}^{\pi}f(x)\sin kx+\frac{a_0}{2\pi k}\int_{-\pi}^{\pi}\sin kx\,dx=-\frac{b_k}{k}$$

Analogously $B_k=\frac{u_k}{k}$ and

$$F(x)=\frac{A_0}{2}+\sum_{k=1}^{\infty}\frac{a_k\sin kx-b_k\cos kx}{k}$$

Taking $x=0$ in the last equality we obtain that

$$0=\frac{A_0}{2}-\sum_{k=1}^{\infty}\frac{b_k}{k}\quad\text{or}\quad\frac{A_0}{2}=\sum_{k=1}^{\infty}\frac{b_k}{k}.$$

Thus,

$$F(x)=\sum_{k=1}^{\infty}\frac{a_k\sin kx+b_k(1-\cos kx)}{k}$$

or by the definition of $F(x)$

$$\int_0^x f(x)dx-\int_0^x\frac{a_0}{2}dx=\sum_{k=1}^{\infty}\frac{a_k\sin kx+b_k(1-\cos kx)}{k}$$

But

$$\frac{a_k\sin kx}{k}=\int_0^x a_k\cos kx\,dx$$

and

$$\frac{b_k(1-\cos kx)}{k}=\int_0^x b_k\sin kx\,dx$$

Substituting the last equalities into (3.7) we have

$$\int_0^x f(x)dx=\int_0^x\frac{a_0}{2}dx+\sum_{k=1}^{\infty}(a_k\cos kx+b_k\sin kx)\,dx.$$

In the last equality the lower boundary can be replaced with any real x_0, i.e. for $-\pi \le x_0 < x \le \pi$ we have

$$\int_{x_0}^{x} f(x)dx = \frac{a_0}{2}(x - x_0) + \sum_{k=1}^{\infty} \int_{x_0}^{x} (a_k \cos kx + b_x \sin kx)\, dx.$$

Let us summarize the results.

Conclusion: The Fourier series expansion of $\int_{x_0}^{x} f(x)dx$ can be obtained on term by term integration of the Fourier series expansion of f(x) and it is independent of the convergence of that Fourier series expansion.

Conclusion: It is possible to calculate $\int_{x_0}^{x} f(x)dx$ knowing only the Fourier series expansion of the function f(x), the function itself can be unknown.

Because of the presence of the term $\frac{a_0}{2}x$ on the right-hand side, this is not clearly a Fourier series expansion of the integral on the left-hand side, the result can be rearranged to be a Fourier series expansion of the function

$$\int_{x_0}^{x} f(x)dx - \frac{a_0}{2}x$$

Example: *Integrate term by term the Fourier series expansion for the square wave*

$$f(x) = \begin{cases} -1 \text{ if } -\pi < x \le \pi \\ 1 \text{ if } 0 < x < x \end{cases}$$

$$f(x + 2\pi) = f(x)$$

Solution: The Fourier series expansion for $f(x)$ is

$$f(x) \approx \frac{4}{\pi} \sum_{k=1}^{\infty} \frac{\sin(2k-1)}{2k-1}$$

If the function is piecewise-continuous over the interval $[-\pi, \pi]$, then care must be taken to ensure that the integration process is carried out properly over the various subintervals. Now we need to consider separately values of x in the intervals $[-\pi, 0]$ and $[0, \pi]$.

If $-\pi < x < 0$ then, integrating term by term, we have

$$\int_{-\pi}^{x} (-1)dx = \frac{4}{\pi} \sum_{k=1}^{\infty} \int_{-\pi}^{x} \frac{\sin(2k-1)}{2k-1} dx$$

that is

$$-(x+\pi) = -\frac{4}{\pi}\sum_{k=1}^{\infty}\left[\frac{\cos(2k-1)x}{(2k-1)^2}\right]_{-\pi}^{x}$$

$$= -\frac{4}{\pi}\left[\sum_{k=1}^{\infty}\frac{\cos(2k-1)x}{(2k-1)^2} + \sum_{k=1}^{\infty}\frac{1}{(2k-1)^1}\right]$$

Replacing the last series in square brackets with $\frac{\pi^2}{8}$, the above simplifies to

$$-x = \frac{\pi}{2} - \frac{4}{\pi}\sum_{k=1}^{\infty}\frac{\cos(2k-1)x}{(2k-1)^2}$$

If $0 < x < x$ then, integrating term by term, we have

$$\int_{-\pi}^{x}(-1)\,dx + \int_{0}^{x} 1\,dx = \frac{4}{\pi}\sum_{k=1}^{\infty}\frac{\sin(2k-1)x}{2k-1}\,dx$$

giving

$$x = \frac{\pi}{2} - \frac{4}{\pi}\sum_{k=1}^{\infty}\frac{\cos(2k-1)x}{(2k-1)^2} \quad \text{if } 0 < x < x.$$

Taking both equations together, we find that the function

$$g(x) = |x| = \begin{cases} -x & \text{if } -\pi < x < 0 \\ x & \text{if } 0 < x < x \end{cases}$$

$$g(x + 2\pi) = g(x)$$

has the Fourier series expansion

$$g(x) = |x| = \frac{\pi}{2} - \frac{4}{\pi}\sum_{k=1}^{\infty}\frac{\cos(2k-1)x}{(2k-1)^2},$$

The last series expansion converges to $g(x)$ at any point despite the fact that the series expansion does not converge to $f(x)$ at the points of discontinuity.

Differentiation of Fourier series

Let $f(x)$ be a continuous 2π-periodic function having an absolutely integrable derivative (with exeption at the finite number of points) and satisfying the condition $f(-\pi) = f(\pi)$. Then

$$f(x) = \int_{0}^{x} f'(x)\,dx + f(0)$$

Thus, by the results of the previous section the Fourier series expansion of $f(x)$ can be obtained from the Fourier series expansion of

$$f'(x) \sim \sum_{k=1}^{\infty} a'_k \cos kx + b'_k \sin kx$$

on term by term integration. The last expansion does not contain the constant term a_0' because

$$a_0' = \frac{1}{\pi} \int_{-\pi}^{\pi} f'(x)\, dx = \frac{1}{\pi} f(x)|_{-\pi}^{\pi} = 0.$$

If the Fourier series expansion of $f(x)$ can be obtained from the Fourier series expansion of $f'(x)$ on term by term integration then the Fourier series expansion of $f'(x)$ can be also obtained from the Fourier series expansion of $f(x)$ on term by term differentiation, i.e. if the function $f(x)$ satisfies the conditions given above and its Fourier series expansion is

$$f'(x) = \frac{a_0}{2} + \sum_{k=1}^{\infty} a_k \cos kx + b_k \sin kx$$

then the Fourier series expansion of $f'(x)$ is

$$f'(x) \sim \sum_{k=1}^{\infty} k b_k \cos kx - k a_k \sin kx$$

There is an " = " sign because under the given conditions the Fourier series expansion of $f(x)$ converges to $f(x)$ but in (25) there is an "~" sign because at the points of discontinuity of $f'(x)$ its Fourier series expansion may not converge to $f'(x)$.

Here it is important that $f(x)$ is continuous everywhere, including the end points $-\pi$ and π. To illustrate this point, consider the Fourier series expansion of the function

$$f(x) = x \text{ it } -\pi < x < x,$$

$$f(x + 2\pi) = f(x)$$

which, from Example, is given by

$$f(x) = 2\sum_{k=1}^{\infty} (-1)^{k+1} \frac{\sin kx}{k}$$

Differentiating term by term, we have

$$f'(x) \sim 2\sum_{k=1}^{\infty} (-1)^{k+1} \cos kx.$$

If the differentiation process is valid then $f'(x)$ must be equal to 1 for $-\pi < x < \pi$. Clearly this is not the case, since the series on the right-hand side does not converge for any value of x.

Notice that the Fourier coefficients of the derived expansion are kb_k and $-ka_k$, so, in contrast to the integrated series, the convergence properties of derived series are worse these of the original series expansion.

Example: Consider the process of differentiating term by term the Fourier series expansion of the function

$$f(x) = x^2 \text{ if } -\pi < x \leq \pi$$

$$f(x + 2\pi) = f(x).$$

Solution: The Fourier series expansion of $f(x)$ is

$$x^2 = \frac{\pi^2}{3} + 4\sum_{k=1}^{\infty}\frac{(-1)^k \cos kx}{k^2} \quad \text{if } -\pi < x \leq \pi.$$

Since $f(x)$ satisfies the conditions of piecewise, we may differentiate above term by term to obtain

$$2x \sim 4\sum_{k=1}^{\infty}\frac{k(-1)^{k+1}}{k^2}\sin kx$$

that is

$$x \sim 2\sum_{k=1}^{\infty}\frac{(-1)^{k+1}\sin kx}{k}$$

which conforms with the Fourier series expansion obtained for the function

$$f(x) = x \text{ if } -\pi < x \leq \pi$$

$$f(x + 2\pi) = f(x)$$

IMPROVEMENT OF CONVERGENCE OF FOURIER SERIES

If we differentiate term by term the Fourier series expansion then the convergence properties are worse. At the same time, for example solving partial differential equations, we have to differentiate the Fourier series expansion more than once. To provide the convergence of the derived series expansions we separate a part of the original series expansion that converges slowly. If we know the sum of the separated slowly convergent part then we can differentiate it immediately. But the coefficients of the remaining part must be decreasing so fast that we could differentiate term by term the remaining part as many times as needed. In other words we have the following problem. For a given trigonometric series, with the sum $f(x)$ which may also be unknown,

$$f(x) = \frac{a_0}{2} + \sum_{k=1}^{\infty} a_k \cos kx + b_k \sin kx$$

we have to separate another series with known sum $\varphi(x)$ so that the coefficients of the remainder series determined by the equality

$$f(x) = \varphi(x) + \sum_{k=1}^{\infty} a_k \cos kx + \beta_k \sin kx$$

should be decreasing sufficiently fast. If this problem is solved then we can replace the function $f(x)$ with the sum of known function $\varphi(x)$ and series with coefficients decreasing sufficiently fast.

The following considerations show that this problem can be solved. For the function $f(x)$ differentiable some times on the interval $[-\pi, \pi]$ its periodic extension is, in general, a discontinuous function or a function with discontinuous derivatives. Thus, it cannot be expected that its Fourier coefficients are decreasing fast. It is easy to see that we can subtract from $f(x)$ a linear function $\varphi(x)$ on the interval $[-\pi, \pi]$ so that the periodic extension of the difference is a continuous function, i.e. the values of this difference are equal at the end points of the interval $[-\pi, \pi]$. For example we can determine the linear function $\varphi(x)$ in the interval $[-\pi, \pi]$ as

$$\varphi(x) = \frac{f(\pi-) - f(-\pi+)}{\pi} x + b$$

where b is any real number. In this case for the difference $g(x) = f(x) - \varphi(x)$

$$g(\pi-) = f(\pi-) - \frac{f(\pi-) - f(-\pi+)}{2\pi}\pi - b = \frac{f(\pi-) + f(-\pi+)}{2} - b$$

and

$$g(-\pi+) = f(-\pi+) - \frac{f(\pi-) - f(-\pi+)}{2\pi} - \pi - b = \frac{f(\pi-) + f(-\pi+)}{2} - b$$

i.e. the periodic extension of the difference $g(x)$ is indeed continuous everywhere. Subtracting from $f(x)$ a certain polynomial we can also guarantee that some derivatives are also continuous at the end points of the interval $[-\pi, \pi]$.

Thus, the function itself and its derivatives can be continuously and periodically extended to the whole set of real numbers. As a result, the Fourier coefficients are decreasing quite fast.

The problem is that a series is given, not the function $\varphi(x)$. We have to find the function $\varphi(x)$ starting from the given series. This difficulty can often be overcome and in this case it is said that the convergence of the series is improved.

Let us consider two methods of improvement of the convergence. The first of these two is based on the well-known fact that the difference of two equivalent infinitely decreasing quantities is the infinitely decreasing quantity of higher order i.e. if $\lim y = 0$ and $\lim z = 0$ then $\lim \frac{y-z}{y} = \lim \frac{y-z}{z} = 0$.

Example: Improve the convergence of the series

$$f(x) = \sum_{k=2}^{\infty} (-1)^k \frac{k^3}{k^4 - 1} \sin kx.$$

Solution: We use the equivalency $\frac{k^3}{k^4-1} \sim \frac{1}{k}$. The difference of these

$$\frac{k^3}{k^4 - 1} - \frac{1}{k} = \frac{1}{k^5 - k}$$

is the infinitely decreasing quantity of higher order. We have

$$f(x) = \sum_{k=2}^{\infty} (-1)^k \left(\frac{1}{k} + \frac{1}{k^5 - k} \right) \sin kx$$

$$= \sum_{k=2}^{\infty} (-1)^k \frac{\sin kx}{k} + \sum_{k=2}^{\infty} (-1)^k \frac{\sin kx}{k^5 - k}.$$

If $-\pi < x < \pi$ then, we have the Fourier series expansion

$$\frac{x}{2} = \sum_{k=1}^{\infty} (-1)^{k+1} \frac{\sin kx}{k} = \sin x + \sum_{k=2}^{\infty} (-1)^{k+1} \frac{\sin kx}{k}$$

$$= \sin x - \sum_{k=2}^{\infty} (-1)^k \frac{\sin kx}{k}$$

that is

$$f(x) = \sin x - \frac{x}{2} + \sum_{k=2}^{\infty} (-1)^k \frac{\sin kx}{k^5 - k}.$$

The order of Fourier coefficients in the last series is $\frac{1}{k^E}$.

A second method of improvement of convergence is based on the representation of Fourier coefficients in the form of the sum

Example: Improve the convergence of the series

$$f(x)=\sum_{k=2}^{\infty}\frac{\sin kx}{k+a}\quad a\ \text{–consant, a > 0.}$$

Solution: Obviously

$$\frac{1}{k+a}=\frac{1}{k}\frac{1}{1+\frac{a}{k}}=\frac{1}{k}\left(1-\frac{a}{k}+\frac{a^2}{k^2}-\frac{a^3}{k^3}+\ldots\ldots\right).$$

Summarizing the geomeric sequence after the term $\frac{a^2}{k^2}$ we have

$$\frac{1}{k+a}=\frac{1}{k}\left(1-\frac{a}{k}+\frac{a^2}{k^2}-\frac{\frac{a^3}{k^3}}{1+\frac{a}{k}}\right)=\frac{1}{k}\left(1-\frac{a}{k}+\frac{a^2}{k^2}-\frac{a^3}{k^2(k+a)}\right)$$

$$=\frac{1}{k}-\frac{a}{k^2}+\frac{a^2}{k^3}-\frac{a^3}{k^3(k+a)}.$$

The given Fourier series expansion can be re-written as

$$f(x)=\sum_{k=1}^{\infty}\frac{\sin k}{k}-a\sum_{k=1}^{\infty}\frac{\sin kx}{k^2}+a^2\sum_{k=1}^{\infty}\frac{\sin kx}{k^3}\sum_{k=1}^{\infty}\frac{\sin kx}{k^3(k+a)}.$$

Using the Fourier series expansions

$$\sum_{k=1}^{\infty}\frac{\sin kx}{k}=\frac{\pi-x}{2},\ \sum_{k=1}^{\infty}\frac{\sin kx}{k^3}=\frac{x^3-3\pi x^2+2\pi^2 x}{12}$$

and

$$\sum_{k=1}^{\infty}\frac{\sin kx}{k^2}=-\int_0^x \ln\left(2\sin\frac{x}{2}\right)dx$$

we have

$$f(x)=\frac{\pi-x}{2}+a\int_0^x \ln\left(2\sin\frac{x}{2}\right)dx+\frac{a^2(x^3-3\pi x^2+2\pi^2 x)}{12}-a^3\sum_{k=1}^{\infty}\frac{\sin kx}{k^3(k+a)}.$$

In the last series expansion the order of coefficiets is $\frac{1}{k^4}$.

COMPLETE SYSTEMS

Let

$$\{\varphi_0(x), \varphi_1(x), ..., \varphi_k(x), ...\}$$

be an orthogonal system of square integrable functions on the interval $[a, b]$. It has been proved that the mean square deviation

$$\delta_n = \int_a^b [f(x) - \sigma_n(x)]^2 dx$$

is minimized if the coefficients of the polynomial are Fourier coefficients of $f(x)$ with respect to the system. This minimal value was denoted by Δ i.e.

$$\Delta_n = \int_a^b \left[f(x) - \sum_{k=0}^{n} c_k \varphi_k(x) \right]^2 dx$$

and using (2.10) gives

$$\int_a^b \left[f(x) - \sum_{k=0}^{n} c_k \varphi_k(x) \right]^2 dx = \int_a^b f^2(x) dx - \sum_{k=0}^{n} c_k^2 \| \varphi_k \|^2 .$$

Definition: The system is said to be *complete* if the sign "$\leq$" in Bessel's inequality can be replaced with the sign of equality i.e. if for any square integrable function $f(x)$ there holds

$$\int_a^b f^2(x) dx = \sum_{k=0}^{\infty} c_k^2 \| \varphi_k \|^2$$

where c_k (k = 0, 1, 2,...) are the Fourier coefficients of $f(x)$ with respect to the system.

The equality is called *Parseval's equality.*

Theorem: *The system is complete if for any square integrable function f(x)*

$$\lim_{x \to \infty} \int_a^b \left[f(x) - \sum_{k=0}^{n} c_k \varphi_k(x) \right]^2 dx = 0$$

where c_k (k = 0, 1, 2...) are the Fourier coefficients of f(x).

Proof: Parseval's equality can be re-written in the form

$$\lim_{x \to \infty} \int_a^b \left| f^2(x) dx - \sum_{k=0}^{n} c_k^2 \| \varphi_k \|^2 \right| = 0$$

and using completes the proof.

If holds it is said that the Fourier series expansion

$$\sum_{k=0}^{\infty} c_k \varphi_k(x)$$

converges in the mean to the parent function *f*(*x*). Now the Theorem states that the system is complete if the Fourier series expansion of any square integrable function *f*(*x*) converges in the mean to this function. This does not necessarily guarantee convergence at any particular point, the convergence in the mean represents a generalization of the convergence at a point. It will be clear by showing that the Fourier series expansion can converge in the mean only to one function. Let us assume that in addition.

$$\lim_{n\to\infty} \int_a^b \left[g(x) - \sum_{k=0}^{n} c_k \varphi_k(x) \right]^2 dx = 0$$

Using the obvious inequality $(a+b)^2 \le 2(a^2+b^2)$ we obtain that

$$0 \le \int_a^b [f(x) - g(x)]^2 dx = 0$$

$$= \int_a^b \left[f(x) - \sum_{k=0}^{n} c_k \varphi_k(x) - g(x) \right]^2 dx \le$$

$$\le \int_a^b [f(x) - g(x)]^2 dx = 0$$

The equalities (4) and (5) give that

$$\int_a^b [f(x) - g(x)]^2 dx = 0$$

i.e. *f*(*x*) = *g*(*x*) with the exception of a finite number of points of discontinuity. From the point of view of Fourier series we need not consider them as different functions because the Fourier coefficients are calculated by integrals whose values do not depend on the values of the function at the finite number of points.

Conclusion: If the system is complete then each square integrable function is determined with its Fourier series expansion with the exception of the finite number of points. In other words - if two functions are different at more than finite number of points then the

Fourier series expansions of these functions cannot coincide.

Properties of complete systems

Proposition: Each continuous function that is orhtogonal to all functions of a complete system is identical to zero

Proof: Since f(x) is orthogonal to any function of the system then

$$c_k = \int_a^b f(x)\varphi_k(x)dx = 0,\ k = 0, 1, 2, ...$$

From Parseval's equation (4.3) we have that

$$\int_a^b f^2(x)dx \sum_{k=0}^{\infty} c_k^2 \| \varphi_k \|^2 = 0$$

and, since $f(x)$ is assumed to be continuous, $f(x) \equiv 0$

Proposition: The Fourier series expansion of each square integrable function f(x) with respect to the complete system can be integrated term by term irrespective of the convergence of this series.

Proof: Let

$$f(x) \sim \sum_{k=0}^{\infty} c_k \varphi_k(x).$$

We have to show that if $a \le x_1 \le x_2 \le b$ then

$$\int_{x_1}^{x_2} f(x)dx = \sum_{k=0}^{\infty} c_k \int_{x_1}^{x_2} \varphi_k(x)dx.$$

Using Cauchy-Bunyakovsky's inequality we obtain that

$$\left| \int_{x_1}^{x_2} f(x)dx - \sum_{k=0}^{n} c_k \int_{x_1}^{x_2} \varphi_k(x)dx \right| \le \int_{x_1}^{x_2} \left| f(x) - \sum_{k=0}^{n} c_k \varphi_k(x) \right| dx \le$$

$$\le \int_a^b \left| f(x) - \sum_{k=0}^{n} c_k \varphi_k(x) \right| dx \le$$

$$\le \left\{ \int_a^b \left[f(x) - \sum_{k=0}^{n} c_k \varphi_k(x) \right]^2 . \int_a^b 1dx \right\}^{\frac{1}{2}} =$$

$$= \sqrt{b-a} \sqrt{\int_a^b \left[f(x) - \sum_{k=0}^{n} c_k \varphi_k(x) \right]^2 dx}.$$

Since the system is complete then by Theorem

$$\lim_{n \to \infty} \sum_{k=0}^{n} c_k \int_{x_1}^{x_2} \varphi_k(x)dx = \int_{x_1}^{x_2} f(x)dx$$

Proposition: Let $\varepsilon > 0$ be any positive number. If for each continuous on the interval [a; b] function F(x) there exists a polynomial

$$\sigma_n(x) = \gamma_0\varphi_0(x) + \gamma_1\varphi_1(x) + \ldots + \gamma_n\varphi_n(x)$$

such that $$\int_a^b [F(x) - \sigma_n(x)]^2 dx < \varepsilon$$

then the system is complete.

Completeness of Trigonometric System

We shall use the following theorem called the Weierstrass First Theorem.

Theorem: Every continuous on the interval $\lfloor -\pi; \pi \rfloor$ periodic function with period 2π satisfying the equality $f(-\pi) = f(\pi)$ can be uniformly approximated by trigonometric polynomials with any accuracy.

Using the Weierstrass First Theorem it is easy to prove.

Theorem: The trigonometric system is complete.

Proof: If f(x) is a continuous periodic function with period 2π on the interval $\lfloor -\pi; \pi \rfloor$ and $f(-\pi) = f(\pi)$ then, by the Weierstrass First Theorem, for any $\varepsilon > 0$ there exists a trigonometric polynomial

$$\sigma_n(x) = \frac{a_0}{2} + \sum_{k=1}^{n} a_k \cos kx + b_k \sin kx$$

such that

$$|f(x) - \sigma_n(x)| < \sqrt{\frac{\varepsilon}{2\pi}}$$

uniformly with respect to $x \in \lfloor -\pi; \pi \rfloor$. It follows that

$$\int_{-\pi}^{\pi} (f(x) - \sigma_n(x))^2 < \int_{-\pi}^{\pi} \frac{\varepsilon}{2\pi} dx = \varepsilon$$

and the completeness of the trigonometric system follows from Proposition 4.6.

Thus, all the properties of complete systems hold also for the trigonometric system:

1. The trigonometric Fourier series expansion of any square integrable function $f(x)$ is convergent in the mean to f(x).
2. If the continuous function $f(x)$ is orthogonal to each function of the trigonomeric system then $f(x)$ is identical to zero.
3. The Fourier series expansion of any square integrable function can be integrated term by term.

SQUARE INTEGRABLE FUNCTIONS

A function $f(x)$ is said to be absolutely integrable on an interval [a; b] if

$$\int_a^b |f(x)|\,dx < \infty .$$

A function f(x) is said to be square integrable on an interval [a; b] if

$$\int_a^b f^2(x)\,dx < \infty .$$

Taking in Cauchy-Bunyakovsky's inequality

$$\int_a^b |f|(x)g(x) < \sqrt{\int_a^b f^2(x)\,dx} . \sqrt{\int_a^b g^2(x)\,dx}$$

$g(x) \equiv 1$, we have that

$$\int_a^b |f(x)|d(x) < \sqrt{\int_a^b f^2(x)\,dx} . \sqrt{b-a} ,$$

i.e. each square integrable function on an interval $[a, b]$ is also absolutcly integrable. If an integrable function is bounded on the interval $[a, b]$ then it is also square integrable. For an unbounded function that is not true, for example the function $f(x) = \frac{1}{\sqrt{x}}$ is integrable on the interval [0, 1] because

$$\int_0^1 \frac{dx}{\sqrt{x}} = 2\sqrt{x}\Big|_0^1 = 2$$

but not square integrable because

$$\int_0^1 \frac{dx}{x} = \lim_{\varepsilon \to 0} \text{In}|x|\big|_\varepsilon^1 = \infty .$$

It also follows that if $f(x)$ and $g(x)$ are two square integrable functions then the product $f(x)g(x)$ is absolutely integrable and therefore integrable. Hence for two given square integrable functions $f(x)$ and $g(x)$ we can define the *inner product*

$$(f; g) = \int_a^b f(x)g(x)\,dx .$$

The inner product defined this way has all algebraic properties known in case of vectors i.e. $(f; g) = (g; f)$, $(\lambda f; g) = \lambda (f; g)$ for any real λ and $(f; g + h) = (f; g) + (f; h)$.

Two functions $f(x)$ and $g(x)$ are said to be orthogonal in an interval $[a; b]$ if $(f; g) = 0$. So the above conditions mean that the functions of the trigonometric system $\{1, \cos x, \sin x, \cos 2x, \sin 2x, ..., \cos kx, \sin kx, ...\}$ are pairwise orthogonal on the interval $[-x, \pi]$. The trigonometric system is said to be an orthogonal system in the interval $[-\pi, \pi]$. In general, the system of square integrable functions

$$\{\varphi_0(x), \varphi_1(x), ..., \varphi_k(x), ...\}$$

is said to be orthogonal in the interval [a; b] if $\varphi_k(x)$ and $\varphi_n(x)$ are orthogonal for each pair of distinct indices k and n; that is, if

$$(\varphi_k; \varphi_n) = \int_a^b \varphi_k(x)\varphi_n(x)\,dx = 0 \text{ for } k \neq n$$

The real number

$$\| f \| = \left[\int_a^b f^2(x)\,dx \right]^{\frac{1}{2}}$$

is called the norm of the square integrable function $f(x)$, i.e. $\|f\| = \sqrt{(f;f)}$. The orthogonal system (2) is said to be orthonormal in the interval $[a; b]$ if $\| \varphi_k \| = 1$ for k = 0, 1, 2,.... The trigonometric system is not orthonormal in the interval $[-\pi, \pi]$ because $\| 1 \| = \sqrt{2\pi}$ and $\| \sin kx \| = \| \cos kx \| = \sqrt{\pi}$ but it is easy to verify that the system of functions

$$\left\{ \frac{1}{\sqrt{2\pi}}, \frac{\cos x}{\sqrt{\pi}}, \frac{\sin x}{\sqrt{\pi}}, ..., \frac{\cos kx}{\sqrt{\pi}}, \frac{\sin kx}{\sqrt{\pi}}, ... \right\}$$

is orthonormal in the interval $[-\pi, \pi]$.

ORTHOGONAL SERIES

The trigonometric system is an example of orthogonal systems we have already considered. Examples of other orthogonal systems that are widely used in practice are Legendre polynomials, Bessel functions, Hermite polynomials, Laugrre polynomials, Jacobi polynomials, Chebyshev polynomials. Let the system be orthogonal on the interval $[a, b]$. Suppose that it is possible to express $f(x)$ as a series expansion of the form

$$f(x) = \sum_{k=0}^{\infty} c_k \varphi_k(x)$$

Our purpose is to determine the coefficients c_k. Assume term by term integration of this series and the series multiplied by $\varphi_n(x)$ on the interval $[a; b]$ to be possible. Multiplying throughout by $\varphi_n(x)$ and integrating term by term, we obtain

$$\int_a^b f(x)\varphi_n(x)\,d(x) = \sum_{k=0}^{\infty} c_k \int_a^b \varphi_k(x)\varphi_n(x)\,dx$$

which, on using (2.3), reduces to

$$\int_a^b f(x)\varphi_n(x)\,d(x) = c_n \int_a^b \varphi_n^2(x)\,dx$$

or

$$\int_a^b f(x)\varphi_n(x)d(x) = c_n \|\varphi_n\|^2$$

which, after replacing n by k again, gives

$$c_k = \frac{1}{\|\varphi_k\|^2}\int_a^b f(x)\varphi_k(x)d(x) \text{ for k} = 0,\ 1,\ 2,\dots.$$

The series is called the Fourier series of $f(x)$ with respect to the system, and the coefficients $c_{k,}$ are called the Fourier coefficients of $f(x)$ with respect to the same system.

MEAN SQUARE DEVIATION, ITS MINIMUM

Let $f(x)$ be an arbitrary square integrable on the interval $[a;\ b]$ function. Consider a polynomial with respect to the system

$$\sigma_n(x) = \gamma_0\varphi_0(x) + \gamma_1\varphi_1(x) + \dots + \gamma_n\varphi_n(x)$$

where $\gamma_0, \gamma_1, \dots, \gamma_n$ are some real constants. Since the functions $\varphi_k(x)$ are square integrable then the difference $f(x) - \sigma_n(x)$ is square integrable also and the integral of square of this difference

$$\delta_n = \int_a^b [f(x) - \sigma_n(x)]^2\, dx$$

is called the mean square deviation between the function $f(x)$ and the polynomial $\sigma_n(x)$. Our purpose is to minimize the mean square deviation, i.e. to determine the coefficients of the polynomial $\sigma_n(x)$ so that mean square deviation δ_n should be minimized. We have

$$\delta_n = \int_a^b f^2(x)dx - 2\int_a^b f(x)\sigma_n(x)dx + \int_a^b \sigma_n^2(x)dx$$

we get

$$\int_a^b f(x)\sigma_n(x)dx = \sum_{k=0}^{n}\gamma_k\int_a^b f(x)\varphi_k(x)dx = \sum_{k=0}^{n}\gamma_k c_k\|\varphi_k\|^2.$$

Further

$$\int_a^b \sigma_n^2(x)dx = \int_a^b\left[\sum_{k=0}^{n}\gamma_k\varphi_k(x)\right]^2 dx = \int_a^b\left[\sum_{k=0}^{n}\gamma_k^2\varphi_k^2(x) + \sum_{\substack{i\neq i\\ 0\le i,j\le n}}\gamma_i\gamma_j\varphi_i(x)\varphi_j(x)\right]dx$$

$$= \sum_{k=0}^{n}\gamma_k^2\int_a^b \varphi_k^2(x)^2\, dx + \sum_{\substack{i\neq i\\ 0\le i,j\le n}}\gamma_i\gamma_j\int_a^b \varphi_i(x)\varphi_j(x)\, dx = \sum_{k=0}^{n}\gamma_k^2\|\varphi_k\|^2$$

because in the second sum all the integrals equal to zero. Now

$$\delta_n = \int_a^b f^2(x)dx - 2\sum_{k=0}^{n} \gamma_k c_k \|\varphi_k\|^2 + \sum_{k=0}^{n} \gamma_k^2 \|\varphi_k\|^2$$

$$= \int_a^b f^2(x)dx + \sum_{k=0}^{n} (c_k - \gamma_k)^2 \|\varphi_k\|^2 - \sum_{k=0}^{n} c_k^2 \|\varphi_k\|^2$$

We have to choose the coefficients γ_k in order to minimize mean square deviation. The first and third term in (8) does not depend on these coefficients. The second term is least if

$$\sum_{k=0}^{n} (c_k - \gamma_k)^2 \|\varphi_k\|^2 = 0,$$

that is for all $k = 0,1,\ldots,n$.

Summarizing, the mean square deviation is minimized if the coefficients of the polynomial are Fourier coefficients of $f(x)$ with respect to the system. In other words - the mean square deviation is minimized by the n-th partial sum of Fourier series. If we denote the minimal value of the mean square deviation by Δ_n, we obtain from in virtue of that

$$\Delta_n = \int_a^b f^2(x)dx - \sum_{k=0}^{n} c_k^2 \|\varphi_k\|^2$$

From the definition of δ_n it follows immediately that Δ_n is non-negative and if n increases then Δ_n can only decrease.

Bessel's inequality

Since in (2.10) $\Delta_n \geq 0$, it follows that

$$\sum_{k=0}^{n} c_k^2 \|\varphi_k\|^2 \leq \int_a^b f^2(x)dx.$$

Hence the sequence

$$s_n = \sum_{k=0}^{n} c_k^2 \|\varphi_k\|^2$$

is bounded and non-decreasing and, consequently, has a finite limit, i.e. the series

$$\sum_{k=0}^{\infty} c_k^2 \|\varphi_k\|^2$$

is convergent, and

$$\sum_{k=0}^{\infty} c_k^2 \| \varphi_k \|^2 \leq \int_a^b f^2(x)\,dx$$

This result is known as Bessel's inequality. For an orthonormal system $\| \varphi_k \| = 1$ and Bessel's inequality takes the form

$$\sum_{k=0}^{\infty} c_k^2 \leq \int_a^b f^2(x)\,dx .$$

The norms of the functions of the trigonometric system {1, cos x, sin x,... cos $k\,x$, sin $k\,x$,...} are $\| 1 \| = \sqrt{2\pi}$ and $\| \cos k\,x \| = \| \sin k\,x \| = \sqrt{\pi}$ for k = 1, 2,.... If $f(x)$ is a square integrable function on the interval $[-\pi\,;\,\pi]$ then for trigonometric system Bessel's inequality takes the form

$$\left(\frac{a_0}{2}\right)^2 \| 1 \|^2 + \sum_{k=1}^{\infty} (a_k^2 \| \cos k\,x \|^2 + b_k^2 \| \sin k\,x \|^2) \leq \int_{-\pi}^{\pi} f^2(x)\,dx$$

i.e.

$$\frac{a_0^2}{4} 2\pi + \sum_{k=1}^{\infty} (a_k^2 \pi + b_k^2 \pi) \leq \int_{-\pi}^{\pi} f^2(x)\,dx$$

or

$$\frac{a_0^2}{2} + \sum_{k=1}^{\infty} (a_k^2 + b_k^2) \leq \frac{1}{\pi} \int_{-\pi}^{\pi} f^2(x)\,dx$$

This is the form of Bessel's inequality used for trigonometric system. The series on the left side of this inequality is convergent.

Theorem: *The series whose terms are the squares of the Fourier coefficients of any square integrable function is always convergent.*

APPLICATION OF FOURIER SERIES

A Fourier series is an expansion of a periodic function $f(x)$ in terms of an infinite sum of sines and cosines. Fourier series make use of the orthogonality relationships of the sine and cosine functions. The computation and study of Fourier series is known as harmonic analysis and is extremely useful as a way to break up an *arbitrary* periodic function into a set of simple terms that can be plugged in, solved individually, and then recombined to obtain the solution to the original problem or an approximation to it to whatever accuracy is desired or practical.

Examples of successive approximations to common functions using Fourier series are illustrated below.

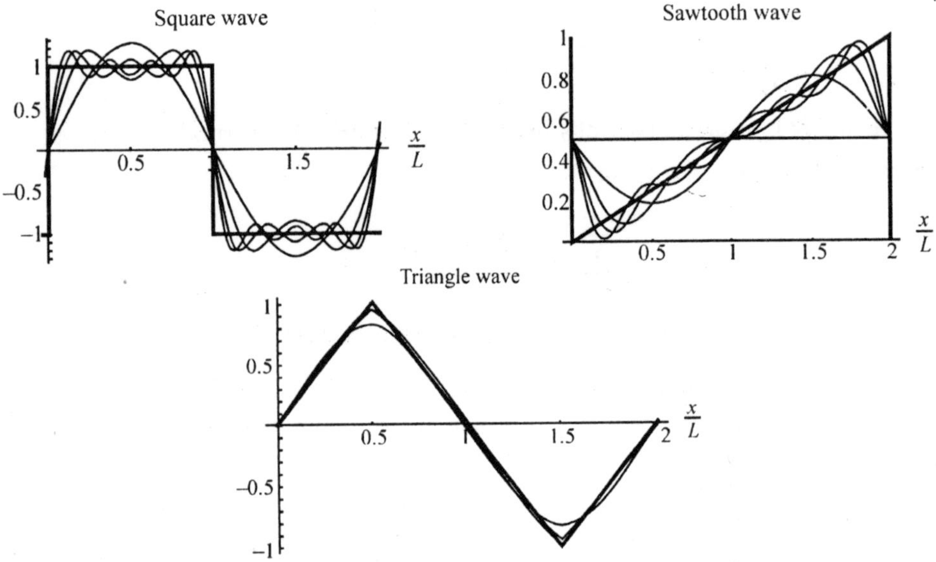

Fig. Successive Approximations to Common Functions

In particular, since the superposition principle holds for solutions of a linear homogeneous ordinary differential equation, if such an equation can be solved in the case of a single sinusoidal, the solution for an arbitrary function is immediately available by expressing the original function as a Fourier series and then plugging in the solution for each sinusoidal component. In some special cases where the Fourier series can be summed in closed form, this technique can even yield analytic solutions. Any set of functions that form a complete orthogonal system have a corresponding generalized Fourier series analogous to the Fourier series. For example, using orthogonality of the roots of a Bessel function of the first kind gives a so-called Fourier-Bessel series.

COMPUTATION OF FOURIER SERIES

The computation of the (usual) Fourier series is based on the integral identities

$$\int_{-\pi}^{\pi} \sin(mx)\sin(nx)\,dx = \pi\delta_{mn}$$

$$\int_{-\pi}^{\pi} \cos(mx)\cos(nx)\,dx = \pi\delta_{mn}$$

$$\int_{-\pi}^{\pi} \sin(mx)\cos(nx)\,dx = 0$$

$$\int_{-\pi}^{\pi} \sin(mx)dx = 0$$

$$\int_{-\pi}^{\pi} \cos(mx)dx = 0$$

for $m, n \neq 0$, where δ_{mn} is the Kronecker delta.

Using the method for a generalized Fourier series, the usual Fourier series involving sines and cosines is obtained by taking $f_1(x) = \cos x$ and $f_2(x) = \sin x$. Since these functions form a complete orthogonal system over $[-\pi, \pi]$, the Fourier series of a function $f(x)$ is given by

$$f(x) = \frac{1}{2}a_0 + \sum_{n=1}^{\infty} a_n \cos(nx) + \sum_{n=1}^{\infty} b_n \sin(nx)$$

where
$$a_0 = \frac{1}{\pi}\int_{-\pi}^{\pi} f(x)dx$$

$$a_n = \frac{1}{\pi}\int_{-\pi}^{\pi} f(x)\cos(nx)\,dx$$

$$b_n = \frac{1}{\pi}\int_{-\pi}^{\pi} f(x)\sin(nx)\,dx$$

and $n = 1, 2, 3, \ldots$. Note that the coefficient of the constant term a_0 has been written in a special form compared to the general form for a generalized Fourier series in order to preserve symmetry with the definitions of a_n and b_n.

A Fourier series converges to the function $\bar{f}$ (equal to the original function at points of continuity or to the average of the two limits at points of discontinuity)

$$\bar{f} = \begin{cases} \frac{1}{2}\left[\lim_{x \to x_0^-} f(x) + \lim_{x \to x_0^+} f(x)\right] & \text{for } -\pi < x_0 < \pi \\ \frac{1}{2}\left[\lim_{x \to \pi^+} f(x) + \lim_{x \to \pi_-} f(x)\right] & \text{for } x_0 = -\pi, \pi \end{cases}$$

if the function satisfies so-called Dirichlet conditions.

As a result, near points of discontinuity, a "ringing" known as the Gibbs phenomenon, illustrated below, can occur.

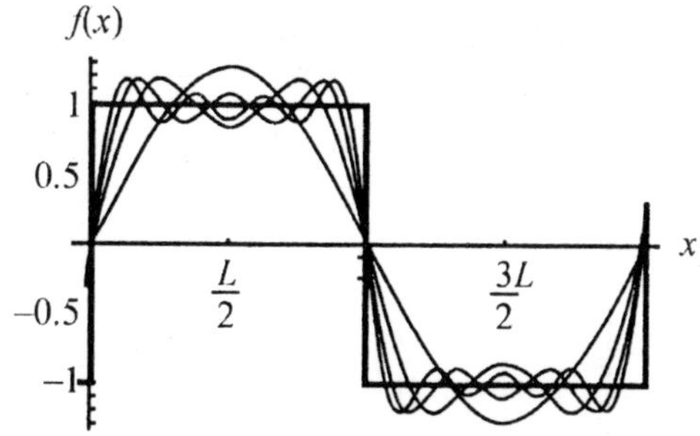

Fig. The Gibbs Phenomenon

For a function $f(x)$ periodic on an interval $[-L, L]$ instead of $[-\pi, \pi]$, a simple change of variables can be used to transform the interval of integration from [–π, π] to [–*L*, *L*]. Let

$$x = \frac{\pi x'}{L}$$

$$dx = \frac{\pi dx'}{L}$$

Solving for x' gives $x' = Lx/\pi$, and plugging this in gives

$$f(x') = \frac{1}{2}a_0 + \sum_{x=1}^{\infty} a_n \cos\left(\frac{n\pi x'}{L}\right) + \sum_{x=1}^{\infty} b_n \sin\left(\frac{n\pi x'}{L}\right)$$

Therefore,

$$a_0 = \frac{1}{L}\int_{-L}^{L} f(x')\,dx'$$

$$a_n = \frac{1}{L}\int_{-L}^{L} f(x')\cos\left(\frac{n\pi x'}{L}\right)dx'$$

$$b_n = \frac{1}{L}\int_{-L}^{L} f(x')\sin\left(\frac{n\pi x'}{L}\right)dx'$$

Similarly, the function is instead defined on the interval [0, 2, *L*], the above equations simply become

$$a_0 = \frac{1}{L}\int_{0}^{2L} f(x')\,dx'$$

$$a_n = \frac{1}{L}\int_{0}^{2L} f(x')\cos\left(\frac{n\pi x'}{L}\right)dx'$$

$$b_n = \frac{1}{L}\int_{0}^{2L} f(x')\sin\left(\frac{n\pi x'}{L}\right)dx'$$

In fact, for $f(x)$ periodic with period $2L$, *any* interval $(x_0, x_0 + 2L)$ can be used, with the choice being one of convenience or personal preference. One of the most common functions usually analyzed by this technique is the square wave. The Fourier series for a few common functions are summarized in the table below.

Function	$f(x)$	Fourier series
Fourier series —saw-tooth wave	$\frac{x}{2L}$	$\frac{1}{2} - \frac{1}{\pi}\sum_{n=1}^{\infty}\frac{1}{2}\sin\left(\frac{n\pi x}{L}\right)$
Fourier series —square wave	$2\left[H\left(\frac{x}{L}\right) - H\left(\frac{x}{L} - 1\right)\right] - 1$	$\frac{4}{\pi}\sum_{n=1,3,5,\ldots}^{\infty}\frac{1}{n}\sin\left(\frac{n\pi x}{L}\right)$
Fourier series —triangle wave	$T(x)$	$\frac{8}{\pi^2}\sum_{\pi=1,3,5\ldots}^{\infty}\frac{(-1)^{(n-1)/2}}{n^2}\sin\left(\frac{n\pi x}{L}\right)$

If a function is even so that $f(x) = f(-x)$, then $f(x)\sin(nx)$ is odd. (This follows since $\sin(nx)$ is odd and an even function times an odd function is an odd function.) Therefore, $b_n = 0$ for all n. Similarly, if a function is odd so that $f(x) = -f(-x)$, then $f(x)\cos(nx)$ is odd. (This follows since $\cos(nx)$ is even and an even function times an odd function is an odd function.) Therefore, $a_n = 0$ for all n.

$$a_n = 0 \text{ for all } n.$$

The notion of a Fourier series can also be extended to complex coefficients. Consider a real-valued function $f(x)$. Write

$$f(x) = \sum_{n=-\infty}^{\infty} A_n e^{inx}$$

Now examine

$$\int_{-\pi}^{\pi} f(x)e^{-imx}dx = \int_{-\pi}^{\pi}\left(\sum_{n=-\infty}^{\infty} A_n e^{inx}\right)e^{-imx}dx = \sum_{n=-\infty}^{\infty} A_n \int_{-\pi}^{\pi} e^{i(n-m)x}dx$$

$$= \sum_{n=-\infty}^{\infty} A_n \int_{-\pi}^{\pi}\{\cos[(n-m)x] + i\sin[(n-m)x]\}dx$$

$$= \sum_{n=-\infty}^{\infty} A_n 2\pi\delta mn$$

$$= 2\pi A_m,$$

so
$$A_n = \frac{1}{2\pi}\int_{-\pi}^{\pi} f(x)e^{-inx}dx$$

The coefficients can be expressed in terms of those in the Fourier series

$$A_n = \frac{1}{2\pi}\int_{-\pi}^{\pi} f(x)[\cos(nx) - i\sin(nx)]dx$$

$$= \begin{cases} \frac{1}{2\pi}\int_{-\pi}^{\pi} f(x)[\cos(nx) + i\sin(nx)]dx & n<0 \\ \frac{1}{2\pi}\int_{-\pi}^{\pi} f(x)dx & n=0 \\ \frac{1}{2\pi}\int_{-\pi}^{\pi} f(x)[\cos(nx) - i\sin(nx)]dx & n>0 \end{cases}$$

$$= \begin{cases} \frac{1}{2}(a_n + ib_n) & \text{for } n<0 \\ \frac{1}{2}a_0 & \text{for } n=0 \\ \frac{1}{2}(a_n - ib_n) & \text{for } n>0 \end{cases}$$

For a function periodic in [$-L/2$, $L/2$], these become

$$f(x) = \sum_{n=-\infty}^{\infty} A_n e^{i(2\pi nx/L)}$$

$$A_n = \frac{1}{L}\int_{-L/2}^{L/2} f(x) e^{-i(2\pi nx/L)} dx$$

These equations are the basis for the extremely important Fourier transform, which is obtained by transforming A_n from a discrete variable to a continuous one as the length $L \to \infty$

APPLICATION OF FOURIER SERIES

We have learnt earlier the way how to use Fourier series. Here, we discuss some application of Fourier series. These series play a vital role in Mathematics as well as in other field.

Simple Fourier Series

Let f be periodic of period 2π, with $f(x) = x$ for x from $-\pi$ to π. Note that this function is a periodic version of the identity function.

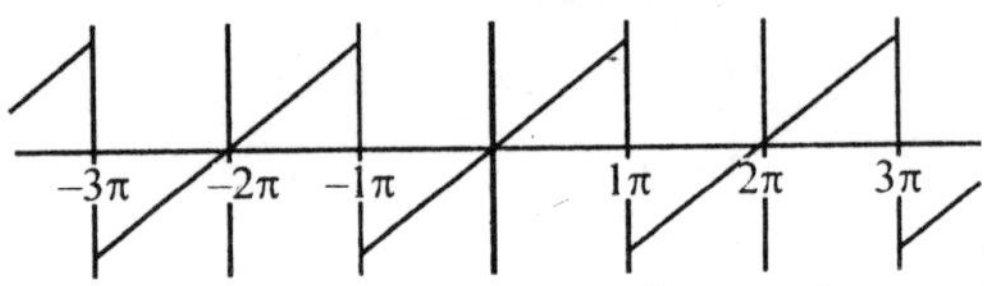

Fig. Plot of a Periodic Identity Function - a Sawtooth Wave.

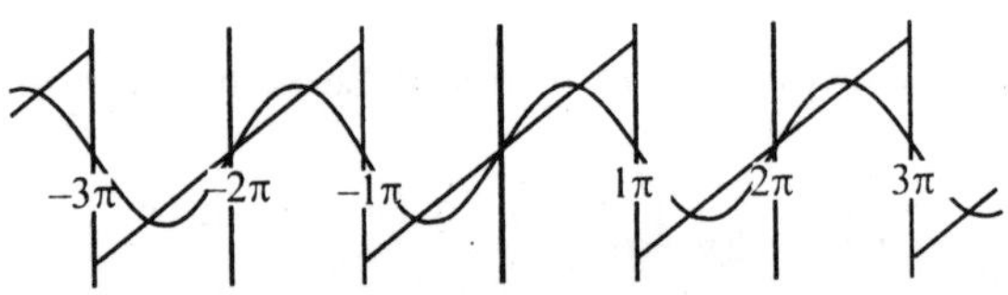

Fig. Plot of the First Five Successive Partial Fourier Series

We will compute the Fourier coefficients for this function.

$$a_n = \frac{1}{\pi}\int_{-\pi}^{\pi} f(x)\cos(nx)\,dx$$

$$= \frac{1}{\pi}\int_{-\pi}^{\pi} x\cos(nx)\,dx$$

$$= a_n = 0 b_n$$

$$= \frac{1}{\pi}\int_{-\pi}^{\pi} f(x)\sin(nx)\,dx$$

$$= \frac{1}{\pi}\int_{-\pi}^{\pi} x\sin(nx)\,dx = \frac{2}{\pi}\int_{0}^{\pi} x\sin(nx)\,dx$$

$$= \frac{2}{\pi}\left(\left[-\frac{x\cos(nx)}{n}\right]_0^{\pi} + \left[\frac{\sin(nx)}{n^2}\right]_0^{\pi}\right)$$

$$b_n = 2\frac{(-1)^{n+1}}{n}$$

Notice that a_n are 0 because the $x \mapsto x\cos(nx)$ are odd functions. Hence the Fourier series for this function is:

$$f(x) = \frac{a_0}{2} + \sum_{n=1}^{\infty}\left[a_n\cos(nx) + b_n\sin(nx)\right]$$

$$= 2\sum_{n=1}^{\infty}\frac{(-1)^{n+1}}{n}\sin(nx), \forall x \in [-\pi, \pi]$$

One application of this Fourier series is to compute the value of the Riemann zeta function at $s = 2$; by Parseval's theorem, we have:

$$\frac{1}{2\pi}\int_{-\pi}^{\pi} x^2 dx = \frac{1}{2}\sum_{n>0}\left[2\frac{(-1)^n}{n}\right]^2$$

which yields: $\sum_{n>0}\frac{1}{n^2} = \frac{\pi^2}{6}$

Wave Equation

The wave equation governs the motion of a vibrating string, which may be fastened down at its endpoints. The solution of this problem requires the trigonometric expansion of a general function f that vanishes at the endpoints of an interval $x = 0$ and $x = L$. The Fourier series for such a function takes the form

$$f(x) = \sum_{n=1}^{\infty} b_n \sin\left(\frac{n\pi}{L}x\right)$$

where

$$b_n = \frac{2}{L}\int_0^L f(x)\sin\left(\frac{n\pi}{L}x\right)dx$$

Vibrations of air in a pipe that is open at one end and closed at the other are also described by the wave equation. Its solution requires expansion of a function that vanishes at $x = 0$ and whose derivative vanishes at $x = L$. The Fourier series for such a function takes the form

$$f(x) = \sum_{n=1}^{\infty} b_n \sin\left(\frac{(2n+1)\pi}{2L}x\right)$$

where

$$b_n = \frac{2}{L}\int_0^L f(x)\sin\left(\frac{(2n+1)\pi}{2L}x\right)dx$$

An Application of Fourier Analysis to the Description and Statistical Analysis of Refractive Error

The description of sphero-cylinder lenses is approached from the viewpoint of Fourier analysis of the power profile. It is shown that the familiar sine-squared law leads naturally to a Fourier series representation with exactly 3 Fourier coefficients, representing the natural parameters of a thin lens. The constant term corresponds to the mean spherical equivalent power while the amplitude and phase of the harmonic correspond to the power and axis of a Jackson cross-cylinder lens, respectively. Expressing the Fourier series in rectangular form leads to the representation of an arbitrary sphero-cylinder lens as the sum of a spherical lens and two cross-cylinders, one at axis 0° and the other at axis 45°. The power of these three component lenses may be interpreted as (x,y,z) coordinates of a vector representation of the power profile. Advantages of this *power vector* representation of a sphero-cylinder lens for numerical and graphical analysis of optometric data are described for problems involving lens combinations, comparison of different lenses, and the statistical distribution of refractive errors.

Various types of finite methods are often used to solve combined conduction and convection problems. These methods are both flexible and accurate but they are also complex to program and relatively slow. At times it can there-fore be advantageous to use an analytical approach. The list of pros and cons for this category of methods is very much a mirror of those for the finite methods. They are easy to program, usually fast but not very flexible. This article is about a Fourier series solution to the problem of calculating the temperature distribution on a convection cooled plate with discrete heat sources. It is by no means the only way to tackle with this problem. At times it is nevertheless the most convenient one. The solution has been known in the heat transfer community for several years. As far as the author knows its origin is to find at the University of Poitiers, France. The purpose of this article is therefore not to explain the how it is derived but to discuss some of its application aspects.

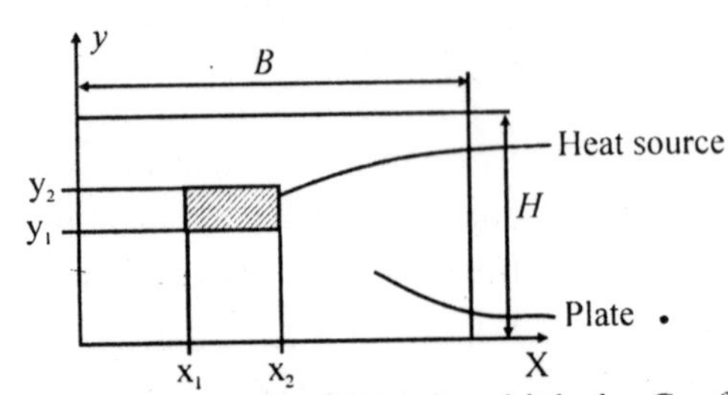

Fig. The 2-D Solution is a Fourier Series in which the Coefficients are given by Another Fourier Series.

$$\begin{cases} \vartheta(x,y) = \dfrac{y}{k^2} \cdot \sum_{n=0}^{N} f(n,y) \cdot A_n \cdot \cos(n \cdot \varpi_x \cdot x) \\ f(n,y) = \sum_{m=0}^{M} B_{mn} \cdot \cos(m \cdot \varpi \cdot y) \end{cases}$$

Where,

$$y = \frac{\dot{q}}{k \cdot \delta} \qquad \chi = \sqrt{\frac{1 \cdot h}{k \cdot \delta}} \qquad \varpi = \frac{\pi}{B} \qquad \varpi = \frac{\pi}{H}$$

$$A_0 = \frac{\varpi_x \cdot (x_2 - x_1)}{\pi}$$

$$A_n = \frac{2 \cdot [\sin(n \cdot \varpi_x \cdot x_2) - \sin(n \cdot \varpi_x \cdot x_1)]}{n \cdot \pi}$$

$$B_{0n} = \frac{\varpi_y \cdot (y_2 - y_1)}{\pi \cdot \left[\left(\dfrac{n \cdot \varpi_x}{\kappa}\right)^2 + 1\right]}$$

$$B_{mn} = \frac{2 \cdot [\sin(n \cdot \varpi_x \cdot x_2) - \sin(n \cdot \varpi_x \cdot x_1)]}{m \cdot \pi \left[\left(\dfrac{n \cdot \varpi_x}{\kappa}\right)^2 + \left(\dfrac{m \cdot \varpi_y}{\kappa}\right)^2 + 1\right]}$$

$$N \geq \frac{3 \cdot E}{x_2 - x_1}$$

$$M \geq \frac{3 \cdot H}{y_2 - y_1}$$

$\dot{q}$ = heat flux ror heat source (W/m^2)

k = thermal condictivity (W/mK)

δ = plate thickness (m)

h = heat transfer coefficient (W/m^{2}K)

The Solution

It is basically a Fourier series in which the coefficients are given by another Fourier series. Double sided cooling has been assumed in this particular case. This can however easily be changed by dividing the heat transfer coefficient with a factor 2, (or by suppressing number 2 in the fourth equation). It should also be noted that the solution is the simplest in a family of solutions that originates from a 3-D solution for a plate with axis dependent thermal conductivities. Since their application aspects are similar the discussion that follows is relevant for all members of this family.

The main advantage with the solution is its simplicity. It can programmatically be fitted into a few procedures, which greatly facilitates code writing. The main disadvantage is that its application window is limited to simple cases. It can sometimes also be slow. As this article will reveal there are nonetheless several tricks that can counterbalance these drawbacks.

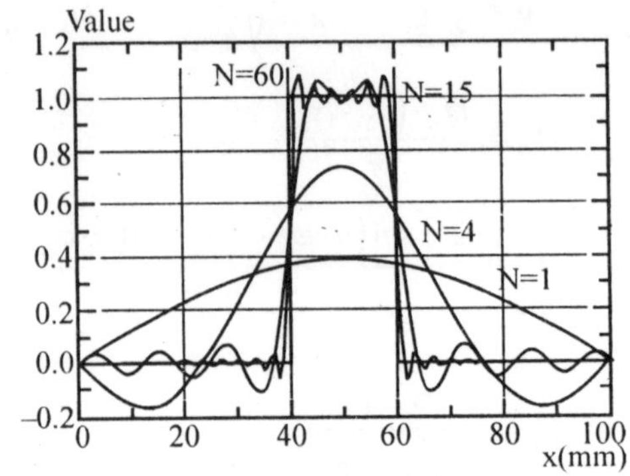

Fig. Fourier Series can be used to Model Square wave Functions. The fit is Strongly Dependent on the Number of terms Included.

To show the total derivation of the solution would require too much space but the basic principle is to apply two perpendicular square functions to model the heat source flux. These functions take the value 1 above the heat source and the value 0 elsewhere. A standard approach for describing such functions is to use Fourier series. These series can be made to model just about any function but their drawback is that the fit only is perfect if an infinite number of terms are included. One must therefore always settle for an approximation.

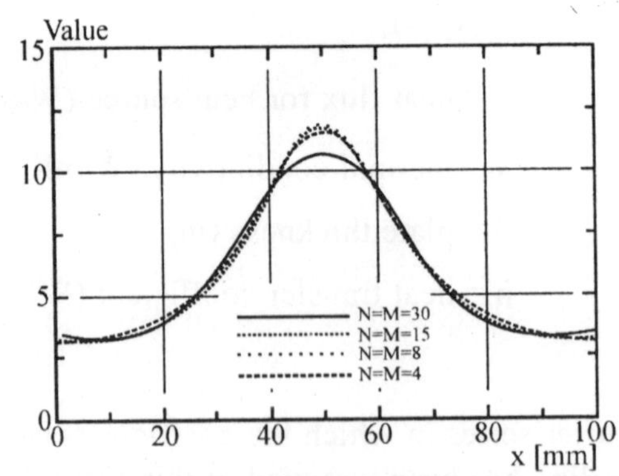

Fig. The Accuracy Impact of the term Count is most Important in the Centre of the heat Source. At some Distance it is Therefore Possible to use half the value Recommended in figure.

To determine the term count needed in a Fourier series is always problematic. The terms do have a tendency to decrease as the series is expanded but not in an orderly and regular manner. It is therefore difficult to use the value of last term as criteria for halting the summation. For the square wave function there is fortunately another possibility. Each term in the modelling series can be looked at as cycloid function with a wavelength that successively decreases. It is evident that a good fit only can be achieved unless the last wavelength is smaller than the width of the square wave. The rule of thumb is to break the summation when the width is covered by 1.5 wavelengths.

The term count has a large impact on the calculation time. It is therefore highly desirable to reduce it as much as possible. Figure shows the impact on the temperature level for the same example as in figure. The accuracy seems to reach saturation for $N > 15$, which confirms that the equations given for M and N in figure are reasonable. When the maximum summation index is reduced to $N = 8$ there is some impact inside the heat source but hardly any on its outside. It therefore seems possible to reduce the recommended values for M and N with a factor 2 for points that are at some distance from the heat source. This is important because it reduces the calculation time with a factor 4.

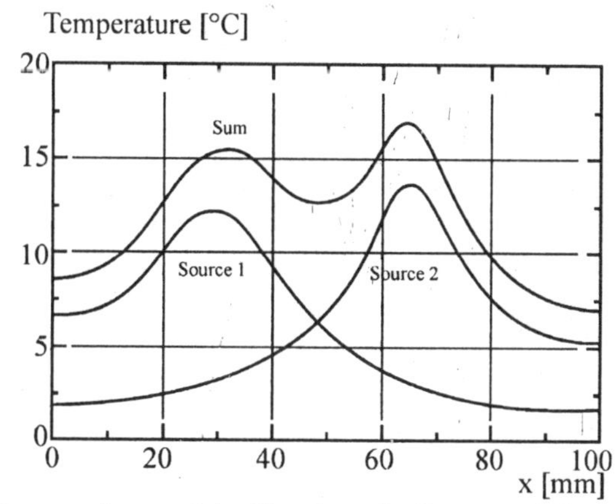

Fig. Superposition can be used to Simulate the Impact of Several heat Sources.

SUPERPOSITION

The super position principle can always be used if the sum of the solutions to an equation also is a solution to that equation. All conduction problems in which the material conductivity is an invariant belong to this category. The superposition principle can therefore be used for almost all conduction problems in thermal design but with a possible exception for half conductors such as silicone. The solution discussed is actually a large scale application of this principle.

Superposition can naturally also be used to simulate the combined impact of several heat sources on a plate. This opens up the possibility for fast studies on how a change of a heat source property in a multi-heat source environment impacts the total temperature field. A procedure that manages this has five steps:

1. Calculate the temperature field for the heat source.

2. Subtract it from the total field.
3. Change the properties of the heat source.
4. Calculate the new temperature field for the heat source.
5. Add it to the total field.

The advantage with this procedure is that the temperature field only has to be recalculated for one heat source at a time. If there are 50 of them this can save a lot of calculation time. This technique can for example be used for rapid layout studies on PCBs. The solution discussed is however still is a bit too slow to make this fully enjoyable when the PCBs are large.

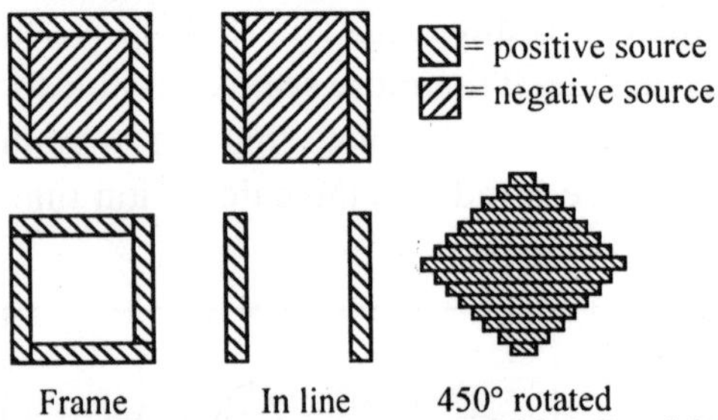

Fig. Various heat Source Shapes can be Created by Combining heat Sources.

The superposition principle can also be used to simulate various heat source shapes. Figure shows a few examples. It should nevertheless be noted that using overlapping, although mathematically elegant, always implies the risk that the two calculation errors combine unfavourably.

Execution Speed

The execution speed is highly dependent on the term count. For small heat sources on large plates the situation is however radically different. A typical 200 X 200 mm PCB with a 10 X 10 mm heat source would for example require 60 terms in each series which adds up to 3600 terms when the two nested series are combined. If the ambition is to create an overview of the temperature field, calculations must further be made on a grid of the order 20 X 20. All in all this results in 1.4 million terms. If there in addition are some 50 components, it is easy to realise that the task is heavy.

There are nevertheless a couple of measures that can be taken to speed up the process. To reduce the term count with a factor 2 for points outside the heat source has already been mentioned. Another possibility is to save the terms in the second summation, f(n,y), in a list and reuse them for all points that have the same y-value. These two measures boost the calculation speed considerably but for large densely packed PCBs it is still difficult to get down to enjoyable execution times.

The situation is much better for heat sinks. They are usually smaller than PCBs and the heat source count is in most cases limited to less than 10. Overviews can therefore be created in a matter of a few seconds, which is a key issue for successful front-end thermal design. Those who not have tired this method can get an idea of how it works on the web site given at the be-ginning of this article. The full potential of the method can however not be exploited unless the fluid temperature increase problem is reasonably well managed.

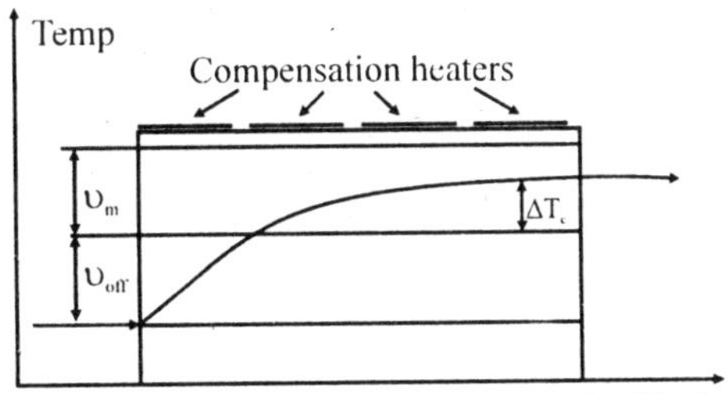

Fig. An Array of Large Heaters that cover the Entire Plate can be used to Compensate for this Error.

$$\dot{Q}_C = h_m \cdot A_c \cdot \Delta T_c$$

where:

$$\begin{cases} \dot{Q} = \text{compensation heat dissipation (VV)} \\ h_m = \text{average heat transfer coefficient (W/m}^2\text{K)} \\ A_c = \text{compensation element surface (m}^2\text{)} \\ \Delta T_c = \text{temperature compensation (K)} \end{cases}$$

Fluid Temperature Increase Compensation

One problem with the Fourier series solution is that it only is valid for uniform fluid temperatures and heat transfer coefficients. This limitation always causes an error. It is often negligible for small heat sinks but for large ones cooled by small flow rates, it can be considerable.

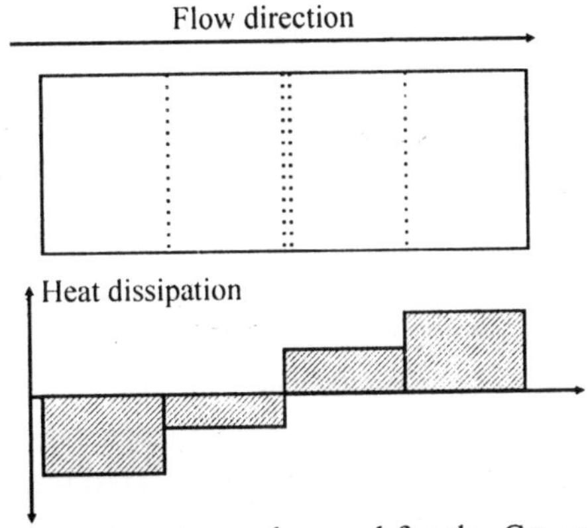

Fig. Different Array Arrangements can be used for the Compensation Heat Sources but the sum of their heat Dissipations must Always be zero.

The above figure shows a case with a high fluid temperature increase. For this case it is obvious the cooling is much more effective at the inlet than at the outlet. This effect is partly caused by a variation of the local temperature difference and partly by a variation of the local heat transfer coefficient.

One way to rectify for this discrepancy is to use compensation heat sources. They are typically only kept alive during the calculation process and they should cover the entire plate. Various array arrangements are possible but the simplest is to use a single inline layout. Another restriction is that the sum of their heat dissipations always must equal zero.

A first level approach, which for many cases is sufficient, is to initially assume that the plate is isothermal. It is a simple task to calculate the corresponding fluid temperature profile. The deviation between the average fluid temperature and the actual fluid temperature determines the strengths of the compensation sources and this is done with a very simple equation. The calculation is performed with the average heat transfer coefficient as parameter and the average fluid temperature as an offset value.

The first level approach basically simulates the fluid temperature profile as if the plate had been isothermal but it is possible to go one step further. The second level approach consists in calculating the average temperature for the compensation sources and use these as a bases for a better approximation of the temperature profile. The basic principle is the same as above but with the complication that the plate now has to be looked at as composed of several isothermal segments. Another difference is that it for this case can make sense to use a two dimensional layout array.

It should also be noted that the compensation sources are large sources that calculate rapidly. Compensating for the fluid temperature increase is therefore not a process that considerably charges the all over execution time.

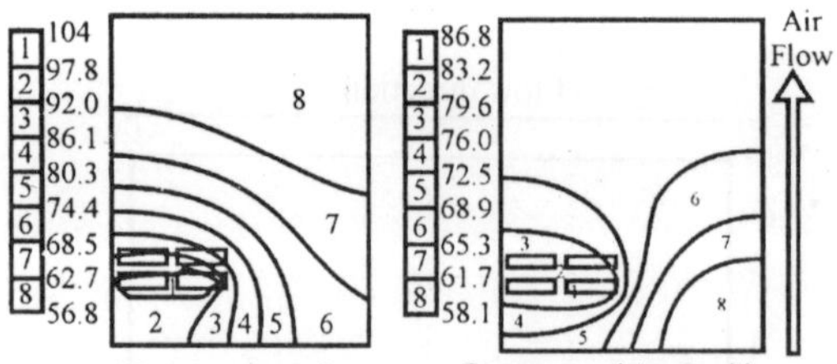

Fig. Comparison of the Temperature Distribution in a 210×260 mm Heat Sink Bottom Plate, with and without Fluid Temperature Compensation.

Above Figure shows an example of the impact of an air temperature increase compensation for rather extreme case. The heat sources are in this case located near the inlet. Their temperature would consequently be calculated too high if no fluid temperature compensation was made.

The result difference is on the 20% level. It is not radical but it considerably expands the usefulness of the solution.